The Solitary Bees

The Solitary Bees

Biology, Evolution, Conservation

Bryan N. Danforth

Robert L. Minckley

John L. Neff

With original artwork by Frances Fawcett

PRINCETON UNIVERSITY PRESS

Princeton and Oxford

Library of Congress Control Number: 2018965627

ISBN: 9780691168982

British Library Cataloging-in-Publication Data is available

Editorial: Alison Kalett and Kristin Zodrow
Production Editorial: Ellen Foos
Text and Jacket Design: Pamela Schnitter
Jacket image: Female *Hesperapis rhodocerata* (Melittidae) visiting a flower
    of *Bahia absinthifolia* (Asteraceae), Willcox, AZ. Courtesy of the authors.
Production: Erin Suydam
Publicity: Matthew Taylor and Julia Hall
Copyeditor: Patricia Fogarty

This book has been composed in Sabon LT Std

Printed on acid-free paper. ∞

Printed in the United States of America

10 9 8 7 6 5 4 3 2

TO OUR FAMILIES,
WHO HAVE MADE THE STUDY
OF SOLITARY BEES
A LOT LESS SOLITARY.

Contents

Acknowledgments

During the preparation of this book, the three authors have been helped by many people (and other animals). First and foremost, we would like to thank Maria Van Dyke, who worked tirelessly over the past two years helping with many aspects of the book. Maria helped assemble and curate our bibliographic database, she helped with table and figure formatting, she proofread the book from start to finish, and she was always ready to offer feedback on all aspects of solitary bee biology. Maria made it possible for the three authors to stay on deadline while at the same time juggling many other demands on her time. Second, we would like to thank Frances Fawcett, who prepared the vast majority of the figures used in this book. Francie is a part-time scientific illustrator who has to balance a busy and demanding life focused on her husband, children, grandchildren, and dog. She worked with starting images of varying quality and rendered them in a consistent, impactful, and aesthetically appealing way. During the preparation of these figures, Francie often worked from original specimens (from the Cornell University Insect Collection) in order to make sure the illustrations accurately captured the true anatomy of the species in question. The book has been greatly enhanced through Francie's careful work on the over 90 images we used throughout the book. Third, we would like to thank the many bee biologists who generously answered our questions and provided key information on bee biology during the preparation of the book (in no particular order): Stephen Buchmann, Avery Russell, Sandra Rehan, Sean Prager, Peter Graystock, Jerry Rozen, Terry Griswold, James Cane, Karl Magnacca, Jakub Straka, Skyler Burrows, and Tom Seeley. We would also like to thank the many bee biologists who published the biological observations that form the foundation of this book. A few people have contributed enormously to our understanding of solitary bee biology through their field work and careful observations, often in remote, harsh, and inaccessible parts of the world. The stories we tell about solitary bee biology are based on work extending back over 80 years by scientists with a deep fascination for these remarkable creatures, including E. Gorton Linsley, Charles Michener, Jerome Rozen Jr., Vince Tepedino, Phil Torchio, Frank Parker, George (Ned) Bohart, Vin Whitehead, Roy Snelling, George Eickwort, Terry Houston, Robbin Thorp, Laurence Packer, Connal Eardley, Ken Walker, Clemens Schlindwein, Jim Cane, Bill Wcislo, Nico Vereecken, and Andreas Müller.

Our deep appreciation goes out to the people who provided us with color photographs and original artwork for the book. Nico Vereecken generously shared with us 12 amazing color photos of bees and their host plants. Nico is an outstanding bee photographer because he is also a PhD-level bee biologist who knows the biology of his subjects better than anyone. His photos capture the intimate partnership between specialist (oligolectic) bees and their host plants, many of which have been studied by Nico and his students and collaborators. Thanks also to Susan Barnett for the photo of sleeping male *Colletes compactus*, Ben Porter for the photo of *Colletes hederae* (benporterwildlife.co.uk), and Birgitte Rubæk for the gouache drawing of *Macropis europaea* (www.behance .net/brubaek).

We are very grateful to John Ascher and John Pickering for making data on bee species richness available on Discover Life (http://www.discoverlife.org/20 /q?search=Apoidea). Accurate estimates of generic, tribal, subfamily, and family-level bee species richness greatly facilitated the preparation of many figures and tables throughout the book.

We would like to thank the many instructors and students of the Bee Course (https://www.thebeecourse.org/). The Bee Course is an annual, 10-day workshop on bee identification and biology that has been offered annually since 1999 at the Southwestern Research Station in Portal, Arizona. It is an immersion course in which the focus is all bees all the time. In writing this book, we often considered the Bee Course students (past, present, and future) as our target audience. Our hope is that this book will help future students arrive in Arizona even more prepared for the course.

Finally, we are very grateful to Alison Kalett, Ellen Foos, Kristin Zudlow, Lauren Bucca, and Dimitri Karetnikov at Princeton University Press for their guidance, advice, and encouragement throughout the development of this book, and the two anonymous reviewers. The reviews significantly improved the quality of the book and the accuracy of our reporting on some details of bee biology.

Bryan Danforth would like to specifically thank people at Cornell who helped in the completion of this book. I am extremely grateful to the students and postdocs who were present in my lab during the writing of this book (Laura Russo, Elizabeth Murray, Heather Grab, Mary Centrella, Kristen Brochu, Margarita López-Uribe, Mia Park, Silas Bossert, Trevor Sless, Erin Krichilsky, and Katherine Urban-Mead). I often shared ideas with them or asked their advice during this time, and they were always willing to provide feedback and suggestions. Their enthusiasm for bee research has been a great source of inspiration and encouragement. I am also grateful to my former graduate students and postdocs who contributed significantly to our understanding of bee phylogeny and evolution (John Ascher, Karl Magnacca, Sedonia Sipes, Jessica Litman, Christophe Praz, Eduardo Almeida, Jason Gibbs, Seán Brady, Sophie Cardinal, and Shannon Hedtke). The Entomology second-floor staff (Cheryl Gombas, Lisa Westcott, Lisa Marsh, Stephanie Westmiller, and Amy Arsenault) have provided a much-appreciated support network during my tenure as Chair of Entomology, and I am very grateful for their support. I am grateful to the staff at Cornell's Mann Library (Mary Ochs, Marty Schlabach, and Sarah Kennedy), who have never failed to track down even the most obscure, ancient publications. Many thanks to the Cornell Pollinator Group (Scott McArt, Katja Poveda, Robert Raguso, André Kessler, Monica Geber, Emma Mullen, and the many students and post-docs who attend) for feedback on various ideas and for reading select chapters of the book prior to publication. I am grateful to the Curator (James Liebherr) and the Collection Manager (Jason Dombroskie) of the Cornell University Insect Collection (http://cuic.entomology.cornell.edu/) for their support of our bee and wasp collection, which was an invaluable resource for the research that went into this book. I am very grateful to my previous department chair (Laura Harrington) and the dean of the College of Agriculture and Life Sciences (Katherine Boor) for granting a sabbatical leave in fall 2016, which allowed me to focus entirely on writing this book. Finally, I am grateful to the National Science Foundation and the US Department of Agriculture for providing research funding that has allowed me to investigate the biology, phylogeny, and ecology of solitary bees.

Jack Neff would like to specifically thank Andy Moldenke, who first introduced him to the wonders of solitary bees. He also would like to mention George (Ned) Bohart, George Eickwort, P. H. Timberlake, Wally LaBerge, and Roy Snelling, who, in various ways, played a role in developing his interest in bees. He also would like to thank the many students at the University of Texas at Austin, especially those in the labs of Beryl Simpson and Shalene Jha, whose questions about bees and pollination have been a refreshing source of intellectual stimulation.

Bob Minckley would specifically like to thank Charles D. Michener for his guidance on bees, and life, and Josiah Austin for unrestricted access to interesting study areas. Also, thanks are due to the librarians at the University of Rochester, the entire Department of Biology at the University of Rochester, and the National Science Foundation for supporting his research.

Finally, we would each like to thank our immediate families, to whom the book is dedicated: Marina Caillaud, and Isabelle and Nicholas Danforth (and many other furry animals small and large) [BND]; Jonathan, Meghan, and Beryl Simpson and our late dog Callie [JLN]; and Shane and Adrian Minckley [RLM]. Our families have been a constant and reliable source of support and inspiration through our many years of research on the biology of solitary bees.

The Solitary Bees

CHAPTER 1

Introduction

This is a book about solitary bees. The idea of a solitary bee may seem strange to many readers. For most people, bees are almost synonymous with social behavior. The word *bee* typically conjures up an image of a colony of hundreds (bumble bees) or thousands (honey bees) of workers. Like a massive factory, the colony hums with activity—each worker performing some key role and all dedicated to selflessly protecting the colony, gathering food and nesting materials, and helping to raise the offspring of the queen.

In reality, the vast majority of bee species on earth live solitary lives. A single female constructs her own nest, defends it against intruders, parasites, and predators, and forages for pollen, nectar, or floral oils as food for her offspring. While some solitary bees nest in dense aggregations of hundreds to thousands of nests, each nest is occupied by just one female, and every female is both worker and queen throughout her life. While they are less conspicuous and well-studied than social bees, solitary bees are fascinating in their own right. Solitary bees exhibit extraordinary diversity in morphology, mating behavior, life history, nest architecture, foraging behavior, and host-plant associations. They are far more abundant than social bees in certain environments (deserts), and they have evolved remarkable adaptations for surviving in these harsh and unpredictable habitats. They are important, but underappreciated, pollinators of many wild and agricultural plants. And they have existed on earth for more than 120 million years. In deciding to write this book, we felt that it was time to provide a modern perspective on the biology of solitary bees.

We had three main goals when we embarked on this project. First, we wanted to excavate the hidden "gems" of solitary bee biology from the specialized scientific literature. The literature on solitary bees is scattered, and many of the most interesting biological stories are buried in the specialized entomological literature. We did not want to only review the well-known stories published in high-impact scientific journals; instead we wanted to bring to light the less well known, but fascinating, studies hidden in less widely read journals, such as the *Journal of the Kansas Entomological Society*, the *Proceedings of the Entomological Society of Washington*, the *Pan-Pacific Entomologist*, and the *American Museum Novitates*. And we found some wonderful stories, like the males of one species of *Anthophora* (Apidae) that collect parsnip perfume to attract females (Chapter 4), the member of the genus *Hylaeus* (Colletidae) that builds an upside-down nest (Chapter 6), and the species of the genus *Lasioglossum* (*Sphecodogastra* spp.) that forage by the light of the moon (Chapter 7).

Second, we wanted to weave together the empirical and natural history studies of solitary bee biology into a modern, comparative framework. The framework we found most useful was phylogeny (Chapter 2). Our understanding of bee (and wasp) phylogeny has changed dramatically over the past 25 years. New data (gene-sequence data) and new phylogenetic methods (model-based methods) have allowed us to more accurately reconstruct the evolutionary origins of bees from hunting wasps as well as the evolutionary relationships within bees. This new view of bee phylogeny has proven extremely useful when interpreting biological patterns across bee families, subfamilies, tribes, genera, and species. In some cases, puzzling observations made when bee phylogeny was less well understood could be reinterpreted and made coherent in light of a more modern view of evolutionary relationships. We make heavy use of this modern view of bee phylogeny throughout the book.

Finally, we wanted to provide a road map for future studies of bee biology. In writing this book, we came across discoveries that had been overlooked or not fully appreciated at the time they were published, and made connections between disparate natural history observations and more recent discoveries based on molecular genetics and genomics. The idea that solitary bees are cultivating fungi within their brood cells is one such idea. Scattered reports dating back to the 1960s of fermentation in the provisions of many distantly related groups of solitary bees are now being reexamined using high-throughput DNA sequencing and stable-isotope analysis. We devote an entire chapter (Chapter 9) to the numerous organisms that inhabit the solitary bee brood cell, including bacteria and fungi, but also annelids, nematodes, and mites. What roles many of these brood-cell inhabitants play is still unclear, but it does appear that bee larvae are benefiting from their presence, either through the protection they provide against other pathogens and predators or through consuming the protein assimilated by these brood-cell inhabitants (Steffan et al. 2019).

ORGANIZATION OF THE BOOK

The chapters of this book are organized in what we hope is a logical and coherent way. We start with an overview of bee phylogeny (Chapter 2). We introduce the currently recognized families and subfamilies and discuss what we know about their phylogenetic relationships. We also use this chapter to define how solitary bees differ from social bees and identify some of the broad ecological differences between these two groups. Taking the phylogeny as a comparative framework, we identify where in the phylogeny of bees social behavior evolved, how many times it evolved, and where and how reversals from sociality to solitary nesting occurred. This chapter is important because it introduces the comparative framework we will use throughout the book as we cover diverse biological attributes of solitary bees.

In Chapter 3, we introduce the basic life-history patterns in bees. We describe the stages that any bee passes through during development. We describe how solitary bees can persist in harsh environments with unpredictable patterns of rainfall and flowering. We explore how diapausing larvae can survive in a state of suspended animation over many years, and what triggers emergence when conditions are suitable for foraging and provisioning. A key theme of this chapter is that, for solitary bees, adult activity is typically very short compared to the period of larval development. Timing is everything for solitary bees.

In Chapters 4 and 5, we turn to males and mating behavior. Male bees intensely compete to mate with receptive females. The strategies that males have adopted to successfully sire offspring are extraordinarily variable. Some males establish territories, others create floral perfumes, and some become highly adept intra-nest assassins. It is common to find multiple reproductive tactics even within a single species—large males can adopt one strategy and small males another. The intensity of male-male competition leads many males to die without ever mating. Life is tough for male bees, and natural selection is constantly honing male mating tactics. We also discuss the surprising roles that males play as occasional, and sometime highly effective, pollinators.

In Chapter 6, we turn to nesting. Solitary bees are creative architects that seem to exploit almost any available site or material for nesting. The vast majority of solitary bees nest underground, in burrows that can be as deep as 5 meters; others modify preexisting cavities by transporting nesting materials; some excavate nests in wood; and some build freestanding nests. Some solitary bees have chosen unconventional nesting substrates, like sandstone, termite nests, snail shells, or the edges of active volcanoes. The brood cells constructed by solitary bees are works of artistry and advanced engineering, and are constructed from diverse materials that range from readily available mud, resin, and leaves to more exotic materials, such as flower petals, floral oils, dung, plant fibers, and mammal hair. Brood cells must protect the developing larvae from desiccation (in deserts), flooding (in the humid tropics), attack by parasites and predators, and infection by pathogenic microorganisms. Some cavity-nesting bees have even adapted to life in urban environments by constructing their brood cells from discarded plastic shopping bags and window-caulking material.

Once a nest is constructed, female solitary bees (excluding the brood parasites; see below) must provision their brood cells with food for developing larvae. In Chapters 7 and 8, we describe the three primary floral resources harvested by female bees—nectar, pollen, and floral oils—and the tools bees use to gather and transport these materials. Many solitary bees have highly specialized mouthparts and legs for accessing hidden floral rewards and effectively transporting them back to the nest. We also describe the various steps in brood-cell provisioning, including how bees learn the landmarks around their nests. We discuss the various intrinsic and extrinsic factors that shape offspring production; female age and body size, floral resource availability, and parasite pressure can all impinge on females as they make day-to-day decisions on whether to produce a male or a female offspring. Finally, we describe a highly specialized group of bees that fly at low light levels. These matinal, crepuscular, and nocturnal species have pushed the boundaries of bee visual and navigational capabilities to the limit.

In Chapters 9, 10, and 11, we cover the diverse organisms that coexist with solitary bees. These include mutualists and commensals, as well as the diversity of parasitoids, predators, and brood parasites that attack solitary bees. One of the most devastating groups of attackers are bees themselves—the brood-parasitic bees, which comprise approximately 13% of all bee species. Like brood-parasitic birds (cuckoos), these "cuckoo" bees exploit their hardworking solitary bee hosts by discreetly (and not so discreetly) depositing their own egg in the host nest and then brutally killing the host egg or larva. A number of ancient lineages of bees have adopted this highly effective strategy.

In Chapter 12, we turn to the topic of bee-plant evolution. How and when did pollen feeding evolve in the first place? How should we look at bee and flowering plant interactions from an evolutionary perspective? Are bees and flowering plants engaged in a love story, an arms race, or something in between? The vast majority of bees are highly specialized herbivores that gather pollen, nectar, and floral oils for their offspring. Bees also show enormous variation in the range of host plants used. There are highly specialized bees that visit just one genus or species of host plant and extreme generalists that collect pollen from many host-plant families. We also discuss the origins and usage of terms like *oligolecty* and *polylecty*. Plants are not passive partners in this relationship. Flowering plants have evolved a diversity of mechanisms to restrict access to their highly valuable floral rewards. Floral morphology, pollen and nectar chemistry, and flowering phenology can all be used to restrict access and impose "honest" visitation. There are both elements of a love story (they both benefit) and of an arms race (they are both seeking mutually assured exploitation) in the coevolution of bees and flowering plants.

In Chapter 13, we turn to the economic value of wild bees as crop pollinators. Solitary bees are fascinating creatures in their own right, but do they impact humans in any really important (economic) way? The answer is clearly yes. Wild (mostly solitary) bees are effective, but underappreciated, pollinators of many economically important crops, including apples and other early-spring flowering trees, blueberries, cranberries, strawberries, watermelons, eggplants, tomatoes, squash and pumpkins, and even coffee. We describe how one calculates the economic value of any single crop pollinator and how wild bees and honey bees compare, in terms of effectiveness, as pollinators. More and more studies are documenting the important role that wild bees play in agricultural pollination.

If solitary bees are important, what are the threats to their long-term viability? In Chapter 14, we explore the diverse threats to solitary bees, including habitat loss, pesticide use, pathogen spillover, loss of genetic diversity, climate change, and invasive species. Solitary bees can be surprisingly resilient in the face of these threats, but there are limits. Highly disturbed, anthropogenically modified habitats generally host a greatly reduced bee fauna of habitat and host-plant generalists. The bee fauna in these habitats can sometimes end up consisting entirely of what we refer to as "trash bees"—generalists that can hang on in the most dramatically altered sites. Needless to say, these habitats cannot support the diversity of specialist and brood-parasitic bees that make more pristine habitats so interesting to a bee biologist.

If you want to gain a glimpse into the extraordinary world of solitary bees, pull up a lawn chair and watch a nesting aggregation of these amazing creatures. You will see females coming and going with nesting materials, pollen, nectar, and other floral rewards. You will see clouds of sex-starved males pouncing on females in an effort to produce offspring. You will see nest parasites, like meloid beetles and bombyliid flies, and brood-parasitic (cuckoo) bees skulking around among the nests in an attempt to lay their eggs. All of this can be viewed in an area smaller than your living room and with nothing more than a lawn chair, good eyes, and patience. It is, as Tennyson wrote, "nature red in tooth and claw." And it is happening in your backyard.

In summary, solitary bees are both biologically fascinating and economically important. They are the "little creatures who run the world," as E. O. Wilson likes to describe insects in general. Yet the study of solitary bee biology has languished in the last few decades. Much of the literature documenting the extraordinary variety of solitary bee life history dates to the 1960s, '70s, '80s, and '90s. This vast literature is essentially inaccessible to most interested naturalists, and observations buried in the primary scientific literature are quickly forgotten. We hope to rekindle interest in solitary bee biology by placing observations of bee natural history into a more modern evolutionary perspective. New insights into bee phylogeny will help us provide an evolutionary framework for understanding variation in behavior and life history. Such a modern synthesis is desperately needed before we lose track of just how remarkable these animals are.

CHAPTER 2

Bee Phylogeny, Bee Diversity, and the Distinction between Solitary and Social Bees

Phylogenies are beautiful things. A phylogeny is a simple, two-dimensional, bifurcating diagram that captures the evolution and diversification of a group of organisms through time.[1] When combined with biogeographic data, phylogenies allow us to infer both dispersal events and vicariance (when a lineage splits due to geological events, such as the breakup of continents, new island formation, or the uplifting of a new mountain range). When combined with data on behavior, life history, ecology, or morphology, phylogenies help us better understand the origin, loss, and parallel evolution of key traits and to predict the traits of unstudied species. And when combined with information on the fossil record, phylogenies can give us a glimpse of the temporal patterns of diversification (i.e., a "time line" of evolution). These simple, elegant, informative stick diagrams are the foundation of comparative biology.

But phylogenies are also hypotheses that can change as we obtain new data or apply new methods of phylogeny reconstruction. Because phylogenies represent historical events that (typically) happened long before humans appeared on earth, we can never actually know if we have the "right" set of relationships. Only when multiple lines of evidence and multiple methods of analysis converge on the same tree can we be confident that the true evolutionary relationships are accurately represented. When phylogenies converge consistently on the same set of topological relationships, a stable system of classification can be developed that encapsulates these relationships.

Our understanding of the phylogeny of bees (and their close relatives, the hunting wasps) has improved considerably over the past 20 years as new molecular data and powerful new methods of phylogenetic analysis have become available. An improved understanding of bee and wasp phylogeny has allowed us to more accurately pinpoint the precise group of hunting wasps from which

---

[1] The importance of phylogenies in modern biology is illustrated by the fact that the one figure Charles Darwin chose for the first edition of the *Origin of Species* was a phylogeny.

bees arose (Sann et al. 2018). An improved understanding of family-level phylogenies in bees has allowed us to reexamine old theories on the historical biogeography of bees (Almeida et al. 2011) and the evolution of some key traits, such as the function of the bifid glossa in Colletidae (Almeida 2008). Molecular phylogenies, when combined with the bee fossil record, have provided new insights into how nest architecture impacts diversification (Litman et al. 2011). Phylogenetic studies have also provided important new insights into the evolution of host-plant associations in bees (Sedivy et al. 2013c, Sipes and Tepedino 2005), the evolution of sociality (Gibbs et al. 2012a, Romiguier et al. 2016), and the evolution of brood parasitism (Cardinal et al. 2010, Sedivy et al. 2013a). In this chapter, we describe the current status of bee (and wasp) phylogeny and classification in order to establish a framework for the future chapters on the biology of solitary bees.

BEE ORIGINS FROM APOID WASPS

What are bees, and more specifically, what are solitary bees? In order to answer this question, we need to step back and consider how bees are related to closely related hunting wasps (Box 2-1). In older classifications (e.g., Brothers 1975), bees (superfamily Apoidea) were thought to be the sister group to a monophyletic group of hunting wasps (superfamily Sphecoidea; Fig. 2-1a). The major difference between bees and hunting wasps is that bees have switched from gathering insects and other arthropods as food for their larvae to collecting protein-rich pollen for larval nutrition; bees are "vegetarian" hunting wasps. Largely based on this biological distinction, it has been assumed that these two groups were closely related, but evolutionarily distinct. Unfortunately, this simple classification subsequently turned out to be wildly incorrect. A series of studies published in the 1980s and 1990s, using morphological data and cladistic methodology, found that bees are actually nested within the "Sphecoidea" (Alexander 1992, Lomholdt 1982, Melo 1999, Prentice 1998, reviewed in Debevec et al. 2012) as sister group to the wasp family Crabronidae (Fig. 2-1b,c). In other words, bees aren't closely related to spheciform wasps; they *are* spheciform wasps. The placement of bees (then called Apoidea) within the "Sphecoidea" required a revised classification. The term Sphecoidea fell into disfavor because it no longer referred to a monophyletic group (see Box 2-2 for a glossary of phylogenetic terms). One simple solution to this problem was to expand the definition of "Apoidea" to include both bees and the four hunting wasp families. The hunting wasps are now typically referred to as the "apoid wasps" (see Box 2-1), and "Anthophila" is now a widely used term to describe the bees.

But the story does not end there. The advent of DNA sequencing opened the door to a whole new approach to resolving insect phylogeny. Initial attempts to resolve the phylogeny of the apoid wasps and bees using molecular data were based on relatively small numbers of genes and taxa (Ohl and Bleidorn 2006, Pilgrim et al. 2008). However, these studies indicated quite strongly that bees are not the sister group to Crabronidae; they are nested *within* the family Crabronidae. Ohl and Bleidorn (2006), based on a single nuclear gene (opsin), found support for placement of bees as the sister group to the crabronid subfamily

BOX 2-1: THE HUNTING WASPS

The term hunting wasp (or apoid wasp) refers to 4 families of parasitic and predatory wasps that were historically placed in the superfamily Sphecoidea (Brothers 1975) but are now placed in Apoidea, along with the bees: Heterogynaidae, Ampulicidae, Sphecidae, and Crabronidae. These 4 families are key to understanding bee origins because they are the closest, non-pollen-feeding relatives of bees. **Heterogynaidae** is a tiny family of just 8 species of enigmatic wasps (Ohl and Bleidorn 2006). They are minute (<5 mm) and mostly black, and while males are winged, females are flightless and brachypterous. We know almost nothing about their biology, but they are possibly parasites of other wasps. **Ampulicidae** (cockroach wasps) includes roughly 200 species of ant-like, fast-moving wasps, often with metallic coloring, that rear their offspring on paralyzed adult cockroaches. **Sphecidae** includes 220 species of large, slender, predatory wasps that prey primarily on spiders, caterpillars, and grasshoppers and their near relatives. The family **Crabronidae** shares many features with bees, including a subterranean nest, central place foraging, and landmark learning. Crabronidae includes nearly 9,000 species (or roughly 90% of apoid wasps) with diverse host preferences (including spiders, springtails, mayflies, grasshoppers and relatives, true bugs, thrips, caterpillars, flies, beetles, and even other Hymenoptera). The vast majority of species are host-specific predators that attack living prey, but some (*Microbembix*) are generalist scavengers that provision their brood cells with dead arthropods. Some crabronids are, ironically, even bee hunters (see Chapter 11). While the vast majority of crabronid wasps are solitary, with a single female occupying each nest, there are some eusocial species in the neotropics, like *Microstigmus* in the Pemphredoninae (Matthews 1968, Ross and Matthews 1989), as well as brood-parasitic species that attack other crabronid wasps (the genus *Stizoides* and the tribe Nyssonini; O'Neill 2001). This family is particularly important for understanding bee origins, because, as we describe in the text, bees are essentially vegetarian crabronid wasps. Good overviews of solitary wasp behavior are provided by Evans and West-Eberhard (1970) and O'Neill (2001).

Philanthinae (Fig. 2-1d). Unfortunately, Ohl and Bleidorn (2006) did not include Pemphredoninae, the subfamily of mostly aphid- and thrips-hunting wasps that Malyshev (1968) hypothesized were closely related to bees. Debevec and colleagues (2012) combined published data from the Ohl and Bleidorn (2006) and the Pilgrim and colleagues (2008) studies with new sequence data for a broader sample of crabronid wasps (including Pemphredoninae) and found support for placement of bees as either the sister group to Philanthinae (the beewolves), the sister group to Pemphredoninae (the aphid hunters), or the sister group to a monophyletic group including both subfamilies (Fig. 2-1e). This helped narrow the possible candidates for the sister group to the bees, but the Debevec and

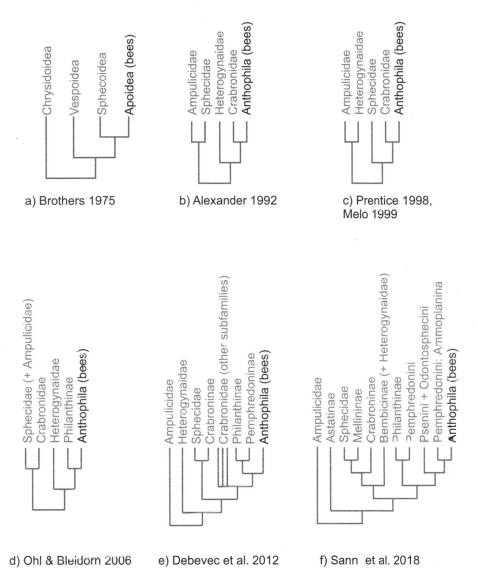

Figure 2-1. Phylogenetic hypotheses for the relationships among bees and hunting wasps: (a) Brothers (1975); (b) Alexander (1992); (c) Prentice (1998) and Melo (1999); (d) Ohl and Bleidorn (2006); (e) Debevec et al. (2012); (f) Sann et al. (2018).

colleagues (2012) study did not include a broad sample of Pemphredoninae and was based on data from just three genes.

A recent study by Sann and colleagues (2018) provides yet another hypothesis for bee origins—that bees are essentially highly derived pemphredonine wasps (Fig. 2-1f). The subfamily Pemphredoninae includes just over 1,000 described species placed into four tribes: Entomosericini, Odontosphecini, Pemphredonini, and Psenini. The prey of Entomosericini and Odontosphecini are unknown. Psenini prey on plant-sucking Homoptera, including Cicadellidae, Membracidae, Cercopidae, and Psyllidae. Pemphredonini (which includes the bulk of the genera

BOX 2-2: A GLOSSARY OF PHYLOGENETIC TERMINOLOGY

Monophyletic group. A taxonomic group that includes all the descendants of a single, common ancestor. Taxonomists always strive to define "monophyletic" groups in their classifications.

Paraphyletic group. A group that includes some, but not all, of the descendants of a single common ancestor. Reptilia is an example of a paraphyletic group because both mammals (Mammalia) and birds (Aves) arose from within Reptilia. Taxonomists generally try to avoid recognizing paraphyletic groups when they establish classifications.

Sister group. When two lineages are each other's closest relatives, they are termed sister groups. Sister groups, by definition, have the same age.

and species of Pemphredoninae) prey on aphids, scales, thrips, and Collembola. Sann and colleagues (2018) analyzed over 195 protein-coding genes and included apoid wasps from all four families, with emphasis on Pemphredoninae[2] and Philanthinae. They found strong support for the grouping of bees as the sister group to the subtribe Ammoplanina—a small group of just 134 species of thrips hunters. These results have huge implications for bee origins (discussed in Chapter 12). Instead of bees being the sister group to all four apoid wasp families, as inferred by Brothers over 40 years ago, bees are actually highly derived, pollen-feeding descendants of a very small group of thrips-hunting wasps (see Chapter 12 for a more in-depth discussion of this topic).

PHYLOGENY OF THE BEES

Since their origin approximately 120 million years ago (mya), bees have diversified into a group of seven families, 28 subfamilies, 67 tribes, 529 genera, and over 20,000 described species. There are five times as many bee species as mammal species, and bees outnumber birds three to one. Fishes are the only vertebrate group that is comparable in size to bees. Our understanding of bee family, subfamily, and tribal-level relationships has changed substantially over the past 20 years. The traditional view of bee phylogeny held that Colletidae, the cellophane bees, were the "basal" or earliest branch of bee phylogeny. However, recent molecular studies have supported a different view—that the family Melittidae represents the basal branch of bee phylogeny, and the Colletidae are highly derived bees that arose later in bee evolution (reviewed in Danforth et al. 2013).

Figure 2-2 presents our current best estimate of the phylogeny of the seven bee families and 28 subfamilies with information on their life history and sociality. We cover the diversity of social behaviors among bees in more detail below, and we provide a short description of each family in the informational boxes that

---

[2] Sann et al. (2018) included representatives of three of the four tribes of Pemphredoninae (Odontosphecini, Pemphredonini, and Psenini). The tribe Entomosericini was not included in their study.

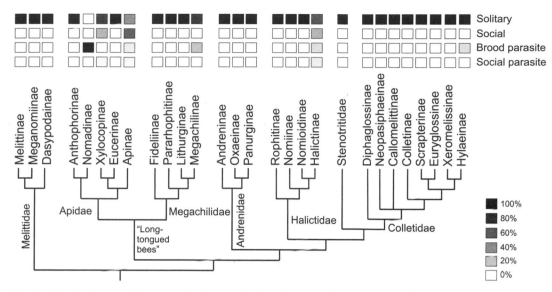

Figure 2-2. Phylogeny of the bee families and subfamilies based on recent morphological and molecular studies. Boxes along the top indicate the relative abundance of solitary, social, brood-parasitic (cleptoparasitic), and socially parasitic species.

accompany this chapter (see Boxes 2-3 to 2-9). This phylogeny is a composite of multiple studies including studies at the family level (Branstetter et al. 2017, Cardinal and Danforth 2013, Danforth et al. 2006a,b, Hedtke et al. 2013) as well as studies focused specifically on subfamily and tribal relationships within families. The phylogeny of Melittidae was analyzed most recently by Michez and colleagues (2009b), Apidae by Cardinal and colleagues (2010) and Bossert and colleagues (2018), Megachilidae by Litman and colleagues (2011) and Gonzalez and colleagues (2012), Andrenidae by Ascher (2004), Halictidae by Gibbs and colleagues (2012b), and Colletidae by Almeida and Danforth (2009).

The phylogeny presented in Figure 2-2 provides the framework we will use for examining variation in behavior, life history, nesting biology, host-plant associations, cleptoparasitism, and even morphology. We will refer back to this tree at various times throughout this book, and readers may wish to flag this figure for future reference.

HOW OLD ARE BEES?

The bee fossil record is fairly extensive, with compression and amber fossils from all bee families except Stenotritidae (Michez et al. 2012). The oldest fossil bee that can be placed in one of the seven extant families is *Cretotrigona prisca* from New Jersey amber (~65 mya; Michener and Grimaldi 1988). This remarkable fossil is strikingly similar to extant stingless bees in the tribe Meliponini, a group of highly social bees that occur throughout the tropical regions. An older fossil (*Melittosphex burmensis*) from Burmese amber (~100 mya) shows a number of bee-like features, including branched hairs, but does not appear to fit into

BOX 2-3: FAMILY MELITTIDAE

Melittidae is one of the smallest bee families, with a total of just 201 described species (Michez et al. 2009b). Melittidae is also an ancient, possibly relictual, bee family that is well-represented in the fossil record back to the Eocene (~53 mya; Michez et al. 2007). Melittid bees are solitary, exclusively ground-nesting, mostly host-plant specialist bees that occur in the temperate, xeric, and Mediterranean climate regions of the Old World and the Nearctic (Michener 1979). The greatest diversity of melittid genera, tribes, and subfamilies occurs in Africa, especially arid regions of southern Africa, where all 3 subfamilies co-occur (Michener 1979). Melittidae is absent from Australia and South America.

Melittidae: *Dasypoda argentata* (Melittidae: Melittinae). Original artwork by Frances Fawcett.

Dasypodainae includes sand-loving, mostly desert bees with narrow host-plant preferences, including *Hesperapis* (in arid North America), *Eremaphanta* (in Central Asia), *Capicola* and *Samba* (in South Africa), and *Dasypoda* (in the Palearctic). Some species (e.g., *H. oraria*) are reported to be monolectic, specializing on a single species of host plant (Cane et al. 1996). **Melittinae** includes *Melitta*, the most widespread genus of melittid bees (Michez and Eardley 2007), and 2 oil-collecting genera: *Rediviva* (22 species) and *Macropis* (16 species). These "oil bees" are treated in detail in Chapter 7. **Meganomiinae** includes large, mostly black-and-yellow, fast-flying species. It is the smallest of the melittid subfamilies, with just 4 genera and 10 described species. This subfamily is largely restricted to sub-Saharan Africa (with one undescribed species reported from Yemen). Unlike the other subfamilies of Melittidae, at least 1 species of *Meganomia* has been shown to be polylectic (Michez et al. 2010).

BOX 2-4: FAMILY ANDRENIDAE

Andrenidae is a large family of nearly 3,000 described species in 3 subfamilies and 8 currently recognized tribes (Fig. 2-2). Andrenidae is a widely distributed family (excluding Australia), with greatest diversity in arid western North America, South America, and the Palearctic. The fossil record of Andrenidae includes several compression fossils from Florissant deposits of Colorado placed tentatively in the subfamily Andreninae (~32 my old; Dewulf et al. 2014) and an amber fossil from the Dominican Republic placed in the subfamily Panurginae (~20 my old; Rozen 1996). All species are solitary, ground-nesting bees. Communal nesting occurs in several genera (including *Andrena*, *Oxaea*, *Panurgus*, *Perdita*, and *Macrotera*). There are no known andrenid cleptoparasites. Many andrenid bees have narrow host-plant preferences, with both behavioral and morphological adaptations to accessing host-plant resources.

Andrenidae: *Protandrena mexicanorum* (Andrenidae: Panurginae). Original artwork by Frances Fawcett.

The subfamily **Andreninae** consists of 6 genera. Of these, 5 genera (*Alocandrena*, *Ancylandrena*, *Megandrena*, *Orphana*, and *Euherbstia*) include a total of just 12 species restricted to arid regions of western North America, Peru, and Chile (Ascher 2004). The remaining genus (*Andrena*) includes over 1,500 species with a mostly Holarctic distribution. Most andrenine species are oligolectic, and the preferred host-plant families of oligolectic *Andrena* include Asteraceae, Apiaceae, Brassicaceae, Ericaceae, Fabaceae, and Rosaceae (Larkin et al. 2006). The subfamily **Panurginae** includes 32 genera and more than 1,300 species. Most panurgines are narrow host-plant specialists. Panurgines are particularly diverse in arid regions of the Western Hemisphere, the southern Palearctic, and Africa. Many species exhibit morphological adaptations to gathering and transporting the pollen and nectar of their preferred host plant. Nearly one-half of panurgine species are in the North American genus *Perdita*, which consists almost entirely of narrow host-plant specialists on an enormous diversity of plant families, including Asteraceae, Papaveraceae, Zygophyllaceae (*Larrea*), Solanaceae, Fabaceae (*Prosopis*), Ericaceae, Boraginaceae, Hydrophyllaceae, and many others (Krombein et al. 1979). The remaining subfamily, **Oxaeinae**, includes 21 species of large, neotropical, fast-flying bees that show a strong preference for flowers with poricidal anthers, such as Solanaceae, some Fabaceae, and Melastomataceae.

BOX 2-5: FAMILY HALICTIDAE

Halictidae is the second-largest family of bees, with nearly 4,500 described species. Halictid fossils are well represented in both Dominican (~23 mya) and Baltic (~42 mya) amber deposits (Engel 2001). However, fossil-calibrated phylogenies suggest that halictids could be much older (between 75 and 96 mya; Cardinal and Danforth 2013).

Halictidae: *Agapostemon angelicus* (Halictidae: Halictinae). Original artwork by Frances Fawcett.

Relationships among the halictid subfamilies are well established (Fig. 2-2). The basal subfamily, **Rophitinae**, is unique among halictid bees in that most species are narrow host-plant specialists. Examples include *Ceblurgus*, which are host-plant specialists on *Cordia* (Boraginaceae); *Xeralictus*, which are host-plant specialists on 2 closely related genera of Loasaceae (*Eucnide* and *Mentzelia*); and *Systropha*, which includes narrow host-plant specialists on Convolvulaceae (*Convolvulus, Ipomoea, Merremia*). Some species may even be monolectic (e.g., *Conanthalictus conanthi* on *Nama hispidum* [Hydrophyllaceae]; Rozen and McGinley 1976). Host-plant preferences in Rophitinae were reviewed by Patiny and colleagues (2007).

Nomiinae, which includes just over 600 species, is a primarily paleotropical group with a diversity of genera in the African and Asian tropics and a small number of genera in Europe and North America. They are absent from South America. Nomiinae includes the only ground-nesting, solitary bee ever managed for commercial pollination: *Nomia melanderi* (Bohart 1972). Nomiinae are a biologically fascinating group, with bizarre and elaborate male morphologies (mostly involving hindlegs [see Ribble 1965 for illustrations] and genitalia) and courtship behaviors involving acoustic communication (Wcislo and Buchmann 1995, Wcislo et al. 1992). Social behavior varies from species that nest solitarily to communal associations (Batra 1966, Vogel and Kukuk 1994, Wcislo 1993, Wcislo and Engel 1996). Some species are host-plant specialists (Minckley et al. 1994), while others are clearly polylectic (Wcislo 1993).

Nomioidinae includes small to tiny metallic blue-green and yellow bees that occur primarily in arid regions of southern Europe, Africa and Madagascar, and central Asia; a single species, *Ceylalictus* (*Ceylalictus*) *perditellus*, occurs in Australia. All nomioidines are ground-nesting, solitary, or communal bees (Danforth et al. 2008).

Finally, the subfamily **Halictinae** is the largest group, with over 3,000 described species, or roughly 80% of all halictid bees. Halictinae is divided into 5 tribes. Augochlorini and Halictini include social taxa (described in more detail in the main body of the chapter). Thrinchostomini and Caenohalictini include solitary bees and a few cleptoparasites, and Sphecodini is exclusively composed of cleptoparasitic species. Within the solitary Halictinae, there are some remarkable bees. The paleotropical genus *Thrinchostoma* includes large bees, some of which are host-plant specialists (e.g., the long-faced, Asian subgenus *Diagonozus* are narrow host-plant specialists on the genus *Impatiens*). Some Caenohalictini (e.g., *Rhinetula*) and some Halictini (e.g., the *Oenothera* specialists within the *Lasioglossum* subgenus *Sphecodogastra*) are matinal, crepuscular, and even nocturnal. Several species of Australian *Lasioglossum* are communal, and some species are known in which males are dimorphic; large-headed flightless males remain within the nest, and small-headed flight-capable males can be collected on flowers (Kukuk and Schwarz 1987, 1988).

BOX 2-6: FAMILY COLLETIDAE

The 2,600 described colletid bees are an important group from the perspective of bee phylogeny. They were originally considered to be "primitive" bees because of the possession of a bifid (forked) glossa that is similar to that of the crabronid wasps. We now know that the bifid glossa of Colletidae is likely a derived trait related to the application of the unique cellophane brood-cell lining that is characteristic of this family (Almeida 2008). All colletid bees are solitary with the exception of 5 species of cleptoparasitic *Hylaeus* (*Nesoprosopis*) in Hawaii (Daly and Magnacca 2003). Colletid bees have their greatest diversity in the Southern Hemisphere continents of Australia (where half of the native bee species are colletids) and South America. Phylogenetic studies (Almeida et al. 2011) have documented frequent interchanges between South America and Australia via Antarctica over the course of colletid evolution. Antarctica would most likely have had a fascinating colletid fauna prior to becoming frozen under miles of ice. Colletids were likely the first group of bees to colonize Australia (approximately 92 mya), well before the arrival of Megachilidae, Apidae, and Halictidae via southeast Asia (Almeida et al. 2011).

Colletidae: *Xeromelissa rozeni* (Colletidae: Xeromelissinae). Original artwork by Frances Fawcett.

The subfamily **Diphaglossinae**, an exclusively New World, mostly tropical group, includes 130 large, fast-flying bees that are often matinal and/or crepuscular foragers. **Colletinae** includes 3 small South American genera (*Hemicotelles*, *Mourecotelles*, and *Xanthocotelles*) plus the large, cosmopolitan, morphologically homogeneous genus *Colletes*. **Neopasiphaeinae** is a morphologically diverse group of bees that are most species-rich in the Australian and neotropical regions, mainly in subtropical and temperate dry biomes. Neopasiphaeinae includes most (but not all) of the genera that were placed previously in the subfamily Paracolletinae. **Callomelittinae** are an enigmatic group of 11 wood-nesting bees restricted to Australia. **Hylaeinae** and **Euryglossinae** are both small to medium-sized, slender, relatively hairless, wasp-like bees that are unusual among bees in transporting pollen internally within the crop. Euryglossinae are endemic to the Australian region, and Hylaeinae have their greatest generic diversity in the Australian region, with a single, cosmopolitan genus (*Hylaeus*) occurring outside of Australia. **Scrapterinae** includes a single genus (*Scrapter*) comprising approximately 60 species that are endemic to Africa, especially the Cape region. *Scrapter* is an enigmatic group that appears to have arrived in southern Africa via long-distance dispersal from Australia approximately 24 million years ago (Almeida et al. 2011). **Xeromelissinae** are small, slender, relatively hairless bees that are restricted to South and Central America. There are 130 described species, with the highest diversity in temperate regions of Chile and Argentina. Members of the subfamilies Diphaglossinae, Neopasiphaeinae, Scrapterinae, Euryglossinae, and Colletinae are mostly ground-nesting bees, Hylaeinae inhabit stems and preexisting cavities, Callomelittinae nest in punky wood, and Xeromelissinae are both stem and ground nesters (Almeida 2008).

BOX 2-7: FAMILY STENOTRITIDAE

Stenotritid bees are an enigmatic group of ancient, solitary, ground-nesting bees restricted to Australia (primarily western Australia). There are just 21 species in 2 genera (*Stenotritus* and *Ctenocolletes*), making this group the smallest family of bees with the most limited geographic range. There are no known stenotritid fossils, but they are estimated to have diverged from Colletidae more than 92 million years ago (Almeida et al. 2011). Stenotritids are large, fast-flying bees, and one species (*C. smaragdinus*) is bright metallic green. They prefer open, sandy, heathland habitats. The evolutionary (phylogenetic) affiliations of Stenotritidae have historically been extremely confusing. Previous hypotheses included their placement as (1) the sister group to all other bee families, (2) the sister group to the andrenid subfamily Oxaeinae, and (3) within Colletidae. Recent phylogenetic studies based on molecular data place Stenotritidae unambiguously as the sister group to Colletidae (Fig. 2-2).

The nesting, mating, and foraging behavior of Stenotritidae have been described in detail by Terry Houston (Houston 1975, 1984, 1987; Houston and Thorp 1984), and Almeida (2008) included Stenotritidae in his review of colletid nesting biology. Stenotritidae are primarily vernal bees, but one species (*C. fulvescens*) is active in late summer and fall. Stenotritids have been collected from flowers of many plant families, but they seem to be primarily associated with Myrtaceae, the predominant flowering-plant family in Australia (Houston 1984).

All are ground-nesting, with nests consisting of a single main burrow and a small number of subterranean brood cells (Almeida 2008). Nests can be extraordinarily deep—up to 3 meters deep in some species (*C. albomarginatus* and *C. nicholsoni*; Houston 1987). Stenotritids do not produce a thick cellophane cell lining, as in Colletidae, and they lack the bifid glossa that characterizes the Colletidae (McGinley 1980). As Houston (1984) described, male mating behavior entails both fast patrolling of potential host plants, patrolling over active nest sites, and hovering in stationary territories. In many species, males and females fly *in copula*, suggesting some form of mate guarding in these bees (Houston 1987).

BOX 2-8: FAMILY MEGACHILIDAE

The family Megachilidae is the third-largest family of bees, with just over 4,000 described species. Megachilidae has a relatively rich fossil record (reviewed by Engel and Perkovsky 2006). There are a diversity of megachilid fossils in Baltic amber (Engel 2001), including 2 distinct tribes of extinct Megachilinae (Glyptapini and Ctenoplectrellini; Gonzalez et al. 2012). *Probombus hirsutus*, the oldest fossil megachilid, is a compression fossil recorded from the late Paleocene (~60 mya). This fossil is clearly a megachilid, but assigning it to any of the extant subfamilies is challenging (Nel and Petrulevicius 2003). Based on the analysis of Litman and colleagues (2011), the Megachilidae are estimated to be between 100 and 120 million years old.

Megachilidae: *Fidelia pallidula* (Megachilidae: Fideliinae).
Original artwork by Frances Fawcett.

Megachilids occur on all continents except Antarctica and occupy a broad range of habitats, from lowland tropical rain forests to deserts. Members of this family use an extraordinary diversity of materials for nest construction, including mud, flower petals, leaves, plant resin, soil, gravel, plant trichomes, and plastic shopping bags (in urban habitats; MacIvor and Moore 2013). They also nest in an amazing diversity of substrates, including walls, stones, and tree branches, and in preexisting cavities in the ground, stems, galls, snail shells, and arboreal termite mounds. Many Megachilidae are host-plant specialists, and cleptoparasitism has arisen repeatedly in the group (19 genera and 668 species are known to be cleptoparasites). This family also includes some of the most important managed pollinators, including *Megachile rotundata* and *Osmia lignaria*.

Subfamily, tribal, and generic relationships have been examined based on both morphological (Gonzalez et al. 2012) and molecular (Litman et al. 2011, 2016) data. The current classification recognizes 4 subfamilies (Fig. 2-2). **Fideliinae** are a fascinating group of relictual, sand-loving, desert bees present in southern Africa, western South America, and Morocco. The split between South America and Africa dates to over 100 million years ago, and fossil-calibrated phylogenies (Litman et al. 2011) have provided solid evidence that Fideliinae pre-dates the breakup of South America and Africa, approximately 35 million years before the extinction of the dinosaurs. **Pararhophitinae**, a distinct group of just 3 species, are also desert bees that range from central Asia to North Africa. **Lithurginae** are widely distributed and nest in wood, stems, and even cattle dung (Sarzetti et al. 2012). Lithurgines are narrow host-plant specialists with preferences for Malvaceae, Cactaceae, Convolvulaceae, and Asteraceae. The subfamily **Megachilinae** includes the majority of species in the family and comprises extraordinarily variable life histories, including mason bees, resin bees, leaf-cutter bees, wool-carder bees, and many brood parasites. Based on an analysis of diversification rates in Megachilidae, the transition from producing unlined brood cells (the primitive condition for the family and the condition for the 3 "basal" subfamilies) to lining brood cells with materials collected from outside the nest led to a significant increase in diversification and a major range expansion from arid, desert habitats currently occupied by Fideliinae and Pararhophitinae. The diverse nesting materials used by Megachilinae may have allowed them to "escape from the desert" (Litman et al. 2011).

any extant bee family (Danforth and Poinar 2011, Poinar and Danforth 2006). Other amber fossils have been described from Oise, France (53 mya; Michez et al. 2007), the Baltic Sea (42 mya; Engel 2001), and the Dominican Republic (23 mya). Extraordinary compression fossils of apid bees have recently been described from Menat, France (60 mya; Dehon et al. 2017, Michez et al. 2009a).

To estimate the antiquity of bees, Cardinal and Danforth (2013) combined a molecular data set for extant taxa with information on the affinities and antiquity of fossil bees to calculate the antiquity of bees as a whole (Fig. 2-3). Their

BOX 2-9: FAMILY APIDAE

Apidae is the largest family of bees, with nearly 6,000 described species and 5 currently recognized subfamilies: Anthophorinae, Nomadinae, Xylocopinae, Eucerinae, and Apinae (Bossert et al. 2018). The fossil record of Apidae extends further back than any other group. In fact, the oldest fossil crown-group bee (*Cretotrigona prisca*), which is estimated to be late Cretaceous, is clearly a member of the extant tribe Meliponini. Fossil-calibrated phylogenies indicate that the family likely arose between 95 and 115 million years ago (Cardinal and Danforth 2013).

Apidae: *Apis mellifera* (Apidae: Apinae). Original artwork by Frances Fawcett.

Apids are by far the most thoroughly studied and familiar bee group. Apidae includes the honey bee (*Apis mellifera*) and 300 species of large, charismatic bumble bees (*Bombus*). Together, honey bees and bumble bees are the most important managed, commercial pollinators. But Apidae also includes an enormously diverse array of solitary and cleptoparasitic lineages, including the long-horned bees (Eucerini), anthophorine bees (Anthophorini), large carpenter bees (Xylocopini), Old and New World oil-collecting bees (*Ctenoplectra, Centris, Chalepogenus, Tetrapedia, Paratetrapedia*, and several others; described in more detail in Chapter 7), and a diversity of brood parasites in 3 subfamilies (Apinae, Nomadinae, and Xylocopinae). In fact, approximately 27% of apid bees are brood parasites—the highest percentage of any bee family.

The phylogeny of Apidae was analyzed most recently by Silas Bossert and colleagues (Bossert et al. 2018). Their revised classification expands the number of apid subfamilies from 3 to 5: Anthophorinae, Nomadinae, Xylocopinae, Eucerinae, and Apinae. **Anthophorinae** includes solitary, largely ground-nesting bees that have been placed in the tribe Anthophorini (e.g., *Anthophora*, *Amegilla*, *Habropoda*, *Pachymelus*, and relatives). **Nomadinae** includes over 1,500 species of exclusively brood parasitic (cleptoparasitic) bees. The various modes of brood parasitism are described in detail in Chapter 10. Nomadinae in our sense includes several tribes of brood-parasitic bees previously placed in the subfamily Apinae (Ericrocidini, Isepeolini, Melectini, Osirini, Protepeolini, Rhathymini, and the genus *Coelioxoides*). This clade of brood parasites has been recovered in previous studies and was previously referred to as the apid "cleptoparasitic clade" (Cardinal et al. 2010). **Xylocopinae** refers to wood- and cavity-nesting groups traditionally placed in Xylocopinae sensu stricto (Xylocopini, Ceratinini, Manueliini, Allodapini), plus 2 tribes of "oil bees": Tetrapediini and Ctenoplectrini (previously placed in Apinae). Xylocopinae includes bees with a diverse range of social behaviors. Tetrapediini, Ctenoplectrini, and Manueliini are solitary; the tribes Ceratinini and Xylocopini are solitary and cooperatively breeding; and the tribe Allodapini includes cooperative breeders and a small number of eusocial species. Social behavior in this group is described in more detail in the main body of this chapter. **Eucerinae** includes the tribes Ancylini, Emphorini, Eucerini, Exomalopsini, and Tapinotaspidini—all solitary, ground-nesting bees. **Apinae** includes the tribe Centridini (solitary, largely oil-collecting bees including the genera *Centris* and *Epicharis*), as well as approximately 1,000 species of "corbiculate" bees: Euglossini, Bombini, Meliponini, and Apini. We discuss this group in some detail in the body of this chapter, under the section on social behavior.

estimates varied based on starting parameters but generally converged on an age of around 125 million years ago. This is an intriguing (but plausible) age for bees. While angiosperm (flowering) plants arose considerably earlier (~140 mya), a well-preserved pollen fossil record indicates the eudicots are precisely 125 million years old. Today, eudicots comprise 75% of flowering plant species, and many are bee-pollinated.

One conclusion to draw from the Cardinal and Danforth (2013) study is that all the bee families (except Stenotritidae) arose well before the mass extinction event that was triggered by the impact of a comet that hit the Yucatán Peninsula (the Cretaceous/Paleogene [K-Pg] mass extinction; Schulte et al. 2010) at the end of the Cretaceous (65.6 mya; Fig. 2-3). This event has been associated with a massive extinction of dinosaurs as well as flowering plants (Labandeira et al. 2002). But did this extinction impact bees? Only one study has examined this question, and the evidence would suggest that it did. Rehan and colleagues (2013) examined diversification rates in xylocopine bees prior to and shortly after the K-Pg. Their analysis documents a slow rate of diversification prior to the K-Pg boundary, followed by a rapid rate increase shortly after the K-Pg boundary. Their results are certainly suggestive that this group of stem- and wood-nesting bees experienced a significant extinction event. It is interesting to

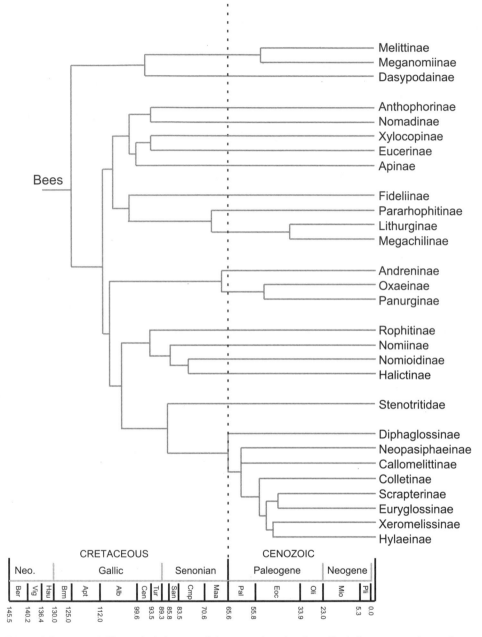

Figure 2-3. A time-calibrated phylogeny of the extant bee families. Clade (crown group) ages from various sources, including Cardinal and Danforth (2013), Branstetter et al. (2017), and Cardinal (2018). The K-Pg boundary is indicated with a dashed line.

consider if some other lineages of bees went extinct at the K-Pg boundary, but in the absence of a much more complete Cretaceous fossil record, it is difficult to evaluate this possibility.

THE OTHER "BEES"—THE VESPID WASP SUBFAMILY MASARINAE AND THE CRABRONID WASP *KROMBEINICTUS*

If "bees" are defined as any lineage of aculeate (stinging) wasp that feeds its offspring with pollen rather than arthropod prey, it might come as a surprise that bees have actually originated more than once on earth. The "bees" familiar to most people are the lineage of 20,000 pollen-feeding solitary, social, and parasitic species that are the focus of this book (also known as the Anthophila). These are by far the most ecologically important and species-rich lineage of pollen-feeding wasps on earth.

But pollen feeding has evolved two other times from within predatory wasps. The wasp subfamily Masarinae (or pollen wasps) arose from within a predominantly predatory group of wasps in the family Vespidae. Predatory Vespidae include the potter wasps (Eumeninae), hover wasps (Stenogastrinae), primitively eusocial paper wasps (Polistinae), and highly eusocial yellow-jackets and hornets (Vespinae). At some point in time (we currently do not have a reliable estimate on the antiquity of pollen wasps), a proto-masarine wasp shifted to provisioning larval cells with pollen rather than arthropod prey, giving rise to what we now call Masarinae. Pollen wasps comprise approximately 300 species in 2 tribes and 14 genera (Gess 1996). They are biologically very similar to bees. Some species build above-ground nests of mud and others excavate below-ground burrows. They are solitary, like the vast majority of bees, and they mass-provision their brood cells with a gooey, semi-liquid mixture of pollen and nectar. Like bees, they are most diverse in arid, Mediterranean climatic regions of western North America, arid South America, the circum-Mediterranean region, western Australia, and southernmost Africa. Southern Africa, in particular the winter rainfall regions of the Fynbos, Succulent Karoo, and Nama Karoo, is the region of highest species richness of Masarinae. Like bees, many species of Masarinae are highly host-plant specific (oligolectic), and they show some of the same host-plant preferences as bees, with numerous species specialized upon the families Zygophyllaceae, Fabaceae, Euphorbiaceae, Malvaceae, Boraginaceae, and Asteraceae (Gess and Gess 2004). Masarines, like bees, have highly modified, elongate mouthparts for accessing floral resources (Krenn et al. 2002). Unlike most bees (except Euryglossinae, Hylaeinae, and some Neopasiphaeinae), masarines carry pollen internally. There are no known cleptoparasitic masarines. For an in-depth treatment of pollen wasps, we recommend Sarah Gess's book *The Pollen Wasps* (1996).

The other group of pollen-feeding wasps are much less well known than the masarines, but they are much more closely related to bees. In a remarkable, but largely overlooked, pair of papers, Karl Krombein and Beth Norden described a crabronid wasp that appears to progressively provision its larvae with pollen (Krombein and Norden 1997a,b). The wasp, *Krombeinictus nordenae* (subfamily Crabroninae, tribe Crabronini), is known from just a handful of specimens studied at one site in Sri Lanka. Female *K. nordenae* nest in the elongate, hollow internodes (stems) of *Humboldtia laurifolia*, a large, tree-like legume. Based on examination of a few active and preserved nests, Krombein and Norden reconstructed the following picture of nesting in *K. nordenae*. Each internode is occupied by a single female wasp and her developing brood. The female appears to rear one offspring at a time and progressively provisions this one larva over the course of several days. Once this larva reaches the last instar, the mother transports it to the bottom of the hollow internode, where the larva spins a cocoon. The mother then lays another egg near the entrance to the cavity, and the cycle repeats itself. Because only one larva is reared at a time, there is no need for cell partitions in the nest. Examination of the pollen collected on the mouthparts of the adult wasp, as well as on the fecal material of the post-defecating larva, led Krombein and Norden to conclude that larvae are reared on pollen from the same plant used for nesting, *H. laurifolia*.

The biology of *K. nordenae* is remarkable in many ways. First, most crabronid wasps are mass provisioners, and progressive provisioning is known to have arisen only a few times within the crabronid subfamily Nyssoninae (Evans 1966). Second, and most remarkably, pollen feeding is a trait that is unknown for crabronid wasps, except for the true bees (i.e., Anthophila). *K. nordenae* is the sole species known within the genus *Krombeinictus*, and it remains to be seen whether other, related genera might also be pollen feeders.

The existence of both the masarine wasps and *K. nordenae*—two additional, independent origins of pollen feeding in the aculeate wasps—demonstrates that the evolutionary transition from carnivory to pollenivory is not unique to bees. One has to wonder why bees, now comprising 20,000 described species, became so much more diverse than either of these two other pollen-feeding wasp lineages. It could just be age—perhaps bees arose well before either the Masarinae or *Krombeinictus* and simply had more time to diversify. Alternatively, perhaps the ability to carry pollen externally on their body was a key innovation that gave bees an evolutionary edge over other pollen-collecting wasps.

WHAT IS A SOLITARY BEE?

This book focuses on solitary bees and their brood parasites. In order to clearly identify the taxa we cover here, we need to first provide some definitions of what *solitary* means in the context of other forms of sociality in bees. The following discussion will illustrate that social categories in bees (and other insects) are not black and white. There is enormous variation in the detailed aspects of social behavior among bees, and bee species or lineages do not always fit conveniently into defined social or life-history categories. It is also important to keep in mind that detailed studies of social behavior have been carried out on only a small proportion of bee species. However, we can use our understanding of bee phylogeny to help infer the likely modes of social behavior given information on related taxa that have been studied. This combination of phylogenetic reasoning, combined with field and laboratory studies of sociality, provides a powerful predictive tool for inferring the likely social behavior exhibited by the over 20,000 bee species on earth.

We will follow a fairly simple lexicon of social and parasitic behaviors (Fig. 2-4a). First, we distinguish solitary, social, and parasitic taxa. Solitary species comprise the vast majority of bees on earth (>75%; Fig. 2-4b). Female **solitary** bees are all capable of producing offspring, they all build and maintain their own nest, and they all forage for the floral resources (primarily pollen and nectar) necessary for provisioning their brood cells. Solitary bees typically live alone, but some form **communal** nests in which multiple reproductively active females share a common main burrow. In these communal bees, each female behaves effectively as a solitary bee, and there is no obvious cooperation among nest mates.

Females of **social** taxa exhibit three key features that, when they occur together, can help define sociality in bees. First, social taxa exhibit reproductive division of labor, meaning some females reproduce (usually described as "queens"), and others build and defend the nest and gather pollen and nectar for the developing offspring (usually described as "workers"). Second, social taxa exhibit cooperative brood care (or alloparental care), meaning females help to rear offspring that are not their own. Cooperative brood care is essentially "helping" to rear individuals that are not your own offspring. Finally, sociality typically involves overlap of generations such that females are long-lived, and mothers and daughters occupy the same nest for an extended period of time.

In our lexicon, eusociality and cooperative breeding share the three defining features of social behavior in bees (described above) but differ in that reproductive division of labor is *permanent* in eusocial societies and *temporary* in cooperatively breeding taxa (Fig. 2-4a). Honey bees, bumble bees, and stingless bees, some halictid bees, and two species of allodapine bees exhibit permanent, caste-based societies. Once a female becomes a non-reproductive worker, she remains a non-reproductive worker for the rest of her life. As we shall see below,

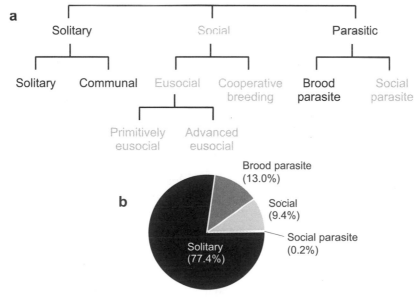

Figure 2-4. A classification of social behaviors in bees (a) and the frequency of these social behaviors across the 20,000 described bee species (b).

the story is more nuanced in the cooperative breeders, because an individual female can assume a role as a worker early in her life (helping her mother produce offspring) and later in life become an egg-laying queen (when her mother dies). Cooperative breeding is widespread in the subfamily Xylocopinae (see below) and may also characterize some Euglossini (especially *Euglossa*). Within Allodapini (a tribe of Xylocopinae), eusociality has evolved twice from cooperatively breeding groups of bees, indicating that cooperative breeding has been a precursor, at least in this group, for eusociality.

Within eusocial lineages, a distinction is often made between primitive eusociality and advanced eusociality. In primitively eusocial bees, queens and workers differ primarily in body size; queens are much larger than workers. New nests are established most often by a single, foundress queen, and nests usually (but not always) last one active season (i.e., they are annual). In advanced eusocial taxa (e.g., honey bees and stingless bees), queens and workers are morphologically distinct, nests last more than one year (i.e., they are perennial), and new nests are founded by a queen accompanied by a group of workers in a process called swarming.

The third category in our lexicon includes parasitic taxa (Fig. 2-4a). Parasitic bees come in many forms, but two broad categories can be recognized: brood parasites (cleptoparasites or "cuckoo" bees) and social parasites. Brood-parasitic bees do not build their own nest or forage for pollen and nectar. Instead, they enter the nests of free-living, pollen-collecting bees and lay an egg in a closed or open brood cell. The adult female or her first-instar larva kills the host egg or larva, and the cleptoparasitic larva then consumes the pollen and nectar provisions of the host. As we shall see in a later chapter, these cleptoparasitic bees have devious ways of attacking their hosts and concealing their eggs. Chapter 10 provides a detailed description of the brood-parasitic bees.

Another category of parasitic bees is the social parasites. Social parasites have a very different mode of parasitism from the brood parasites described above. Social parasites attack only social bee species. A female social parasite enters the nest of a social host and effectively replaces the resident queen as the egg-laying female, while, at the same time, co-opting the resident workforce to rear her own offspring. In some cases, the host queen is killed, and in others, she remains within the nest but ceases to lay eggs. In either case, the social parasite usurps the resident workers for her own devious objectives—to raise more social parasites. In both modes of parasitism, parasitic taxa often attack closely related hosts, and this is especially true for social parasites. However, for brood parasites, there are some very clear exceptions to this "rule" (Cardinal et al. 2010). Like virtually all the terms in our lexicon, the distinction between brood parasite and social parasite can be blurred. Social parasites attack only social hosts, but some brood parasites (e.g., *Sphecodes*) attack both solitary and social hosts.

The lexicon described above, while not perfect, allows us to characterize the social behavior of the 20,000 described bee species (Table 2-1). Phylogenetic studies, combined with information on social behavior for those species that have been studied in detail, indicate that over 15,000 bee species (or just over 77%) are solitary and/or communal (Fig. 2-4b). Social taxa (including the cooperatively breeding bees in the Xylocopinae and the eusocial taxa in the Halictinae

Table 2-1. Species richness across families and life history traits.
Species richness data compiled from *Discover Life* (Ascher & Pickering 2018).

Family	Subfamily	Solitary	Brood parasite	Social	Social parasite	Total
Melittidae	Melittinae	97				97
	Meganomiinae	10				10
	Dasypodainae	94				94
Apidae	Anthophorinae	742				742
	Apinae	496	9	748	31	1,284
	Eucerinae	1,194				1,194
	Nomadinae		1,565			1,565
	Xylocopinae	788	1	240	15	1,044
Megachilidae	Fideliinae	17				17
	Pararhophitinae	3				3
	Lithurginae	63				63
	Megachilinae	3,348	668			4,016
Andrenidae	Andreninae	1,539				1,539
	Oxaeinae	22				22
	Panurginae	1,396				1,396
Halictidae	Rophitinae	261				261
	Nomiinae	619				619
	Nomioidinae	94				94
	Halictinae	2,135	371	905	15	3,426
Stenotritidae		21				21
Colletidae	Diphaglossinae	130				130
	Neopasiphaeinae	421				421
	Callomelittinae	11				11
	Colletinae	516				516
	Scrapterinae	59				59
	Euryglossinae	389				389
	Xeromelissinae	130				130
	Hylaeinae	955	5			960
	Totals	15,550	2,619	1,893	61	

and Apinae) account for approximately 10% of all bee species. Brood parasites account for just over 12% of bee species. In some families (e.g., Apidae), clepto-parasites account for more than 25% of the species. By focusing on the solitary bees and their numerous brood parasites and excluding the social and socially parasitic taxa, this book covers the biology of 90% of all bee species.

It is important to identify where in the phylogeny social taxa occur in order to exclude these social taxa from our treatment of solitary bees. Social bees occur in three subfamilies (Fig. 2-2): Apinae, Xylocopinae, and Halictinae. In each of these subfamilies, the social taxa account for only between 10%

and 30% of the described species. We describe below, for each of these three subfamilies, the phylogenetic distribution of social bees and their different life histories (Fig. 2-4a).

Apinae

Seven hundred of the over 1,200 described species of Apinae (sensu Bossert et al. 2018) are eusocial. Eusociality in Apinae is restricted to a well-defined, monophyletic group referred to as the "corbiculate" bees (Fig. 2-5). Female corbiculate bees (excluding those that are brood or social parasites) have a highly modified hind tibia, which forms a concave, shiny structure (the corbicula) for carrying pollen and plant resins (Martins et al. 2014). Corbiculate bees include four tribes: Euglossini (orchid bees), Bombini (bumble bees), Meliponini (stingless bees), and Apini (honey bees). Most Euglossini are solitary or communal, but some species of *Euglossa* exhibit features typical of a more social bee, including multi-female nests, overlap of generations, and temporary division of labor in which some females forage and others guard the nest (reviewed in Cardinal and Danforth 2011). The vast majority of the 260 species of bumble bees are primitively eusocial, but the group also includes at least 30 social parasites. All species of Apini and Meliponini are advanced eusocial with morphologically distinct castes, long lives, perennial colonies, and obligate swarm founding. There are no known social parasites in Apini and Meliponini.

The phylogeny of the corbiculate tribes has been a remarkably controversial topic (reviewed by Cardinal and Danforth 2011, Cardinal and Packer 2007). The most recent analyses of transcriptomic (Romiguier et al. 2016) and genomic (Bossert et al. 2017) data support the tree shown in Figure 2-5. According to this phylogeny, the largely solitary Euglossini is the sister group to the remaining eusocial tribes. In this book, we include the mostly solitary Euglossini and exclude the eusocial tribes Bombini, Meliponini, and Apini.

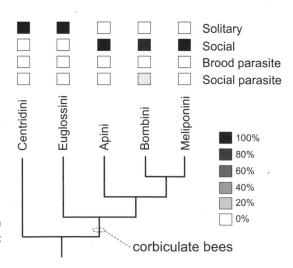

Figure 2-5. Tribal relationships within the subfamily Apinae (based on Bossert et al. 2017).

Xylocopinae

Xylocopinae refers to wood- and cavity-nesting groups traditionally placed in Xylocopinae sensu stricto (Xylocopini, Ceratinini, Manueliini, Allodapini) plus two tribes of solitary "oil bees": Tetrapediini and Ctenoplectrini (previously placed in Apinae; Fig. 2-6). Classifying the social behavior of Xylocopinae sensu stricto is particularly challenging. The tribe Manueliini includes just three species in a single genus (*Manuelia*), two of which are known to be solitary (Daly et al. 1987, Flores-Prado et al. 2008). The tribes Ceratinini and Xylocopini both include species that are truly solitary as well as species that are cooperatively breeding (Rehan et al. 2012, Groom and Rehan 2018). Sociality in these groups never entails lifelong castes. An individual female may behave as a worker in her mother's nest before eventually achieving reproductive status later in life when her mother dies (Rehan et al. 2012).

Bees in the tribe Allodapini are bizarre in all kinds of ways. Except in the genus *Halterapis*, they nest in stems but do not produce distinct brood cells. Offspring are reared through progressive provisioning, and nests consist of groups of reproductively active females that cooperate in provisioning and nest defense. Many species have well-defined behavioral castes (Schwarz et al. 2007). Allodapines all fall into our "social" category, with most species exhibiting cooperative breeding and two species (*Hasinamelissa minuta* and *Exoneurella tridentata*) exhibiting eusociality with distinct morphological castes (Schwarz et al. 2011). There is also a diversity of social parasites with evocative names like *Inquilina*. Social parasitism is estimated to have arisen 11 times within Allodapini (Tierney et al. 2008b).

Rehan and colleagues (2012) demonstrated that weak sociality (i.e., cooperative breeding) is the ancestral state for the Xylocopinae sensu stricto and that cooperative breeding has persisted in this group for over 100 million years.

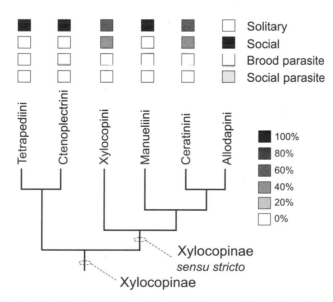

Figure 2-6. Tribal relationships within the subfamily Xylocopinae (based on Rehan et al. 2012 and Bossert et al. 2018).

Given that Xylocopini and Ceratinini include many truly solitary species and that Manueliini, Tetrapediini, and Ctenoplectrini are exclusively solitary, it is appropriate to include these five tribes in a book on solitary bees. However, Allodapini is a group almost exclusively comprised of cooperative breeders, with some that are eusocial (Fig. 2-6). They differ from solitary bees in so many ways that we do not treat them here.

· Halictinae

The third subfamily of bees that includes social taxa is Halictinae (Fig. 2-7). This subfamily includes over 3,000 described species in five tribes, and nearly half of these species are thought to be eusocial. Eusocial taxa are restricted to two of the five tribes: Augochlorini and Halictini. The other tribes are predominantly solitary (Thrinchostomini and Caenohalictini) or exclusively cleptoparasitic (Sphecodini). Gonçalves (2016) recently analyzed phylogenetic relationships within Augochlorini and demonstrated that the largely eusocial genera (including *Augochlora*, *Augochlorella*, and *Pereirapis*) form a monophyletic group. A total of 61 species out of roughly 600 described Augochlorini are presumed or known to be eusocial. We do not treat these social taxa in this book but do include the roughly 540 species of solitary augochlorines, some of which have been studied in detail (e.g., *Augochlora pura*). Halictini includes the vast majority of eusocial halictid bees (Gibbs et al. 2012b). More than half of the bee species in Halictini are eusocial, and these social taxa are concentrated in just two speciose genera: *Halictus* and *Lasioglossum* (Fig. 2-7). *Halictus* consists of primarily eusocial species, but one is solitary (*H. quadricinctus*; Sitdikov 1988), one is a social parasite (*H. chalybaeus*; Pauly 1997), and many are socially polymorphic (Eickwort et al. 1996, Richards 1994, Richards et al. 2003).

Approximately one-third of the 1,800 *Lasioglossum* species are believed to be eusocial. The subgeneric classification of *Lasioglossum* is notoriously complex, but recent taxonomic and phylogenetic work by Jason Gibbs has provided a more stable classification that can help with classifying species as either solitary or social. Two main clades exist within *Lasioglossum*: the *Lasioglossum* series (also called the "strong-veined" group) and the *Hemihalictus* series (also called the "weak-veined" group; Fig. 2-7). Each "series" includes many subgenera. Subgenera within the *Lasioglossum* series are typically solitary, but a small number are cleptoparasitic (nested within the subgenus *Homalictus*). There are no known social species. However, in the *Hemihalictus* series, many (but not all) subgenera are eusocial. The main eusocial subgenera include *Dialictus*, *Evylaeus*, and *Sphecodogastra* (sensu Gibbs et al. 2013). We do not treat the eusocial *Lasioglossum* here, but we do include the solitary species, some of which have been studied in great detail (such as the *Sphecodogastra* species that specialize on *Oenothera*; McGinley 2003).

Excellent books and review papers exist on the social bees. For a broad overview of social insects, Wilson's (1971) text is still an excellent, but slightly dated, source. For an overview of the social bees, see Michener (1974), and for a more recent treatment of select groups, see Choe and Crespi (1997). Halictine and

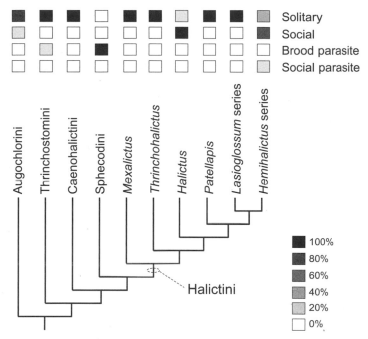

Figure 2-7. Tribal (and generic-level) relationships within the subfamily Halictinae, showing the phylogenetic affinities of the genera *Halictus*, *Lasioglossum*, *Mexalictus*, *Patellapis*, and *Thrincohalictus* within the largely eusocial Halictini (based on Gibbs et al. 2012b).

allodapine social behavior has been reviewed by Schwarz and colleagues (2007), and the evolution of social behavior in allodapines was analyzed by Tierney and colleagues (2008b) and Schwarz and colleagues (2011). Social behavior in bumble bees is summarized nicely in Kearns and Thomson (2001). The nesting biology of stingless bees has been reviewed by Roubik (2006), and a recently published book by Vit and colleagues (2013) contains a diverse assemblage of chapters on stingless bee biology. Finally, for the tribe Apini, there are numerous books on the biology of the domesticated honey bee, including Winston (1987), Caron and Connor (2010), and Tautz (2008). For a treatment of Asian honey bees, see Oldroyd and Wongsiri (2009), and for a fascinating book on honey bee decision making, see Seeley (2010).

ECOLOGICAL DIFFERENCES BETWEEN SOLITARY AND SOCIAL BEES

There are some fundamental ecological differences between solitary and social bees (Table 2-2) that will be important to keep in mind as we describe the various aspects of solitary bee biology in subsequent chapters. First, the period of adult activity in solitary bees is much shorter than it is for eusocial bees (as we will see in Chapter 3). A typical solitary bee is an adult for just a few weeks to a month, and this active period often coincides with floral resource availability, especially for narrow host-plant specialists. By contrast, most social bees are active for an extended period of time (spanning months), because (1) they are

Table 2-2. Fundamental differences in life history between social and solitary bees.

Life history trait	Social	Solitary
Period of adult activity	Over a long period of time (several months)	Over a very short period of time (sometimes weeks)
Generations per year	Always multivoltine	Usually univoltine
Host-plant specificity	Typically broadly polylectic	Broadly polylectic to narrowly oligolectic
Information sharing among nest mates	Yes (in highly social Apidae)	No (never)
Highest species richness	Humid tropics	Arid regions (deserts)
Buffered from environmental stresses	Yes (by virtue of colony size)	No

multivoltine and (2) they generally have very broad host-plant preferences compared to solitary bees. While solitary bees have both narrow and broad host-plant preferences, social bees are almost universally polylectic.[3]

Second, the ability to communicate among colony mates (either by the dance language in honey bees or the use of odor trails in stingless bees) allows colonies of some advanced eusocial bees to rapidly exploit resources that are kilometers from the nest site. By contrast, an individual solitary bee must rely on her knowledge of the floral resource availability in the immediate vicinity of her nest. As we shall see in a later chapter, the foraging ranges of solitary bees are far smaller than those of eusocial bees—sometimes on the order of 300–500 meters from the nest. A larger worker force, the ability to communicate about the location of resources, and larger home-range size enables social bee species to respond more rapidly to local changes in resource abundance (Crone 2013).

Finally, the biogeographic distribution of solitary and social bees is conspicuously different. The species richness of solitary bees is highest in arid, Mediterranean climate regions, such as arid western North America (the Chihuahuan and Sonoran deserts), the circum-Mediterranean region (including southern Europe, the Middle East, and northern Africa), western Australia, arid southern Africa (the Greater Cape Floristic Region), and xeric regions of Chile and Argentina (Michener 1979). These regions host a diverse solitary bee fauna with high levels of endemism, unique relictual lineages, and typically very high levels of host-plant specialization (Moldenke 1979). These are also regions of extraordinarily high plant diversity and highly seasonal rainfall patterns. By contrast, social bee diversity is highest in the humid tropics, where highly eusocial Meliponini and Apini predominate, and in the north temperate, boreal regions, where primitively eusocial Bombini predominate.

Why such a striking difference in geographic peaks of diversity between solitary and social bees? A number of hypotheses have been proposed. First, solitary

---

[3] There is one exception to this general rule. *Bombus gerstaeckeri* (Apidae) is a eusocial bee that appears to be narrowly host-plant specific (Ponchau et al. 2006). Females forage exclusively on monkshoods (*Aconitum*, Ranunculaceae). This narrow host-plant preference is exceptionally rare among eusocial bees.

bees, because they are largely ground-nesting, may not do well in the tropics because moist, humid soils may allow fungi to invade and overcome their brood cells. Second, and probably more importantly, solitary, univoltine, host-plant specialists may be ideally suited for desert environments where flowering occurs in a very short temporal window. In desert environments, solitary bees time their emergence to coincide with peak flowering of their host plant and can delay emergence to avoid catastrophic mortality in drought years (Danforth 1999, Minckley et al. 2013). The ability to track host-plant resources gives solitary bees an advantage in desert environments where rainfall is rare and unpredictable (see Chapter 3). Social bees, in contrast, require floral resources over a much longer period of time (months) in order to rear the multiple worker generations essential for offspring production. Hence, deserts, with a short, unpredictable, but abundant period of flowering, are ideal habitats for solitary bees but do not provide social bees with floral resources over a long enough period for them to rear multiple worker broods. Whatever the reason, the geographic distribution of species richness in solitary and social bees indicates that these two groups may be evolutionarily adapted to highly divergent environmental conditions (see Groom and Rehan [2018] for an exploration of this hypothesis in the tribe Ceratinini).

ORIGINS AND REVERSALS IN SOCIAL BEHAVIOR AMONG BEES

While the ancestor of bees was undoubtedly a solitary, ground-nesting species, sociality has originated in multiple bee lineages. Within Halictidae there are likely two (and possibly three) origins of eusociality (Gibbs et al. 2012b). Cooperative breeding appears to have evolved once in the common ancestor of Xylocopinae sensu stricto (Rehan et al. 2012). Based on the most recent analyses of corbiculate relationships (Fig. 2-5), eusociality appears to have had a single origin in this clade (Romiguier et al. 2016). We can therefore infer that sociality (including cooperatively breeding and eusociality) evolved independently between four and five times in bees.

But phylogenetic studies have also revealed numerous "reversals" back to solitary nesting in the Halictini (within *Lasioglossum*; Danforth et al. 2004, Gibbs et al. 2012b) and in the Augochlorini (Danforth and Eickwort 1997, Gonçalves 2016), both lineages of predominantly social bees. These "secondarily solitary" taxa demonstrate that the evolution of sociality in bees is not unidirectional; solitary lineages can give rise to social descendants and social lineages can give rise to solitary descendants (Wcislo and Danforth 1997). Reversals from primitive eusociality to solitary nesting are common. Within the genus *Lasioglossum* alone, there are as many as 12 reversals from eusociality back to solitary nesting (Danforth 2002, Danforth et al. 2003a). This high rate of reversal in social behavior within primitively eusocial bees is likely explained by the fact that primitively eusocial bees pass through a solitary foundress phase for at least part of their life cycle. In contrast, in advanced eusocial lineages (stingless bees and honey bees), there is no solitary phase in the life cycle, and these lineages have never given rise to solitary descendants. Advanced eusocial taxa seem to have reached a "point of no return" (Wilson and Hölldobler 2005).

BOX 2-10: A GLOSSARY OF SOCIAL TERMINOLOGY

Solitary. A single, adult female builds and occupies each nest. She constructs her own brood cells, provisions them with pollen and nectar, guards her own nest, and lays her own eggs. Approximately three-quarters of all bees are solitary (Fig. 2-4).

Communal. Adult, reproductively active females share the same nest entrance but otherwise behave as solitary females. Each female forages for pollen and nectar and lays eggs in a brood cell she constructs. Lateral tunnels branching from the common main tunnel may be occupied by different females. There may be benefits in terms of nest defense from sharing a common nest entrance, but there is no obvious cooperation among females and no reproductive division of labor or cooperative brood care.

Cooperatively breeding. This form of sociality is largely restricted to stem- and wood-nesting members of the subfamily Xylocopinae. Mothers and their female offspring remain together in the same nest for an extended period of time. There is reproductive division of labor, with the mother doing most of the reproduction and the daughters remaining in the nest as non-reproductive guards or even foragers. In cooperatively breeding societies, there is also caste polymorphism; the daughters serve exclusively as guards at the nest entrance, while the mother constructs additional brood chambers, provisions them with pollen and nectar, and lays eggs. What makes this system different from true eusociality is that the associations among mothers and daughters are *temporary*. Mothers will eventually die and leave the nest to their reproductively active daughters. Guarding (and non-reproduction) in these bees is a temporary state rather than a permanent condition.

Eusocial. Eusocial taxa exhibit reproductive division of labor, cooperative brood care, and overlap of generations. However, unlike cooperatively breeding societies, reproductive division of labor and caste status remain constant throughout the lifetime of an individual. Once a female bee becomes a worker, she remains a worker. Eusocial taxa are typically divided into two broad categories: primitive eusociality and advanced eusociality.

Primitively eusocial. In these societies, queens and workers differ only in body size. This type of eusociality is exhibited by most eusocial Halictini and Augochlorini, Bombini (bumble bees), and two species of Allodapini. Many primitively eusocial bees have given rise to secondarily solitary descendants, suggesting that primitive eusociality is an evolutionary labile state that can easily revert to solitary nesting.

Advanced eusocial. Queens and workers in these societies are morphologically distinct. In addition, colony founding in advanced eusocial bees is generally achieved through swarm founding, in which the queen and a large number of workers leave the nest together. Queens typically cannot found nests on their own. The tribes Meliponini (stingless bees) and Apini (honey bees) exhibit this kind of sociality.

Socially polymorphic. These are species in which there are both eusocial and solitary populations. Sociality in socially polymorphic species typically varies with latitude (at low latitudes populations can be eusocial, whereas at higher latitudes they switch to solitary nesting) and altitude (low-elevation populations can be eusocial, whereas they are solitary or communal at higher elevations). Halictids exhibit substantial social polymorphism in some species (*Halictus rubicundus*, *Lasioglossum albipes*, and *Megalopta genalis*). The existence of social polymorphism clearly indicates how sociality in bees can be a labile trait.

Brood parasitic or cleptoparasitic. In brood-parasitic bees, females do not build nests or collect pollen and nectar for larval nutrition. In fact, these bees lack the structures for gathering, manipulating, and carrying pollen. Instead, they enter the nests of pollen-collecting bees and lay their eggs in either open or closed brood cells. The adult female or her first-instar larva kills the host egg or larva, and the brood parasite then consumes the pollen provisions of the host bee. Female brood parasites are often heavily armored to defend themselves against the attack of the host female. Brood parasites are often, but not always, closely related to their hosts. This life history is described in detail in Chapter 10.

Socially parasitic. Social parasites enter the nests of social bees and kill or replace the host female as the primary egg layer. These bees attack only social hosts and are often closely related to their hosts (e.g., the subgenus *Psithyrus* in bumble bees).

Robbers. Robbing (Box 8-3), or stealing the pollen/nectar provisions of other bees, is a fairly rare phenomenon. It is a way of life for stingless bee species in the genera *Lestrimelitta* (in the New World; Sakagami et al. 1993) and *Cleptotrigona* (in the Old World; Portugal-Araújo 1958); both rob from other species of stingless bees. Intra-specific robbing has also been described in several species of *Xylocopa* (Hogendoorn 1991, Hogendoorn and Velthuis 1993, Lucia et al. 2017).

In a fascinating modeling study, Fu and colleagues (2015) demonstrate mathematically that while solitary nesting is a low-risk, fairly conservative strategy, eusociality is a high-risk but very high-reward strategy. This is because the initial investment in worker production by eusocial species effectively delays the production of reproductives. If colonies die off during the worker phase, they are reproductive dead ends and go extinct. This means that eusocial species, especially in an unpredictable environment and early on in their evolutionary history, are more prone to extinction than a solitary species inhabiting the same environment. However, once eusociality is established, it can become a more effective reproductive strategy than solitary nesting. The modeling study of Fu and colleagues (2015) parallels what we see empirically when we map sociality onto phylogenies of various bee groups. Eusociality has arisen rarely in bees, consistent with the idea that there is a high rate of extinction. But reversals to solitary nesting occur repeatedly within virtually every lineage of eusocial bees. When lineages do overcome this threshold to become eusocial, these lineages tend to be the numerically most abundant bees in the environment. Think stingless bees or honey bees in the tropics, where the biomass of these eusocial species is extremely high relative to the biomass of solitary bee species in the same geographic region. So, rather than thinking of eusociality as the "pinnacle of bee evolution" to which all bees are striving, think of eusociality as a high-risk, high-reward investment strategy. Putting all your money down on a small, high-tech startup company might yield big returns down the road, but it could also mean a catastrophic loss of your investment capital. Solitary bees have adopted a more conservative investment strategy: choose low, but stable, long-term investment returns, and hold your stocks for a very long time.

In this chapter, we have attempted to lay some preliminary groundwork for the rest of this book. The phylogeny of the bee families and subfamilies (Fig. 2-2) provides a framework for interpreting all aspects of bee biology, including social and solitary nesting, as well as the diversity of traits we will discuss in relation to the biology of solitary bees. A basic knowledge of the seven families of bees will be useful to readers as we discuss the broad features of bee nesting biology, mating behavior, host-plant preferences, and life history. We have attempted to make a clear distinction between social and solitary bees both from the perspective of behavior (how they differ) and from the perspective of phylogeny (how they are related to each other). We can now move on to explore in detail the fascinating and complex lives of the solitary bees. The starting point for such a treatment is to describe the various life-history patterns observed in solitary bees.

CHAPTER 3

The Solitary Bee Life Cycle

For solitary bees, timing is everything. Unlike social bees that are active for several months, most solitary bees have a very narrow temporal window of activity. Because of year-to-year variation in weather and the timing of host-plant flowering, collection records made over many years usually indicate a much broader period of activity than actually occurs in any single year. As we shall see below, the actual period of adult foraging can be exceptionally short—10 days to two weeks in many ground-nesting, solitary bees (Cameron et al. 1996, Minckley et al. 1994). In temperate regions of the world where there are pronounced seasons, individual solitary bee species show a progression of activity from those that are active in the early days of the spring to species active in midsummer and species active in early fall. Stevens (1948, cited in Linsley 1958) documented the turnover of *Andrena* species through the year in a site in North Dakota (Fig. 3-1). Each species is active for, on average, one month, but different species are active from the earliest days of spring (in early April) to the early fall (late September). Such narrow windows of adult activity are partially driven by host-plant association, especially for narrow host-plant specialists (Minckley et al. 1994). For the North Dakota *Andrena,* there are early-spring species (e.g., *A. frigida, A. mariae, A. illinoiensis*, and *A. nigrae*) that specialize on early-spring-flowering Rosaceae and Salicaceae, midsummer species (e.g., *A. ziziae, A. miranda*, and *A. thaspii*) that specialize on Umbelliferae, and late-summer species (e.g., *A. nubecula, A. helianthi, A. canadensis*, and *A. hirticincta*) that specialize on late-blooming Asteraceae.

In desert environments—areas that host the highest species richness and phylodiversity of solitary bees—the temporal synchrony between solitary bee activity and flowering time is especially key to offspring production. Desert species have been shown to track host-plant availability between years (Minckley et al. 2013) and to use some of the same cues as their host plants (soil humidity) to trigger emergence (Danforth 1999). In arid environments, oligolectic bees are also far better at tracking host-plant flowering than polylectic bees (Minckley et al. 2000). The close synchrony between bee emergence and host-plant flowering is a remarkable aspect of solitary bee biology that has received very limited attention.

If solitary bees are active for just a few weeks per year, what are they doing for the rest of the year? Most solitary bees have an extended period of adult or larval diapause. Whether bees overwinter as adults or as last-instar larvae appears to be driven partially by habitat and partially by phylogeny. Northern

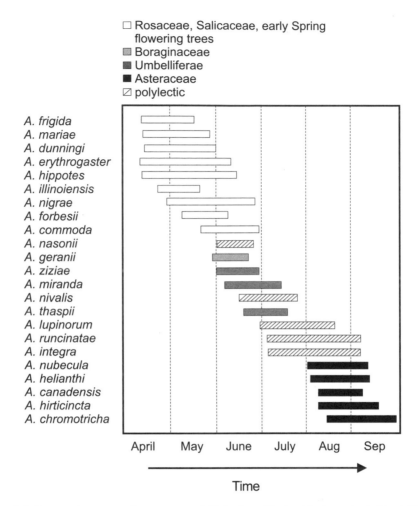

Figure 3-1. Temporal patterns of emergence of 23 *Andrena* (Andrenidae) species in North Dakota. Data from Stevens (1948). Figure redrawn from Linsley (1958). Bars are colored according to host-plant preferences.

Hemisphere, cold-temperate bee groups, such as *Andrena* and many *Osmia*, overwinter as adults. However, there are some *Andrena* that overwinter as last-instar larvae (e.g., *A. accepta* and *A. haynesi*; Neff and Simpson 1997), and some species even show a mixture of adult and larval overwintering (e.g., *A. rud-beckiae*; Neff and Simpson 1997). Adults are likely less resistant to desiccation than last-instar larvae, but adult overwintering allows early-spring solitary bees to emerge, construct nests, and forage as soon as the weather permits. In arid, desert environments, most species overwinter as prepupae. Prepupae have a thick, impermeable outer integument that is resistant to desiccation and allows these larvae to overwinter for an extended period of time (multiple years; see below). As we shall see below, extended diapause in desert bees is an adaptation to the short, highly unpredictable rainfall patterns that characterize the arid regions of the world (Davidowitz 2002).

There are exceptions to the general rule outlined above. In two autumnal species of *Colletes*, *C. halophilus* in Europe (Sommeijer et al. 2012) and *C. collaris* (Goukon and Maeta 2016) in Asia, the overwintering stage appears to consist of second- and third-instar larvae that are slowly feeding on the liquid pollen/nectar provisions. This would seem to be a bizarre way to spend the inactive time of year. Feeding larvae are much more vulnerable than last-instar larvae, prepupae, or adults. But in these two species, and presumably other related species, feeding larvae have adopted a novel mode of overwintering for reasons we do not fully understand.

FROM EGG TO ADULT—THE GROWTH AND DEVELOPMENT OF SOLITARY BEES

Bees, like all arthropods, have an exoskeleton, which means that development occurs through a series of discrete stages punctuated by molting, or shedding of the exoskeleton. Bees are also holometabolous insects, which means that at one point during development, they undergo metamorphosis—a dramatic change in morphology—from larva to a resting stage called the pupa and ultimately to the adult. A bee starts its life as an egg and then passes through a series of five larval stages before pupating and then emerging as a fully formed adult (Fig. 3-2).

Bee eggs

All bees begin life as an egg. Bee eggs vary widely in morphology, especially among brood-parasitic bees (described in more detail in Chapter 10). In pollen-collecting bees, eggs are more or less similar in shape, color, and texture. Pollen-collecting bees all produce an elongate, gently curved, sausage-shaped, whitish egg that is placed on, within, or beneath the pollen/nectar provision mass (Fig. 3-3;

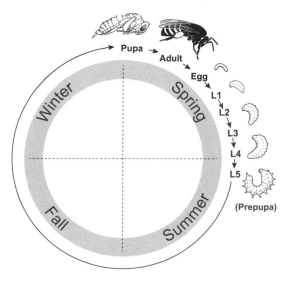

Figure 3-2. Life cycle of a typical solitary bee through the calendar year. Adults are active for a short period of time, during which they emerge, mate, construct and provision nests, and then die. Bees have five larval instars that develop quickly within the brood cell. The last larval instar (the prepupa) undergoes a prolonged diapause that can last most of the year. Pupation takes place early in the season when conditions are favorable, and adults emerge shortly thereafter.

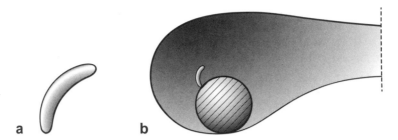

Figure 3-3. Brood-cell, egg, and pollen provisions in *Perdita nuda* (Andrenidae): (a) egg (redrawn from Torchio 1975, Fig. 3) and (b) egg located on pollen-provision mass in brood cell (redrawn from Torchio 1975, Fig. 2).

described in more detail in Chapter 8). While egg shape is fairly consistent across pollen-collecting bees, egg size is not. Egg size does, in general, scale with adult body size; larger bees have larger eggs. However, some bee eggs are enormous relative to their body size, while other bee eggs can be minute compared to adult body size.

Iwata and Sakagami (1966) reviewed egg size in both pollen-collecting and brood-parasitic bees and found that egg size ranges from less than 1 millimeter (in some brood-parasitic bees) to over 2 centimeters (in carpenter bees, Xylocopini). They developed a measure of egg size relative to adult body size—the "egg index"—that has been used widely in the bee literature. The "egg index" is simply egg length divided by adult thoracic width. Typical pollen-collecting bees have an egg index of approximately 1.0 (egg length = thoracic width), whereas some groups (the subfamily Xylocopinae) can approach an egg index of 2.0 (egg length is 2x thoracic width), and eggs of brood-parasitic bees fall in the range of 0.3–0.5 (egg length is one-third to one-half thoracic width).

Why do carpenter bees produce gigantic eggs and brood parasites produce tiny eggs? For brood-parasitic bees, the explanation seems fairly clear. Brood parasites produce large numbers of small eggs rapidly because they have the potential to parasitize many brood cells in a short period of time—especially when they encounter a large nesting aggregation or a communal nest of the host (Alexander and Rozen 1987). This also helps explain why brood parasites often have an increased number of ovarioles relative to their pollen-collecting hosts (a topic we cover in more detail in Chapter 10). Why carpenter bees have enormous eggs is less clear. One hypothesis is that bee eggs placed in a wood (rather than soil) nest may be prone to dehydration. Perhaps carpenter bees produce large eggs (which have a relatively low surface-to-volume ratio) in order to prevent egg desiccation prior to hatching.

Egg hatching generally occurs shortly after oviposition (typically within hours). Two different modes of egg hatching have been documented by Rozen and colleagues (Rozen et al. 2017). In the most widespread mode of egg hatching, first-instar larvae bear hatching spines along the lateral surface of their body that rasp open the egg chorion along its long axis. This mode of egg hatching has been documented in a wide variety of solitary and social bees. However, an alternative mode appears to have evolved in the cleptoparasitic apids. In this latter mode, the larva exits through an apical opening at the anterior end of the egg. Larvae in these groups lack the hatching spines. This novel mode of egg hatching appears to be a synapomorphy of the apid subfamily Nomadinae (as recently defined by Bossert et al. 2018).

The larval stage

Most solitary bees spend the vast majority of their lives as larvae. Larvae are the primary feeding stage of bees and the stage in which most pollen, nectar, and floral oils are consumed. As a consequence, the larval stage is the stage in which bees must acquire the nutrients to reach adulthood; but in this stage, they also must detoxify the secondary plant compounds that are likely present in pollen, nectar, and possibly floral oils. The larva is therefore a crucial, but sometimes underappreciated, stage in the life cycle of a solitary bee.

Bee larvae are typically rather nondescript, soft-bodied, legless, eye-less, grub-like creatures with a distinct head capsule bearing a pair of small antennae, mandibles, maxillae, and 1 posterior labium (Fig. 3-4b,c). Actively feeding larvae have a soft, flexible body with distinct spiracles on 10 of the 11 body segments. Many bee larvae bear dorsal tubercles that assist in movement within the tight confines of the brood cell (Fig. 3-4a). Pollen-collecting bees have rather uniform larvae, but as we shall see in a later chapter, the larvae of some brood-parasitic (cleptoparasitic) bees are highly modified, with enlarged, sickle-shaped mandibles for dispatching their host egg or larva.

We know a great deal about larval development in solitary bees from two different types of studies. First, nests excavated or examined in the field can allow the investigator to piece together a picture of larval development through examination of brood cells with larvae in various stages of development. If the brood cells examined span the full range of larval instars, one can reconstruct a very complete picture of larval growth, pollen feeding, defecation, pupation, cocoon formation, and overwintering. One can also identify brood-cell parasites and understand how they interact with the host inside the brood cell. A complete picture of development may also require examination of brood cells at various times throughout the year (early, peak, and late season, and during the period of larval or adult diapause). Studies of this type form the basis of much of what we know about solitary bee life history, nesting, and brood development, and a vast literature exists dating back to the earliest studies by naturalists, such as Jean-Henri Fabre in the mid-19th century. No one has studied more bees in this way and across more regions of the world than Jerome G. Rozen Jr. at the American Museum of Natural History, in New York City. Rozen's studies span every family of bee and have been conducted on virtually every continent (except Australia). We have relied heavily on Rozen's published work for much of this chapter—indeed, much of this book.

A more detailed picture of larval development can be obtained through studies of bees in a greenhouse setting or in observation nests that can be opened periodically and examined. Torchio and colleagues documented larval (and embryonic) development using observation nests of bees such as *Colletes* (Torchio et al. 1988), *Hylaeus* (Torchio 1984a), *Anthophora* (Torchio and Trostle 1986),

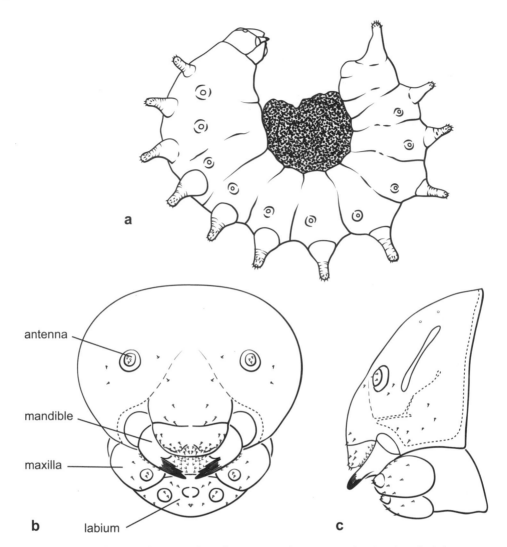

Figure 3-4. Larval morphology in solitary bees: (a) *Perdita octomaculata* (Andrenidae), last-instar, post-defecating larva (redrawn from Eickwort 1977, Fig. 12); (b) *P. octomaculata*, frontal view of head capsule (redrawn from Eickwort 1977, Fig. 13); (c) *P. octomaculata*, lateral view of head capsule (redrawn from Eickwort 1977, Fig. 14), showing antenna, labrum, mandible, maxilla, and labium.

Osmia (Torchio 1989a), and *Megachile* (Trostle and Torchio 1994), and the brood parasites associated with these bees, including *Stelis* (Torchio 1989b), *Xeromelecta* (Torchio and Trostle 1986), and *Epeolus* (Torchio and Burdick 1988). These studies provide an extraordinarily in-depth look at larval development, feeding, movements within the brood cell, and interactions between host and parasite.

Like their solitary wasp ancestors, all solitary and brood-parasitic bees pass through five larval instars (Evans and West-Eberhard 1970). In bees, the first larval instar is non-feeding. Molting of the first instar takes place within the chorion of the egg, and the first larval skin and chorion are shed at the same time (Torchio et al. 1988). The second through fifth larval stages are the actively feeding stages; larvae use their mandibles to consume the pollen/nectar provisions stored by their mother. Larval development in solitary bees is usually rapid. Most bee larvae complete feeding in between one and three weeks, but larval development can sometimes take much longer (two months in *Colletes cunicularius*; Malyshev 1923).

Bee larvae exhibit a variety of approaches to consuming the pollen/nectar provisions. In all families except Colletidae, the larvae rest with their ventral surface appressed to the pollen/nectar provisions during feeding. Colletid bees are unique in that the larvae lie on their sides on the surface of the soupy, pollen/nectar provisions (Torchio 1984a). Larval colletids do a kind of sidestroke across the liquid provisions, feeding as they swim over the pollen/nectar soup. Feeding while lying on the side appears to be widespread in Colletidae but absent from the closely related family Stenotritidae (Torchio 1984a). In Diphaglossinae, which feed on very liquid provisions, Rozen (1984b) described how larvae float on their sides on the surface of the provisions and periodically dip their heads downward into the liquid pollen/nectar provisions as they feed, apparently stirring up the pollen from the bottom of the brood cell. Larval feeding has been observed in a number of other colletid bee groups, and in all cases, the larvae lie on their side as they feed (Eickwort 1967, Michener 1960, Rajotte 1979, Rozen and Michener 1968, Sarzetti et al. 2013, Sommeijer et al. 2012, Torchio 1984a, Torchio et al. 1988).

Based on Torchio's embryological studies, the unusual position of the colletid larvae is due to a major difference in embryological development. Most bees (in fact, most insects) undergo a 180° rotation during embryonic development. This rotation positions the ventral surface of the larva downward (toward the pollen/nectar provisions) when the first-instar larva exits the egg. In Colletidae, larvae rotate only 90°, positioning them on their side as they exit the egg. This unique feature of colletid embryonic development is clearly a derived trait that unites Colletidae as a monophyletic group.

All other bees feed with the ventral surface appressed to the pollen/nectar provisions. Larvae feeding on solid or semi-liquid provisions (e.g., Anthophorini, Eucerini) initially float on the pollen/nectar surface and consume the provisions as they move around the periphery of the brood cell. In *Anthophora urbana*, larvae gradually reduce the pollen/nectar provisions to a more solid column with a distinct feeding groove in which they move as they graze from the surface of the pollen provisions (Torchio and Trostle 1986). Bees that feed on solid provision masses usually consume the pollen provisions by gradually feeding on the pollen mass while remaining in the same position (above the pollen provisions), or they move around the pollen provisions, gradually reducing the pollen provision while doing so (e.g., *Exomalopsis*; Rozen 1984a; Fig. 3-5). In Panurginae (Andrenidae), which feed on a nearly perfectly spherical pollen ball, the larvae start feeding on top of the pollen provisions and then do a flip in the last larval instar such that the larva ends up beneath the pollen provisions with the provisions cradled on their ventral surface (Danforth 1991a, Rozen 1967a; Fig. 3-6). Because these larvae also have well-developed dorsal tubercles (Fig. 3-4a), this seems like a strategy designed to limit fungal or microbial growth by keeping the pollen provisions away from the humid brood-cell wall. In Megachilidae, larvae

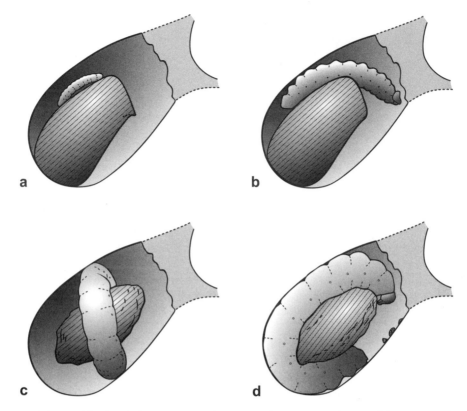

Figure 3-5. Larval feeding behavior in Exomalopsini (Apidae): (a) *Exomalopsis solidaginis*, small larva resting on pollen provisions (redrawn from Rozen 1984a, Fig. 20); (b) *E. solidaginis*, third-instar larva feeding on pollen provisions (redrawn from Rozen 1984a, Fig. 22); (c) *E. solidaginis*, fourth-instar larva circling grooved pollen loaf (redrawn from Rozen 1984a, Fig. 23); (d) *E. solidaginis*, early fifth instar completing pollen loaf and starting to defecate (redrawn from Rozen 1984a, Fig. 24).

do not move from the site where the egg was initially deposited until the last larval instar, at which point they migrate around within the brood cell (Rozen pers. comm.). Most pollen consumption in solitary bees takes place during the last larval instar.

A novel mode of pollen/nectar consumption was reported in *Colletes kincaidii* (Colletidae) by Torchio and colleagues (1988 p. 619). In its second through fifth instars, a larva aspirates the soupy pollen/nectar provisions into its anus and rectal cavity, then brings the tip of its abdomen to its head and extrudes the pollen/nectar provisions into its mouth. This bizarre mode of feeding has been observed only in this one study, so it is difficult to know how widespread it is among colletid, or other, bees. Torchio and colleagues (1988) hypothesize that the rectal transfer of pollen/nectar provisions might be related to the passage of glandular secretions into the mouth along with the pollen/nectar provisions.

In a few cases, bee larvae have also been reported to eat the waxy brood-cell lining. This behavior has been reported in *Anthophora abrupta* (Batra and Norden 1996, Norden et al. 1980) and in *Amegilla dawsoni* (Houston 1991a).

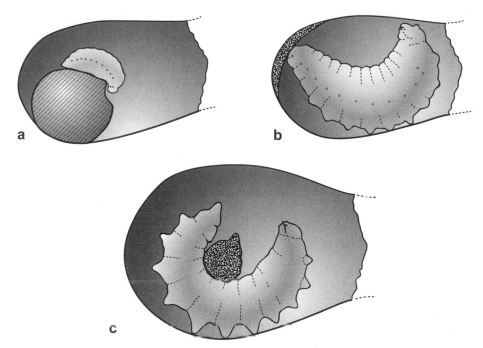

Figure 3-6. Larval feeding and defecation in Panurginae (Andrenidae): (a) *Panurginus potentillae*, young larva feeding on pollen provisions (redrawn from Rozen 1967a, Fig. 12); (b) *P. potentillae*, mature prepupa after defecation on the rear of the brood cell (redrawn from Rozen 1967a, Fig. 13); (c) *Perdita zebrata*, mature prepupa with fecal material on venter of abdomen; note larval tubercles that project below the larval body (redrawn from Rozen 1967a, Fig. 14).

We discuss this phenomenon in more detail in Chapter 6. Suffice it to say that the brood-cell lining in Anthophorini is highly modified relative to the typical brood-cell lining known from most other bees and appears to serve as an additional nutritional source for developing larvae.

In the last (fifth) larval instar, feeding larvae begin to defecate. Defecation can take place at the latter stages of pollen consumption or after the pollen/nectar mass has been entirely consumed. Defecation normally precedes diapause, but in one brood-parasitic bee (*Sphecodes* near *fragaria*), defecation takes place at the end of prepupal diapause (Torchio 1975). Defecation presents a particular challenge for an insect larva confined to a small, hermetically sealed brood chamber: How to dispose of the fecal material? Typically, bees apply the fecal material to the back (Panurginae; Rozen 1967a), sides (Exomalopsini; Rozen 1984a), or front of the brood cell. However, some bees apply the fecal material as a ring around the equator of the brood cell (*Eremapis*, Apidae; Neff 1984) or retain the fecal material on the venter of the larva (*Perdita*, Andrenidae; Eickwort 1977). In bees that spin cocoons, the fecal material is generally deposited before or during cocoon construction, and fecal matter is deposited either outside the cocoon or incorporated into the cocoon (e.g., in Exomalopsini [Rozen 1977b] and some Megachilini [Neff and Simpson 1991]; see also Figs. 29–32, 34–37, 39 in Radchenko and Pesenko 1994). In Diphaglossinae, which consume extremely liquid provisions, the larvae first chew a hole in the cellophane lining at the bottom of the brood cell and then defecate a very liquid fecal material that drains out through the hole. Only after they have allowed the fecal liquid to drain out of the brood cell do they finalize the construction of the cocoon (Rozen 1984b).

Prepupae, in bees that overwinter in this stage, typically have a thick, rigid, water-resistant cuticle that allows them to resist desiccation over the long period of inactivity. A number of studies have hinted at the possibility that last-instar larvae produce a larval exudate that spreads over the surface of the prepupa. Torchio (1975) described a "ferrous-colored, lipid-like material" that was exuded from glands on the surface of the body of the prepupa of *Sphecodes* (Halictidae). This material gradually spread over the surface of the prepupa and eventually hardened, forming what Torchio believed to be an antimicrobial coating to the prepupa. Rozen (2016a) described a similar fluid in the larvae of *Hesperapis rhodocerata* (Melittidae), a bee that has no brood-cell lining. These studies suggest the tantalizing possibility that bee prepupae have chemical strategies hidden up their sleeves for dealing with fungal or other microbial attack.

In many bee groups, the last-instar larva spins a thick, water-resistant cocoon that forms an effective barrier against desiccation and pathogen or parasite attack. All bee larvae possess paired salivary openings at the apex of the larval labium (Michener 1953a; Fig. 3-4b). However, in larvae that spin a cocoon, these salivary openings protrude forward as well-developed, slit-like salivary lips. The cocoon consists of silken fibers embedded in a water-resistant, impermeable matrix that is an effective barrier against desiccation and a physical barrier against parasite and predator attack (Rozen et al. 2011a, Rozen and Hall 2011). Cocoons are expensive (nutritionally and energetically) to produce. Cocoons average 49.3% of adult body weight in *Osmia cornuta* (Bosch and Vicens 2002) and 47.5% of adult body weight in *O. ribifloris* (Neff pers. obs., cited in Neff 2008) and therefore represent a significant investment for the developing larva.

Cocoon production is believed to be a primitive state in solitary bees because larval silk glands and cocoon production are widespread in the hunting wasps from which bees arose (Michener 2007, Radchenko and Pesenko 1994). However, cocoon production has apparently been lost repeatedly in bee evolution because only some bee subfamilies and tribes produce cocoons (Table 3-1). The subfamilies Melittinae, Meganomiinae (Melittidae), Rophitinae (Halictidae), Diphaglossinae (Colletidae), and all Megachilidae produce cocoons. The fact that cocoon production is in the "basal" subfamilies of Halictidae and Colletidae is consistent with the idea that the cocoon is a plesiomorphic (ancestral) trait in bees. Within Apidae, there is substantial variation in which groups do or do not produce a cocoon. Many pollen-collecting apids produce a cocoon (Table 3-1), including all Euglossini, Ctenoplectrini, Exomalopsini, Tapinotaspidini, Ancylini, Eucerini, and Emphorini. However, there is variation within some tribes. In Centridini, species of *Centris* produce cocoons, but species of *Epicharis* do not. In many Exomalopsini and Emphorini, cocoon production appears to be facultative, meaning that larvae that emerge in the same year do not spin a cocoon, whereas those that are going to enter diapause and overwinter for an extended period of time do (Neff and Simpson 1992; Rozen 1977b, 1984a; Rozen and Snelling 1986). No Anthophorini produce a cocoon, and most members of the subfamily Xylocopinae do not spin a cocoon.

Table 3-1. Distribution of cocoon production across the families, subfamilies, and tribes of solitary bees.

Family	Subfamily	Tribe	Glandular or oil-based brood-cell lining	Cocoon formation	References
Melittidae	Melittinae	Melittini	Dufour's gland	+	Celary 2006
		Macropidini	floral oils	+	Rozen & Jacobson 1980
	Dasypodainae	Hesperapini	No	-	Rozen & McGinley 1991
		Dasypodaini	No	-	Lind 1968
	Meganomiinae		Dufour's gland	+; unusual, porous cocoon	Rozen 1977c
Andrenidae	Andreninae		Dufour's gland	-	
	Panurginae		Dufour's gland	-	Rozen 1967a, 1989b
	Oxaeinae		Dufour's gland	-	Rozen 2018
Halictidae	Rophitinae		Dufour's gland	+, except *Conanthalictus*	Rozen & Özbek 2008
	Nomiinae		Dufour's gland	-	Rozen 2008a
	Nomioidinae		Dufour's gland	-	Rozen 2008a
	Halictinae		Dufour's gland	-	Rozen 2008a
Colletidae	Diphaglossinae		Dufour's gland	+	Rozen 1984a
	Neopasiphaeinae		Dufour's gland	-	Almeida 2008
	Callomelittinae		Dufour's gland	-	Almeida 2008
	Colletinae		Dufour's gland	-	Almeida 2008
	Scrapterinae		Dufour's gland	-	Almeida 2008
	Euryglossinae		Dufour's gland	-	Almeida 2008
	Xeromelissinae		Dufour's gland	-	Almeida 2008
	Hylaeinae		Dufour's gland	-	Almeida 2008
Stenotritidae			Dufour's gland	-	Houston & Thorp 1984
Megachilidae	Fideliinae		No	+	Litman et al. 2011
	Pararhophitinae		No	+	Litman et al. 2011
	Lithurginae		No	+	Litman et al. 2011
	Megachilinae		No; use p ant products instead	+	Litman et al. 2011
Apidae	Anthophorinae	Anthophorini	Dufour's gland	-	Brooks 1983
	Apinae	Centridini	Floral oils	+ *Centris*/- *Epicharis*	Rozen 1965a
		Euglossini	Dufour's gland	+	Rozen 2016b

(continued)

Table 3-1. (Continued)

Family	Subfamily	Tribe	Glandular or oil-based brood-cell lining	Cocoon formation	References
	Eucerinae	Ancylini	Dufour's gland	+	Straka & Rozen 2012
		Emphorini	Dufour's gland	+/facultative	Neff & Simpson 1992
		Eucerini	Dufour's gland	+	Rozen 1965a
		Exomalopsini	Dufour's gland	+/facultative	Rozen 1984b
		Tapinotaspidini	Floral oils	+	Rozen et al. 2006
	Nomadinae	Ammobatini	n/a; cleptoparasites	-	Rozen 1991
		Ammobatoidini	n/a; cleptoparasites	-	Rozen 1991
		Biastini	n/a; cleptoparasites	-	Rozen 1991
		Brachynomadini	n/a; cleptoparasites	-	Rozen 1991
		Caenoprosopidini	n/a; cleptoparasites	-	Rozen 1991
		Coelioxoidini	n/a; cleptoparasites	-	Alves-dos-Santos et al. 2002
		Epeolini	n/a; cleptoparasites	-	Rozen 1991
		Ericrocidini	n/a; cleptoparasites	+	Rozen & Buchmann 1990
		Hexepeolini	n/a; cleptoparasites	-	Rozen 1991
		Isepeolini	n/a; cleptoparasites	+	Rozen 2003a
		Melectini	n/a; cleptoparasites	+ most/- Thyreus/some facultative Xeromelecta	Rozen 1969a
		Neolarrini	n/a; cleptoparasites	-	Rozen 1991
		Nomadini	n/a; cleptoparasites	+?	Rozen 1991
		"Osirini"	n/a; cleptoparasites	+	Bogusch 2005, Michelette et al. 2000, Rozen et al. 2006, Straka & Bogusch 2007
		Protepeolini	n/a cleptoparasites	facultative	Rozen et al. 1978
		Rhathymini	n/a cleptoparasites	+?	Rozen 1969a
		Townsendiellini	n/a; cleptoparasites	-	Rozen 1991
	Xylocopinae	Ceratinini	No	-	Sandra Rehan pers. comm.
		Ctenoplectrini	Floral oils	+	Rozen 2010
		Manueliini	No	-	Flores-Prado et al. 2008
		Tetrapediini	Floral oils	-	Alves-dos-Santos et al. 2002 Rocha-Filho & Garófalo 2016
		Xylocopini	No	-	Sean Prager pers. comm.

Among the brood-parasitic apids, there is substantial variation in cocoon production as well. Like pollen-collecting apids, many brood-parasitic apids produce cocoons, but some do not (Table 3-1). In many cases, host and parasite are similar with regard to cocoon production. *Centris* produce cocoons, as do their brood parasites (*Ericrocis*) (Fig. 3-7a,b). Even facultative cocoon production is something shared by unrelated hosts and brood parasites. *Leiopodus singularis* (Protepeolini) shows facultative cocoon production, just like its host, *Diadasia ochracea* (Rozen et al. 1978). However, host and parasite can also differ. Anthophorini do not produce cocoons, whereas their brood parasites (Melectini) mostly do (*Thyreus* is the sole exception).

Rozen and Hall (2011) and Rozen and colleagues (2011a) examined the ultrastructure of the cocoon in *Osmia chalybea* (Megachilidae: Osmiini; Fig. 3-8a) and its brood parasite *Stelis ater* (Megachilidae: Anthidiini; Fig. 3-8b) in order to understand how the "nipple" of the cocoon functions as a barrier to parasites and predators but also allows sufficient gas exchange for larval respiration. While the cocoon nipples of the two species differ externally, they appear to consist of the same functional elements, which simultaneously provide a barrier to parasites and water loss but also allow for gas exchange. In *O. chalybea*, the nipple consists of a complex series of rigid structures (the cap and disk; Fig. 3-8a) embedded in a matrix of cocoon fibers. Below the cap is a porous outer screen that shows minute punctures through which gas exchange likely takes place. Similarly, below the disk there is an inner screen that is also porous

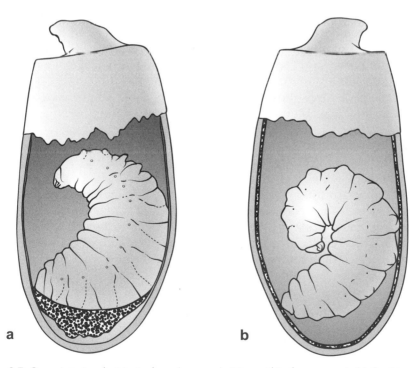

a   b

Figure 3-7. Overwintering, last-instar larva (prepupa) sitting within the cocoon in (a) *Centris caesalpiniae* (redrawn from Rozen and Buchmann 1990, Fig. 15) and (b) *Ericrocis lata* (redrawn from Rozen and Buchmann 1990, Fig. 16).

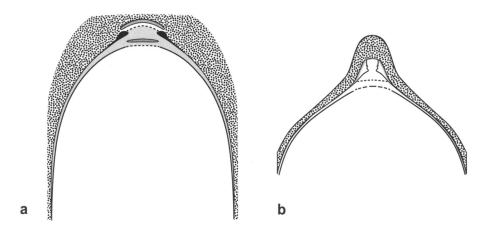

Figure 3-8. Cocoon ultrastructure in two megachilid bees: (a) ultrastructure of the cocoon in *Osmia chalybea*, showing configuration of the cocoon nipple (redrawn from Rozen and Hall 2011, Fig. 17) and (b) ultrastructure of the cocoon in *Stelis ater*, a brood parasite of *O. chalybea*, showing configuration of the cocoon nipple (redrawn from Rozen and Hall 2011, Fig. 43).

and seemingly permeable to the passage of gases. Together, the rigid cap and disk function as a barrier, whereas the outer and inner screens appear to function as filters. Measurements of gas exchange across different parts of the cocoon confirmed that gas exchange takes place through the nipple but not through the more rigid, impermeable posterior portions of the cocoon (Rozen et al. 2011a). These findings indicate clearly that the cocoon in solitary and brood-parasitic bees is a highly adapted device for simultaneously protecting the prepupa from parasites and desiccation while allowing the effective exchange of gases for larval development.

The metabolic rates of overwintering prepupae and adults have only rarely been quantified. This is unfortunate because many solitary bees show remarkable adaptations for long-term, multiyear diapause in extremely harsh environments (see section "Bet-hedging in desert bees," below). We would expect these desert bees to exhibit extremely low metabolic rates as compared to cold-temperate bees, which have received the most study. The best data we have on cold-temperate solitary bees comes from a study by Kemp and colleagues (2004) in which they compared the metabolic rates of *Megachile rotundata* (which overwinters as a prepupa) and *Osmia lignaria* (which overwinters as an adult). Their results help explain why so many solitary bees overwinter as prepupae rather than adults. In *M. rotundata*, the prepupal metabolic rates averaged between 0.1 and 0.2 ml/g*hr over the period of prepupal diapause, or approximately one-sixth of the metabolic rate of feeding larvae and one-tenth the metabolic rate of a recently emerged, adult bee. Most remarkably, the body weight remained unchanged during the roughly eight months of prepupal diapause.

For *O. lignaria*, the overwintering adult metabolic rate was considerably higher than in *M. rotundata*: 0.25–0.5 ml/g*hr. The metabolic rate of overwintering adults was only slightly less than feeding larvae and one-sixth that of a recently emerged bee. Perhaps most significantly, there was a steady decline in adult body weight over the six months of diapause, suggesting that adult diapause may be an ineffective way to survive extremely long periods of dry conditions.

Based on the Kemp and colleagues (2004) study, there are clear advantages to prepupal diapause. Prepupal metabolic rates are lower than diapausing adult metabolic rates, and prepupae do not lose body weight at the same rate as overwintering adults. *O. lignaria* and *M. rotundata* are convenient experimental models for examining diapausing metabolic rates in bees, but the real insights will come when we examine metabolic rates in desert bees that experience multiyear diapause in extremely harsh, arid environments.

The pupal stage

After bee larvae have reached the last larval stage (the prepupa), they can either remain in that stage for an indefinite period of time (in a state of diapause) or continue their development toward adulthood. In either case, the prepupa eventually receives the hormonal signals that trigger metamorphosis—the transition from last-instar larva to the much more adult-like pupal stage. The pupal stage is, superficially, very much like the adult stage, but with greatly reduced wings. The pupa has a clearly defined head (with compound eyes, paired antennae, and the full component of adult mouthparts), a mesosoma (with three pairs of legs and wing buds), and a metasoma that allows, at this stage, the distinction between male and female offspring. There are also features unique to the pupal stage, such as an apical spine that allows the pupa to rotate around its long axis (Torchio 1975), mesosomal and metasomal tubercles that support the pupa while it rests on its back in the brood cell, and various spines on the legs, wings, and even head that limit contact with the brood-cell walls (Torchio and Trostle 1986; Fig. 3-9). The pupae of some bees are remarkably active. Torchio and Trostle (1986) describe 360° flips made by *Xeromelecta californica* pupae resting within the brood cells. These flips are so powerful that Torchio and Trostle reported feeling the movement through the wall of the brood cell. Unlike the last-instar larval stage, the pupal stage is a vulnerable stage and is rarely the stage in which bees overwinter for extended periods of time (but see Herbst 1922, Haider et al. 2014a, and Chiappa et al. 2018). In most solitary bees, the pupal stage is short—on the order of one to two weeks.

Emergence of adult bees from the pupal exoskeleton is the ultimate, final stage of bee development. Once the adult bee has emerged from the pupal skin,

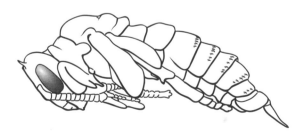

Figure 3-9. Lateral view of female pupa in *Perdita octomaculata* (Andrenidae) (redrawn from Eickwort 1977, Fig. 20).

she must expand her wings and harden her exoskeleton. At this stage, the fully formed adult is ready to emerge from the ground, stem, or wood nest constructed by her mother. In some bees, the adult will remain in the brood cell for an extended diapause (e.g., early-spring *Andrena*, *Colletes*, and *Osmia*) or emerge immediately if the weather and flowering conditions are favorable for foraging and reproduction. Bees use their mandibles and legs to dig out of their mother's nest.

THE ANNUAL LIFE CYCLE

We describe below a number of different annual life cycles that have been described in solitary bees. In some life cycles, bees spend the majority of their time as last-instar larvae, whereas in other life cycles, bees spend the majority of their time as adults. In some bees, the last larval stage (prepupa) is the overwintering stage that allows bees to pass through an unfavorable period of the year (the dry season in a desert, the wet season in the humid tropics, or the winter in the cold-temperate regions). In other bees, overwintering takes place in the adult stage; very rarely, overwintering can take place in the pupal stage (Chiappa et al. 2018). Some bees overwinter as adults within the brood cells, while others overwinter as adult family groups in the mother's nest. Which bee groups exhibit which life cycle is determined to some extent by phylogeny; closely related groups tend to show similar life history. However, these life cycles are clearly also shaped by natural selection and represent unique solutions to survival in particular habitats and in the face of particular environmental challenges.

Univoltine; prepupal diapause—the likely "ground-plan" for bees

Most solitary, ground-nesting bees conform to this life history (Box 3-1). Univoltine life cycles with prepupal diapause occur in Melittidae, Andrenidae, Colletidae, Stenotritidae, some Halictidae, most Megachilidae, and most ground-nesting Apidae. Adults are typically active for a relatively short period of time; they spend the rest of the year as a last-instar, overwintering larva (prepupa) within the brood cell. *Calliopsis pugionis* (Andrenidae: Panurginae), a solitary, ground-nesting bee, is typical of this type of life cycle. *C. pugionis* occurs in the coastal sage scrub habitats of cis-montane southern California and the adjacent Colorado Desert (Shinn 1967). This species is active from late March to early June. Like many *Calliopsis*, it is oligolectic. Females collect pollen and nectar from closely related genera of Asteraceae (*Encelia*, *Coreopsis*, *Chaenactis*, and *Hemizonia*; Shinn 1967).

Larvae spend approximately nine months of the year in diapause underground (Fig. 3-10). Larvae pupate in early spring (March), and adults emerge in late March to early April. Based on emergence data, this species is protogynous (females emerge before males). Protogyny is typical of many panurgine bees, in which females mate repeatedly throughout their lives and males are in a "race to be last" (see Chapter 4 on male behavior). *Calliopsis pugionis* forms

BOX 3-1: *MELITTA LEPORINA* (MELITTIDAE)—THE LIFE CYCLE AND NESTING BIOLOGY OF A "GROUND-PLAN" BEE

Phylogenetic studies of bees have indicated that the family Melittidae, a small, relictual group including mostly narrow host-plant specialists, is the basal family of bees (Chapter 2). We might therefore look to this family to give us a picture of the most basal, "ground-plan" life history of solitary bees. *Melitta leporina*—a solitary, ground-nesting, univoltine, oligolectic bee from the western Palearctic—represents just such a "ground-plan" bee (Celary 2006). *M. leporina* is active in summer, from late June to mid-August. Like most solitary bees, *M. leporina* is protandrous (males emerge before females). Mating takes place on flowers. In the evening, because males do not have nests, males gather together to form sleeping clusters on flowers. Each female constructs a subterranean burrow in which to rear her offspring. Nests range from 25 to 40 centimeters in depth, and each brood cell is constructed at the end of a short lateral tunnel. Females line the brood cell with a hydrophobic coating derived from the Dufour's gland (located in the metasoma). The brood cell coating provides a barrier against microbes as well as an impermeable, waterproof layer for maintaining a stable level of humidity within the brood cell.

Female *M. leporina* are hard workers. They start foraging at approximately 8:00 a.m. and forage until nearly 8:00 p.m. Each pollen/nectar provision mass requires approximately 15 foraging trips. Foraging trips average between 20 and 30 minutes, and females quickly (<1 minute) deposit their pollen loads within the brood cell before venturing out on another foraging trip. Once the full complements of pollen and nectar have been collected, the female sculpts the pollen into a conical mass and lays a single egg on top. She then closes the brood cell with soil (which takes approximately 20 minutes). Female *M. leporina*, like many solitary bees, make one last trip of the day to collect pollen for their own nutritional needs. This last "feeding trip" likely provides an essential source of protein for future egg production. Like many solitary bees, *M. leporina* is a narrow host-plant specialist that restricts foraging to legume genera such as *Medicago* (alfalfa), *Melilotus* (sweet-clover), *Trifolium* (clover), and *Vicia* (vetch). The foraging range of this bee is very small—25–30 meters from the nest, based on Celary's observations, and almost certainly not more than 100 meters. Within the closed brood cell, larval development is rapid. Eggs hatch in 3–4 days, and larvae complete feeding in just 2 weeks. Larvae then defecate and spin a protective cocoon. They spend the next 9 months as last-instar larvae (prepupae) in a state of diapause within the brood cell. Like most solitary bees, lifetime fecundity is low (9–12 offspring per female). And like almost all solitary bees, *M. leporina* also has enemies, such as brood-parasitic bees in the genus *Nomada* (Apidae).

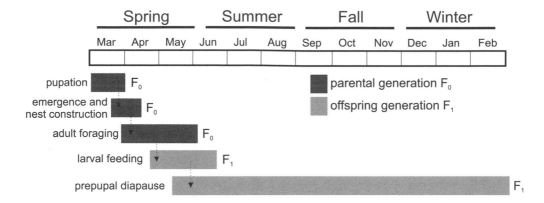

Univoltine; larval diapause (e.g., *Calliopsis pugionis*)

Figure 3-10. Life cycle of a typical, early-spring, univoltine solitary bee with prepupal diapause (e.g., *Calliopsis pugionis*, Visscher and Danforth 1993). F_0 = parental generation, F_1 = first offspring generation.

large, dense nesting aggregations with as many as 500 nests per square meter at some sites (Visscher and Danforth 1993). Adult activity peaks in the period April–May, when females are actively provisioning, on average, one brood cell per day. Larvae consume the pollen provisions fairly quickly (6–9 days in a closely related panurgine species, *Macrotera portalis*; Danforth 1991a), defecate, and then enter diapause as a prepupa by mid-June. *C. pugionis* is clearly univoltine, with no evidence of multiyear diapause based on excavations of nesting sites shortly after adult emergence (Visscher et al. 1994). Despite the apparently long period of adult activity indicated by museum collection records (Shinn 1967), Visscher and Danforth (1993) estimated that, on average, female *C. pugionis* produce just 6.4 cells per nest. If females make just one nest during their lifetime, this would suggest a very low rate of annual offspring production (see below).

Univoltine to multivoltine; adult diapause

Other groups of univoltine bees overwinter as adults rather than last-instar larvae. These are most often cold-temperate, Holarctic groups, such as *Osmia*, *Andrena*, some *Colletes* and *Megachile*, and *Protosmia rubifloris* (Griswold 1986). *O. lignaria* (Megachilidae: Megachilini: Osimiini)—among the most thoroughly studied of all solitary bees—exhibits this type of life cycle (Bosch et al. 2008; Fig. 3-11). Emergence of adult *Osmia* takes place in the early spring (March/April in North America). While male and female emergence periods overlap, males typically emerge 2 to 4 days before females (Bosch et al. 2008); hence, they are protandrous. Mating typically takes place as females leave the natal nest. The timing of emergence is dependent on temperature and can vary widely from year to year.

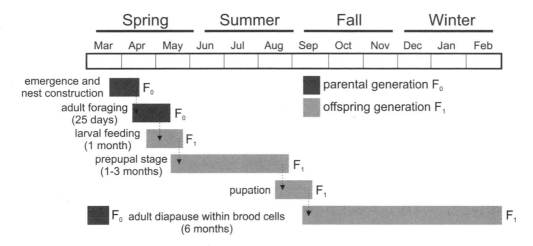

Univoltine; adult diapause (e.g., *Osmia lignaria*)

Figure 3-11. Life cycle of a typical, early-spring, univoltine solitary bee with adult diapause within the brood cells (e.g., *Osmia lignaria*, Bosch et al. 2008). F_0 = parental generation, F_1 = first offspring generation.

Following mating, females enter a short (2–3 day) pre-nesting period, during which ovarian development takes place. Once ovarian development is complete, females seek out a suitable cavity for nesting and then commence foraging for mud (for brood-cell partitions) and pollen/nectar (for brood-cell provisions). In *O. lignaria*, females typically complete one, and occasionally two, cells per day (Torchio 1989a). Individual females are active for up to 25 days, and a single female can construct multiple nests. Eggs hatch after approximately one week, and completion of the five larval instars typically takes one month. Once larval feeding has finished, the last larval instar defecates and then spins a cocoon. Both egg and larval development are temperature-dependent. In early summer, larvae enter a dormant, prepupal stage, which can last from one to three months. Pupation typically takes place in late summer. Adults eclose in early fall and then enter a six-month overwintering diapause triggered by cold temperature. Overwintering adults remain within their brood cells until they emerge the following spring. Most other lineages of Megachilidae overwinter as last-instar larvae (Bosch et al. 2001).

A variant on this form of life history is referred to as "parsivoltinism" by Torchio and Tepedino (1982). In several species of *Osmia* (*O. iridis*, *O. montana*, and *O. californica*), Torchio and Tepedino documented an unusual life cycle in which some larvae remain as prepupae through the first winter and then molt to adulthood for a second year of overwintering within the cocoon. Parsivoltinism can be viewed as a form of bet-hedging (see below) because not all larvae from the same generation emerge in year one; some persist on to year two, but as adults overwintering within the cocoon. This life cycle appears to be unique to Megachilidae. Additional species of *Osmia* that are reported to have a parsivoltine life cycle include *Osmia latreillei* (Krombein 1969), *O. latisulcata* (Parker 1984), *O. texana* (Tepedino and Frohlich 1984), *O. tanneri* (Torchio

1984b), and *O. ribifloris* (Sampson et al. 2004). Species of *Megachile* that are reported to have a parsivoltine life cycle include *M. assumptionis* (Almeida et al. 1997), *M. flavipes*, *M. lanata*, and *M. cephalotes* (Sihag 1983).

Univoltine; adult diapause with overwintering in adult family groups (e.g., Manueliini, Ceratinini, and Xylocopini)

Members of the subfamily Xylocopinae sensu stricto exhibit an unusual life cycle involving long-term overlap of mothers and their developing brood and overwintering of adult males and females in small family groups. This kind of life cycle is unlike that in any other group of bees, because adult females remain in the nest with their developing brood throughout the period of larval development and even after emergence of the adult offspring from their brood cells. As a result, there is far more interaction between adults and their fully developed adult offspring in Xylocopinae than in any other group of bees. This helps explain why so many xylocopines exhibit weak forms of sociality, such as cooperative breeding (see Chapter 2). One advantage of this life cycle is that mothers provide long-term protection to their developing, and even fully developed, offspring by guarding the nest until the onset of winter (or the inactive period). Mothers rarely abandon their brood; orphaning rates can be exceptionally low (2%) in some species of *Ceratina* (Rehan and Richards 2010).

Manuelia postica (Apidae: Xylocopinae: Manueliini) provides a good example of this kind of bee life cycle (Fig. 3-12). The genus *Manuelia* is a relictual

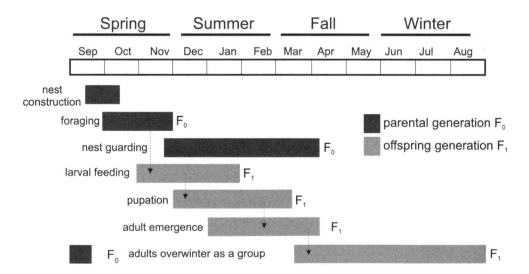

Univoltine; adults overwinter as a group (e.g., *Manuelia postica*; from Flores-Prado et al. 2008)

Figure 3-12. Life cycle of a typical univoltine xylocopine bee that overwinters in adult family groups (e.g., *Manuelia postica*; Flores-Prado et al. 2008). F_0 = parental generation, F_1 = first offspring generation.

group of just three species that occur in southern Chile and Argentina (Michener 2007). These stem-nesting bees resemble species of *Ceratina* (Ceratinini)—a much more widespread group. *M. postica* was studied by Luis Flores-Prado and coauthors (2008) in the Altos de Lircay National Park in Chile for six years. Adults spend the winter months (April–September) in hibernating assemblages of adult males and females. In the spring, adults disperse from these hibernaculae and initiate nesting in the hollow, pithy stems of *Chusquea* (Poaceae), *Aristotelia* (Elaeocarpaceae), and *Rubus* (Rosaceae). Females excavate an opening along the length of the stem and then construct brood cells both above and below the opening. Foraging and brood-cell provisioning take place from early October through November, and females rear multiple larvae in discrete brood cells separated by a partition made of wood particles (see Fig. 1 in Flores-Prado et al. 2008). Larvae consume the pollen/nectar provisions through the summer, pupate, and then emerge as adults. By early fall (March/April), adult offspring start to form overwintering assemblages of adult males and females. During the entire period of offspring development, the mother remains with her brood and protects the nest from intruders by blocking the entrance with her metasoma. Most nests consist of a single adult guarding the brood, but nests of up to three adult females were recorded in the study. Orphaned nests (nests with no adult female) were observed from November to February, indicating that there is some mortality of adults during brood development. Whether these nests suffer significantly higher brood mortality than guarded nests is not clear.

Uni- or multivoltine; late-summer mating with overwintering, fertilized females (e.g., solitary and social Halictinae)

In both solitary and social members of the halictid subfamily Halictinae, only fertilized, adult females overwinter (as in bumble bees). Mating takes place late in the season of adult activity, and males die before the onset of winter. In these bees, fertilized foundresses (gynes) found new nests in the spring. Fertilized females typically overwinter in specialized hibernaculae in the soil, wood, or stem in which they nest (Michener 2007).

Augochlora pura, a common, solitary, polylectic bee in eastern North America, illustrates this kind of life cycle (Fig. 3-13). *A. pura* was studied by Stockhammer (1966) based on observations of natural nests in the field as well as artificial, observation nests established in a flight room at the University of Kansas. Female *A. pura* behave essentially like ground-nesting bees, but instead of utilizing soil for their nests, they excavate burrows in pulpy, rotten wood in the understory of deciduous hardwood forests. Overwintering, fertilized females emerge in spring (mid- to late April) and found new nests. They forage and rear a brood of offspring that emerge as adults in early June. It is difficult to assess the fate of the overwintering foundresses at this stage, but Stockhammer speculates that they are unlikely to live much beyond the emergence of their first brood offspring. The number of subsequent generations is highly variable and difficult to establish. Egg to adult development time was approximately three weeks in the laboratory colonies, suggesting that *A. pura* is capable of producing at least one generation per month. Stockhammer

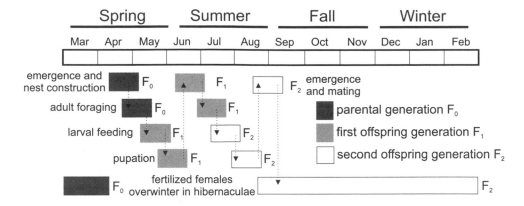

Bivoltine; adult diapause, females overwinter as
fertilized gynes (e.g., *Augochlora pura*)

Figure 3-13. Life cycle of a typical uni- or multivoltine halictine bee (Halictidae) with overwintering of fertilized females (gynes) (e.g., *Augochlora pura*; Stockhammer 1966). F_0 = parental generation, F_1 = first offspring generation, F_2 = second offspring generation.

estimated that, under natural field conditions, *A. pura* produces two to three generations during the summer months. In the artificial conditions of the flight room, they produced up to six generations during the course of his studies, so generation number is obviously not a fixed trait in these bees. In Figure 3-13, we have shown two adult summer generations for simplicity. Late-summer females emerge, mate, but do not construct nests or start provisioning brood cells. Instead, these females construct a specialized chamber in pulpy wood for overwintering. These hibernaculae are built off of existing burrows in the wood and extend downward to near the soil surface. Here, females construct a chamber that is roughly the size of a brood cell and spend the winter in hibernation. Females do not leave the hibernacula until the following spring. Males die off in the late summer/early fall.

This life cycle has been documented in other solitary halictines, including species of *Agapostemon* (Roberts 1969), *Lasioglossum* (*Sphecodogastra*) *comagenensis* (Batra 1990), *L.* (*Sphecodogastra*) *lusorium* (Bohart and Youssef 1976), and *L.* (*Lasioglossum*) *majus* (Boesi et al. 2009). A variant of this life cycle was reported by Field (1996) for *L.* (*Sphecodogastra*) *fratellum*. In this low-fecundity bee, some adult females overwinter a second year and continue to provision brood cells as two-year-old adults. These bees are exceptionally "lazy" foragers; they do not forage on days that would appear to be suitable for foraging and provisioning brood cells. Extended longevity may be an adaptation to producing offspring in the cold-temperate habitats that these bees inhabit.

Given that sociality has evolved repeatedly in the Halictinae, one wonders if this unusual life cycle—in which only fertilized, adult females overwinter—might provide some pre-adaptation for social life. Perhaps the fact that there are no males available in the early spring, when the female's first brood offspring emerge, gives her a unique opportunity to "enslave" her daughters as workers.

Partially bivoltine; prepupal diapause

A lifestyle known as partial bivoltinism has been described in a number of solitary, ground-nesting bees, especially those in the apid tribe Emphorini (including *Diadasia*, *Ptilothrix*, and other genera) but also in the Nomiinae (Hannan et al. 2013). In this life cycle, a first generation of adult bees produces offspring that either emerge immediately and start nesting within the same year as their mother or enter prepupal diapause until the following year. The second-generation adults that emerge and start to produce offspring can sometimes have very low reproductive success, especially if host-plant flowering is diminished late in the season. However, occasionally when host-plant resources are sufficient, these second-generation adults can successfully produce a second brood of offspring within the same season. These larvae then invariably enter diapause and remain in the ground until the following year.

A bee that exhibits this life history is *Diadasia rinconis* (Apidae: Eucerinae: Emphorini), a ground-nesting, cactus specialist that is active in the early spring in the southwestern United States (Fig. 3-14). This species was studied over a 10-year period at Pedernales Falls State Park in Texas by Neff and Simpson (1992). *D. rinconis* is active in April, May, and early June, roughly coincident with the flowering of its cactus host plant (various species of *Opuntia*). Females of the first generation were capable of provisioning up to 3 brood cells per day, with approximately 10 foraging trips/cell. Larvae develop rapidly—egg to adult development takes just over 2 weeks—and some second-generation females emerge to begin a second round of mating, nest construction, and brood-cell provisioning. Other larvae enter diapause and wait until the following year. Interestingly, larvae that emerge in the same year do not spin a cocoon, whereas those that enter diapause do. The second generation of foraging females

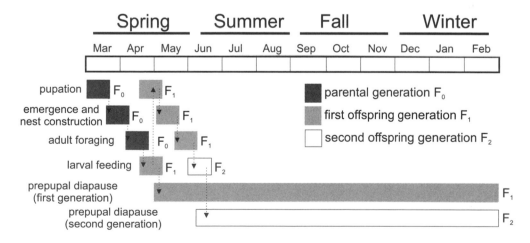

Partial bivoltine; larval diapause; temperate life cycle (e.g., *Diadasia rinconis*)

Figure 3-14. Life cycle of a temperate, partially bivoltine bee with prepupal diapause (e.g., *Diadasia rinconis*; Neff and Simpson 1992). F_0 = parental generation, F_1 = first offspring generation, F_2 = second offspring generation.

emerged when floral resources were in decline, and they were not as productive as first-generation females. Foraging trip times of second-generation females were nearly twice as long as those of the first-generation females, suggesting that second-generation females had to work considerably harder to fully provision a brood cell. While emerging second-generation females may produce some offspring in good years, they may be a reproductive dead end in years of reduced or early cactus flowering.

This kind of life history can be viewed as a form of bet-hedging (see below). Producing a second generation within the same year may boost reproductive success but is also highly risky. First-brood offspring that enter diapause until the next year can be viewed as an insurance policy against catastrophic loss of the second spring generation. Neff and Simpson (1992) were able to determine that all the offspring of a single female either emerged in the same year as their mother or entered diapause to the next year. This suggests that there may be a genetic component to the choice of whether to develop directly to adulthood or to enter diapause and wait until the next year.

Distinct seasonal patterns of emergence are not unique to desert bees. In tropical regions of the world, the alternation between rainy season and dry season can also establish discrete periods that are optimal for solitary bee foraging and reproduction, and periods that are not favorable (Martins and Antonini 1994). The life cycle of *Ptilothrix plumata* (Apidae: Eucerinae: Emphorini) has been studied extensively by Rogério Martins and colleagues (1996, 1999, 2001; Fig. 3-15). *P. plumata* is a ground-nesting, oligolectic bee that was studied at the Ecological Station of the Campus Pampulha of the Universidade Federal de Minais Gerais over a six-year period. Females are narrow host-plant specialists on various genera of Malvaceae, including *Pavonia*, *Sida*, and *Herissantia* (see Box 8-1). Females build shallow, single-celled nests in hard-packed soils. Peak adult activity

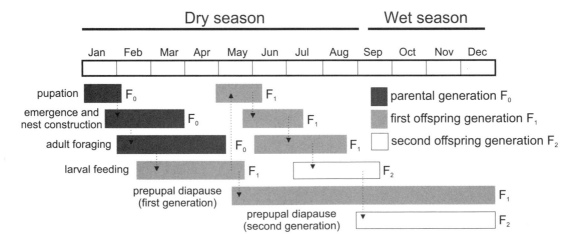

Partial bivoltine; larval diapause; tropical life cycle (e.g., *Ptilothrix plumata*)

Figure 3-15. Life cycle of a tropical, partially bivoltine bee with prepupal diapause (e.g., *Ptilothrix plumata*; Martins et al. 1996, 2001). F_0 = parental generation, F_1 = first offspring generation, F_2 = second offspring generation.

occurs during the dry season (January–August). *P. plumata* exhibits a partially bivoltine life cycle similar to the life cycle of *D. rinconis*, but extended over a longer period of time. Martins and colleagues (1996, 2001) discovered that larvae fell into two discrete categories: same-season larvae that skip diapause and emerge within the same year as their mother, and next-season larvae that enter diapause and emerge in the following dry season. Same-season larvae show an egg to adult development time of 30–120 days, whereas next-season larvae enter diapause for between 300 and 390 days. The probability of producing a next-season larva increases over the course of the dry season. Larvae produced early in the dry season (January–February) invariably emerge as adults the same year. Larvae produced in the middle of the dry season (March–May) had an equal probability of emerging as entering diapause. And larvae produced late in the dry season (June–August) invariably entered diapause. This strategy represents a spreading of risk across two years that is similar to what we saw in *Diadasia rinconis*. In fact, the two life cycles are identical, except that the period of adult activity is considerably longer in *P. plumata* than in *D. rinconis*.

Why do *P. plumata* (and another univoltine, tropical emphorine, *Diadasina distincta*; Martins and Antonini 1994) avoid the wet season? The authors hypothesize that brood cells containing pollen/nectar provisions may be especially prone to fungal infection during the humid wet season and that females *P. plumata* benefit from avoiding any provisioning during this period.

Partial bivoltinism has been documented in other solitary bee species: *Macrotera portalis* (Danforth 1991a), *Microthurge corumbae* (Mello and Garófalo 1986), *Megachile rotundata* (Tepedino and Parker 1988), *Anthophorula nitens* (Rozen and Snelling 1986), and *A. chionura* (Rozen and Macneill 1957). Hannan and colleagues (2013) describe a form of partial multivoltinism in a tropical Japanese bee, *Nomia chalybeata*. In this bee, there are multiple generations per year, and a mix of emerging and overwintering prepupae are produced in each generation. However, the proportion of overwintering prepupae increases over time, so that in the last generation nearly all prepupae enter diapause until the following year.

Bivoltine

In arid regions of the southwestern United States, there are two discrete periods of flowering—early-spring flowering that is triggered by the winter rainfall and late-summer/early-fall flowering that is triggered by the late-summer "monsoon" rains. Many bees are active for just one of these periods, but some species clearly have both a spring and a late-summer emergence, each of which produces one brood of offspring. *Anthophorula sidae* (Apidae: Eucerinae: Exomalopsini) is one such bee that was studied by Barbara and Jerome G. Rozen Jr. in the vicinity of Willcox, Arizona (Rozen 1984a; Fig. 3-16). *A. sidae* is a ground-nesting, communal bee with up to eight adult females per nest. Females are narrow host-plant specialists on Malvaceae, especially the genus *Sida*. Adult *A. sidae* are active in both the spring and the late summer, and excavations of nests in the late summer revealed that the spring generation does not spin a cocoon, whereas the late-summer generation does. The variation in cocoon spinning in *A. sidae*

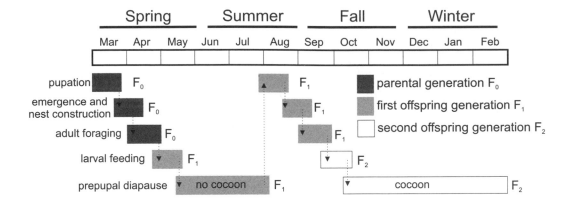

Bivoltine; larval diapause (e.g., *Anthophorula sidae*)

Figure 3-16. Life cycle of a desert, bivoltine bee with prepupal diapause (e.g., *Anthophorula sidae*; Rozen 1984a). F_0 = parental generation, F_1 = first offspring generation, F_2 = second offspring generation.

parallels the pattern seen in *Diadasia rinconis* (Fig. 3-14); larvae that spend a short time in diapause do not invest in cocoon spinning, whereas larvae that overwinter for several months do. The cocoon is costly to produce (see below), and it makes sense that bee larvae do not spin a cocoon except when one is essential for larval survival. *Perdita coreopsidis* is another ground-nesting, communal, southwestern desert bee that clearly shows a bivoltine life cycle (Danforth 1989a). Even some brood parasites of bivoltine bees have a bivoltine life cycle (Rozen et al. 1978).

BET-HEDGING IN DESERT BEES

Arid regions of the world are the places where we find the greatest species richness and diversity of solitary bees, and it is in deserts that we find some of the most fascinating examples of life-history adaptations in bees. Deserts are harsh environments. Rainfall is limited, and annual precipitation is unpredictable and highly variable from year to year (Davidowitz 2002). Flowering plants in desert regions adopt a number of strategies for coping with the highly seasonal and unpredictable rainfall. First, they use cues, such as rainfall, to time key events, such as flowering and seed germination. Rainfall is often the cue that triggers flowering (in perennial plants) or germination (in annual plants). Second, to insure that they do not lose all their reproductive effort in one bad year, many desert annuals produce seeds that germinate over multiple years. Some seeds germinate in year one, others in year two, and still others in subsequent years. By spreading their offspring production over many years, annual plants are effectively "hedging their bets" against catastrophic losses in a very dry year. The combination of rainfall-induced emergence and delayed (multiyear) germination allows desert annual plants to survive in this harsh environment.

It turns out that desert bees do the same thing. Delayed emergence, or bet-hedging, has been documented in a number of desert-adapted bee species (Table 3-2). In one extreme case, larvae of an Australian desert bee, *Amegilla dawsoni*, were observed to remain in diapause for up to 10 years (Houston 1991a). In some cases, demonstrations of delayed emergence happened more or less by accident. Jerome Rozen Jr. kept overwintering prepupae of *Pararhophites* in a desk drawer at the American Museum of Natural History for five years before an adult emerged from the cocoon (Rozen 1990). Other cases have been documented through repeated visits to the same site over multiple years. A nest site of *Hesperapis rhodocerata* (Melittidae) was discovered south of Animas, New Mexico, in 2010 and visited annually for five years before another emergence was reported in 2015 (Rozen 2016a). In other cases, authors monitored nesting aggregations through drought years and documented the presence of viable, overwintering prepupae in the soil even though there were no adults flying in the drought year (Rust 1988). In the case of *Macrotera portalis*, prepupae excavated early in the season were reared under seemingly ideal temperatures and humidity, yet only 50% emerged in any one year (Danforth 1999). Even brood parasites (*Xeromelecta californica*) of desert hosts (*Anthophora pueblo*) show delayed, multiyear diapause (Orr et al. 2016). So solitary desert bees appear to be capable of spreading their risk across multiple years by adopting the same tactics as desert annual plants—bet-hedging. But do they also use rainfall as a signal for emergence?

Anecdotal (Hurd 1957) and experimental (Danforth 1999) evidence suggests that rainfall is a cue for pupation in desert bees. Hurd (1957) documented emergence of *Hesperapis fulvipes* in the fall of 1951 in association with heavy summer rains in the Mojave and Colorado deserts of southern California. *H. fulvipes* is normally a spring bee that specializes only on desert sunflower (*Geraea canescens*). However, the heavy summer rains of 1951 brought *Geraea* into bloom in the fall and also triggered the emergence of the specialist *Hesperapis* at a time of year when they are not normally active. As far as we know, Hurd was the first to hypothesize that specialist bees are using the same cues (rainfall) as their host plants to trigger emergence.

A more direct demonstration of rainfall-induced emergence was made by Danforth (1999) for *Macrotera portalis*, a late-summer, host-plant specialist on *Sphaeralcea* (Malvaceae). Overwintering prepupae excavated from nests in southern Arizona and New Mexico were assigned randomly to two groups: early rainfall and late rainfall. The early-rainfall group was exposed to high humidity on day one of the experiment, whereas the late-rainfall group was exposed to high humidity two weeks later. Only 60% of the bees in either group pupated and emerged as adult bees, but bees in group two emerged precisely two weeks later than those in group one, indicating that humidity (or rainfall) is the proximate cue used by these bees to trigger emergence. Prepupal body weight also predicted likelihood of pupation. Small, underweight prepupae were far more likely than larger, heavier prepupae to pupate in any given year. It seems likely that fat and water reserves may predict whether a prepupa will delay emergence to another year or emerge in the current year. Such conditional dependence is also reported for the seeds of annual plants.

Faunistic studies of desert bee communities also suggest that solitary bees are tracking rainfall, especially for species that are host-plant specialists on

Table 3-2. Documented cases of delayed emergence (multiyear diapause) in solitary bees.

Family	Subfamily	Species	Maximum recorded length of diapause (years)	Stage of diapause	Evidence of rainfall-induced emergence	Reference
Andrenidae	Andreninae	Andrena spp.	2	prepupa	yes	Thorp 1979
Andrenidae	Panurginae	Calliopsis larreae	2	prepupa	yes	Rust 1988
Andrenidae	Panurginae	Macrotera portalis	3	prepupa	yes	Danforth 1999
Andrenidae	Panurginae	Perdita floridensis	2	prepupa	no	Norden et al. 2003
Andrenidae	Panurginae	Perdita nuda	3	prepupa	yes	Torchio 1975
Apidae	Anthophorinae	Amegilla dawsoni	10	prepupa	no	Houston 1991a
Apidae	Anthophorinae	Anthophora pueblo	4	prepupa	yes	Orr et al. 2016
Apidae	Eucerinae	Melissodes robustior	3	prepupa	no	MacSwain 1958
Apidae	Nomadinae	Xeromelecta californica	4	prepupa	yes	Orr et al. 2016
Megachilidae	Megachilinae	Chelostoma philadelphi	2	prepupa	no	Krombein 1967
Megachilidae	Megachilinae	Osmia spp.	2	prepupa year 1, adult year 2	no	Torchio & Tepedino 1982
Megachilidae	Megachilinae	Trachusa sp.	2	prepupa	yes	Thorp 1979
Megachilidae	Pararhophitinae	Pararhophites orobinus	5	prepupa	no	Rozen 1990
Melittidae	Dasypodainae	Hesperapis regularis	3	prepupa	no	Stage 1966
Melittidae	Dasypodainae	Hesperapis rhodocerata	5	prepupa	yes	Rozen 2016a

ephemeral desert annuals. Minckley and colleagues (2013) documented specialist and generalist bee abundance and species richness across a three-year period in mixed Chihuahuan Desert habitat along the United States–Mexico border. Years one and three were typical years in terms of rainfall, but year two was a drought, and annual plants greatly reduced their flowering. Using a combination of net collecting and pan trapping, the researchers found that generalist bees emerged in large numbers in the drought year, but specialists on annual plants and rainfall-sensitive perennial shrubs (e.g., creosote bush, *Larrea tridentata*) did not emerge in the drought year. However, specialists on mesquite (*Prosopis velutina*), which blooms even in drought years, did emerge. Asynchrony of most generalists and synchrony of the entire specialist bee community with rainfall and host-plant flowering is strong evidence that specialist bees are tracking the same environmental cues (most likely rainfall) as their associated host plants.

Further evidence for the importance of bee-plant synchrony comes from faunistic studies across the deserts of the southwestern United States. Minckley and colleagues (2000) documented patterns of bee species composition across four distinct desert regions: Chihuahuan, Mojave, Upper Sonoran, and Lower Sonoran. They focused exclusively on the bee fauna of creosote bush (*L. tridentata*), which hosts a rich fauna of both generalist and specialist bees. They found the greatest diversity, abundance, and biomass of specialist bees in the regions with the lowest overall rainfall and highest inter-annual variance in rainfall (the Mojave and Lower Sonoran deserts). Specialists, rather than generalists, appear to be capable of surviving in the harsh conditions of the Lower Sonoran Desert by virtue of the fact that they can track host-plant bloom better than generalists. By effectively tracking host-plant bloom, specialist bees render a highly unpredictable environment (from the perspective of a generalist) predictable.

It seems that solitary bees and seeds really are a lot more similar to each other than anyone would have predicted. Not only do solitary bees show rainfall-induced emergence, like annual plants; they also show multiyear diapause and condition-dependent emergence. Just as desert annual plants have a "seed bank" that persists through dry, unfavorable years, so too bees have a "bee bank" that allows desert bee populations to persist in some of the most inhospitable regions of the world (Minckley et al. 2000).

LIFETIME FECUNDITY AND DETERMINANTS OF ADULT LIFE SPAN IN SOLITARY BEES

Given the typically short window of time solitary bees are active as adults, it is worth considering what we know about lifetime reproductive output in solitary bees. Few studies have estimated life expectancy or lifetime fecundity in solitary bees, but both life expectancy and lifetime fecundity appear to be low for many solitary bee species. Minckley and colleagues (1994) estimated female life expectancy to be 14 days in *Dieunomia triangulifera*, a solitary, ground-nesting, oligolectic halictid. For *Andrena agilissima* (Andrenidae), Giovanetti and Lasso (2005) estimated that adult females live between 4 and 30 days, and an average female was estimated to live for just 16 days. Cameron and colleagues (1996) estimated that female *Melissodes rustica* (Apidae) actively provision brood cells for just 10 days (assuming females occupy just one nest, which may not be the case). The life expectancy for most oligolectic solitary bees seems to be surprisingly short.

In addition, daily offspring production is documented to be low in many solitary bees. *Andrena*, in particular, seem to produce brood cells at a remarkably slow rate. Giovanetti and Lasso (2005) summarized data on the rate of offspring production for various *Andrena* species and found that most *Andrena* provision less than one brood cell per day. Franzén and Larsson (2007) made similar observations when studying *A. humilis* in Germany. They summarized total lifetime reproductive success in various *Andrena*, *Panurgus*, and *Panurginus* (Andrenidae) and found that a short adult life span in these bees and a low rate of cell provisioning (0.4 to 1.4 cells per day) means that many *Andrena* have a surprisingly low rate of total lifetime fecundity. Most species that they surveyed produced less than 10 offspring over their entire lifetime. Low offspring production could be due to a number of factors, including low resource availability or low rates of egg production. Egg limitation may put an upper limit on reproduction in many solitary bees because eggs are costly to produce and require a stable source of plant protein for the production of vitellogenins (egg proteins). Egg production, more than floral resource availability for provisioning brood cells, may be the primary limiting factor in solitary bee reproductive success (Neff 2008; see Chapter 8 for a more in-depth discussion).

Lifetime reproductive success in *Andrena* appears surprisingly low. But is this true of other solitary ground-, stem-, wood-, or cavity-nesting bees? In order to examine lifetime reproductive success across a wider array of species, we assembled data on the lifetime fecundity of 12 species of solitary bees in 4 of the 7 bee families from published accounts (Table 3-3). A few caveats need to be taken into consideration when examining Table 3-3. First, these values do not account for brood mortality. These are simply estimates of the total number of brood cells produced and provisioned, not the number of offspring that actually survived to adulthood. Because brood-cell mortality can be high in solitary bees and wasps (Danks 1971), these measures of lifetime reproductive output should be viewed as "best case" estimates. We discuss brood-cell mortality due to both brood-parasitic bees and various non-bee parasites and predators in later chapters in this book (see Chapters 10 and 11). Second, it is important to keep in mind that quantifying total reproductive success is not easy in many solitary bees, because cells cannot always be accurately associated with a single burrow (in ground-nesting bees) and females may make more than one nest over their lifetime.

Note that lifetime reproductive success is highly variable, from as few as 6.7 offspring in *Panurgus banksianus* (Andrenidae), to as many as 35 offspring in some mason bees (*Osmia*, Megachilidae). Low lifetime fecundity seems to be true for many ground-nesting oligolectic bees, but is also true for wood-nesting bees (*Xylocopa*) and stem-nesting bees (*Ceratina*). Low fecundity does not seem to be limited to bees with short life spans. Species of *Ceratina* and *Xylocopa*, which live for more than a year as adults and in which provisioning takes place over an extended period of time, also have low lifetime fecundity. Data assembled by Sean Prager on the lifetime fecundity of 22 species of both tropical and temperate *Xylocopa* shows that these bees have lifetime fecundity of just 3–9

Table 3-3. Lifetime reproductive success in solitary bees.
Based on various sources.

Species	Family	Mean lifetime offspring procuction	Range lifetime offspring production	N	Reference
Panurgus banksianus	Andrenidae	6.7 ± 2.1	4–10	7	Münster-Swendsen 1970
Ceratina calcarata	Apidae	6.9 ± 3.0	1–14	110	Rehan & Richards 2010
Ceratina calcarata	Apidae	9.25 ± 3.27	4–15	20	Johnson 1990
Xylocopa violacea	Apidae	7.43 ± 7.58	3–14	65	Vicidomini 1996
Xylocopa virginica	Apidae	7.7 ± 5.4	0–16	9	Prager 2008
Nomia chalybeata	Halictidae	11.9 ± 4.7	5–20	14	Hannan et al. 2013
Anthidium septemspinosum	Megachilidae	27.1 ± 4.5	19–31	14	Sugiura 1994
Heriades carinata	Megachilidae	10.0 ± 4.5	5–16	7	Matthews 1965
Osmia bicornis	Megachilidae	18.6 ± 6.3	10–27	5	Raw 1972
Osmia bruneri	Megachilidae	19.0 ± 6.7	9–27	16	Frohlich & Tepedino 1986
Osmia cornifrons	Megachilidae	37.2 ± 12.6	9–60	20	Sugiura & Maeta 1989
Osmia lignaria	Megachilidae	22.7 ± 13.2	1–48	32	Tepedino & Torchio 1989
Osmia lignaria	Megachilidae	28.9 ± 12.2	5–49	17	Tepedino & Torchio 1982
Megachile pacifica	Megachilidae	35.4 ± 11.3	14–62	42	Kim 1997

offspring (Prager 2008). One general trend that does emerge from Table 3-3 is that megachilid bees have unusually high lifetime reproductive success compared to other ground- and stem-nesting bees. As we shall see in a later chapter, foraging and provisioning in megachilid bees is strikingly different from those behaviors in other bees. Megachilids typically make many, short, rapid foraging trips, unlike other bees. This trait may be related to the use of a metasomal scopa that puts limits on the amount of pollen that can be carried per trip.

It is also possible that our data are biased. Many of the megachilids listed in Table 3-3 are managed pollinators or invasive species. In both cases, these kinds of species may show higher rates of offspring production and increased lifetime reproductive output than the average mason or leaf-cutter bee. We are cautious about concluding that megachilids have universally higher lifetime fecundity than other bees—but the trends are suggestive. What does appear to be true is that many solitary ground-, stem-, and wood-nesting bees have low lifetime fecundity. This has important implications for bee conservation, which we discuss in more detail in Chapter 14.

What factors impact life expectancy in solitary bees? Straka and colleagues (2014) examined the factors that determine life expectancy in two European solitary bee species: *Anthophora plumipes* (Apidae; a host-plant generalist) and *Andrena vaga* (Andrenidae; a host-plant specialist). Females were individually marked at emergence, and then Straka and colleagues monitored foraging activity (time spent outside the nest), temperature, precipitation, pollen availability, and, ultimately, adult life span. In both species, adult life span was short—roughly two weeks—consistent with the accounts cited above for other solitary, ground-nesting bees. However, within that short period of time, females were only actively foraging on between 5 and 6 days, or the equivalent of 42–44% of their adult life span. This is a remarkably short period of adult foraging activity. For both bee species, there was a clear impact of activity (proportion of adult life span spent foraging) on life expectancy. Bees that foraged more frequently died earlier. This is not too surprising given that increased activity can lead to oxidative damage to proteins and lipids in controlled laboratory studies of fruit flies (Magwere et al. 2006). What it does suggest is that adult bees may have a finite number of days for foraging and brood-cell provisioning. Bees that are active foragers appear to suffer in terms of life expectancy. Other determinants of life span included the date of first foraging. Bees that started foraging earlier in the season had higher life expectancy than bees that started foraging later in the season. This effect could be due to an overall decrease in floral resources over the course of the season. If floral resources are limited, bees may need to work harder to collect them. Finally, both the temperature and precipitation on days that bees were active impacted their life span; bees foraging on cooler, wetter days lived longer than bees foraging on warmer, drier days. The impact of precipitation was more pronounced than the impact of temperature. This may be explained if bees are approaching some thermal maximum when foraging, as has been suggested for some desert bees (Chappell 1984). The responses of these two species to both intrinsic as well as extrinsic factors were remarkably similar, suggesting that these patterns may be generally applicable to many solitary bees. This study provides a clear demonstration that foraging activity can reduce life expectancy. This may help explain why "lazy" bees are so often reported in studies of solitary bee nesting. Bees that do not forage when conditions appear to be suitable for foraging may be wisely adopting a strategy to extend their own longevity, or they could just be waiting for their next mature oocyte to develop.

In this chapter, we have outlined the developmental trajectory of bees from egg through five larval instars to the pupal stage and, finally, to adulthood. We have emphasized the fact that, for most solitary bees, the period of adult activity is extremely short, and there is a very high premium on timing emergence with host-plant flowering. Most solitary bees spend the vast majority of time as overwintering, last-instar larvae (prepupae), but some groups overwinter as adults in the brood cell or family groups of adults resting in the natal nest. We have described the various life-history patterns in solitary bees, including univoltine, bivoltine, partially bivoltine, and variations on these themes. We have described how solitary desert bees adopt a bet-hedging strategy of multiyear diapause and how rainfall appears to be a trigger for emergence, especially in those bees with narrow host-plant preferences. In the remaining chapters of this book, we will be focusing largely on adult behavior, including mating, nest construction, foraging, provisioning, and defense against parasites and predators.

CHAPTER 4

Alternative Male Mating Tactics: The Race to Be First and the Race to Be Last

"The birds and the bees" is a common euphemism for sex. While almost all of us eventually become familiar with the intricacies of human sex, far fewer spend much time contemplating the sex lives of birds, and even fewer still, those of bees. The sexual behavior of bees is remarkably diverse. All male solitary bees are capable of mating with multiple females, but relatively few attain this goal. This drive is sharpened by the fact that many males will not mate at all even under the best of circumstances. This is because in most solitary bee species, females mate only once, and there are more males ready to mate than receptive females. Male-biased sex ratios are an expected outcome in outbreeding systems when males are smaller than females, as is usually the case among solitary bees (Stubblefield and Seger 1994). This means some males are doomed to a life of involuntary chastity as those males that are able to mate with more than one female reduce the supply of receptive females.

Male solitary bees employ a wide array of tactics in their frantic attempt to mate and leave offspring. At the same time, females are not passive players in this game, and there is considerable evidence of female choice in many species. Bee sexual interactions range from quick, secretive trysts to extended amorous bouts involving long periods of courtship, prolonged copulations, and extended periods of post-copulatory signaling or mate guarding. For males, tactics range from a lifetime of gathering bouquets of scent to court a female to fierce or deadly battles for emerging virgins. How males find mates, what females have to say about it, and related topics will be dealt with in this chapter.

The structure of bee mating systems depends on the distribution of receptive females in space and time (Alcock et al. 1978, Alcock 2013a, Emlen and Oring 1977). When females mate only once or have a brief receptive period, the narrow window of female availability favors a frantic race to be first among males. On the other hand, when females have extended or even unlimited receptivity and mate repeatedly, there is no benefit to a male in being first to mate. Rather there will be a race to be last, since in this case, the victory in the sexual sweepstakes goes to the male that mates just before a female oviposits. This phenomenon is driven largely by a phenomenon called sperm precedence.[1] The tactics that males use in order to encounter females will depend, to a large extent, on where receptive females occur, and dramatically different, alternative male tactics may exist within the same species. Individual males may also vary their mating tactics in response to varying female densities, levels of receptivity, and the abundance of conspecific males.

We know remarkably little about the mating frequency of female solitary bees. For the vast majority of species, there are no published data. It is commonly assumed that most females mate only once or are receptive only briefly. Mating in most species is rarely observed beyond a week or so after female emergence, and most detailed studies indicate that females mate for a brief period of time early in life (Eickwort and Ginsberg 1980). Another reason to assume that most female solitary bees mate once is that there are only limited benefits, and a distinct downside, to multiple mating. Male solitary bees produce thousands of sperm (Garófalo 1980), but a female solitary bee typically lays no more than 20 eggs in her lifetime (see Chapter 3). Of these eggs, only half or fewer will be the fertilized eggs that produce daughters, so her sperm requirements are quite low. One benefit of mating multiply is the higher genetic diversity multiple mating likely conveys to her offspring (Alcock et al. 1978), but even here the optimal number of mates is likely to be small. On the negative side, repeated mating or mating attempts take valuable time that could be used to rear more offspring (Rossi et al. 2010, Stone 1995). Unless there are extenuating circumstances, females will usually be better off getting the sex business out of the way early, ignoring the horny males, and proceeding with provisioning nests.

MONANDRY AND THE RACE TO BE FIRST

The race to be first is presumably why males of most solitary bees emerge as adults and commence activity before females (protandry) (Robertson 1918). When looking for bee virgins, it is almost certainly better to be a few days early to the party and spend some time without any female company than to be late and find that many females have already found mates and are off the market. The predominance of protandry among solitary bees in turn explains the standard distribution of the sexes in the linear nests typical of cavity-nesting bees (see Chapter 6): females in the inner cells and males in the outer, an arrangement that means early-emerging males will not be blocked by later-emerging females (Gerber and Klostermeyer 1970). While many solitary bee species are truly protandrous, some cases are more ambiguous. Careful observations found that while male *Andrena erythronii* (Andrenidae) are commonly observed flying well before females, male emergence actually commences on the same day

---

[1] Sperm precedence occurs in many insects and arises from the fact that the female sperm storage organ (the spermatheca) is a blind-end sac. The last sperm to enter the spermatheca is likely the first sperm used when females fertilize their eggs. When females mate multiply (polyandry), this creates a significant incentive to be the last male to mate prior to oviposition. While sperm precedence is likely to be the case for many bee species, it has been demonstrated using genetic markers in only one case (*Anthidium manicatum*; Lampert et al. 2014).

as female emergence, or at most a day or two earlier. Apparently in this species most females mate almost immediately upon reaching the soil surface and then quickly dig into the soil, delaying foraging until several days later and giving the impression of a significant difference in emergence times (Michener and Rettenmeyer 1956). Such behavior may confound estimates of protandry based solely on collections, particularly collections at flowers.

Mating at the nest site upon female emergence

As indicated above, if females are monogamous, selection will favor males that are the first to locate receptive females at sites where the probability of encountering receptive virgins is highest (Alcock 2013b, Paxton 2005). For species that nest gregariously, emergence sites are an obvious place to encounter numerous virgin females. Females of most bee species emerge in a small window of time early in the season. In such species, it is common to see hundreds or even thousands of males patrolling a large nest site when females are first emerging. Such behavior has been reported for the vast majority of gregariously nesting species.

Males that patrol emergence sites typically fly low to the ground, and they periodically drop to inspect the surface or enter holes made by emerging females. When they sense an appropriate cue, males dig to the emerging females (Alcock et al. 1976, Cane 1994a, Cane and Tengö 1981). A common sight at these emergence sites is clusters of males attempting to grasp and mount an emerging female (Fig. 4-1). These "mating balls" may persist for several minutes, usually dispersing after a victorious male and his newfound partner depart for

Figure 4-1. Cluster of male *Colletes hederae* (Colletidae) attempting to mate with a newly emerged female. Photo by Ben Porter (benporterwildlife.co.uk).

the margins of the nest site to complete their nuptials. Fighting in these mating balls is usually just a matter of harmless grappling, but in *Amegilla dawsoni* it can reach lethal intensity (see Box 4-1). In this species, males may slash one another with their tibial spines, sometimes severely damaging wings or other body parts (Alcock 1996a). Even the females are occasionally injured or accidentally decapitated by the powerful struggling males (Alcock 1996a). Mating balls that contain no females sometimes form probably through simple errors on the part of desperate males ready to mate with anything, or when a male picks up compounds used by females to signal receptivity and thus has become, at least temporarily, a pseudo-female as perceived by the antennae of his fellow suitors. A common, although not universal, feature of these mating clusters is that bigger is often better. Larger males tend to dominate these competitions and, in many but not all cases, do most of the mating (Alcock 1996b, 2013b).

BOX 4-1: *AMEGILLA DAWSONI* (APIDAE)— THE MYSTERY OF THE MIDGET MALES

Amegilla dawsoni is a large, univoltine, monandrous, ground-nesting bee found on the arid plains of western Australia. Adults are active from July to September, the drier half of the local rain season. Although some females construct isolated nests, most females join massive nesting aggregations of over 10,000 nests in areas of bare soil (Houston 1991a). Females show limited size variation, but males vary greatly, with a continuous but strongly bimodal size distribution (Houston 1991a). The large males are about the same size as females, but the small males in the most abundant size class average only half the weight of their big brothers. Large males are powerfully built fighters that patrol emergence areas and attempt to mate with virgin females as they emerge from their natal nests. Although they do not dig down to reach emerging females, as has been reported in some other species with large males (Alcock 1997, Cane and Tengö 1981), they do cluster about emergence holes and regularly form mating balls in which they fight vigorously to control emerging virgins. These fights can include biting and slashing with their tibial spurs that can result in extensive wing damage (Alcock 1996a). The largest males dominate these mating clusters and do almost all of the mating at the nest site. The smaller males are relegated to patrolling flowers and vegetated areas at the margins of the emergence sites, where they search for the few females that escape the emergence sites without being mounted by the large males. Late in the day, the density of males at the emergence site commonly decreases, as the number of emerging females decreases and the large males retire to rest. At this time, some smaller males will patrol the emergence site in search of virgins (Alcock 1997). Since females lose receptivity soon after copulating, few females are potential mates for the smaller males (Simmons et al. 2000).

While large males do most of the mating, small males outnumber them by a ratio of around 4 to 1 (Alcock 1996c,b). Exactly why females produce so many small sons despite their obvious fitness disadvantages is not obvious, but a number of factors seem to contribute. For one, females initially provision cells producing females and then switch to a mix of females and large males. It is only in the latter half of the flight season that they switch to mainly producing small males. By this time, floral resources are limited, so that it takes significantly longer to collect a given amount of nectar and pollen. Since the value of an incomplete cell is zero, it may be better to produce small cells and small males than risk failure by attempting to produce a large male (Alcock et al. 2005). Another factor may be related to phenology. Although small male cells are provisioned at the end of the previous season, small males are the first to emerge, both within a season and within a given day. As such, they may at least temporarily be escaping competition from the large males. Also, small males do the hard work of digging an emergence tunnel and as such may be doing a service for their larger brothers and sisters (Alcock 1996a). Finally, large size is not without its disadvantages. Large males have a shorter adult life span than small males, probably due to the wear and tear of their combative habits plus their increased exposure to emergence-site predators like butcher birds (Alcock 1996a; see also Chapter 11).

An interesting evolutionary path found in a number of species that compete for females at dense nest aggregations has been the evolution of fighting morphs, sometimes called metanders (Moure 1963), betas (Chemsak 1985), or majors (Alcock 1996b, 1997). These are found in several subgenera of *Centris* (*Xerocentris*, *Trichocentris*, *Centris* s. str., and *Wagenknechtia*, Apidae), where they have evolved at least twice, and again in *Amegilla* (Apidae) (Box 4-1). Unlike the flightless, macrocephalic males that occur in some *Lasioglossum* (Halictidae) and *Macrotera* (Andrenidae) (see below), they are not truly dimorphic, as the size distributions of males in these species are continuous. Nevertheless, the largest of these "metandric" morphs can differ so greatly in shape and coloration from a "normal" male that several have been described as new species (Snelling 1984) and, in one case, were the basis for a new subgenus (Snelling 1956). Besides being much larger than "normal" males, large metandrics have a more robust form and larger, more heavily muscled legs (Neff and Simpson 1992, Snelling 1984), useful features when struggling to grasp a female or dislodge a rival in a mating ball. They can also differ strikingly in coloration. Metandric males of *Centris* s. str. have more extensive maculation on their metasoma than normal males. In other metandric *Centris*, the metasoma of the metanders is also covered with pale pubescence rather than being dark-haired as in normal males.

Small males in these species take a very different approach to finding mates (Alcock et al. 1977b). Rather than fighting with the larger males for emerging females, the smaller, faster males patrol the periphery of the emergence area or nearby flower patches for virgin females that have eluded the dominant metandrics that patrol the emergence area. In one study, fighting among large males provided the opportunity for 11% of emerging *Amegilla dawsoni* females to leave the emergence area unmated (Alcock 1997). In the same study, it was observed that some small males did patrol the emergence site, but did so either early or late in the day, avoiding the optimal midday hours when most females had emerged but also when large males were most active. *A. dawsoni* is a cold-season bee, and small males that fly early and late at the emergence site appear to be making the "best of a bad lot" by searching when they are at a thermal disadvantage.

While the development of distinct castes of fighting males that dominate gregarious emergence sites is an interesting evolutionary story, it is not a typical one. For starters, most bees do not nest gregariously, so the advantages of being a large male fighting for emerging females in a small, easily searchable area vanishes. Second, with the exception of the examples mentioned above, most bees that nest in large aggregations have not evolved distinct large, fighting male morphs. The usual pattern at gregarious emergence sites is that most males initially patrol nesting areas and later shift to patrolling nearby shrubs, bushes, or flowers where conspecific females forage for pollen and nectar (Neff 2003). In some cases, such as *Diadasia rinconis*, larger males seem to be at an advantage in gaining access to emerging virgins (Neff and Simpson 1992), but in others, there is no clear large male advantage (Seidelmann 1999, Shimamoto et al. 2006). As the season progresses and fewer receptive females emerge at the nest site, males typically disperse and switch to patrolling or defending other sites where they are likely to encounter virgin females, usually flowers.

Pre-emergence mating within communal nests

Males can also monopolize females if they are aggregated in communal nests, as occurs in some species of *Andrena* (Andrenidae). Think of a large, communal nest of foraging, reproductively active females as a nesting aggregation with a single nest entrance. Some *Andrena* are known to form communal nests with hundreds of females. Careful studies have found that high proportions (71–97%) of the females in these species mate in their natal nests before emergence, presumably with nest mates since there is a high degree of inbreeding (Paxton and Tengö 1996, Paxton et al. 1999a). Exactly what goes on inside these nests is unknown. Post-emergent males of these species can be found patrolling flowers and occasionally copulating with females. Since some of these females bear pollen loads, there apparently is some degree of polyandry. As we shall see below, intra-nidal mating takes on a whole new dimension in some communal species with multiple-female mating.

Mating at flowers and flower patches

Most bees do not nest gregariously, which greatly diminishes the opportunity for males to monopolize receptive females as they emerge from the nest. However, females must forage for floral resources, and males can take advantage of this opportunity if the host plants are clumped. Flowers are where males of most solitary species look for mates (Linsley 1958). This is particularly true for oligolectic bees in which females are concentrated in areas where the host plant is in flower. Male *Systropha* (Halictidae) establish mating territories for up to 13 consecutive days around patches of *Convolvulus* (Convolvula-ceae), the sole source of pollen and nectar for foraging females. They strike any conspecific male that enters their territory with the sclerotized tips of T7 and S8,[2] often rendering the victim groggy but otherwise undamaged. Other flower-visiting insects are not attacked (Fraberger and Ayasse 2007). Male *Pachymelus limbatus* (Apidae) exhibit a similar mating strategy. Males in this species aggressively defend scent-marked patches of flowers, chasing away all potential flower visitors except virgin conspecific females (Nilsson and Raba-konandrianina 1988).

In some species, not all males adopt the same strategy, with some males holding territories and others behaving as wanderers. Females of the small bee *Protodiscelis palpalis* (Colletidae) are oligolectic on *Hydrocleys martii* (Limno-charitaceae), an aquatic herb occurring in small ephemeral ponds in the arid Caatinga region of northeastern Brazil (see Box 12-2). In this system, the spatial distribution of flower patches varies from day to day. Some males fight each day to establish new territories over flower patches, while others are nonterritorial wanderers that randomly patrol flowers in the ponds. Territory holders de-fend only a small core area in the heart of their territories and ignore the non-aggressive wanderers that pose no serious threat to their territorial dominance. In this system, territory holding appears to be superior to wandering in terms of male reproductive success (Oliveira et al. 2012, 2013).

Male mating tactics around flower patches can also change facultatively. Males of *Hylaeus alcyoneus* (Colletidae) vary greatly in body size, with the larg-est males weighing nearly five times the mass of the smallest males. Smaller males are non-aggressive patrollers, while the larger males fight to establish and hold territories on the individual spikes of *Banksia* (Proteaceae) with the greatest resources. When territory holders are experimentally removed, they are quickly replaced, sometimes by patrollers, indicating that here patrolling and territoriality are flexible alternatives, not fixed strategies (Alcock and Houston 1987). Large males have a curiously enlarged and protuberant S3 that they use to bash intruding males; small males do not. Similar structures are found in other territorial Australian Hylaeinae (Colletidae; Alcock and Houston 1996). Males of *Ptilothrix fructifera* (Apidae) range greatly in size but show no discrete size classes. Large males defend territories of flowering *Opuntia* (Cactaceae) plants, the female host plant, while smaller males patrol non-flowering plants between the *Opuntia* patches and the nest site. When flowers were added to previously flowerless patrolled areas, some of the smaller males became territo-rial, clearly demonstrating behavioral flexibility in mating tactics (Oliveira and Schlindwein 2010).

---

[2] Throughout the book, we use commonly accepted abbreviations to refer to the metasomal terga and sterna of bees. Terga and sterna are numbered consecutively along the length of the metasoma so that T2 refers to the second metasomal tergum, T3 refers to the third metasomal tergum, S7 refers to the seventh metasomal sternum, and so forth.

What to do when females are dispersed?

Flowers and emergence areas are not the only sites where males encounter receptive females. Typically, parasitic bees are irregular flower visitors and do not occur in high densities at nest sites; thus alternate mating rendezvous sites are needed. Male *Nomada* (Apidae) patrol routes they mark with volatile compounds produced from their cephalic glands that serve to attract females (Alcock et al. 1978, Bergström 2008, Tengö and Bergström 1976, 1977). These male-produced odors may also play a role in helping female *Nomada* gain access to the nests of their *Andrena* hosts (see Chapter 10).

Other bees are known to establish and defend territories where there are no flowers or other resources that would attract receptive females, yet males defend these sites aggressively from conspecific males. This non-resource-based form of territoriality is common in *Xylocopa* (Apidae) (discussed below). Similar strategies are found in other bees, such as males of some *Centris* (Apidae) species that hover in the crown of trees (which may or may not be flowering) and attract females with scents produced by glands in their swollen hind tibiae (Coville et al. 1986, Stort and Cruz-Landim 1965, Vinson et al. 1996). Males of *Protoxaea gloriosa* (Andrenidae) have enlarged eyes and hover within territories they establish in patches of flowering plants, darting out after any approaching females. Larger males typically hold territories with more flowers (Alcock 1990), but inexplicably, some territories are established in patches of non-rewarding flowers (Cazier and Linsley 1963). Besides *Protoxaea* and some *Xylocopa*, similar hovering-with-darting behavior is found in other species with enlarged eyes, such as *Caupolicana* (Colletidae) and *Melitturga* (Andrenidae) (Rozen 1965b, Stephen et al. 1969). Such mating tactics presumably reflect some combination of conditions related to prolonged receptivity among females, dispersed nesting sites, and widespread floral resources.

POLYANDRY AND THE RACE TO BE LAST

Multiple mating by females (polyandry) is relatively rare among bees, occurring mainly in the Anthidiini (Megachilidae) and Panurginae (Andrenidae), and sometimes among communal species. Although uncommon, polyandry is of interest because it reverses many of the rules males must follow in finding mates. As noted earlier, the race changes from being the first to mate with a given female to being the last to mate before she oviposits. This may put a premium on extended mate guarding since a male has no guarantee his most recent mate will not mate once again soon after he releases her. The advantage of protandry is also diminished since there is less urgency to emerge early and spend time and energy searching for females before they emerge. In fact, to do so may reduce one's life span at the other end when mating is still possible. By the same token, there is no need for the nearly universal arrangement of males in the outer cells and females in the inner cells of linear nests, given that males don't have to emerge early. Linear nests seem to be rare among polyandrous species, but it is probably not a coincidence that in linear nests of *Anthidium* (Megachilidae), a genus in which polyandrous behavior appears to be universal, females are in the outer cells and males in the inner (Krombein 1967, Parker 1987, Sugiura 1991).

Male bees in polyandrous species follow multiple paths to maximize reproductive success. In *Anthidium* (Megachilidae), dominant males defend territories in patches of flowers attractive to females, a strategy referred to as resource defense polygyny. There is considerable intra-specific variation in body size in male *Anthidium*, with large males typically as large as or even larger than females, contrary to the usual pattern in bees. As one might expect, large males usually hold territories, with small males relegated to the periphery. Territory holders typically attempt to mate with any conspecific females but are not always successful. Females can reject male advances entirely or copulate briefly with the territory holder and return to foraging. Territorial male *A. maculosum* spend most of their time interacting with conspecifics, either chasing off smaller satellite males, defending against intruding males attempting to take over their territory, or pursuing and/or mating with visiting females (Alcock et al. 1977a). Most of these interactions end up without any actual damage to any of the participants, but not all male *Anthidium* are so restrained. Box 4-2 provides a detailed account of the biology of *A. manicatum*, one of the most aggressive of all bees.

Frequent brief copulations by territory-holding males are not the only path to reproductive success in polyandrous bees. Panurgines (Andrenidae) are a group in which polyandry is widespread. Many species have very prolonged copulations, with females flying with the attached male from flower to flower or from flower to nest (Fig. 4-2). Since the aim of the game is to be the last to mate before a female oviposits, theory predicts males should adjust their behavior as the probability of oviposition changes. Just such a change was observed in *Calliopsis puellae* (Rutowski and Alcock 1980). Females in this species are oligolectic on cichorioid composites, such as *Malacothrix fendleri* (Asteraceae), whose flowers open in the morning and close around noon. Females are believed to provision one cell per day, so ideally a male would want to mate with her during her last pollen-foraging trip. Unfortunately for the males, there is no precise way to determine which trip is the last one, so they adjust their mating tactics over the course of the day. Early in the morning, when females are unlikely to be finished provisioning, copulations are relatively brief. As the morning progresses, the duration of copulations increases so that pairs fly first from flower to flower and eventually from flowers back to the nest. Females are able to forage while *in copula*, and since males assist in flying, extended mating apparently does not place a great foraging burden on the female. These prolonged copulations almost certainly have little to do with sperm transfer and primarily serve as a form of mate guarding.

A feature that is widespread among polyandrous species is extreme allometric variation in male head size and shape. This is particularly common in the following tribes of Andrenidae of the New World: Calliopsini (*Arhysosage* spp.; Rozen 2013), Protandrenini (*Psaenythia*, *Rhophitulus*, *Psaenythisca*; Michener 2007, Ramos and Rozen 2014), and Perditini (*Perdita* and *Macrotera*; Danforth and Desjardins 1999), but also in some Asian Halictidae (*Lasioglossum* [*Sudila*] spp.; Sakagami et al. 1996) and an Australian Colletidae (*Leioproctus* [*Ottocolletes*] *muelleri*; Houston and Maynard 2012). Males in these species

BOX 4-2: *ANTHIDIUM MANICATUM* (MEGACHILIDAE)— THE MURDEROUS BULLY

Anthidium manicatum is a large, handsome, yellow-and-black bee native to much of Europe, western Asia, and northern Africa. The females are polylectic, with an apparent preference for blue flowers of the Lamiaceae, Fabaceae, and Scrophulariaceae (Severinghaus et al. 1981). Many of these flowers are commonly planted as ornamentals, so sightings of *A. manicatum* in gardens are frequent. Females of *A. manicatum* nest in preexisting cavities, often far above the surface of the ground. Probably aided by its cavity-nesting habits, it has been inadvertently introduced into North and South America and New Zealand. Following these introductions, it has rapidly expanded, and now it is one the world's most widespread bees (Strange et al. 2011). The biology of female *A. manicatum* is not particularly remarkable, but the males have attained considerable notoriety due to their extremely aggressive mating tactics. They have even been accused of contributing to honey bee decline.

In *A. manicatum*, copulation entails neither prolonged courtship nor postcoital guarding. Copulatory bouts are quite brief, lasting only 10–14 seconds (Severinghaus et al. 1981). Territory-holding males tend to be larger than wanderers, and males holding larger territories are larger than those in smaller territories. However, these roles are not fixed, since when a large male was removed from a territory, the replacing bee was often smaller (Starks and Reeve 1999). This form of territoriality, called resource defense polygyny, is probably practiced by males of all *Anthidium* species, but the males of *A. manicatum* step it up a notch. As in all male *Anthidium*, the apex of the T7 is trifid, and there may be large, lateral-hooked processes on T6. In most male *Anthidium*, these modified terga have a functional role in mating. During copulations, T7 pushes down on the female S6, while the female T6 fits into depressions on the male S6, facilitating genitalic entry (Toro and Rodriguez 1997). In *A. manicatum*, these processes have lost their marital functions and play no role in copulation (Severinghaus et al. 1981). Instead, they have assumed martial ones. While most *Anthidium* males defend their territories against other males with nonlethal contests, the males of *A. manicatum* violently defend their territories against all other flower visitors (with the exception of conspecific females), frequently attacking bees much larger than themselves, such as bumble bees. Honey bees are a common target. They do this by ramming them at high speed and simultaneously curling their metasoma to strike them with the spines of T5 and T6, often resulting in damaged or even detached wings, a virtual death sentence for a foraging bee (Wirtz et al. 1988). This aggressive behavior was noted by Ward (1928), who commented, "The strange thing is that a male aculeate should act like this for the males of this group are a byword for uselessness and laziness." The result of this aggressive behavior is that, by excluding other flower visitors, male *A. manicatum* make their territories more attractive to females. For females, the price of admission is only 10 seconds or so of quick sex.

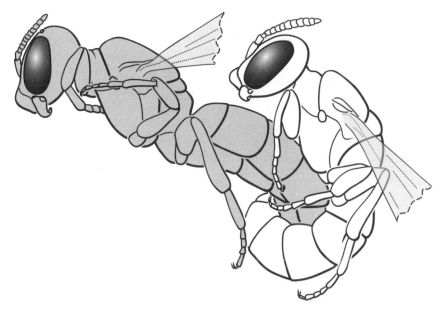

Figure 4-2. Mating pair of male and female *Callonychium chilense* (Andrenidae) flying *in copula* (redrawn from Toro 1985, Fig. 1).

vary widely in body size, but their head size scales strongly allometrically—meaning that smaller males have small heads, but larger males have grotesquely enlarged, massive heads (Danforth and Neff 1992) (Fig. 4-3). Males with larger heads have greatly expanded mandibular muscles, longer mandibles, and mandibles that are more widely spaced (Danforth and Neff 1992). One possible advantage of a larger head and stronger mandibles is greater facility in grasping females. This can be especially advantageous when males and females fly *in copula* (Rozen 2013). A greatly expanded head may also be advantageous in male-male combat over receptive females. In *Macrotera* (*Macrotera*) *texana*, a panurgine bee with strong positive head allometry, large-headed males defeated small-headed males in battles on *Opuntia* flowers, were more often found *in copula* than small-headed males, and mated more frequently, suggesting a strong reproductive advantage to being a large-headed male (Danforth and Neff 1992).

An extreme form of the kind of macrocephaly described above involves the evolution of truly dimorphic males (Fig. 4-4a,b). This phenomenon has evolved independently in a few communally nesting North American panurgines and Australian halictines. In North America, male dimorphism occurs in *Macrotera* (*Macroteropsis*) *portalis* and *Macrotera* (*Macroterella*) *mellea* (Rozen 1989b), while in Australia, dimorphic males have been found in three species of *Lasioglossum* (*Chilalictus*) (Halictidae; Knerer and Schwarz 1976, Kukuk and Schwarz 1988). Male dimorphism appears to have evolved from the extreme male head size allometry described above (Danforth and Desjardins 1999). One can easily imagine how dimorphism could arise from positive head allometry; elimination of males of intermediate body size would yield a population of males with two discrete, widely divergent head morphologies. Male dimorphism seems particularly common in communal bees in which large numbers of receptive females co-occupy the same nest. Large communal nests represent a reproductive

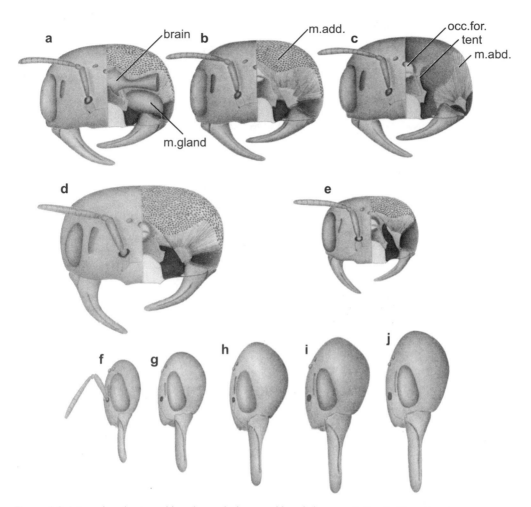

Figure 4-3. Internal and external head morphology and head-shape variation in *Macrotera texana* (Andrenidae): (a–c) internal anatomy of typical male head, showing location of brain, mandibular gland (m.gland), mandibular adductor muscles (m.add.), mandibular abductor muscles (m.abd.), occipital foramen (occ.for.), and tentorium (tent); (d–e) large and small males, showing relative magnitude of adductor muscles; (f–j) lateral views of male heads, showing size variation (from Danforth and Neff 1992, Fig. 1).

bonanza for any male that can monopolize some or all of the resident female population. The morphology and behavior of dimorphic males in *Macrotera portalis* is described in Box 4-3.

Strategies involving chemical signaling

An interesting contrast between resource-based and non-resource-based male mating strategies is found among the carpenter bees (*Xylocopa* spp., Apidae). In many *Xylocopa* species, the males have large mesosomal glands that produce mixtures of volatile compounds (Minckley 1994). These species are also strongly

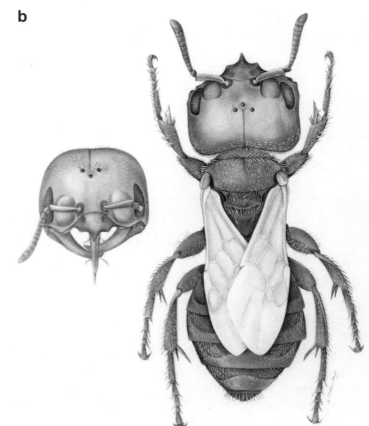

Figure 4-4. Dimorphic males in *Macrotera portalis* (Andrenidae): (a) small-headed, flight-capable male and (b) large-headed, flightless male. Original artwork by Natalia Florenskya.

BOX 4-3: *MACROTERA PORTALIS* (ANDRENIDAE)— A REMARKABLE BEE WITH "DIMORPHIC" MALES

Macrotera portalis is an extraordinary bee. This small, oligolectic, polyandrous andrenid bee inhabits the Chihuahuan Desert of North America (Danforth 1996). Males and females are active in the late summer (August to October) in synchrony with the monsoon rainy season. Up to 30 adult females construct a shallow (~10 cm deep) nest in silty soils devoid of vegetation, and nests are reused for many years. Females are host-plant specialists on flowers in the mallow genus *Sphaeralcea* (Malvaceae). Populations are highly structured, with significant levels of inbreeding within populations and significant genetic differentiation among populations, almost certainly a result of the unusual mating system in this species (Danforth et al. 2003b).

What makes this species so remarkable is that males occur in two discrete, non-overlapping forms: small-headed, flight-capable males (Fig. 4-4a) and large-headed (macrocephalic), flightless "walking heads" (Fig. 4-4b) (Danforth 1991b). Small-headed males occur at *Sphaeralcea* plants, where they spend the day in the cup-shaped flowers and attempt to mate with foraging females. At night, they rest within the closed *Sphaeralcea* flowers. Their morphology and behavior are typical of other members of the genus *Macrotera* and many other panurgine bees.

In contrast, the macrocephalic, flightless form occurs only within active, occupied nests, almost certainly their natal nests. As many as 26 males have been found within a single nest. Macrocephalic males are highly modified relative to their small-headed brothers, with enormously enlarged heads that are packed with massive mandibular muscles, elongate, robust mandibles, reduced compound eyes, a pair of well-developed genal projections, and a greatly expanded pair of glandular facial foveae. Soft regions of the body, such as the proboscidial fossa and the membranous neck, are shielded by flanges and ridges. They also have noticeably reduced wings and have entirely lost the indirect flight musculature that typically fills the mesosoma of an adult, flight-capable bee. It appears that they have "traded in" flight musculature for the expanded mandibular muscles.

The behavior of these macrocephalic males is as remarkable as their morphology. Intra-nest behavior was studied by Danforth (1991a,b), who established observation nests of this species at a site near Rodeo, New Mexico. Nests were made by sandwiching soil between two plexiglass plates and placing these observation nests in a below-ground chamber covered with a plywood lid. With the aid of a flashlight and a voice-activated recorder, the behavior of both males (Danforth 1991a) and females (Danforth 1991b) were observed over a two-year period.

These flightless males cannot forage and are not fed by trophallaxis by the females, so it appears that they receive their nourishment by snacking on provisions deposited by females in partially provisioned brood cells. Macrocephalic males do two things within nests that help explain why they are equipped with an enormous, heavily armored head. First, they fight with other macrocephalic males within the same nest. Males attack each other with mandibles agape, appearing to target the soft, membranous regions of the head. These fights can last hours and often end in the death of one of the males. Second, males mate with females within the nest, and these matings take place immediately prior to oviposition. Sacrificing mobility, these macrocephalic males are guaranteeing that they will be the last to mate and thus should have the highest probability of paternity. In this "race to be last," the macrocephalic males win first prize because they are likely supplanting the sperm of previous matings with small-headed males on flowers. Being a macrocephalic male does have its drawbacks—you just might get killed by your brother.

sexually dimorphic in coloration and structure. In one of the best-studied species, *X. varipuncta* (Fig. 4-5), males are known to establish small territories at non-resource-related sites, such as ridgelines or the tops of trees, where they hover and release the volatiles in the late afternoon. Territory location changes frequently, probably as a reaction to changing wind conditions and subsequent attempts to optimize scent dispersal (Alcock and Smith 1987, Minckley and Buchmann 1990). Bioassays have shown that female *X. varipuncta* are attracted to the isolated dominant compounds in these male-produced mixtures, indicating that the compounds alone are probably the primary female attractant (Minckley et al. 1991). Females regularly approach the hovering males, but only a very small fraction of these encounters result in copulation (Alcock and Smith 1987).

Not all *Xylocopa* males rely on chemical lures. Some take the more traditional path of establishing territories at patches of flowers that they aggressively defend and where they attempt to mate with any receptive females they encounter. Such males typically have enlarged eyes and reduced or absent mesosomal glands. In addition, they are monomorphic; males are the same color as conspecific females. Not all *Xylocopa* fit neatly into this dichotomous scheme, since some males have large glands but practice resource defense. In other cases, there may be mixed strategies. *X. micans* is bivoltine and has relatively large mesosomal glands. First-generation males mainly practice resource defense and have reduced glandular activity, while second-generation males have increased glandular activity and do not defend resource-based territories (McAuslane et al. 1990). Such mixed strategies may explain at least some, but probably not all, of the apparent mismatches of glandular structure and mating behavior seen in some *Xylocopa* species (Leys and Hogendoorn 2008).

Probably the most famous, and most unusual, of the non-resource-related mating systems is found among the euglossines, the so-called orchid bees (Apidae; Fig. 4-6). Rather than searching for females, male euglossines spend their long lives gathering volatile compounds from a variety of sources, including fungi, feces, damaged plant tissue, and rotting fruit and flowers in at least 16 plant families (Ramírez et al. 2002, Roubik and Hanson 2004, Whitten et al. 1993). The majority of the volatile compounds used by male orchid bees come

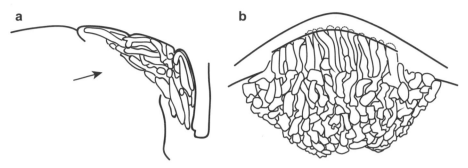

Figure 4-5. Male mesosomal gland in *Xylocopa varipuncta* (Apidae): (a) sagittal section of posterior part of the mesosoma and (b) interior view of the glandular reservoir from angle indicated by the arrow in (a). M=metanotum, S=scutellum, P=propodeum (redrawn from Minckley 1994, Figs. 5E,F).

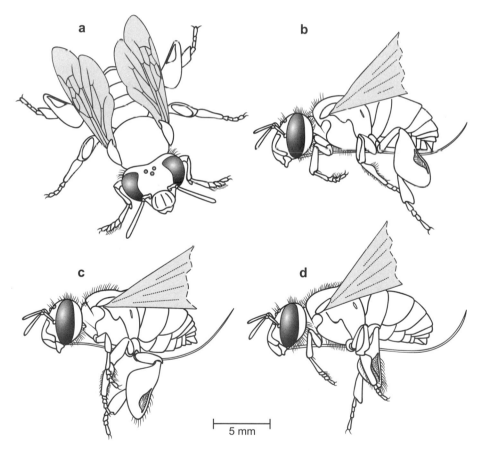

5 mm

Figure 4-6. Movements of male orchid bees (*Euglossa imperialis*) while collecting and storing floral fragrances: (a) male bee mopping floral oils from the surface of a flower with expanded foretarsi; (b) hovering; (c) in-flight transfer of floral fragrances from foretarsus to midtarsus; (d) in-flight transfer of floral fragrances from midtarsus to hind-tibial storage organ (redrawn from Kimsey 1984, Figs. 3–6).

from over 600 species of orchids, or roughly 10% of the orchid flora of the neotropics (Ramírez et al. 2011). Using a form of enfleurage involving lipids produced by their labial glands, males harvest the volatiles with brushes of specialized hair on their foretarsi (Fig. 4-6a) (Eltz et al. 2007, Whitten et al. 1989). The volatiles are then transferred to their greatly expanded hind tibiae (Fig. 4-6c,d), where they are stored in an internal fibrous matrix (Vogel 1966). Analysis of the aromatics produced by orchids and other scent sources demonstrated that they are primarily mixes of common terpenes and volatile esters. Subsequent work showed that males were attracted to individual compounds when they are presented as baits and that different species had different preferences for the various compounds.

Exactly what males were doing with the volatiles has long been a mystery. Hypotheses have included that they function as intoxicants (van der Pijl and Dodson 1969), as a sort of elixir of life that prolongs male life span (Dodson et al. 1969), or as male anti-predator defenses (Roubik 1989). None of these found much support. Scent collection was long thought to have something to do with mating behavior, but exactly how the scents were used was unclear, since females were not attracted to natural (or artificial) scent sources (Schemske and Lande 1984, Vogel 1966). Were these scents used as nuptial gifts (Roubik 1989), as pheromones for attracting other males to form leks that in turn attract females (Kimsey 1980), or as precursors for female-attracting pheromones produced by the males (Williams and Whitten 1983)? While plausible, none of these hypotheses were supported either. It now appears that no modification of the volatiles is needed and that it is the mix or blend of compounds the males collect that drives the system. Males attract females by perching on trees, extracting the compounds from their hindlegs, transferring them to the felt-like tibial pads, and dispersing them with fanning actions of their wings (Bembé 2004, Eltz et al. 2005). Females follow the odor plumes and closely approach the fanning male. If the bouquet passes muster, she lands and mating ensues. If not, she leaves, and the male will either have to hope a less picky female accepts him, or he'll have to upgrade his chemical mix. Details of exactly what females are using to evaluate male suitability remain unclear, but it seems likely that the ability to accumulate a complex bouquet is an indicator of fitness (Whitten et al. 1989).

In a remarkable, but little known, case of convergent evolution, one species of *Anthophora* is also reported to collect floral exudates, combine them with glandular secretions, and use them to attract females. Norden and Batra (1985) studied *A. abrupta* at a farm near Baltimore, Maryland, in the early 1980s. Males were observed chewing the leaves, petioles, and stems of parsnip (*Pastinaca sativa*, Apiaceae) plants in partial shade. Damage inflicted by the male chewing behavior caused the parsnips to exude a thick, smelly sap that was collected by males on specialized "mustaches" of dense, closely spaced, apically flattened hairs on the lower margin of the clypeus. Mandiblar gland secretions were added to the plant exudate, and then males marked surfaces upwind of nesting females, including grass stems, a honeysuckle bush, and even a car bumper! Female *A. abrupta* are polylectic but were not observed to visit parsnip flowers for nectar or pollen. Morphological adaptations to collecting parsnip exudates (e.g., the "mustache") are not present in other *Anthophora* or other genera of Anthophorini (Norden and Batra 1985). The behavior documented by Norden and Batra is particularly puzzling because parsnip is not native to North America.

FEMALE CHOICE AND COURTSHIP BEHAVIOR

Often overlooked, but important to recognize, is the central role females have in the mating game. Males do most of the work, but females signal their receptive state in a variety of ways. In some cases, the signals are visual. For example, in *Hesperapis* and *Macropis* (Melittidae), foraging females hold their hindlegs up over their bodies when males approach, both signaling their non-receptive state and making it difficult for a male to mount (Cane et al. 1983a, Stage 1966). Foraging females of *Anthophora occidentalis* (Apidae) reject overzealous males by falling to the ground or curling their metasoma under and, like the melittids, raising their hindlegs (Batra 1978a). However, by far the most common signals used by females are chemical. Virgin female bees in many species have been shown to produce volatile compounds that are attractive to males (Ayasse et al. 2001, Cane and Tengö 1981, Fraberger and Ayasse 2007). In species with females that become unreceptive shortly after mating, such as *Centris pallida* (Apidae), there is a period of post-copulatory display that involves many of the behaviors used in courtship, such as wing whirring and leg and antennal stroking before becoming unreceptive (see Box 4-4 for a detailed description of courtship behaviors in *Osmia bicornis* [Megachilidae]). If prematurely interrupted by an observer, some females retained their receptivity and later mated with another male even if sperm was transferred. Under natural conditions, such interruption may serve as an indirect method of female choice as it favors mating with a dominant male (Alcock and Buchmann 1985).

BOX 4-4: *OSMIA BICORNIS* (MEGACHILIDAE)— COURTSHIP AT ITS FINEST

Osmia bicornis (long known as *O. rufa*) is an abundant, monandrous, univoltine, polylectic bee that occurs throughout most of Europe, northern Africa, and western and southwestern Asia. Like most bees, the males of *O. bicornis* are significantly smaller than females, and as one might expect based on sex allocation theory, males outnumber females nearly two to one (Seidelmann 1999, Seidelmann et al. 2010). Females are sexually receptive as soon as they emerge, and males search for virgins at flowers and emergence sites (Raw 1976). Males do not defend territories, and male-male interactions are uncommon. No large male advantage has been detected (Seidelmann 1999).

Mating behavior in *O. bicornis* is quite complex. After locating a virgin, mating typically begins with an attempt to copulate as soon as the male mounts her. Such initial attempts are almost always rejected as some period of courtship is usually required to win her favor. Courtship consists of a series of active phases, with antennal tapping, leg waving, and buzzing sounds. These are interspersed between rest periods when essentially nothing happens. The duration of the courtship phase is highly variable, averaging 14.1 ± 14.5 (0.08–51.5) minutes. Eventually the frequency of the leg motions and the antennal drumming increases, as does the buzzing, which becomes a continuous whirr. The male then spreads his wings, slips backward on the female, and attempts to insert his genitalia. If successful, copulation ensues. This involves a series of metasomal pumping motions accompanied by back swings of his antennae. The copulatory phase is relatively brief, averaging only 0.9 ± 0.3 minutes (0.5–1.5), with the female initiating termination via an escalating series of pushing motions with her legs (Seidelmann 2014). The majority of the time, these copulatory attempts are unsuccessful, so he moves to a rest phase before resuming courtship. Even then, success is not guaranteed, as only 37% of mating attempts result in copulation (Seidelmann 2014). There is some evidence that one of the criteria female *O. bicornis* use to judge male suitability is the duration of the vibrational signaling (buzzing), a possible indicator of male vigor (Conrad and Ayasse 2015). Following a successful copulation, a male switches to a post-copulatory display phase. This involves a complex mix of antennal strokes, buzzing noises, and whole-body vibrations and averages 13.1 ± 10.7 minutes (2.8–53). Upon completion of the post-copulatory display, the female typically has lost receptivity and will reject any subsequent male attention. However, if the male is removed prematurely, the female retains receptivity and will accept the attentions of subsequent males. Curiously, if the removed male is placed on a virgin female, he will continue the post-copulatory display, and she will lose receptivity, even though no semen has been transferred. Even odder, just being in the vicinity of a pair involved in a post-copulatory display will increase the probability of a female losing receptivity (Seidelmann 2014). Although female *O. bicornis* very rarely mate after experiencing the joys of the post-copulatory display, male *O. bicornis* have a backup plan to further insure female fidelity (male fidelity not being part of the game). During copulation, they implant a mating plug, which effectively blocks any sperm that might be transferred during a later copulation. Mating plugs are known in some social bees, but this is the only known occurrence among solitaries (Seidelmann 2014).

Bee mating often has an auditory component. Buzzes and squeaks have been reported during the mating of a variety of bees (Conrad and Ayasse 2015, Larsen et al. 1986, Torchio 1990, Toro and Riveros 1998, Wcislo and Buchmann 1995, Wcislo et al. 1992). In most cases, male bees lack special organs for sound production, and the sounds are produced by vibrations of the flight muscles, with or without wing movement. An exception is *Meganomia binghami* (Melittidae), in which males have stridulatory areas on the pregradular area of S4, S5, and S6 that produce loud rasping sounds during mating (Rozen 1977c). Exactly what female bees hear is hard to say, as they lack tympanal ears capable of detecting sound over long distances. Since the mating sounds are always produced while the male grasps the female, it is not clear if these sounds are transmitted through the air or are vibrational signals transmitted via body contact. The importance of these signals in courtship and female choice deserves further attention.

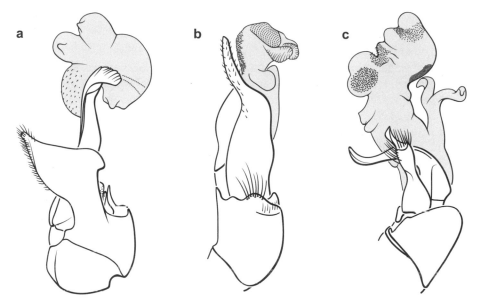

Figure 4-7. Male genital capsule with membranous endophallus everted: (a) *Perdita albipennis* (Andrenidae); (b) *Arhysosage flava* (Andrenidae); (c) *Epanthidium bertonii* (Megachilidae). (Redrawn from Roig-Alsina 1993, Figs. 4–6).

More subtle forms of female choice may also be at work in solitary bees. Male genitalia in bees consist of a sclerotized genital capsule (and associated S7 and S8) and an unsclerotized, eversible intromittant organ (the endophallus). The sclerotized genitalic components are highly variable in bees. They are often so morphologically variable that accurate male identification cannot take place without careful examination of the sclerotized genitalic components. Hence the lurid fascination that bee taxonomists take in drawing and describing male genitalia. The endophallus also has a surprising level of morphological complexity. In a clever study that involved inflating the endophallus of a wide variety of bees, Roig-Alsina (1993) found that endophallus morphology was correlated with mating system. In many bee groups (including Colletidae, Oxaeinae, and Andreninae), the endophallus is a simple eversible or non-eversible sac (Roig-Alsina 1993). However, in bee groups with polyandrous mating (Panurginae, Anthidiini, and the genus *Apis*), the everted endophallus can be wildly variable and complexly lobed (Roig-Alsina 1993) (Fig. 4-7). The function of an eversible, elaborately lobed endophallus is unclear. One possibility is that it enhances clasping ability, something that may be important in bees with prolonged copulations, such as in *Calliopsis puellae*, discussed above (Roig-Alsina 1993). This would seem an unlikely explanation in other polyandrous bees, such as *Anthidium* (Megachilidae), where copulation is usually brief (<30 seconds, Michener and Michener 1999). Another possibility is internal signaling and cryptic female choice, in which females might be using genital structure to evaluate male quality (Eberhard 1991, Roig-Alsina 1993). The correlation between polyandry and elaborate male endophallus morphology suggests that cryptic female choice may be an important driver of genitalic diversity in bees.

ANATOMY: MALE SECONDARY SEXUAL STRUCTURES

Beyond the mundane features distinguishing male from female bees, such as the number of flagellar or metasomal segments, many male bees sport bizarre body modifications. For example, in the appropriately named *Hylaeus* (*Gongyloprosopis*) *preposterous* (Colletidae), the antennal scape is grotesquely enlarged and globose (Fig. 4-8) (Snelling 1982). Similarly enlarged scapes are found in a number of other *Hylaeus* species (Michener 1965, Snelling 1966). The antennae of male bees are always at least as long as those of their respective females, and often much longer. Some bees take this to extremes. Many male Eucerini (Apidae), the so-called long-horned bees, have antennae nearly as long as their bodies. Male *Ctenioschelus* (Apidae), a genus of brilliantly metallic, parasitic bees, have extremely long antennae that give them the appearance of cerambycid beetles (Michener 2007). Male antennae sometimes take even stranger forms. The males of *Trichocerapis mirabilis* (Apidae: Eucerini) have elongate antennae whose basal segments are "normal," but segments 8, 9, and 10 are threadlike, and the apical 11th segment is expanded into a flattened black disk, resulting in a bola-like structure (Michener 2007). Similar, albeit shorter, bola-like antennae are found in *Spatunomia filifera* (Halictidae: Nomiinae), although here only the apical segment is modified. Probably the strangest antennae are found in *Cladocerapis bipectinatus* (Colletidae: Neopasiphaeinae), whose males sport bipectinate flagellar segments (Houston 1991b). Merely pectinate flagellar segments are found in the related *Leioproctus macmillani* (Houston 1991b). The function of these antennal

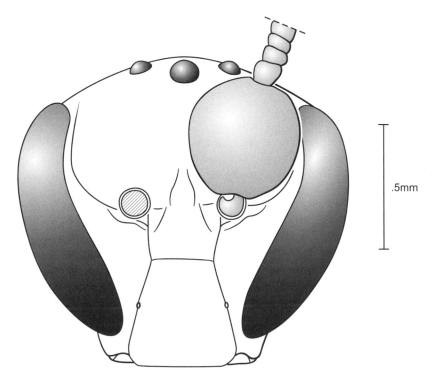

.5mm

Figure 4-8. Grossly enlarged antennal scape in *Hylaeus* (*Gongyloprosopis*) *preposterous* (Colletidae) (redrawn from Snelling 1982, Fig. 25).

oddities remains largely speculative, but since virtually everything male bees do is centered around mating, a sex-related function is assumed. One study found that the elongate antennae of male *Eucera berlandi* have 10 times as many pore plates and three times as many olfactory neurons as those of their females, suggesting an enhanced role of chemoreceptors for mate location (Streinzer et al. 2015). The enlarged scape of *H. preposterous* could have a glandular function; the bola-like antennae of *T. mirabilis* and *S. filifera* might be visual signals held in front of the eyes of the female during courtship, while the pectinate antennae of *C. bipectinatus* could play some role in mate location. Unfortunately, no observations of mating behavior have been reported in these bees.

In other cases, the functions of male structural oddities are known. The highly modified hindlegs of many male nomiine bees (Halictidae: Nomiinae) are known to play a role in grasping the female during courtship and mating (Wcislo and Buchmann 1995), and a similar function is expected for the similarly modified legs of males in other genera like *Tetrapedia, Caenonomada, Ancyloscelis, Anthophora, Xylocopa* (all Apidae), *Geodiscelis* (Colletidae), *Meganomia* (Melittidae), and *Megachile* (Megachilidae). While the males of many familiar genera like *Andrena, Colletes,* and *Lasioglossum* lack distinctive arrays of secondary sexual adaptations, others, like *Megachile willughbiella*, are virtual mobile assemblages of the bee equivalent of bondage gear (Wittmann and Blochtein 1995). Males of *M. willughbiella* use the spines on their forecoxae to pinion the female's head, while their modestly modified midlegs pin the female's wings, and the hindlegs grasp and lift her metasoma (Fig. 4-9). The male's mandibles have a prominent

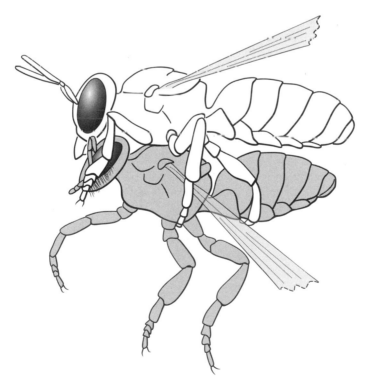

Figure 4-9. Male and female *Megachile willughbiella* (Megachilidae) *in copula* (redrawn from Wittmann and Blochtein 1995, Fig. 1).

ventral process that holds the female's antennae in a groove on his gena. Once the antennae are in place, the male then inserts the female's flagellum into a grooved structure on his highly modified forebasitarsi. With this, her antennae are in close contact with glandular pores on his basitarsi. At the same time, his expanded and fringed basitarsi serve as blindfolds, covering her eyes and creating a double whammy of visual and chemical signals. Various combinations of spines, glandular structures, and modified legs are found in many other male *Megachile*, as well as some *Xylocopa* (Anzenberger 1977, Osten 1989, Wittmann and Blochtein 1995). The elaborate ways some male bees, like those of *M. willughbiella*, constrain the movements of females during courtship and mating could be interpreted to mean these acts are very one-sided and male-dominated. However, the presence of chemical signaling strongly indicates elements of female choice, so that the various ways males constrain female movements help males deliver signals to the female, but ultimately the choice to mate or not is hers.

CHAPTER 5

The Surprising Utility of Males

Male bees are often maligned as slackers, layabouts unworthy of respect. This ill-deserved reputation is based entirely on the behavior of male honey bees, a group that, in contrast to virtually all other male bees, really are slackers. Male honey bees are known as drones, a name that has become synonymous with an indolent lifestyle based on the labors of others. While drone honey bees are rarely encountered, male solitary bees are a common feature in most natural landscapes. As discussed in Chapter 4, a large portion of the lives of male solitary bees is dedicated to a frantic search for mates. However, their lives are not just sex, sex, sex all the time, so here we examine some other aspects of male bee behavior, such as where they sleep and what role they play as pollinators. Nonetheless, even here sex isn't far out of the picture, since in several famous pollination systems, plants have harnessed the sex drives of male bees to their own devious ends.

MALE SLEEPING SITES

Whether or not bee "sleep" is the same as the sleep of vertebrates is a matter of semantics, but bees certainly spend considerable amounts of time in a "sleep-like" state (Kaiser 1995), so the term seems appropriate. Non-parasitic female solitary bees almost always spend their nocturnal rest periods in their nests, but male solitaries—since they do not construct nests, nor, with few exceptions, is their presence tolerated in the nests of females—must sleep elsewhere. The most commonly observed sleep sites are flowers (Fig. 5-1). These are not necessarily the most common male sleep sites, but they are the ones most likely to be noticed. In many cases, males, singly or in small groups, sleep on open flowers or flower heads or in the shelter of tubular corollas (Rau and Rau 1916, Vereecken and McNeil 2010). Males are not restricted to open flowers as sleep sites. Male squash bees (*Peponapis* and *Xenoglossa* spp.) are early risers, so usually if one wants to see them in action, one needs to be out at dawn. However, if one likes to sleep in, an easy way to find male squash bees is to go out later in the day and squeeze old, closed squash (*Cucurbita* spp.) blossoms, since male squash bees often spend the night in them. Although squeezing closed pricklypear blossoms is not recommended, males of several *Macrotera* and *Diadasia* species use closed pricklypear (*Opuntia*, Cactaceae) flowers as sleep sites, entering in the evening

Figure 5-1. Male *Macrotera portalis* (Andrenidae) emerging from a closed *Sphaeralceae* (Malvaceae) flower in the early morning. Males of many bee species sleep within closed flowers during the night. Photo by Bryan Danforth.

while the flowers are open and forcing their way out between the wilted petals the following morning (Neff and Danforth 1991, Neff and Simpson 1992). Similarly, male bees of a variety of species commonly spend the night in the large, nectar-less flowers of various prickly poppy species (*Argemone* spp., Papaveraceae), flowers whose petals close for the night. These flowers produce no nectar and are not visited by these males during the day. Males moving between such flowers looking for suitable sleep sites may lead to some pollination, but floral records based on the use of flowers as sleep sites can give a misleading view of floral visitation patterns. The effects of males using flowers as sleep sites are not always benign; the case of *Hexantheda missionica* and *Calibrachoa elegans* is a good example (Box 5-1).

Besides flowers, many males (and most females of parasitic species) sleep while grasping twigs or small branches with their mandibles (Rau and Rau 1916) (Fig. 5-2). Such sleeping bees commonly adopt characteristic postures that may vary from species to species (Kaiser 1995, Linsley 1962). One of the more interesting features of many twig or branch sleepers is the formation of sleeping aggregations of tens or even hundreds of individuals in close proximity, forming a swarm-like mass (Fig. 5-3). Males that spend the day competing with one another in the search for females put aside their differences and come together, perhaps for mutual protection via a dilution effect (Alcock 1998). These sleeping aggregations may persist for many days at the same site, and aggregations have been reported for many species (Linsley 1962,

BOX 5-1: *HEXANTHEDA MISSIONICA*—
THE NOT-SO-HELPFUL MALES

Hexantheda (Colletidae: Neopasiphaeinae) is a small genus of drab little South American bees. Morphologically, *Hexantheda's* main claim to fame is the presence of extra segments on their labial palpi, as many as five extra for a total of nine in *H. enterriana* (Urban and Graf 2000). All the *Hexantheda* species are believed to be specialists on species of either *Petunia* or its close relative, *Calibrachoa* (Solanaceae) (Almeida and Gibran 2017). The extra labial palpi protrude well beyond the tip of the glossa and presumably help these "short-tongued" bees to reach the nectar hidden in the narrow basal tubes of the trumpet-shaped *Calibrachoa* and *Petunia* flowers.

Although some *Calibrachoa* species appear to be bird-pollinated, most are bee-pollinated shrubs (Fregonezi et al. 2013). The pollination biology of most *Calibrachoa* species has not been studied in any detail, but *C. elegans* is an exception. *C. elegans* is a self-incompatible annual with magenta flowers, each of which remain open for up to five days (Stehman and Semir 2001). At the sites in Minas Gerais, Brazil, studied by Stehman and Semir, individuals of *H. missionica* were the only visitors to *C. elegans* flowers. Females of *H. missionica* collected nectar and pollen from *C. elegans* and, as they carried large pollen loads and regularly contacted the stigmas, almost certainly acted as pollinators. Fruit set from hand outcrosses (81%) was higher than that of open visited controls (59%), suggesting the possibility of pollen limitation. *H. missionica* scopal loads consisted solely of *C. elegans* pollen, indicating a high degree of dietary specialization. However, since the bee is far more widespread than the plant, it must be using other floral hosts elsewhere.

Male *H. missionica* patrol *Calibrachoa* flowers in search of females and occasionally enter the flowers for nectar. The latter activity may result in some pollination, although Stehmann and Semir doubt they are doing much because of the small amounts of pollen on their bodies, something they attribute to frequent grooming. The males also regularly spend the night in *Calibrachoa* flowers. By itself, this is nothing special, but with *Hexantheda* there is a twist. *Calibrachoa* flowers aren't ideal sleep sites, as the style and anthers occupy much of the center of the outer corolla tube. Male *Hexantheda* solve this problem by using their mandibles to cut and remove the style and anthers, leaving a roomy sleeping space; this is a nice trick for the bee, but in the process the flower is sterilized. This seems to be an odd interaction in which the female bees are indispensable pollinators, while their males are, from the plant's viewpoint, highly dispensable floral predators.

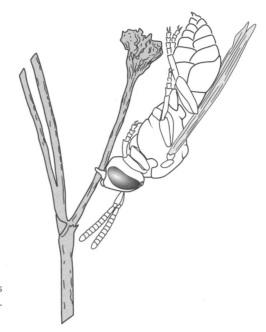

Figure 5-2. Sleeping male *Nomada* (Apidae) with mandibles clamped on a twig and body in rigid sleeping position (redrawn from Kaiser 1995, Fig. 1C).

Figure 5-3. Sleeping aggregation of male *Colletes compactus* (Colletidae) beneath leaves of a mint (Lamiaceae). Photo by Susan Barnett.

Alves-dos-Santos et al. 2009 and references therein). In most cases, these sleep-ing aggregations are monospecific, but mixed aggregations have been reported (Stephen et al. 1969). Exposed male sleep aggregations are the most commonly encountered types, but subterranean or otherwise hidden sleep aggregations also are known for some species, like *Nomia melanderi* (Halictidae; Stephen et al. 1969). An odd variant on the usual exposed male, hidden female sleep pattern is found in *Centris burgdorfi* (Apidae), a gregariously nesting species of coastal Brazil. In this species, males fight for access to emerging females at the nest site during the day but form sleeping clusters in old burrows at the nest site, while females sleep exposed in groups on the plants they use as floral oil hosts (Sabino et al. 2017).

Various other cavities—such as small cavities in the soil, crevices, spaces un-der stones, or cavities in wood—are used as sleep sites by males, but given their scattered and cryptic nature, they are hard to evaluate. Some males, such as those of *Hesperapis regularis* (Melittidae; Burdick and Torchio 1959) and *Col-letes thoracicus* (Colletidae; Batra 1980), construct shallow sleeping burrows in the ground. Other males, such as those of *Andrena perplexa* (Andrenidae), are known to at least occasionally spend the night in the burrows of their fe-males (Stephen 1966). Of course, some of the macrocephalic males mentioned in Chapter 4 spend their nights in nests, but in these cases, the nests are both sleep sites and boudoirs.

MALES AS POLLINATORS

Male bees are often considered to play a minor role as pollinators. In a world where bee equals honey bee, this makes perfect sense, since male honey bees don't visit flowers. In the broader world of solitary bees, the story is consider-ably more complex. There, the roles of male bees as pollinators range from being, at most, minor contributors to being absolutely essential. Nonetheless, a lack of respect for the pollinating prowess of male bees is not without some basis. Male bees do not provision nests, so their overall requirements for pollen and nectar are far less than those of non-parasitic females. However, males typically spend more time on the wing than their respective females, so they may consume more nectar than their females, although personal usage is almost always dwarfed by the amount that goes into nest provisions. Male bees do consume some pollen, but their pollen consumption is trivial com-pared to that of females (Cane et al. 2017). Lower pollen and nectar require-ments mean they need to visit far fewer flowers, and with fewer flower visits comes a reduced potential for a role as a pollinator. In addition, male bees rarely, if ever, visit certain types of flowers. Chief among these are nectar-less pollen-rewarding flowers, particularly those with functionally poricidal an-thers, such as are found in *Solanum, Senna, Chamaechrista*, and many other genera (Buchmann 1983; see also Chapter 7). While male bees may patrol these flowers in search of mates, no male bee is known to be a vibratile pol-len harvester, so they almost never land and work them in a manner that would lead to pollination. Since male bees do not construct nests, males are not normally associated with resin-producing flowers, such as those of *Clusia* or *Dalechampia*.[1]

Another group of normally bee-pollinated flowers in which male bees play a very limited role are the oil flowers. With a couple of notable exceptions, males of oil-collecting taxa lack the structures for floral oil collection and transport; accordingly, they very rarely visit these flowers. The exceptions are males of *Tetrapedia* and *Paratetrapedia*. Despite their similar names, *Tetrapedia* and *Paratetrapedia* (sensu lato) are not particularly closely related (Danforth et al. 2013). In these genera, males have a foretarsal oil-collecting apparatus that is virtually identical to those of their respective females, as well as areas of specialized branched hairs where they store the oils they collect (Aguiar and Melo 2007, Cappellari et al. 2012, Neff and Simpson 1981). In *Paratetrapedia* sensu lato, the putative oil-storage organs are scopa-like patches of branched hairs on the hindlegs (Aguiar and Melo 2007), while in *Tetrapedia*, the males have both branched hairs on the hindlegs and mats of densely branched hairs on their distal terga (Alves-dos-Santos et al. 2009, Cappellari et al. 2012). These tergal mats are unique, as they have no morphological counterpart among their respective females. Oil collection has been documented for male *Tetrapedia* (Singer and Cocucci 1999), but there seem to be no direct observations for male *Paratetrapedia*. Males and females of *Tetrapedia* are strongly dimorphic, facilitating behavioral observations, but males and females of *Paratetrapedia* are remarkably similar, so distinguishing them in the field can be quite difficult. In the absence of reports of activity at flowers, oil collection by male *Paratetrapedia* can be inferred from the presence of floral oils in the branched hairs of their hindlegs (JLN pers. obs.).

Unfortunately for the sake of a tidy tale, we do not currently know why these males collect floral oils. It is highly unlikely that they are collecting the oils for their own consumption. Presumably the oils play some role in mating, perhaps as a nuptial gift (which would be unique among bees), or perhaps they function as some type of female attractant or have some other, currently unknown function. Similarly, the role of these oil-collecting males as pollinators is unclear. Male *Paratetrapedia*, which visit male malpighiaceous floral hosts in a "normal" manner, probably are effective pollinators, while male *Tetrapedia*, which, like their females, collect oil from beneath the flower without regularly contacting the flower's sexual organs, are likely to be far less effective (Cappellari et al. 2012). A male *T. diversipes* was observed collecting oil from flowers of *Gomesa paranensoides* (Orchidaceae, as *Oncidium paranaense*), and as it had a pollinarium on its clypeus, it appeared to be an effective pollinator (Singer and Cocucci 1999).

There are remarkably few studies comparing the relative effectiveness of male and female bees as pollinators, and the few available tell a mixed story. A few studies have shown that on an individual-visit basis, males of some species are equally effective as pollinators as their respective females (Herrera 1987, Neff and Simpson 1990), but more typically, females have been found to be more effective pollinators than males (Cane 2002, Cane et al. 2011, Larsson 2005, Motten et al. 1981, Neff and Simpson 1990, Parker 1981, Tepedino 1981). However,

---

[1] There is one exception to this statement. While most species of *Dalechampia* (Euphorbiaceae) are pollinated by resin-collecting female bees (including female orchid bees), pollination by scent-collecting male orchid bees has arisen repeatedly in the genus (Armbruster et al. 2009).

just because males of a given species are less effective than their respective females does not mean they are not significant pollinators. Several studies have found that male bees were more effective pollinators than many other visitors in the studied systems (Cane 2002, Herrera 1987). In addition, one needs to be specific in making comparisons. Male *Megachile rotundata* were found to be as effective as their nectar-collecting females at pollinating blueberries, but they were significantly less effective than pollen-collecting females of their own or any other studied species (Javorek et al. 2002). The story is also more complicated than simple comparisons of single-visit effectiveness. Males typically make longer inter-flower movements than females, even on small scales (Herrera 1987, Ne'eman et al. 2006). This reduces the possibility of geitonogamy (movement of pollen between flowers on a given individual plant) and enhances the bee's role in promoting gene flow. This could be enhanced further, since in many species male behavior involves patrolling flowers in search of females, with only infrequent actual floral visits. This suggests that, although difficult to measure, the proportion of long inter-floral moves by males may be high, and the role of males as pollinators—particularly on plants with large floral displays where the probability of geitonogamous movements by females is high—may often be underestimated. Remarkably little is known about the foraging ranges of male solitary bees.

Some male bees regularly return to a particular site to sleep (Alcock 1998, Linsley 1962), but unlike females, males are not true central-place foragers, whose activities are centered upon a nest. A strong positive correlation has been demonstrated between body size and foraging range in female bees (Greenleaf et al. 2007), so it seems reasonable to assume that the same size-dependent relationship holds for males. Nonetheless, actual foraging ranges of males may be either much smaller or larger than that predicted by size alone. For example, the large fighting male morphs often found in territorial species probably have foraging ranges much smaller than would be predicted by size alone. On the other hand, we know some males move much farther than would be expected for nest-based females. In a study using miniaturized radio transmitters weighing 300 milligrams, a seemingly hefty load for a 600-milligram bee, male *Exaerete frontalis* were found to forage in core areas averaging 45 hectares, or the equivalent of a foraging radius of 4.2 kilometers (Wikelski et al. 2010). In a record-setting extreme, a male *Euglossa viridissima* was collected 12 days after being marked, having traversed a remarkable 95 kilometers from the marking site (Pokorny et al. 2015). This could give long-distance pollen movement a whole new meaning.

Another factor is phenology. Many bee species are protandrous, so for many plants, particularly those that flower in the early days of spring, male bees may be the most important floral visitors before the females emerge and begin foraging (Motten et al. 1981, Tepedino 1981). Beyond making longer inter-floral moves or simply being the only game in town, male bees may offer advantages as pollinators, since they are often "cheaper" pollinators from the plant's perspective than females, at least in terms of pollen usage. Pollen is the very stuff of pollination, that which must be moved about to maximize plant male fitness, and many features of floral morphology and pollen presentation have been interpreted as means to reduce pollen loss to pollen-foraging bees (Harder and Barclay 1994; see also Chapter 12). Since male bees do not forage for pollen, or at least spend very little time doing so to meet their very limited individual needs, male bees should be far less costly in terms of pollen loss than females.

Besides their standard role as pollinators as a by-product of their foraging behavior, males of some bee species are involved in a number of distinctive pollination systems unique to male insects. Probably the oddest of these are the sexual-deception or pseudocopulatory systems. These are sometimes called instances of Pouyannian mimicry after Maurice-Alexandre Pouyanne, a Frenchman who first discovered the phenomenon while observing orchids in Algeria (Pouyanne 1917). The rewardless flowers in these systems use combinations of morphology, color, tactile cues, and scent to directly harness the sexual drives of male insects to their benefit by pretending to be sexually receptive females. Males encounter these flowers while searching for mates and attempt to mate with them, in the process picking up or depositing pollen. While these bizarre pseudocopulatory pollination systems are uncommon among angiosperms, they have multiple independent origins (Gaskett 2011). Long thought to be restricted to the orchids, pseudocopulatory pollination systems have also been discovered in the Asteraceae and Iridaceae (Ellis and Johnson 2010, Vereecken et al. 2012). While our concern is with male solitary bees, they are not the only players in these systems. Male thynnid wasps are the major participants in Australia (Stoutamire 1983), while in the Mediterranean region, the list of sexually deceived visitors to flowers of *Ophrys* (Orchidaceae) includes male scarab beetles, syrphid flies, and scoliid and crabronid wasps. However, the primary pollinators of most *Ophrys* species are male solitary bees (Fig. 5-4). One recent compilation listed roughly 117 bee species as *Ophrys* pollinators; the list is dominated by species of *Eucera* and *Andrena* but also includes species of *Anthophora*, *Bombus*,

Figure 5-4. Male *Andrena morio* (Andrenidae) approaching an *Ophrys incubaceae* (Orchidaceae) flower. Photo kindly provided by Nico Vereecken.

Chalicodoma, Colletes, Eupavlovskia, Melecta, Osmia, Tetralonia, Tetraloni-ella, and *Xylocopa* (the number is imprecise as some of the "species" appear to be *nomen nuda*) (Gaskett 2011).

Ophrys are often called "bee orchids" because of their supposed resemblance to bees. The labellum of an *Ophrys* flower is typically elongate, vaguely resembling a bee body, and often has hairy lateral projections. The colors of the labellum are quite variable; the most common patterns are reddish brown with various blue, purple, or white maculae. To the casual human observer, these flowers are rather odd-looking and only vaguely bee-like. Of course, what constitutes being bee-like in this context is only relevant to a bee, and we are not bees. A myriad of studies, dating back to the pioneering work of Kullenberg (Kullenberg 1956, Kullenberg and Bergström 1976) and Borg-Karlson (1979) have shown that what really matters in attracting male insects to an *Ophrys* flower is its scent. These scents are functionally allomones, mixtures of compounds that match the pheromones of the models—in this case, sexually receptive female bees. The pheromones are species-specific and typically consist of a mixture of n-alkanes, n-alkenes, and other compounds. The orchid allomones are produced in about the same amounts as the bee-produced pheromones and closely mimic the chemical composition of those pheromones.[2]

Perfection in the similarity of the mimic to its model is often expected to be the ideal in mimetic systems. However, this does not seem to be the case in at least some pseudocopulatory systems. A study of the interactions of male *Colletes cunicularius* and *Ophrys exaltata* found that chemical imperfection on the part of the mimic can be adaptive (Vereecken and Schiestl 2008). While the scents of the model (female sympatric *Colletes*) and the mimic (the orchid) are mixes of identical compounds, the ratios of the various constituents differ slightly between bees and orchids and between female bees of sympatric and allopatric populations. In a series of choice tests, male *C. cunicularius* were found to prefer orchid scents over those of sympatric female *Colletes*, and to prefer the scents of allopatric females over sympatric females. Some males have a preference for the exotic in their sexual activities. Who knew?

Pseudocopulatory systems are very one-sided; the plants get all the benefit (pollination), and the male bees get nothing but wasted time (and perhaps lost sperm, although actual ejaculation on the flowers appears to be quite rare). Another male-driven pollination system, this time exploiting the sleeping habits of male bees, is not so one-sided. As noted earlier, male bees use a wide array of sites as sleeping sites: flowers, plant stems, shallow sleeping burrows, and various small cavities. Although it seems like an odd strategy, several plants with non-rewarding flowers are known to be pollinated primarily by male bees that use them as sleeping sites. These flowers are dull in color and have corollas modified to form a short, dark tube suggestive of a burrow or bee nest. Male bees, particularly male *Eucera*, but sometimes other insects, use these as sleep sites. Jostling between males in the tubes often results in inter-floral movements that may lead to pollination. These so-called shelter mimics are best known

[2] A *pheromone* is a chemical substance produced and released into the environment by an animal that affects the behavior of members of the same species. An *allomone*, on the other hand, is a chemical that is produced and released by an individual of one species that impacts the behavior of a member of another species, usually to the benefit of the sender but not the receiver.

in the orchid genus *Serapias* and the royal irises (*Iris* section *Oncocyclus*; Dafni et al. 1981, Sapir et al. 2006, Vereecken et al. 2013). Beyond structural mimicry, these systems have a chemical component; their floral scents perhaps mimic the scent of bee nests (Pellegrino et al. 2012). Interestingly, shelter mimicry appears to be the evolutionary precursor of the sexual-deception pollination systems of *Ophrys*. That story has been further complicated by an independent origin of sexual deception, from shelter mimicry in the royal irises and a secondary reversion to shelter mimicry from pseudocopulation in *Ophrys* (Vereecken et al. 2012).

An additional odd male-only pollination system has been suggested for certain oncidiine orchids. Male *Centris* were observed holding territories near inflorescences of *Oncidium hyphaematicum* and *O. pardothrysus*, orchids whose flowers are vaguely bee-like (although they also might be considered to be mimics of flowers of the Malpighiaceae). The males drove away insects that entered their territories and, when the flowers were moved by the wind, struck at the orchid flowers, regularly removing pollinaria. Although pollination levels in the orchids were high, fruit set was very low, suggesting a combination of self-incompatibility and geitonogamy (Dodson and Frymire 1961). Similar behavior was observed by male *Centris insularis* (as *C. insulans*) at flowers of *Tolumnia henekenii* (Orchidaceae), although in this case, some of the bee behavior was consistent with a pseudocopulatory system (Dod 1965).

The largest of the pollination systems dominated by male bees is that involving scent collection by male euglossines (Dressler 1968). Male euglossines spend most of their lives collecting volatile compounds that they store in their highly modified hind tibiae (Dressler 1982, Roubik and Hanson 2004) (Fig. 4-6). Ultimately, these compounds are used during courtship; the amounts and variety apparently are important indicators of male quality to female euglossines (Eltz et al. 2015a). These volatiles are collected from many sources, including fungi, feces, plant wounds, and rotting fruit, as well as the vegetation and flowers in at least 16 families of angiosperms (Ramírez et al. 2002, Whitten et al. 1993). In many cases, male euglossines are collecting scents at flowers without playing any significant role as pollinators, and in some cases their scent-collecting activities may actually be destructive (Cappellari et al. 2009). Scent-collecting male euglossines are known to be significant pollinators of species in at least six angiosperm families, most notably the Araceae, Euphorbiaceae, and Orchidaceae (Ramírez et al. 2002, Roubik and Hanson 2004, Whitten et al. 1993). Of course, the fame of the euglossines and the source of their common name (orchid bees) come from their association with the Orchidaceae. Between 600 and 700 species of neotropical orchids are totally dependent on scent-collecting male euglossines as pollinators. These specialized flowers lack nectar, and like most orchids, their pollen is presented as pollinaria, which are not useful as bee food (Harder and Johnson 2007). As they do not seem to qualify as general floral mimics, female euglossines very rarely visit them and play no role in their pollination. Reproductive isolation among these orchids is maintained by the species-specific mixes of volatile compounds that attract different bees and the morphological complexity of the flowers, which leads to species-specific placement of pollinaria on particular parts of the bees' bodies (Roubik and Hanson 2004, van der Pijl and Dodson 1969). Some of these orchid flowers are remarkably complicated. In the bucket orchids (*Coryanthes* spp.), male euglossines land on the hypochile (the basal part of the floral lip) to collect scents. A male may fall into the liquid-filled, bucket-shaped epichile, from which it cannot easily escape because of specialized slippery cells lining the bucket and wetting agents in the liquid that keeps the bee from flying out (Gerlach 2011). The only exit is through a tunnel-like structure in which, if it is of the proper size, a pollinaria is deposited or picked up from its metasoma (Crüger 1864, Dodson 1965).

The male euglossine-orchid interaction was long held to be a fine example of close coevolution, with the bees dependent on orchids for scent sources and the orchids dependent on the bees for pollination (Barth 1985, O'Toole 2013). The latter half of that relationship is certainly true, but the former has frayed considerably. Molecular studies have shown that there is little correspondence between euglossine and orchid phylogenies, ruling out narrow coevolution, and that male euglossines were probably collecting scents at least 12 million years before the euglossine scent-collection pollination syndrome arose in the orchids (Ramírez et al. 2011). In a more modern demonstration of the lack of male euglossine dependence on orchids, *Euglossa dilemma*, following an apparently accidental introduction from Central America, has successfully colonized southern Florida, despite the absence of scent-producing orchids from its native range (Skov and Wiley 2005). It now collects scents in Florida from such non-orchid sources as leaves of basil, allspice, and *Melaleuca* (Pemberton and Wheeler 2006).

Male euglossines have long been believed to be the only bees involved in scent-collection pollination systems, but a recent study indicates a second, independent origin of the phenomenon. Some male *Paratetrapedia* have dense mats of short, branched hairs on S3 (Aguiar and Melo 2011). Another recent study has shown that males of *P. chocoensis* use their sternal brushes to harvest scents (volatile oils) from the flowers of *Anthurium acutifolium* (Araceae), pollinating the flowers in the process (Etl et al. 2017). As noted earlier, these males have arrays of setae on their foretarsi that are identical to those of their oil-collecting females. Etl and colleagues (2017) speculate that the males are harvesting fatty oils from oil flowers, transferring them to the sternal brushes, and then using them as an aid in collecting and storing the scents, a process analogous to the use of endogenously produced lipids as an aid in oil collection by male euglossines.

CHAPTER 6

Nest Architecture and Brood-Cell Construction

For many species, the nest represents the largest investment (in terms of time and energy) that a female bee makes in her lifetime. Not only must a female choose an appropriate site, she must construct a nest in soil, wood, or other substrates and form the brood cells, which she will provision with pollen, nectar, or floral oils, and lay her eggs (Chapter 8). The brood cell will be the home of her offspring for one to many years (Chapter 3). How the nest is constructed and the brood cell prepared will profoundly impact her reproductive success. A poorly constructed brood cell or a sloppily constructed nest can expose her offspring to freezing, flooding, desiccation, or attack by both brood-parasitic bees (Chapter 10) and non-bee parasites and predators (Chapter 11).

Solitary bees construct their nests in a wide variety of substrates. Most bees construct their nests in the soil, but many bees excavate nests in wood, pithy stems, *Agave* stalks, or unexpected locations, like active termite nests. Other bees construct freestanding nests or use any of a wide array of preexisting cavities, such as old beetle burrows, abandoned bee or wasp nests, snail shells, and abandoned fly galls. Most bees line their brood cells with glandular secretions, but others use floral oils, and many megachilid bees use various combinations of sand, pebbles, and plant products (leaves, leaf pieces, leaf masticate, petals, trichomes, and resin).

The bewildering diversity of nesting substrates and materials used by bees makes it very difficult to develop a universally acceptable classification of nesting types. For simplicity, we distinguish among four broad categories of nest construction described below and shown in Figure 6-1.

Soil excavators comprise the vast majority of solitary bees. Cane and Neff (2011) estimated that 64% of non-parasitic bees are ground-nesting, soil-excavating bees. Virtually every bee subfamily includes at least some ground-nesting soil excavators (Fig. 6-1), and the earliest bees were likely ground-nesting.

A second category includes the wood excavators. This category includes not only the bees that excavate nests in solid wood, such as many *Xylocopa* species (the large carpenter bees), but also taxa that excavate in rotten or soft wood and pithy stems (Fig. 6-1). We also include here oddball taxa that excavate their nests in horse and cattle dung (Cane 2012, Sarzetti et al. 2012).

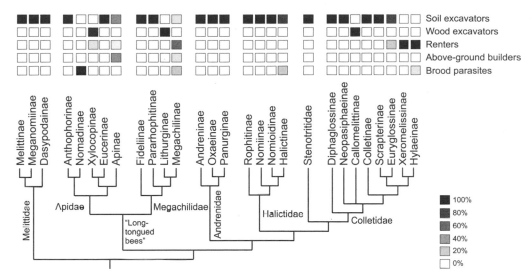

Figure 6-1. Phylogeny of the bee subfamilies and families, with the proportion of species in each that are either soil excavators, wood excavators, renters, or above-ground builders. We have also included the category of brood parasites for completeness.

A third broad category includes the renters—bees that construct nests in pre-existing cavities (Iwata 1976). Renting is the second-largest nesting category, and the renters are a taxonomically diverse assemblage (Fig. 6-1). Renters, while they do not excavate a nest, spend considerable time and effort preparing the nest and brood cells. Some collect and transport foreign materials to the nest (Megachilinae; Table 6-1), others add glandular secretions to line and water-proof the brood cell (Colletidae), and a third group use floral oils (sometimes mixed with other materials) in brood-cell preparation (Centridini, Tetrapediini, and Ctenoplectrini). Renters inhabit a diversity of cavities, including holes in the ground, cavities in stone, gaps in masonry, beetle burrows, snail shells, vacated fly galls, and virtually any human material that looks and feels like a stem, tube, or hollow cavity (including the hydraulic tubing of small aircraft; Eickwort and Rozen 1997).

A final nesting category includes the above-ground builders. This category includes bees that construct freestanding nests of various materials (including mud, resin, and plant trichomes). Taxa included here are the euglossine bees that construct their nests with resin, bark, and plant fibers (Apinae) and diverse groups of Megachilinae that construct freestanding nests of mortar (mud and pebbles glued together with glandular secretions), plant resins (alone or combined with other materials), and plant trichomes (sometimes combined with mammal fur and bird feathers) (Table 6-1).

Not all bees fit neatly into one of the four broad categories listed above. Some bee groups (e.g., Hylaeinae, Xeromelissinae, Tetrapediini, Ceratinini, and others) fall into a gray zone between renting and pith nesting, sometimes initiating nests in preexisting cavities in pithy stems but then secondarily enlarging them. Many of the bees that nest in arboreal termite nests are also hard to classify, be-cause they must first excavate to get into the termite nests, but once inside, they

Table 6-1. Variation in the nesting substrates used by bees in the family Megachilidae.

The classification used here is based on a variety of sources. For subfamily and tribal-level classification, we follow Gonzalez et al. (2012). For generic classification within tribes, we follow Praz et al. (2008a; Osmiini), Litman et al. (2016; Anthidiini), Trunz et al. (2016; Megachilini). Cleptoparasitic and fossil lineages have been excluded. The tribal placement of two genera of Megachilinae (*Pseudoheriades* and *Afroheriades*) is not well established so we have excluded them from the table. The biology of *Afroheriades* is unknown while *Pseudoheriades* makes resin-lined nests in pre-existing cavities in wood (Gess & Roosenschoon 2017).

Subfamily	Tribe	Group	Genus	Soil excavators	Wood excavators	Renting below-ground	Renting above-ground	Above-ground builders	Materials used
Fideliinae			*Neofidelia*	X					n/a
			Fidelia	X					n/a
Pararhophitinae			*Pararhophites*	X					n/a
Lithurginae			*Austrothurgus*		X				n/a
			Lithurgus		X				n/a
			Lithurgopsis		X				n/a
			Microthurge		X				n/a
			Trichothurgus		X				n/a
Megachilinae	Aspidosmiini		*Aspidosmia*					X	resin + pebbles
	Anthidiini	Anthidium	*Afranthidium*				X-snail		fiber
			Anthidioma						fiber?
			Anthidium	X-rare		X	X		fiber
			Indanthidium						fiber?
			Pseudoanthidium			X	X	X	fiber
			Serapista					X	fiber
		Anthodioctes	*Anthodioctes*				X		resin
			Aztecanthidium						resin
			Duckeanthidium				X		resin
			Epanthidium					X	resin
			Hypanthidioides				X	X	resin

		Hypanthidium					
		Notanthidium				X	resin
		Acedanthidium					
		Anthidiellum			X	X	resin
		Bathanthidium			X	X	resin
		Benanthis					
		Cyphanthidium					fiber?
	Dianthidium	Dianthidium		X?		X	resin
		Eoanthidium					
		Icteranthidium					resin
		Pachyanthidium			X	X	resin + fiber
		Paranthidium	X				resin
		Plesianthidium				X	fiber
		Rhodanthidium		X?	X-snail		fiber or resin
		Apianthidium					
	Trachusa	Trachusa	X				resin
		Trachusoides					
Osmiini	Chelostoma	Chelostoma			X		mud
	Heriades	Heriades			X		resin
		Hofferia			X		resin
		Ochreriades			X		mud
		Othinosmia		X?		X	resin, pebbles
		Protosmia			X-snail		resin
		Stenoheriades			X?		resin?
	Osmia	Ashmeadiella	X	X	X		leaf masticate, mud

(continued)

Table 6-1. (Continued)

Subfamily	Tribe	Group	Genus	Soil excavators	Wood excavators	Renting below-ground	Renting above-ground	Above-ground builders	Materials used
			Atoposmia			X	X		leaf masticate, mud
			Haetosmia	X					leaf masticate with mud and pebbles
			Hoplitis	X	X	X	X-snail	X	mud, petals, leaf masticate, pebbles, pith particles, resin
			Osmia	X	X	X	X-snail	X	resin, mud, plant masticate, plant pith, resin
			Wainia				X-snail		petal pieces, sand
	Megachilini		*Gronoceras*						
			Heriadiopsis						
			Matangapis						
			Megachile	X	X	X	X	X	resin, mud, leaves, petals, leaf pieces, leaf masticate, pebbles, wood fragments
			Noteriades						

often incorporate termite-created cavities into their nests (Carrijo et al. 2012). We will be the first to admit that the nest architecture of solitary bees does not follow such a single rigid classification.[1]

Whereas some bee groups are relatively uniform in their methods of nest construction and nest-site choice, there are other groups that seem to have diversified into virtually every nesting substrate imaginable. This is true especially within the Apidae and Megachilidae. Species of *Centris* (Apidae), for example, nest in a remarkable diversity of substrates, including horizontal ground, vertical clay banks, mud or adobe walls, abandoned mud dauber (*Sceliphron* or *Trypoxylon*, Sphecidae) nests, old bee cells and burrows, and arboreal termite mounds (Coville et al. 1983). *Centris* in the subgenera *Hemisiella* and *Heterocentris* are renters and cavity nesters that use holes in wood, logs, stumps, and linear trap nests. Likewise, within the subfamily Megachilinae (Megachilidae), there are genera (*Megachile*, *Hoplitis*, and *Osmia*) with species that fit each of our four categories above (Table 6-1; Cane et al. 2007, Michener 2007, Trunz et al. 2016).

SOIL EXCAVATORS

Excavating a below-ground nest in the soil is by far the most common nesting strategy among solitary bees (Fig. 6-1). The families Melittidae, Andrenidae, and Stenotritidae are composed exclusively of soil excavators. Most Halictidae are soil excavators, as are the majority of colletids. Within the Colletidae, five of the eight subfamilies consist primarily of soil excavators.[2] A large proportion of the pollen-collecting (i.e., non-parasitic) Apidae are soil excavators, and this is almost certainly the primitive state for the family. In the Megachilidae, the basal subfamilies (Fideliinae and Pararhophitinae) consist exclusively of desert-dwelling, sand-loving, ground-nesting bees.

Soil-excavating bees are capable of performing extraordinary engineering feats (Fig. 6-2). While the maximum cell depth for most bee nests averages a relatively modest 35 centimeters (Cane and Neff 2011), some bee nests are much deeper. Nests of some *Ctenocolletes* (Stenotritidae) attain depths greater than 3 meters (Houston 1984, 1987). Most deep nests are multicelled, but *Habropoda pallida* (Apidae), a sand-dune specialist, constructs single-celled nests that extend to nearly 2 meters (Bohart et al. 1972). A cell of another sand specialist, *Andrena* (*Callandrena*) *haynesi*, was found at a depth of 2.7 meters (Parker and Griswold 1982). The deepest recorded nest of any solitary bee is made by *Exomalopsis aureopilosa* (Apidae). As reported in Ronaldo Zucchi's (1973) PhD thesis (cited in Rozen 1984a), it extended downward to just over 5 meters. The extraordinary depth of this nest may be due to the fact that this is a communally nesting species and the nest in question was an exceptionally large

---

[1] Other classifications of bee nest architecture exist. For alternative treatments of this topic, we refer readers to classic books and papers by Malyshev (1935), Iwata (1976), Sakagami and Michener (1962), Stephen et al. (1969), and Radchenko and Pesenko (1994, 1996).
[2] In Colletidae, soil nesting occurs exclusively in Diphaglossinae, Neopasiphaeinae, and Scrapterinae, almost all Colletinae, and a majority of Euryglossinae, but only rarely in Xeromelissinae and Hylaeinae (see Almeida 2008, Michener 2007).

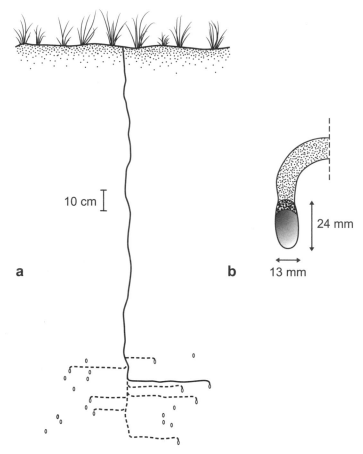

Figure 6-2. Nest of *Oxaea flavescens* (Andrenidae): (a) side view of the open burrow (solid line), filled burrows (dashed lines), and cells (ovals) (redrawn from Roberts 1971, Fig. 9) and (b) side view of brood cell (redrawn from Roberts 1973, Fig. 11).

one, with 884 adult females and 46 males. It is common for communal nests to persist for many generations; the main tunnel is progressively extended downward with each new generation. Even small bees can produce exceptionally deep nests. Ramos and Rozen (2014) documented a 2-meter-deep nest in a small panurgine bee (*Psaenythisca wagneri*). Females of this species are just 5 millimeters in length, which means the nest depth was 400 times their own body length. A six-foot-tall human would need to excavate down 2,400 feet to construct a tunnel of comparable size. The Chilean miners trapped in the Copiapó mine in 2010 were 2,300 feet below the surface, and it took three months of drilling with heavy equipment to liberate them.

Nest aggregations of solitary, ground-nesting bees can be gigantic and can appear seemingly overnight. The largest nesting aggregation ever reported consisted of over 12 million nests of *Dasypoda plumipes* (Melittidae; Blagoveschenskaya 1963). Nesting aggregations of over 100,000 individual nests have been reported in Apidae, Halictidae, and Melittidae (Table 6-2). What triggers the initiation of nesting is not clear, but rainfall seems like a logical possibility in arid

Table 6-2. Variation in aggregation size for gregariously nesting solitary, ground-nesting bee species.

Total number of bee nests were estimated in a variety of ways.

Species	Family	Number of individual bee nests	Literature cited
Dasypoda plumipes	Melittidae	12,000,000	Blagoveschenskaya 1963
Nomia melanderi	Halictidae	5,300,000	Cane 2008
Centris caesalpiniae	Apidae	423,000	Rozen & Buchmann 1990
Anthophora edwardsii	Apidae	180,000	Cane 2008
Dieunomia triangulifera	Halictidae	155,000	Minckley et al. 1994
Anthophora bomboides	Apidae	133,000	Hanson & Ascher 2018
Hesperapis peninsularis	Melittidae	100,000	Stage 1966
Mesoxaea texana	Andrenidae	80,000	Cockerell 1933
Andrena postomias	Andrenidae	28,000	Maeta et al. 1988
Eremapis parvula	Apidae	8,000	Neff 1984

regions of the world (see Chapter 3 for a further discussion of rainfall-induced emergence). Rainfall may make soils more malleable and easily manipulated by solitary, ground-nesting bees as well.

Ground-nesting bee aggregations can persist at the same site for decades (Table 6-3). A combination of appropriate soil texture and large, stable populations of suitable host plants may explain the long-term persistence of bee aggregations. Some ground-nesting bees are regularly found in aggregations, but the aggregations move from site to site en masse from one year to the next (e.g., *Centris caesalpiniae*, Rozen and Buchmann 1990; *Diadasia rinconis*, Neff and Simpson 1992). How they recruit to these newly established aggregations remains a mystery, although it is likely that chemical cues play a role (Cane 1996).

Ground-nesting bees, especially those that nest in large aggregations, are capable of moving considerable amounts of soil. Cane (2003) calculated the volume of soil moved by nesting female *Nomia melanderi* (Halictidae) in the Touchet Valley of southeastern Washington. While individual nest tumuli are small, these bees nest in extremely high densities (>200 nests/m²). Individual nesting aggregations can number more than 1 million nests. Across the entire Touchet Valley, Cane estimated a population of 9.5 million nesting female bees in 2001. Based on average nest densities, Cane calculated that the population of *N. melanderi* moved 96 tons (87,500 kg) of soil in a single year. *N. melanderi* is a medium-sized bee that constructs nests of no more than 0.5 meter in depth. Ground-nesting bees, like earthworms, contribute significantly to soil aeration and rejuvenation.

One hazard for a ground-nesting bee is flooding. Aggregations of soil-nesting bees, especially those nesting along watercourses, can be vulnerable to catastrophic flooding. Fellendorf and colleagues (2004) documented a flooding event in Germany that did severe damage to a massive nesting aggregation of *Andrena*

Table 6-3. Variation in the longevity of continuously occupied nesting aggregations of solitary, ground-nesting bees.

Species	Family	Duration	Literature cited
Andrena agilissima	Andrenidae	11 years	Andrietti et al. 1997
Panurginus polytrichus	Andrenidae	20 years	Neff 2003
Anthophora abrupta	Apidae	20 years	Cane & Norden pers. obs., cited in Cane 2008
Colletes kincaidii	Colletidae	>20 years	Torchio et al. 1988
Crawfordapis luctuosa	Colletidae	>20 years	Roubik & Michener 1984
Habropoda laboriosa	Apidae	>20 years	Cane pers. obs., cited in Cane 2008
Colletes inaequalis	Colletidae	>20 years	Cane pers. obs., cited in Cane 2008
Trachusa byssina	Megachilidae	22–35 years	Friese 1923
Dieunomia triangulifera	Halictidae	22–35 years	Minckley et al. 1994
Trachusa perdita	Megachilidae	24–27 years	Thorp pers. obs., cited in Cane 2008
Trachusa gummifera	Megachilidae	24–27 years	Thorp pers. obs., cited in Cane 2008
Anthophora bomboides	Apidae	40 years	Hanson & Ascher 2018
Macropis nuda	Melittidae	40 years	Rozen & Jacobson 1980
Anthophora pueblo	Apidae	40 years	Orr et al. 2016
Nomia melanderi	Halictidae	50 years	Cane 2008
Peponapis pruinosa	Apidae	50 years	K. Brochu pers. comm.
Anthophora plumipes	Apidae	>50 years	O'Toole & Raw 1991
Andrena vaga	Andrenidae	60 years	Ulrich 1956

vaga along the Rhine River. However, some bees seem to be able to survive flooding (see Box 6-1). The ground-nesting, oil-collecting bee *Epicharis zonata* (Apidae) nests in lowland, sandy, coastal habitats in French Guiana. These bees are active as adults during the dry season, but they nest in habitats that are flooded for up to nine months during the rainy season (Roubik and Michener 1980). Brood cells are located up to 0.5 meter in depth, and larvae remain hermetically sealed in their brood cells while completely immersed under water. The thick, water-repellent brood-cell lining and a double cell closure appear to be adaptive features that allow these bees to survive in seasonally flooded habitats. Other bees have sufficiently water-resistant brood cells to remain under water for extended periods of time. Species of *Colletes* (Colletidae; Albans et al. 1980), *Dasypoda* (Melittidae; Cane 1991), and *Calliopsis* (Andrenidae; Visscher et al. 1994) have been recorded nesting in seasonally flooded sites (reviewed in Fellendorf et al. 2004). *Perdita floridensis* (Andrenidae), a bee adapted to seasonal flooding in the southeastern United States, is remarkable in that the prepupae migrate out of the brood cells while underwater; these are truly "benthic" bees (Box 6-1).

BOX 6-1: *PERDITA FLORIDENSIS*—A REMARKABLE "BENTHIC" BEE

Perdita floridensis (Andrenidae: Panurginae) is an early-spring host-plant specialist that occurs in the eastern United States from North Carolina to Florida. Like most *Perdita*, *P. floridensis* is a small (~0.5 cm in length), solitary, ground-nesting bee. *P. floridensis* was studied in detail at the Archbold Biological Station (ABS) in central Florida by Norden and colleagues (Norden et al. 2003). At the ABS, female *P. floridensis* form massive nesting aggregations in low-lying sandy areas that form seasonally flooded ponds in the rainy season (June–September). During the brief period of adult activity (mid-April to mid-May), females mate, construct their subterranean burrows, and provision their brood cells. Brood cells are constructed at a depth of between 15 and 45 centimeters, and females provision their brood cells exclusively with pollen and nectar collected from gallberry (*Ilex glabra*, Aquifoliaceae), a common shrub growing at the site. Females produce between 5 and 8 brood cells per nest, and after consuming the pollen/nectar provisions, larvae molt to the prepupal stage. Nesting sites become submerged in early July and remain under knee-deep water for as much as six months (until January). The prepupae of *P. floridensis*, unlike any other bee that we are aware of, migrate out of their brood cells and wiggle through the saturated sand by curling and uncurling their bodies. Like all *Perdita* prepupae, the prepupae of *P. floridensis* have elongate dorsal tubercles that appear to function well to propel them through the soupy, saturated soils. They move as far as 5 centimeters from their natal brood cell. Why they migrate through the sand is not clear, but Norden and colleagues (2003) speculate that it allows the prepupae to move into areas in which algal photosynthetic activity elevates the oxygen content of the surrounding water. Movement through the soil may be an adaptation for avoiding anoxic regions of the saturated soil matrix. While there are a number of reports of solitary bee larvae surviving under water in seasonally flooded habitats, no bee other than *P. floridensis* is known to leave the safety and comfort of the brood cell to migrate through the saturated soil matrix. This remarkable "benthic" species illustrates just how adaptable solitary bees are to highly seasonal and variable environments.

Nest location

What factors determine the locations of the nests of soil-burrowing bees? It is clear that soil texture, aspect, and ground cover all play a role in where bees choose to nest. Cane (1991) analyzed the texture of the soils used by 32 ground-nesting bees from diverse habitats in North America. In this small sample, soil type was not correlated with the taxonomic relations of the bees, and most bees were using soils with either a high proportion of sand or sand mixed with silt or clay. Many bees are known to nest preferentially in particular substrates. Sand specialists (psammophiles) are probably the largest group of edaphic specialists, but other bees are known to specialize in particular substrates, including sandstone (see Box 6-2), vertical clay cliffs (Michener and Lange 1958), or even the edges of active volcanos (Erenler et al. 2016). A surprising diversity of bees excavate nests in arboreal termite nests (Carrijo et al. 2012) (see Box 6-3). Soil specialists (such as those that nest in dunes) are of particular concern from a conservation perspective (see Chapter 14).

BOX 6-2: BEES THAT NEST IN STONE

Among the most highly specialized of ground-nesting bees are those that have evolved the ability to nest in sandstone cliffs and outcroppings. These "stone-nesting bees" are a small, but heterogeneous group. One of the most intensively studied of these stone nesters is *Colletes daviesianus* (Colletidae), a widespread Eurasian species that often nests gregariously in burrows excavated in sandstone cliffs (Friese 1912, Mader 1980). Mader has documented hundreds of such sites in Germany, including in houses built with sandstone blocks (Mader 1999). Strictly speaking, *C. daviesianus* isn't a sandstone specialist, since it is known to excavate its nests in areas of bare, sloping soil or even rotten mortar in masonry structures. Nonetheless, it seems to prefer vertical sandstone exposures when they are available. Similar nesting habits have been reported for *C. kincaidii*, a polylectic North American species (Torchio et al. 1988). Although widespread, reports of its biology are available for only a small area near Bonny Doon, California, where it nests gregariously in sandstone cliffs. Nest reuse was common at the Bonny Doon site, yet some females initiated new nests in the hardened sandstone cliff faces. Interestingly, nest parasitism by *Epeolus compactus* was four times higher in reused nests than in newly initiated ones (Torchio and Burdick 1988), a possible explanation for why females would go to all the work of initiating nests in such a hard substrate.

Anthophora pueblo (Apidae) is a recently described sandstone specialist that occurs at a handful of sites across Nevada, Arizona, Utah, Colorado, and New Mexico (Orr et al. 2016). Females prefer to excavate in the softer sandstone outcroppings at sites where rocks of varying hardness occur, yet significant mandibular wear in older females indicates there is a real cost to nesting in sandstone. What might be the benefits of this remarkable behavior? Orr and colleagues (2016) speculated that brood cells excavated in sandstone may be better protected from the myriad parasites and predators that attack solitary bees. In addition, sandstone could offer some protection against microbial pathogens.

Macrotera opuntiae (Andrenidae), one of the best-known American examples of a sandstone bee, turns out to be something of a cheater. As its name suggests, the bee is a pricklypear cactus (*Opuntia*) specialist. The host plants are widespread, but nests of *M. opuntiae* are known only from the White Rocks, a sandstone outcrop near Boulder, Colorado. Clarence Custer reported on the biology of *M. opuntiae* in a paper with the intriguing title "The bee that works in stone; *Perdita opuntiae* Cockerell" (Custer 1928). He found that the bees excavated communal nests with multiple entrances and that cells were excavated either in the soft interior sandstone or in large, sand-filled cavities within the rocks. He assumed that the cavities were the result of the extensive tunneling by the bees into solid stone (Custer 1928, 1929). A subsequent study at the same site reached a different conclusion: nest entrances into the sandstone were mostly through preexisting cracks (Bennett and Breed 1985). While the bees did construct some entrances through solid stone, these were secondary additions after the bees had gained access to the softer interior of the rock. It thus appears that while it shares some traits with the other sandstone bees, the nest biology of *M. opuntiae* may be more similar to that of a typical ground-nesting bee, albeit involving accommodations to a very unusual substrate.

BOX 6-3: BEES THAT NEST IN ACTIVE TERMITE MOUNDS

A surprising number of solitary bees utilize above-ground, arboreal termite nests as substrates for their nests (reviewed in Carrijo et al. 2012). The stingless bees (Meliponini) are frequently recorded using above-ground *Nasutitermes* nests. Among solitary bees, there are clearly some highly specialized termitophiles that specialize in nesting only within the above-ground nests of termites. Termitophiles have been reported in the Euglossini, Eucerini, and Centridini (Apidae), Neopasiphaeinae (Colletidae), and Megachilini (Megachilidae). Termite subfamilies used by solitary bees include Apicotermitinae, Nasutitermitinae, Syntermitinae, Termitinae, and Rhinotermitinae. Within the genus *Centris* (Apidae), the subgenus *Ptilotopus* appears to include numerous obligate termitophiles associated with Termitinae and Syntermitinae. *Ptilotopus* even have a highly specialized cleptoparasite (*Acanthopus*, Apidae) that specializes on these arboreal solitary bees as hosts. Female *Ptilotopus* construct their tunnels and brood cells just beneath the surface of the termite nest. Burrows enter the walls of the termite nest horizontally and then extend vertically beneath the hard exterior of the termite nest. Bee tunnels are isolated from the chambers and tunnels of the termites, and the two species coexist without any signs of hostility or aggression (Coville et al. 1983). *Megachile pluto* (Megachilidae), a bee first collected by Alfred Russel Wallace and the largest bee on earth (Vereecken 2018), is also an obligate termitophile. Messer (1984) described the nests of *M. pluto* in Indonesia. Females of this rare, massive bee nest exclusively in the nests of *Microcerotermes amboinensis*. In Australia, Maynard and Rao (2010) reported an apparent obligate relationship between the colletid bee *Leioproctus nigrofulvus* and the above-ground nesting termite *Coptotermes lacteus* (Rhinotermitidae). While rare, these relationships suggest that bees benefit from the association for a variety of reasons. Arboreal termite nests are solid structures that provide a barrier against parasite and predator attack. They are also likely beneficial for soil-nesting bees, because they provide an above-ground, well-insulated, thermostable environment that is unlikely to be flooded in tropical environments where soils are humid and prone to inundation. Finally, when termites have aggressive nest defense (e.g., in the Nasutitermitinae), the associated bees may benefit from the aggressive behavior of the termite workers. The termites may also benefit from the presence of stinging or biting bees, so this is likely a mutualistic relationship.

This rare, but remarkable, mutualism between solitary bees and arboreal termites indicates how evolutionarily flexible bee-nesting habitats can be. When soil-like materials are available, even when the soil is up in a tree, certain bee species have evolved the ability to exploit these materials for nesting.

Most solitary bees have a marked preference for nesting where their mother nested, a phenomenon referred to as philopatry. Philopatry may be an adaptation to insuring that female bees find the right soil texture for nesting, but long-term use of the same site can lead to an accumulation of parasites over time. In some communal species, females reuse their mother's nest over many generations, which might be viewed as an extreme form of philopatry.

It is difficult to conclusively document philopatry without careful, long-term studies of marked females. Yanega (1988) conducted a remarkably detailed five-year study of dispersal dynamics in *Halictus rubicundus* (Halictidae) in the front yard of his family's home in Queens, New York. By following over 220 individually marked females, Yanega found that most initiated nests within 0.5 meter of their natal nest. A mere four females during the five-year study dispersed to a second aggregation just 3 meters away. *H. rubicundus* is a socially polymorphic species, but this surprising level of local philopatry likely holds for many solitary bees.

Nest excavation and structure

Bees that nest in soil are masters of excavation. Morphological adaptations to soil nesting include well-developed basitibial plates on the hind tibia that allow bees to move rapidly within the nest tunnels and a well-developed pygidial plate on the female T6 that is used to pack soil along the length of the burrow and within the brood cells. Think of the pygidial plate as a trowel used to tamp the soil as bees excavate their tunnels and brood cells. The triangular impressions of the pygidial plate can sometimes be seen in recently completed brood cells of soil-nesting bees. Many wood- and cavity-nesting bees lack the specialized pygidial and basitibial plates of ground nesters, indicating that these morphological traits are intimately associated with movement within the nest and the manipulation of soil. Even some ground-nesting bees that have recently evolved from wood-nesting ancestors (e.g., *Proxylocopa* or *Cnemidorhiza*, Apidae) have well-developed pygidial and basitibial plates (Minckley 1994).

Ground-nesting bees use a variety of methods for excavating their nests. If the soil is compact, they use their mandibles to dig into the soil surface (Torchio and Youssef 1968, Torchio et al. 1988). If the soil is loose, they use their mandibles and legs in combination (Stage 1966). Breaching the soil surface can be a major obstacle for soil-nesting bees. Depending on conditions, nest initiation can sometimes take more time than excavating the remainder of the nest. This may explain why some soil nesters (e.g., *Dieunomia triangulifera*) initiate nesting after a heavy rain (Minckley et al. 1994) or when the soil is experimentally moistened by helpful entomologists (Wuellner 1999). Some bees that nest in compressed, hard-packed soils, such as *Diadasia*, use regurgitated nectar to soften the soil surface. Others gather water to soften hard-packed soils. Female *Ptilothrix* have an amazing ability to alight on the surface of ponds, cattle tanks, and puddles as they load up on water for new nest construction (Rust 1980).

Once a female has broken through the soil surface, she then faces the problem of excavating the main tunnel and disposing of the excavated soil. Excess soil can be shuttled to the nest entrance, packed into lateral tunnels after cells are constructed, or used to reinforce the tunnel walls. Females dig with their head toward the bottom of the nest and remove soil by raking it as they back up, using the forelegs and the head (Neff and Simpson 1991, Wuellner 1999). Alternately, a combination of legs and the metasoma can be used to move soil, either as a packet of soil between the curled metasoma and the hindlegs or by the bee pushing the soil to the nest entrance with the dorsum of the metasoma, like a plow or bulldozer (Hurd and Powell 1958).

Bees use a variety of methods to dispose of excess soil at the surface. The majority of bees simply pile the loose soil at the nest entrance to form a mound or tumulus. The first sign of an active nesting aggregation of ground-nesting bees is often the appearance of volcano-like tumuli made of freshly excavated soil. These tumuli can be quite conspicuous when deeper soils differ in color or texture from surface soils. Depending on the species and the orientation of the nest on the ground, the nest entrance can be centered in the tumulus or positioned to one side (Stephen et al. 1969).

Some bees dispose of excavated soil by spraying it widely from the nest entrance with vigorous kicking motions of the legs. Bees in the Fideliinae, a basal branch of the Megachilidae, and the genus *Hesperapis* (Melittidae) have modified hindlegs that allow them to spray soil a considerable distance from the nest entrance (Rozen 1970, 1973; Stage 1966). Such behavior is widespread among ground-nesting crabronid wasps (O'Neill 2001), suggesting that soil-kicking behaviors in the basal Megachilidae and Melittidae may be ancestral behaviors held over from their wasp relatives (Litman et al. 2011).

Another approach to soil disposal is to carry the soil away from the nest on the wing. Female *Megachile frontalis* (Megachilidae) in New Guinea carry pellets of mud in their mandibles a short distance from the nest (150–300 mm) and deposit them seemingly at random (Michener and Szent Ivany 1960). Such behavior presumably makes the nest entrance less conspicuous to parasites and predators (see Chapter 11).

Various ground-nesting bees form tubes of cemented soil or turrets (Fig. 6-3) at the nest entrance. Conspicuous freestanding turrets are built by most emphorines and some anthophorines (both Apidae) and nomiines (Halictidae), but hidden turrets of more loosely consolidated soil are found within the tumuli of many other bees, particularly those nesting in sandy soils (Wcislo and Engel 1996, Neff and Simpson 1997). For reasons that are not entirely clear, some bees

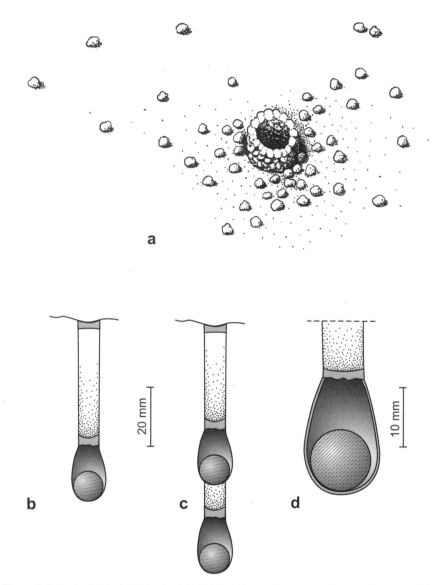

Figure 6-3. Nest of *Ptilothrix* (Emphorini: Apidae), illustrating very shallow, single- or double-celled nests: (a) nest entrance of *Ptilothrix plumata*, showing turret structure (redrawn from Martins et al. 1996, Fig. 1b); (b) single-celled nests of *P. bombiformis* (redrawn from Rust 1980, Fig. 2); (c) double-celled nest of *P. bombiformis* (redrawn from Rust 1980, Fig. 3); (d) brood cell, provision mass, and egg of *P. bombiformis* (redrawn from Rust 1980, Fig. 4). Note that the egg is placed below the pollen/nectar provisions in *Ptilothrix*.

even "decorate" their turret with brightly colored pollen grains (e.g., *Diadasia afflicta*; Neff et al. 1982).

Turrets may have many different functions in solitary bees. Turrets prevent sand and soil particles from falling into the burrow entrance (North and Lilly-white 1980), and they may provide a barrier to flooding in humid environments. Turrets may also provide an obstacle to parasite attack, especially when parasites are hunting along the soil surface (e.g., Mutillidae). Linsley (1958) pointed out that turrets are often made by species that nest in dense aggregations, suggesting that they play a role in nest recognition. When turrets are impregnated with glandular secretions, they may have a unique odor that allows individual female bees to find their nest among hundreds of other nests in dense aggregations (Rajotte 1979, Shimron et al. 1985). In some solitary bees (*Diadasia afflicta*, Neff et al. 1982; *D. rinconis*, Neff and Simpson 1992; and *Ptilothrix plumata*, Martins et al. 1996), the turret is broken down by females when nests are completed, and the material is reused to construct a solid plug at the nest entrance. Finally, the turret may play a role in thermoregulation. Female *Anthophora abrupta* nest in vertical banks and construct a long (>7 cm) downward-oriented turret that can take as long as six hours to complete. Norden (1984) found that, on cool mornings, the nests with turrets were substantially warmer than nests without turrets and that females in nests with turrets initiated foraging 15–30 minutes earlier than those without turrets.

The below-ground architecture of bee nests is incredibly variable. A typical bee nest consists of a vertical burrow, with lateral tunnels leading to brood cells. Whereas most ground nesters make multicelled nests, some bee species regularly make single-celled nests (e.g., *Perdita maculigera*, Michener and Ordway 1963; some *Ptilothrix*, Rust 1980, Fig. 6–3; *Diadasina distincta*, Martins and Antonini 1994; and *Samba spinosa*, Kuhlmann and Timmermann 2011). Lateral tunnels vary greatly in length. Some bee species build brood cells immediately off the main burrow (e.g., some Halictidae; Sakagami and Michener 1962), whereas others construct elongate, winding lateral tunnels (e.g., *Hesperapis*; Stage 1966) that can be exceedingly difficult for a melittologist to follow when backfilled with soil by the resident female. Cells in ground nesters may be arranged singly, in clusters, or in linear series. There are additional variations on these themes. In bank-nesting *Anthophora abrupta*, the nest is essentially just a linear chain of cells built off a larger chamber near the soil surface (Norden 1984; Fig. 6-4), while in *Diadasia afflicta*, each nest is a short, vertical main burrow ending in a cluster of either single cells or short cell series (Neff et al. 1982). Some of the most complex below-ground nests are constructed by females in the tribe Augochlorini (Halictidae). These bees build clusters of cells suspended by pillars within subterranean chambers (Sakagami and Michener 1962).

WOOD AND PITH EXCAVATORS

A relatively small group of bees excavate their nests in wood (both solid and heavily decaying), pithy stems, and structurally similar materials, such as horse dung (Fig. 6-1). This group includes not only the bees that excavate nests in solid wood, such as most *Xylocopa* species (the large carpenter bees, Apidae),

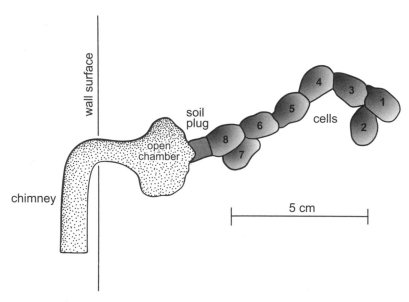

Figure 6-4. Nest of *Anthophora abrupta*, a bank-nesting bee, showing the open chamber from which a more or less linear series of brood cells extend into the adobe wall (redrawn from Norden 1984, Fig. 3a).

but also taxa that excavate in rotten or soft wood, such as many Lithurginae (Megachilidae); members of the colletid subfamily Callomelittinae; some halictid bees, such as *Augochlora pura* and *Megalopta* species; various *Megachile* species, particularly in the subgenera *Megachile* s. str.; and *Xanthosarus* (Stephen 1956, Katayama 2001). This group also includes bees that excavate in pithy stems, like *Ceratina* and *Manuelia* (both Xylocopinae), as well some osmiine megachilids, like *Hoplitis producta*. We also include here oddballs like *Trichothurgus bolithophilus* and *Osmia integra*, bees that excavate their nests in dry dung (Cane 2012, Sarzetti et al. 2012).

The nest tunnels and brood cells of wood- and pith-nesting bees are constructed entirely with the mandibles (Gerling et al. 1981), and accordingly, the mandibles of females in these groups bear specializations for excavating in wood (Michener and Fraser 1978). The mandibles of Xylocopini (Apidae) and Lithurginae (Megachilidae) are stouter than those of ground-nesting bees and bear an enlarged median tooth. Superficially very similar mandibles are found in *Anthophora (Clisodon)* spp. (Apidae; Michener and Fraser 1978) and *Osmia pilicornis* (Megachilidae; Prosi et al. 2016), both wood nesters. Wood-nesting Halictidae (*Augochlora pura*, *Megalopta* spp.) and Colletidae (Callomelittinae) also show mandibular modifications associated with excavating in solid or decaying wood (Michener and Fraser 1978).

Most of the wood nesters treated here make their cell partitions and nest closures with "sawdust" or wood fragments produced during nest excavation. With no wood particles handy, *Trichothurgus bolithophilus* uses dung scrapings for its nest plugs and partitions, while *Osmia integra* brings in a mix of sand and leaf masticate to make its cells (Cane 2012). Pith-nesting *Hoplitis* species have been reported using soil (Davidson 1896) or leaf masticate (Rau 1928) for their nest partitions and, in one case, alternating layers of soil and leaf masticate for the nest plug (Davidson 1896). *O. pilicornis* constructs its nest partitions and plugs with leaf masticate, while the wood-excavating *Megachile* species construct their cells with leaf pieces.

Xylocopinae sensu stricto—a clade of mostly wood- and pith-nesting bees

The subfamily Xylocopinae (Fig. 6-1) includes a diversity of wood-, pith-, and cavity-nesting species. The tribe Xylocopini (the large carpenter bees) includes a single, large genus (*Xylocopa*) that occurs worldwide and nests in a variety of substrates ranging from robust, erect flower scapes to solid wood (Fig. 6-5). As one might expect, these bees have specialized mandibles for excavating wood, and their enormous heads are packed with powerful mandibular muscles. *Xylocopa* species that nest in flower scapes prefer certain plants for their nests. In North America, they use *Agave, Yucca,* or *Dasylirion* (Asparagaceae); in southern Africa, they prefer *Aloe*; and in Australia, they use *Xanthorrhoea* (both in the family Asphodelaceae). The scapes of these plants have a tough exterior with a pithy core, so once the outer layer is breached, the nest can be excavated quickly. A disadvantage of these substrates is that they are relatively short-lived. The narrow scapes restrict most nests to two long branches that extend up and down from the nest entrance. Cells are provisioned one branch at a time, which places similarly aged larvae together in the same region of the nest. This arrangement means that when newly eclosed adults excavate their way out of their cells, they do not damage their much younger siblings (Gerling et al. 1981).

Xylocopa that nest in solid wood are derived from those described above (Minckley 1998; Leys et al. 2000, 2002; Blaimer et al. 2018). Unlike nests in linear stalks, nests in solid wood can have many short branches, with brood cells at the end of each branch, much like a ground-nesting bee. Excavating a nest in solid wood involves a significant investment in time and energy. *Xylocopa* that nest in sound wood benefit from slow decay of the nesting substrate, and nests can persist for many years.

There have been few detailed observations of nest construction by wood-nesting bees. Gerling and colleagues (1981) studied the intra-nest activities of *X. pubescens* using X-ray imaging. They found that the females made their

Figure 6-5. Nest of a wood-nesting bee (*Xylocopa pubescens*), showing an adult female at the nest entrance and three teneral adult offspring in individual brood cells (redrawn from X-ray images in Blom and Velthuis 1988, Fig. 1).

perfectly round tunnels by slowly rotating along their long axis as they chewed, with the sawdust and pith accumulating beneath and behind them. Eventually, any sawdust not needed for construction of the cell partitions was pushed out the nest entrance. In these and many other *Xylocopa* nests, the nest entrance is only slightly wider than the bee, allowing the bees to defend their nests by blocking the entrance with the head or metasoma (Watmough 1983). The broader interior tunnels allow nest mates to pass each other in the nest.

One enigmatic subgenus of *Xylocopa* (*Proxylocopa*) constructs subterranean, soil nests (Gutbier 1914). As the name suggests, *Proxylocopa* was once thought to be a "primitive" *Xylocopa* (Malyshev 1913, cited in Michener 2007), but phylogenetic analyses now indicate the group arose from a wood-nesting ancestor (Leys et al. 2000, 2002; Minckley 1998). *Proxylocopa* occur in Old World deserts, which may account for the shift back to soil nesting (Hurd and Moure 1963). A detailed study of nesting in one species (*X. xinjiangensis*) indicates that females build two kinds of nests: a deep overwintering nest in which females spend the cold winter months, and a much more shallow breeding nest when they are actively foraging and provisioning brood cells (He and Zhu 2018). These bees nest in vertical banks, cliff faces, and eroded mounds of soil.

The remaining tribes of Xylocopinae sensu stricto, Manueliini and Ceratinini (the small carpenter bees; Fig. 2-6), all nest in soft wood or pithy stems. Manueliini consists of a single genus, *Manuelia*, with just three species known from Chile and southern Argentina. *M. postica* and *M. gayatina* are biologically similar to the scape-nesting *Xylocopa* in that they burrow into the sides of unbroken stems to create a linear nest that extends above and below the entrance hole (Claude-Joseph 1926, Flores-Prado et al. 2008). *M. gayi* is unique in that females nest in rotten logs. Multiple, linear nests commonly branch from a common entrance (Claude-Joseph 1926). As in most *Xylocopa*, individual cells in all three species are barrel-shaped and are separated by pith partitions.

Ceratina (Ceratinini), a large worldwide genus, takes a different approach. These small carpenter bees excavate simple, linear nests that begin where a broken stem exposes its pithy interior. Plants preferred by *Ceratina* for nest sites include *Rubus* (raspberries and blackberries, Rosaceae), *Lantana* and *Verbena* (Verbenaceae), *Bambusa* (bamboo, Poaceae), *Xanthorrhoea* (Asphodelaceae), and *Stachys* (Lamiaceae) (Daly 1966, Michener 1962, Okazaki 1992, Rau 1928). The linear series of cells are separated by partitions made of compacted pith (Fig. 6-6). Cell partitions are smooth and convex on their outer face (toward the entrance) but rough and flat on the inner face (Michener 1962). There is considerable flexibility in nest structure. Female *Ceratina* sometimes leave empty cells between provisioned cells, and they occasionally omit partitions between completed provision masses (Sakagami and Laroca 1971). Most reports indicate that the nests are unlined, but *C.* (*Euceratina*) *dallatorreana* lines its brood cells with a thin, transparent, hydrophobic, wax-like material that is produced by glands in the second and third metasomal sterna (Daly 1966; Fig. 6). Nests often have a "turning chamber" (a region of expanded diameter near the nest entrance) that allows females to rotate around within the confines of the nest tunnel (Okazaki 1992).

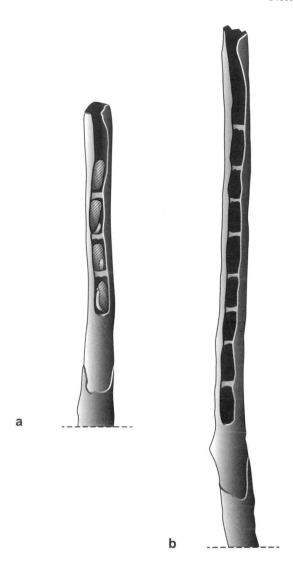

Figure 6-6. Twig nests of *Ceratina dallatorreana*: (a) small, active nest with pollen and nectar provisions (redrawn from Daly 1966, Fig. 3); (b) large, active nest with pollen provisions removed (redrawn from Daly 1966, Fig. 4).

Lithurginae—a clade of wood-nesting megachilid bees

The subfamily Lithurginae represents an early branch of the family Megachilidae that nests in wood (Fig. 6-1; Table 6-1). *Lithurgus* and some species of *Microthurge* and *Lithurgopsis* (all Megachilidae) excavate their nests only in rather soft wood (Brach 1978, Garófalo et al. 1981, Malyshev 1930, Parker and Potter 1973). *Lithurgopsis apicalis* nests in the same species of *Agave* used by some large carpenter bees, but initiates its nests in places where a crack in the stem exposes the soft interior, suggesting that females have difficulty penetrating the tough outer layers (Rozen and Hall 2014). Reuse of the same entrance can lead to the formation of communal nests in this species. Similar communal behavior and nest reuse has been observed in *Microthurge corumbae* (Garófalo et al. 1992).

Some lithurgine bees use seemingly bizarre substrates. *Trichothurgus bolithophilus* (Megachilidae), a bee of the treeless Patagonian plains of southern Argentina, excavates its nests in dried horse manure (*bolitos*) (Sarzetti et al. 2012). Since horses weren't introduced into Patagonia until the 16th century and there have been no native large mammals in that region since the demise of the giant ground sloths 10,000 years ago, it seems likely that grass clumps or similar materials originally served as nest substrates for this bee prior to the arrival of horses.

Other wood-nesting bees

Besides the two main lineages of wood and pith nesters described above, wood and pith nesting has arisen in other bees. The Australian colletid, *Callomelitta picta* (Colletidae: Callomelittinae) was reported to nest in wood by Rayment in a classic publication with the poetic title *A Cluster of Bees* (1935). More recent observations by Terry Houston (2018) have confirmed that female *C. picta* nest in soft, pithy wood. In the Halictidae, a temperate species, *Augochlora pura*, nests in rotten logs (Stockhammer 1966), and members of the tropical genera *Megalopta* and *Xenochlora* excavate nests in rotting lianas or dead branches (Sakagami and Moure 1965, Tierney et al. 2008a). Other wood excavators include *Osmia pilicornis* (Megachildae), a rare osmiine known to excavate nests in solid wood (Prosi et al. 2016). Species of *Clisodon*, a small subgenus of *Anthophora* (Apidae) usually excavate their nests in rotting logs but have also been reported to use pithy stems (Medler 1964), while most, but not all, species of *Alcidamea*, a large Holarctic subgenus of *Hoplitis* (Megachilidae), excavate their nests in pithy stems.

RENTERS—BEES THAT NEST IN PREEXISTING CAVITIES

Renting, or constructing brood cells in preexisting cavities, is a widespread strategy in both solitary bees and wasps (Iwata 1976, O'Neill 2001). Renting is particularly common in the colletid subfamilies Xeromelissinae and Hylaeinae (Almeida 2008), and the megachilid subfamily Megachilinae (Litman et al. 2011). However, renting also occurs in Apidae (e.g., *Ctenoplectra, Tetrapedia,* and some *Centris),* and the colletid subfamilies Euryglossinae and Colletinae (Fig. 6-1). The distinction between renting and nesting in pithy stems is sometimes blurred. We treat Hylaeinae and cavity-nesting Euryglossinae and Xeromelissinae as renters here, but we discuss pith nesters, such as *Ceratina*, with other bees that nest in solid or punky wood (see previous section).

Cavity-nesting bees use a variety of substrates to house their nests. Nest sites include preexisting cavities in the ground, cracks in stone, terrestrial snail shells, earthworm burrows, hollow stems, beetle burrows, chloropid fly galls, and the vacated nests of other bees and wasps. They can also utilize many human materials, including straws (both paper and plastic), fencing materials, hydraulic tubing, and gaps in window frames. Virtually any hollow tube of the right diameter can serve as a domicile for at least some species of cavity-nesting bees.

While nesting in a preexisting cavity might seem like a simpler solution than constructing a nest de novo, cavity nesting presents its own set of challenges.

First, above-ground, cavity-nesting bees experience much greater fluctuations in temperature and humidity than most soil-nesting bees. Above-ground nests are scarcely buffered from the ambient temperatures and humidity in the environment, and their larval or adult overwintering stages must be able to survive much more substantial extremes than ground-nesting bees confront. This may be one reason that cavity-nesting bees do not delay emergence in the face of extreme drought, whereas ground-nesting bees do (Minckley et al. 2013). Second, above-ground, cavity-nesting bees experience a unique set of brood parasites and predators that specialize on the nests of above-ground nesting bees and wasps (see Chapter 11). Above-ground cavity nesters have the advantage of escaping from parasites in three dimensions (Wcislo 1996), but once discovered, the parasites can attack from all sides. Finally, because cavity-nesting bees often build their brood cells in linear series, they are faced with a challenge. Most bees are protandrous, meaning males emerge earlier in the season than females (see Chapter 4). If male and female cells were randomly arranged along the length of the nest, early-emerging males could damage their female nest mates as they vacated the nest. Renters typically put female cells on the back of the nest and male cells in the front in order to avoid this outcome.

A common feature of many cavity-renting bees is that they often have a slender, elongate body form. This body shape is essentially universal in the Hylaeinae and Xeromelissinae (Colletidae) and is found in many cavity-renting Osmiini, Anthidiini, and Megachilini (Megachilidae). An elongate, slender body reduces cross-sectional area, allowing a larger bee to use a smaller-diameter cavity than would be possible with a "normal" bee shape. Most species of cavity-nesting bees either carry pollen internally (Hylaeinae) or on the metasomal venter (Xeromelissinae and Megachilinae). This puts a limit on how much pollen these bees can carry on a single foraging trip but again reduces cross-sectional area.

Unlike ground-nesting bees, there is substantial evidence that finding an appropriate cavity can limit populations of renting bees. Nest usurpation seems particularly common in cavity-nesting bees (Eickwort et al. 1981, Krombein 1967), and observations of fighting among females in the vicinity of suitable nesting substrates suggests that competition for nest sites may be intense (Černá et al. 2013b, Field 1992, Krombein 1967).

Above-ground, cavity-nesting species seem to do particularly well in small, fragmented urban and suburban habitats. Studies in both desert (Cane et al. 2006) and grassland sites (Hinners et al. 2012) found cavity nesters were more common in small, fragmented urban habitats than in the surrounding "natural" habitat. This was not true for ground-nesting bees. The explanation seems to be that nesting substrates provided by humans (homes, sheds, fences, and walls) allow populations of cavity nesters, but not ground nesters, to benefit from human development.

Colletid renters—the cellophane bees

In the family Colletidae, renting is most common in the subfamilies Hylaeinae and Xeromelissinae (Fig. 6-7) but has also been reported in Euryglossinae

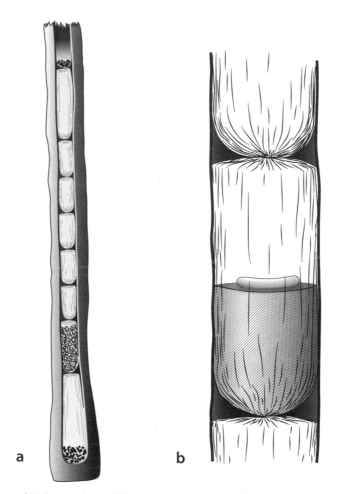

a   b

Figure 6-7. Nest of *Chilicola ashmeadi* (Xeromelissinae): (a) overall architecture of the stem nest and (b) details of the brood cell (redrawn from Eickwort 1967, Figs. 12 and 13). Note the cellophane lining of the brood cell and the semi-liquid provisions with the egg floating on top.

(Almeida 2008, Eickwort 1967; Fig. 6-1). Most renting colletids use hollow stems for their nests, but there are some exceptions. Females of *Hylaeus pectoralis*, a species that ranges from England to Japan, nest exclusively in galls produced by chloropid flies (*Lipara lucens*) on the common reed (*Phragmites australis*) (Bogusch et al. 2015). *H. pectoralis* occurs only in wetland habitats where the host plant and fly galls are present. The narrow substrate requirements of this bee helps explain why it is currently listed as critically endangered in the Czech Republic (Bogusch et al. 2015).

Other *Hylaeus* have similarly specialized nesting requirements. *H. variegatus* uses the empty burrows of earthworms and ground-nesting halictine bees (Malyshev 1935), and *H. tricolor* (and other members of the subgenus *Hylaeopsis*) use the vacated mud nests of *Trypoxylon* wasps (Sphecidae; Hook

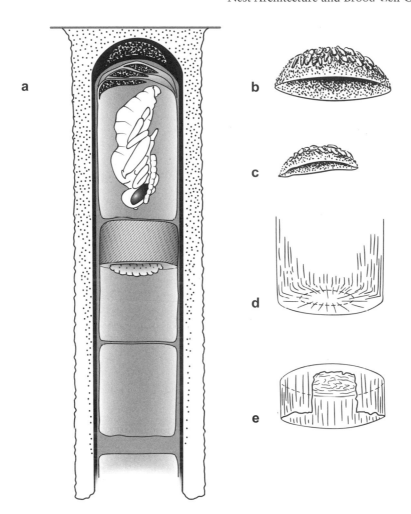

Figure 6 8. Highly specialized nest of *Hylaeus* (*Hylaeopsis*) *tricolor* located in the vacated brood cells of *Trypoxylon* wasps: (a) completed nest with two brood cells and a vestibular cell; the topmost brood cell shows a fully developed pupa, the second brood cell shows a fully provisioned cell with a small, developing larva, and the bottommost cell is a vestibular cell; (b) remains of wasp meconium; (c) meconium of pupated *Hylaeus*; (d) cellophane brood-cell lining of *Hylaeus*; (e) cellophane cell cap with closed ring (redrawn from Sakagami and Zucchi 1978, Fig. 7).

et al. 2010, Sakagami and Zucchi 1978). These bees produce a short, linear series of brood cells that face *downward* from the mud dauber nest (Fig. 6-8). Why the pollen/nectar provisions do not simply drain out of the brood cells is a mystery.[3]

---

[3] Actually, it is probably just a combination of surface tension and viscosity that keeps the provisions from running out of the brood cell. *Hylaeus tricolor* is one of the very few bees that make downward-facing brood cells.

Megachilid renters—leaf-cutters and relatives

By far, the largest and most diverse group of renting bees is found in the megachilid subfamily Megachilinae (Fig. 6-1). This subfamily dwarfs all other megachilid subfamilies, with a total of over 4,000 described species (Chapter 2). By comparison, the three basal megachilid subfamilies include fewer than 90 species. One reason these bees have undergone such impressive diversification may be their widespread use of plant products for brood-cell construction. Plant-derived materials provide both a waterproof barrier that resists desiccation as well as an effective barrier against microbes. Based on work by Jessica Litman and colleagues, the use of plant-derived products has allowed the subfamily Megachilinae to "escape from the desert" and diversify into what is now a cosmopolitan and species-rich subfamily (Litman et al. 2011).

Many renting megachilines are generalized in their choice of a nest site. Female *Megachile brevis* were observed making nests in plants (living and dead), holes in a garage door, under cow dung or mats of prairie grass, and in preexisting cavities in the ground (Michener 1953b). *Osmia bicornis*, another cavity-nesting generalist, has been found nesting inside locks, cracks in window frames, old *Anthophora* nests, and even a fife left in a garden shed (Raw 1972).

However, there are also much more highly specialized cavity nesters. Nesting exclusively in terrestrial snail shells has evolved repeatedly in the Megachilidae (Table 6-1). Many Osmiini nest exclusively in terrestrial snail shells (Table 6-1). In genera such as *Osmia*, *Protosmia*, and *Wainia*, snail-shell nesting is widespread (Kuhlmann et al. 2011, Müller 2018). Snail-shell nesting is less common in the genus *Hoplitis*, but reports of snail-shell nesting in three separate subgenera suggest it has arisen repeatedly in this genus (Müller and Mauss 2016, Sedivy et al. 2013a). Two genera of Anthidiini (*Afranthidium* and *Rhodanthidium*) are also reported to nest in snail shells (Table 6-1). Some snail-shell nesters do little to hide the shells, but others bury them or conceal them with layers of plant fragments (Müller et al. 1997).

Some renting megachilines reuse vacated nests of ground-nesting bees, and although most are opportunists, some are much pickier. Female *Megachile assumptionis* nest exclusively in the shallow, vacated, one-celled nests of a single species: *Ptilothrix plumata* (Apidae; Fig. 6-9). Brood-cell construction is a complex process involving many steps. Female *M. assumptionis* line the empty *Ptilothrix* brood cell with a thin layer of leaf masticate, provision the cell, oviposit, and then close the cell with a plug of sand, leaf masticate, and a glandular material. After that, the vertical burrow is filled with cut leaf pieces, and a final closure is made with small pebbles and mud; the mud is made with collected water added to soil gathered from near the nest (Martins and Almeida 1994). *M. assumptionis* nesting activity peaks during the period when vacated *Ptilothrix* nests are most abundant.

The materials used in brood-cell construction in Megachilinae are remarkably variable. Mud, pebbles, sand, wood chips, leaves, flower petals, plant trichomes, and resins are all used in nest construction by renting bees, and some renters have even adapted to using human materials. In Toronto, Canada, trap-nest studies uncovered cut plastic as a substitute for leaves in the nests of the

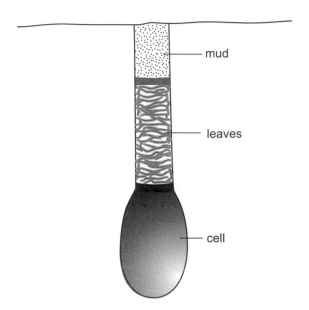

— mud

— leaves

— cell

Figure 6-9. Nest of *Megachile assumptionis*, a specialist cavity-nesting leaf-cutter bee that builds its nests only in the vacated nests of *Ptilothrix plumata* (redrawn from Martins and Almeida 1994, Fig. 3).

leaf-cutter bee *Megachile rotundata* and a polyurethane-based exterior building sealant in place of plant resins in nests of the resin bee *M. campanulae* (MacIvor and Moore 2013). In Dubai, Gess and Roosenschoon (2017) found *M. patellimana*, which normally constructs its nests from pieces of grass leaves, collecting narrow, grasslike pieces of plastic.

The tools used by megachilines to harvest and transport materials for nest construction are highly varied. Bees that collect mud or sticky resins often have concavities on the clypeus or labrum (or both) for carrying balls of nest-making materials back to the nest. Bees that cut leaves have highly modified mandibles, with sharp edges for cutting the tough leaf material. Bees that specialize on cutting more tender tissues (like flower petals) and bees that use leaf masticate typically lack these specialized cutting mandibles. Bees that collect leaf trichomes (the wool-carder bees, the "*Anthidium* group" of Litman et al. 2016; Table 6-1) have finely serrated, comb-like mandibular teeth for scraping and transporting leaf trichomes.

Construction of a single megachiline brood cell is a time-consuming process. The complex cell partitions of *Megachile pugnata* each require more than 30 collecting trips to complete—and that is before any brood-cell provisioning can take place. Females make 15 trips for leaf masticate, 15 trips for soil, and a final 3 or 4 trips for oval leaf pieces to complete a single cell partition (Frohlich and Parker 1983). *M. brevis* (a leaf-cutter bee) typically requires 28 individual leaf pieces to construct a cell, and each leaf piece is carefully chosen based on texture, size, and pliability (Michener 1953b). Female *M. brevis* use stiff leaves (from *Rosa*) for the outer layers of the brood cell and then use more pliable leaves to construct the inner layers. Leaves for the brood-cell walls were also larger than leaves placed on the ends of the brood cell. Thin leaves and flower petals are used to fill gaps and extra space.

Apid renters

A number of apid "oil bees" are renters. In *Tetrapedia*, an above-ground, renting "oil bee," the brood-cell partitions are constructed of sand mixed with floral oils (Alves-dos-Santos et al. 2002, Camillo 2005). The brood cell is coated with floral oils and then provisioned with pollen mixed with floral oils. Each brood cell can take up to 50 hours to construct, provision, and close (Camillo 2005). *Ctenoplectra*, another cavity-nesting "oil bee," constructs brood cells within pre-existing galleries in wood or in the mortar nests of other bees (*Megachile* [*Chalicodoma*]; Rozen 1978, Sung et al. 2009). In *Centris*, another predominantly oil-collecting group, most females excavate their own nests, but some (members of the subgenera *Hemisiella*, *Heterocentris*, and *Xanthemisia*) are renters. They are odd for renters in that they essentially build a ground nest in a cavity in wood by bringing in filler (wood chips) and soil to construct a cell that, unlike that in most other cavity nesters, does not conform to the shape of the cavity (Aguiar and Garófalo 2004, Jesus and Garófalo 2000). Cavity-nesting "oil bees" (such as *Centris*, *Tetrapedia*, and *Ctenoplectra*; Apidae) transport sand and floral oils in the highly modified scopal hairs of their hindlegs (see Chapter 7).

ABOVE-GROUND BUILDERS—BEES THAT BUILD FREESTANDING NESTS

A relatively small proportion of solitary bee species build freestanding, above-ground nests. These bees are restricted to the apid tribe Euglossini and diverse members of the megachilid subfamily Megachilinae (Fig. 6-1; Table 6-1). The materials used to build freestanding nests are varied and include mud, resins and resins mixed with other materials, and plant trichomes sometimes mixed with feathers and mammal fur (Fig. 6-2).

The "true" mason bees

A number of bees in the subfamily Megachilinae (*Megachile* subgenus *Chalicodoma* and the several osmiine genera; Table 6-1) build above-ground nests from mud and pebbles mixed with glandular secretions. These are the "true" mason bees described by Jean-Henri Fabre (Fabre 1914) as part of his famous collection of essays entitled *Souvenirs Entomologiques*. Species of *Megachile* (*Chalicodoma*) make their nests on exposed surfaces, often vertical walls, with pebbles cemented together with a mortar of soil mixed with "saliva," although sometimes they are just mortar without the pebbles. Studies on the mason bee *M. (C.) sicula* have shown that the "saliva" is a mix of long-chain hydrocarbons, mainly hentriacontene and tritriacontene, produced by the labial glands (Kronenberg and Hefetz 1984a). The glandular mix is used not only to form the mortar but also is spread over the nest exterior to render it hydrophobic. Similar nests are built by bees of *Hoplitis* (*Hoplitis*), a group also called mason bees (Eickwort 1975). While some mason bees, like *M. parietina*, have only one female per nest, others, like *M. pyrenaica* and *H. anthocopoides* (Fig. 6-10), are communal,

Figure 6-10. Partially completed above-ground mortar nest of *Hoplitis anthocopoides* (Megachilidae) (redrawn from Eickwort 1975, Fig. 7). A thick mud layer will be added to complete the nest.

with up to six females per nest (Eickwort 1975, Fabre 1914). In the communal species, each female builds her own brood cells, but multiple females cooperate to create the outer hydrophobic mortar layer (Eickwort 1975, Fabre 1914). Interestingly, not all is peaceful cooperation in these communal species, as intraspecific brood parasitism is very high. When nesting gregariously, over 50% of the brood cells may be opened by females other than the original mother. The intruder removes the original egg, replaces it with one of her own, and recloses the cell (Eickwort 1975). In situations like this, the benefits of communal nesting may easily be outweighed by the downsides.

The resin bees

The "resin bees" in the tribe Anthidiini (including the "*Anthodioctes*" and "*Dianthidium*" groups of Litman et al. 2016) include genera that make above-ground nests constructed exclusively of resins or resin combined with stones and other materials (Table 6-1). Freestanding nests can be attached to the undersides of leaves and on rocks, trees, plant stems, and other substrates. *Dianthidium* species typically build freestanding clusters of resin cells that are covered with an outer layer of small pebbles or shell fragments so that individual cells are not recognizable (Clement 1976, Frohlich and Parker 1985). In *Anthidiellum*, members of the subgenera *Anthidiellum* s. str. and *Loyolanthidium* construct odd single-celled, pot-like resin nests with a prominent beak or spout (Grigarick and Stange 1968). The nests are attached to a variety of exposed surfaces, usually singly, but sometimes in series (Schwarz 1928). The function of the spout is unclear, but it may serve as an air-exchange mechanism in what is otherwise a seemingly impermeable nest. Similar, but even odder resin cells are constructed by *Plesianthidium volkmanni*. In this species, females construct clusters of resin cells with very long spouts or necks inside large cavities (Gess and Gess 2014). *Pachyanthidium bicolor* builds clusters of individual short, spouted cells from a mix of "resin" and plants hairs that it attaches to leaves, although in this case, the "resin" may actually be *Euphorbia* latex (Michener 1968). Freestanding, pebble-crusted resin cells similar to those of *Dianthidium* are built by *Othinosmia globicola*, an African heriadine megachilid (Michener 1968, Kuhlmann and Timmermann 2009).

The nest and brood-cell construction of one resin bee, *Dianthidium ulkei* (Megachilidae: Anthidiini), was studied in detail by Frohlich and Parker (1985), who used plexiglass observation nests in a greenhouse. They provided the nesting females with three species of Asteraceae for pollen and nectar, and pieces of pine bark for resin. Female *D. ulkei* build nests in a variety of settings, but they can be induced to nest in hollow tubes or plant stems, making it possible to observe the details of brood-cell construction. Female bees gathered and transported the resin back to the nest, held between the forelegs and mandibles. Within the nest, females mixed abdominal secretions into the resin. These secretions rendered the resin more liquid and therefore more malleable. Softened resins were used to line the nest tunnel, construct brood-cell partitions, and close the nest entrance. Females incorporate additional materials into the nest closure, including pebbles, dirt, vermiculite, paper, small sticks, pith, and bark (Frohlich and Parker 1985). A single-celled freestanding nest observed by Hicks (1933) was estimated to have required 1,000 trips to complete since it contained 880 pebbles and 40 small twigs. In contrast, a *D. ulkei* nest constructed in an old *Anthophora bomboides* nest consisted almost entirely of resin and would have required many fewer trips (Hicks 1933).

Resins are an ideal material for the construction of a freestanding nest. They are soft and pliable when first collected but can harden to a solid, water-repellent, highly resistant material. Resins also exhibit antimicrobial properties that may prevent pollen and nectar provisions from spoiling (Messer 1985). The role of glandular secretions in either hardening resins or making them more pliable is not well understood.

The wool-carder bees

The wool-carder bees in the tribe Anthidiini (the "*Anthidium*" group of Litman et al. 2016) include genera that construct remarkable, freestanding, above-ground nests composed of plant and animal fibers (Table 6-1). *Afranthidium* (Michener 1968) and *Serapista* (Gess and Gess 2014) both construct above-ground, "wool" nests they attach to plants (Fig. 6-11). *Serapista* nests are highly variable in shape, even within a species (Gess and Gess 2014). A single nest entrance on the upper surface of the woolen mass provided access to the brood cells within (Fig. 6-11). Female *S. rufipes* also build an above-ground nest similar in size and shape to *S. denticulata*, but these bees will also nest in preexisting cavities in the ground. The below-ground nests consist of the same woolen material, composed primarily of plant fibers, but the wool is used to line the nest. The "wool" of the wool-carder bees usually refers to the plant trichomes collected by the bees. However, some *Serapista* species incorporate bits of mammal fur into their nests, so they are true "wool-carders."

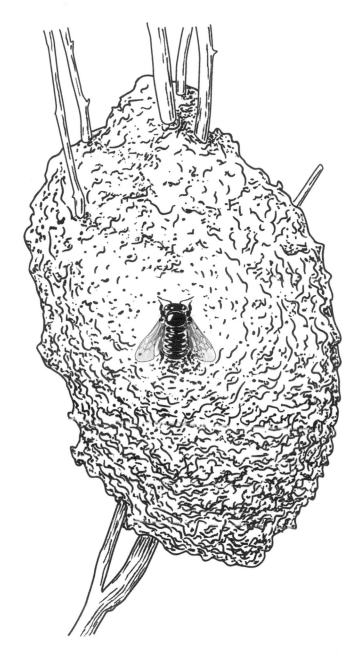

Figure 6-11. Above-ground nest of *Serapista rufipes* (Megachilidae) with a single female resting on the exterior of the nest. The nest is constructed from plant fibers. *S. rufipes* is unusual in that it can construct nests both above-ground (as in this figure) and below-ground, in preexisting cavities (drawing based on photograph [Fig. II.4.23a] in Gess and Gess 2014).

Aspidosmia—an ancient, enigmatic group of above-ground builders

The southern African genus *Aspidosmia* was recently placed in a separate tribe (Aspidosmiini; Table 6-1) within the Megachilinae (Gonzalez et al. 2012). This enigmatic group combines morphological features shared with both the Osmiini and Anthidiini. It appears to be most closely related to extinct genera known only from Baltic amber fossil deposits (*Glaesosmia*, *Glyptapis*, and *Ctenoplectrella*). *Aspidosmia* thus represents a unique, relictual, perhaps ancient lineage of bees building freestanding nests. *Aspidosmia* nests are multicelled structures constructed of small pebbles embedded in a matrix of macerated plant material and are built on the underside of loose stones (Brauns 1926).

Orchid bees (Euglossini: Apidae)

Among the most striking freestanding nests made by any solitary bee are those of certain orchid bees (Euglossini). Most euglossines construct their nests in preexisting cavities, but some construct freestanding nests on the undersides of leaves or attached to plant stems. Some, like those of *Euglossa cybelia*, are relatively simple clusters of cells constructed with resin, bark, and plant fibers (Gonzalez et al. 2011). More typically, the cell cluster is enclosed by a resin involucrum. These may be dome-like, as in *E. championi* or *E. dressleri* (Roubik and Hanson 2004), or nearly spherical, as in *E. dodsoni* (Riveros et al. 2009). The nests of *E. hyacinthina* and *E. turbinifex* are particularly striking. Here, the involucrum is a ridged, top-like structure (Young 1985, Wcislo et al. 2012). As the tip of the top points down, this shape may function like a leaf drip-tip, a design facilitating the shedding of moisture (Roubik and Hanson 2004). The resins used for nest construction are gathered from wounds on plants in the families Burseraceae, Fabaceae, and Clusiaceae, as well as the resin flowers of *Dalechampia* (Euphorbiaceae) and *Clusia* (Clusiaceae) (Armbruster 1984, Roubik and Hanson 2004). In some cases, the bees intentionally inflict wounds to the plant to induce resin production (Schwarz 1948, Johnson 1983).

THE BROOD CELL

A feature shared by all bees that build nests (i.e., bees that are not brood parasites) is the construction of a brood cell. The brood cell houses the pollen, nectar, or floral oils collected by adult females and serves as a home for the developing and overwintering larvae or adults. Virtually all bees line their brood cells in some way to provide a waterproof, microbe-resistant barrier against the harsh and variable external environment (Fig. 6-12). Brood-cell linings can serve as barriers to desiccation in arid regions, as barriers to flooding in seasonally flooded environments, and as hermetically sealed refuges from the various microbes, parasites, and predators that attack the brood cells of bees. We describe below the main strategies for brood-cell preparation in solitary bees.

Brood cells of most wood-excavating or cavity-renting solitary bees are cylindrical, while those of most ground-excavating species are ovoid or teardrop-

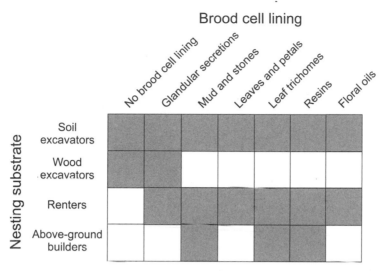

Figure 6-12. Bees construct their nests in a variety of substrates and line their brood cells with a variety of materials. This figure illustrates the various combinations of nesting substrate and brood-cell lining that are known to exist among extant bee lineages. Note that bees that nest in pithy stems and wood generally do not line their brood cells, whereas those that nest in soils and preexisting cavities do.

shaped chambers (Malyshev 1935). Brood cells can be oriented horizontally, vertically, or at virtually any angle with respect to the horizontal. Horizontal brood cells are typical for ground-nesting bees, but vertical brood cells are constructed in a number of lineages.[4] Many groups that construct vertical cells also have fairly liquid provisions, but there are exceptions to this rule. Emphorini (Apidae) make vertical brood cells but have relatively dry provisions (Neff et al. 1982, Neff and Simpson 1992). Brood cells in soil-excavating bees are not just excavated cavities in the soil. Most soil excavators prepare the brood cell by packing fine soils or sand into the walls of the brood-cell chamber, and some bees even bring soils in from outside the nest (Michener 1964). Female *Pararhophites orobinus* (Megachilidae) construct a discrete "receptacle" of soil mixed with nectar that holds the provision mass at the back of the brood cell (McGinley and Rozen 1987; Fig. 8-5d). Brood-cell shape can also vary within species, especially in wood and pith nesters. *Ceratina dentipes* construct barrel-shaped cells when nesting in stems with soft and abundant pith, but they build more cylindrical cells when pith is scarce (Okazaki 1992).

Most solitary bees, regardless of where they build their nests, line their brood cells in some way (Fig. 6-12; Table 3-1). Soil-excavating bees generally line their brood cells with glandular secretions, but they can also use plant-based materials (including leaves, flower petals, and leaf trichomes), and a highly specialized group of bees use floral oils (Fig. 6-12; see Chapter 7 for a more detailed treatment of the "oil bees"). Wood excavators either do not line their brood cells or line them with glandular secretions derived from the Dufour's gland, the

[4] Groups that make vertical brood cells include Eucerini, Emphorini, Anthophorini, soil-nesting Centridini, Oxaeinae, Diphaglossinae, and Stenotritidae.

mandibular gland, or glands located in the metasoma. Most cavity renters line their brood cells with either glandular secretions or plant materials (Fig. 6-12). Brood-cell linings are widespread in bees (Table 3-1), but rare in their closely related crabronid wasp relatives. The widespread use of cell linings among bees and the rarity of such structures in hunting wasps that provision their young with immobilized, but living, insects suggests that lining of the brood cell is an adaptation to storing the relatively liquid pollen/nectar provisions of bees (Michener 1964).

Brood cells lined with glandular secretions

Most ground-nesting bees smooth the cell wall and then line it with secretions from exocrine glands that may or may not be combined with collected materials such as resins, mud, or fine soils brought from other parts of the nest.

The primary source of secretions for the cell lining of solitary bees in the families Andrenidae, Halictidae, Colletidae, and Apidae is the Dufour's gland (Fig. 6-13). This gland is located near the poison sac or venom gland in the metasoma of most stinging Hymenoptera (Aculeata) and is likely derived from the colleterial gland in other insects (Abdalla and Cruz-Landim 2001, Mitra 2013). The histology and gross morphology of the Dufour's gland has been described across a wide variety of bees by Lello (Lello 1971a–d, 1976). In general, the gland is composed of a single layer of cells opening into a central reservoir. At the tip of the metasoma, the reservoir narrows to a duct connecting to the sting chamber (Hefetz 1987).

The Dufour's gland varies widely in size and shape (Fig. 6-14). In Colletidae (Fig. 6-14a), Andrenidae (Fig. 6-14b), and Halictidae (Fig. 6-14c), the Dufour's gland is a large, sausage-shaped gland located within the metasoma, typically positioned above the reproductive organs and digestive tract. In Halictinae, it

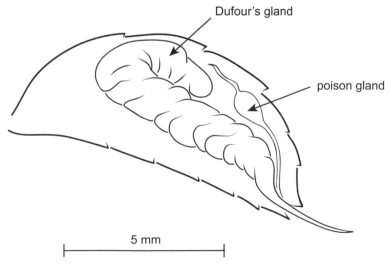

Figure 6-13. Dufour's gland in *Colletes* (Colletidae) (redrawn from Almeida 2008, Fig. 4).

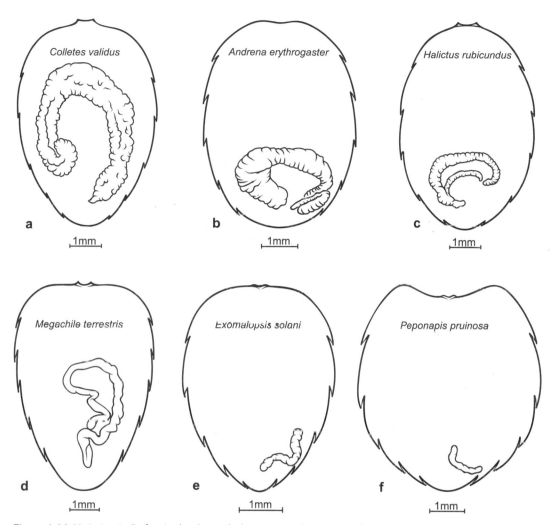

Figure 6-14. Variation in Dufour's gland morphology among bees: (a) *Colletes validus* (Colletidae); (b) *Andrena erythrogaster* (Andrenidae); (c) *Halictus rubicundus* (Halictidae); (d) *Megachile terrestris* (Megachilidae); (e) *Exomalopsis solani* (Apidae); (f) *Peponapis pruinosa* (Apidae) (redrawn from figures in Lello 1971a–d).

can be bi- or trilobed, and in most Apidae, it is a filiform structure (Fig. 6-14e,f). The particularly thick cell lining of many Colletidae (see below) is produced by a Dufour's gland (Fig. 6-13, 6-4a) that can represent 10% of live body weight (Duffield et al. 1984) and occupy as much as 20–50% of the metasomal cavity (Batra 1980). Cleptoparasitic bees, which do not build nests, have greatly reduced Dufour's glands.

The secretions produced in the Dufour's gland consist primarily of large, polar molecules that provide a waxy, hydrophobic coating to the brood cell. The exact composition of the Dufour's gland products varies among groups (Cane 1981, Hefetz 1987). Major Dufour's gland products are macrocyclic lactones in the Colletidae and Halictidae, terpenoid esters and long-chain hydrocarbons in most Andrenidae, aliphatic hydrocarbons (essentially paraffin waxes) in the Calliopsini (Andrenidae; Cane 1981), and triglycerides in some Apidae (*Anthophora*) (see below). Application of the Dufour's gland products to the brood-cell wall typically involves deposition of the glandular products on the bottom of the brood cell and then brushing movements of the paraglossae or glossa to spread the products evenly over the inner walls of the brood cell (Torchio 1984a, Torchio et al. 1988, Danforth 1991a).

The chemistry of the cell linings of Colletidae are particularly complex and well-studied (see review by Almeida 2008). In most colletids, females apply a thick, cellophane-like coating to the brood cell using the brush-like, bifid glossa (Almeida 2008). In some colletids, this cellophane-like coating extends along the full length of the burrow and can even extend outside the nest (e.g., *Meroglossa torrida*; Michener 1960). The colletid brood-cell lining is composed of macrocyclic lactones produced by the Dufour's gland (Hefetz et al. 1979), which are likely polymerized by labial enzymes to form a unique polyester lining (Albans et al. 1980, Torchio et al. 1988). The colletid brood-cell lining is resistant to most organic solvents and may represent a novel biological polymer that Hefetz and colleagues (1979) called a "laminester." Albans and colleagues (1980) showed that it slowed desiccation and appears to have antimicrobial properties. Not all Colletidae make the same large investment in the cell lining. The subfamily Diphaglossinae produces just a thin, fragile cell lining (Rozen 1984b).

Norden and colleagues (1980) described a remarkable case in which the Dufour's gland products serve a dual role—as brood-cell coating and as larval food. Various observers had noted that *Anthophora* species have a thick, white material coating the brood cells that smelled cheesy (Malyshev 1925, Rozen 1969b). Norden and colleagues (1980) studied an aggregation of *A. abrupta* in Maryland that nested in vertical clay banks; they were able to coerce females to nest in glass tubes. The observation tubes allowed them to document adult deposition of the cell lining and larval-feeding behavior within cells. Liquid triglycerides derived from the Dufour's gland were used by females initially to line their brood cells and later were added to the provision mass. Once applied to the burrow walls, the liquid triglycerides convert to solid diglycerides that have a noticeably "cheesy" odor. The thick, white, fatty layer coating the cell surface is consumed by the larva after it finishes the pollen provisions, clearly indicating that the coating serves as larval food. The Dufour's gland in these bees is enormous, occupying approximately half of the volume of the metasoma (Norden et al. 1980), and is also highly modified, with a large, sac-like reservoir and multiple fingerlike projections (Fig. 6-15). Norden and colleagues (1980) and Cane and Carlson (1984) reported triglycerides in the Dufour's gland contents of other *Anthophora* species and species in the closely related genus *Habropoda* (as *Emphoropsis*). A "cheesy odor" was also noted from cells of *H. laboriosa* (Cane 1994a). Consumption of the "waxy" cell lining has also been reported for larval *Amegilla dawsoni* (Houston 1991a). These reports suggest that members of the tribe Anthophorini (*Anthophora*, *Habropoda*, and allied genera, Apidae) use triglycerides as both a cell lining and a nutritional additive to the pollen/nectar provision mass. It is rare in bees for adult females to provide food other than pollen and nectar to their developing offspring. It is even rarer when larval food is derived from glandular secretions of the adult females. The production

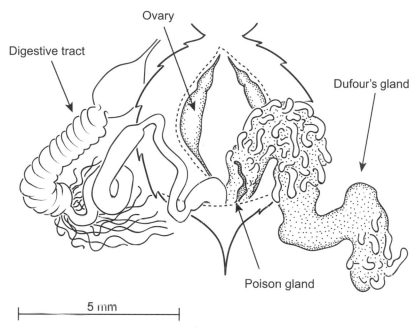

Figure 6-15. Dufour's gland morphology in *Anthophora abrupta* (Apidae). Note the size and elaborated morphology of the Dufour's gland (redrawn from Norden et al. 1980, Fig. 2).

of "royal jelly" by honey bees would be a similar, but phylogenetically unrelated, example. Female anthophorines make a substantial contribution to larval growth and development through the production of the calorie-rich secretions of the Dufour's gland.

In some Megachilidae, Dufour's gland secretions are apparently mixed into the provisions (but not the cell partitions). Williams and colleagues (1986) reported a "goat-like" smell in the provisions of *Megachile integra* and *M. mendica* and traced this odor to Dufour's gland secretions. Whether these secretions serve as larval food or protection of the provision mass was not clear. Functions of Dufour's gland secretions other than for the cell lining in other bees include nest and nest-mate recognition in *Andrena* (Andrenidae; Bergström and Tengö 1974) and *Eucera* (Apidae; Shimron et al. 1985), and as trail-marking pheromones in *Xylocopa* (Apidae; Vinson et al. 1978). Hefetz (1987) mentions that Dufour's gland secretions of *A. vetula*, a species that nests in the total darkness of caves, are essential for nest recognition. Female *A. vetula* "fly into the cave as long as the light enters; they then alight and walk toward their nests while tapping their antennae on the ground" (Hefetz 1987). That Dufour's gland secretions are used for marking nest entrances in gregariously nesting bees means that they must show substantial intra-specific variation. Indeed, this has been shown in one social bee, *Lasioglossum zephyrum* (Smith et al. 1985, Smith and Wenzel 1988).

Some *Xylocopa* (Apidae) line their nests with a thin coating of glandular material, primarily aliphatic hydrocarbons, and have been observed licking the nest walls shortly after nest construction (Gerling et al. 1983). The source of this lining was thought to be the Dufour's gland, but chemical studies have shown that only the ground-nesting *X. olivieri* uses Dufour's gland secretions to line its nests (Kronenberg and Hefetz 1984b). *X. sulcatipes*, and probably other wood-nesting *Xylocopa*, line their nests with compounds from the metasomal yellow glands. These paired glands, located beneath the ventrolateral margins of the sterna, get their name from their bright yellow color when they are active during the nesting season (Gerling et al. 1989).

Likewise, females of *Ceratina (Euceratina) dallatorreana)* have glands located on the second and third metasomal terga that produce waxy secretions that are apparently used to waterproof the interior surfaces of the brood cell (Daly 1966).

The salivary and mandibular glands have also been proposed as sources of secretions used in the cells, but these are far less studied than those of the Dufour's gland. Mandibular glands have been shown to produce compounds that have antimicrobial properties (Cane et al. 1983b).

Brood cells lined with plant materials

Many Megachilidae and some Apidae use material derived from plants to line their cells in place of Dufour's gland secretions (Table 6-1). These plant-derived materials include whole or chewed leaves, resins, trichomes, and floral oils.

Resin is a plant product that is widely used by solitary bees for brood-cell construction (Table 6-1). Resins have some obvious advantages for nest construction: they are waterproof, have antimicrobial properties, and can be used to form structures, which harden with tremendous structural integrity (Armbruster 1984, Lokvam and Braddock 1999, Messer 1985). A diverse group of bees in the Megachilidae and Apidae collect resins from flowers (primarily *Clusia* [Clusiaceae] and *Dalechampia* [Euphorbiaceae]) or a wide array of non-floral sources (Armbruster 1984, Simpson and Neff 1981). One might expect resin to be difficult to manipulate and carry back to the nest. Orchid bees (Euglossini: Apidae) carry resins in their hindleg corbicula, and resin-collecting Megachilidae transport resins between their paired mandibles and elongate labrum. Glandular secretions are likely added to the resins to render them more malleable (Frohlich and Parker 1985; see above). Resins produced from vegetative sources typically dry within a day or so, but floral resins may remain malleable for a week or more. Many bees that collect resins incorporate other materials (including mud and feces) into the nest, but they invariably line the brood cell solely with resin.

Leaves are another plant material that solitary bees use for brood-cell construction. More than one thousand species of *Megachile* and various Osmiini (Megachilidae) use leaves to construct their brood cells (Fig. 6-16). Leaves are used in their entirety, cut into leaf pieces, or chewed into a paste (leaf masticate). Leaf masticate is used only rarely for cell linings but is mixed with sand and other materials to form cell partitions and nest plugs in a variety of cavity-renting Osmiini and Megachilini (Megachilidae). It seems likely that secondary plant compounds present in leaves help to protect provisions and developing offspring from bacterial and fungal attack. However, few studies have examined the chemical ecology of leaf-cutter bees, and we know very little about how leaf choice and chemistry impact microbial growth in the pollen/nectar provisions.

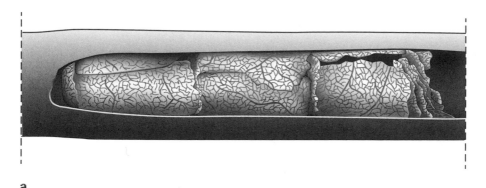

a

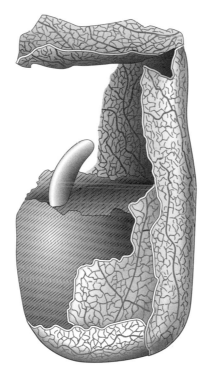

b

Figure 6-16. Nest architecture and brood-cell construction in leaf-cutter bees (*Megachile*: Mega-chilini: Megachilidae): (a) intact nest with three consecutive, leaf-lined brood cells and (b) partially opened brood cell with pollen/nectar provisions and recently laid egg (*Megachile rubi*; redrawn from Eickwort et al. 1981, Fig. 7).

Flower petals are an alternative to leaves for cell construction. Some bees use petal pieces interchangeably with leaf pieces for cell construction (*Megachile* [*Litomegachile*] *brevis*; Michener 1953b), while others use petals exclusively (*Megachile* [*Megachiloides*] spp. [Bohart and Youssef 1972] and *Osmia avosetta* [Rozen et al. 2010a]), or alternate petals and mud (*M.* [*Chrysosarus*] spp., *Osmia*

[*Tergosomia*] spp., *Hoplitis* [*Anthocopa*] spp.; Zillikens and Steiner 2004). *Osmia (Tergosmia)* cells typically have a mud layer amid the petal layers forming the cell wall, while in *Hoplitis (Anthocopa)*, the petal-lined cells are separated by mud partitions (Müller 2018 and references therein). The nests of *Osmia (Tergosmia) avosetta* are particularly striking, since petals are used to line the nest tunnel and tubes of colorful petals often protrude above the burrows' entrances. Rozen and colleagues (2010a) speculated that petals are used by *O. avosetta* because they are pliable (and therefore easily manipulated) and that they increase humidity for the developing larvae in an arid environment. Rozen and colleagues (2010a) provide a review of petal use in other osmiine bees.

A number of groups in the Apidae (Centridini, Tetrapediini, Tapinotaspidini, and Ctenoplectrini) and Melittidae (*Macropis*, *Rediviva*) use floral oils as brood cell linings and closures (Fig. 6-17). Because these oils are also used as larval food, we discuss them in Chapter 7. Many bee groups that have adopted floral oils as a brood-cell lining have a reduced Dufour's gland (Table 3-1; Cane et al. 1983a, Kuhlmann 2014).

A monophyletic group of six genera (*Afranthidium*, *Anthidioma*, *Anthidium*, *Indanthidium*, *Pseudoanthidium*, and *Serapista*) in the tribe Anthidiini (Megachilidae)—the wool-carder bees—construct their brood cells from plant hairs (trichomes) that they collect using highly specialized, comb-like mandibles (Litman et al. 2016; see also Table 6-1). Plant trichomes alone may provide some defense against desiccation or microbial attack, but wool-carder bees also add additional plant products to the wool (Eltz et al. 2015b, Müller et al. 1996). Female *Anthidium manicatum*, for example, gather glandular exudates from *Pelargonium* (Geraniaceae), *Antirrhinum* (Scrophulariaceae), and *Crepis* (Asteraceae) during brood-cell preparation (Müller et al. 1996). Glandular products

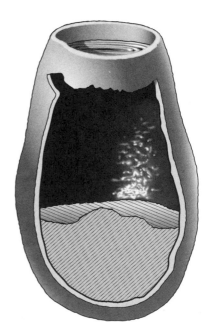

Figure 6-17. Brood-cell and pollen provisions of a ground-nesting bee, *Epicharis zonata*. The shiny brood-cell lining is derived from floral oils (redrawn from Roubik and Michener 1980, Fig. 5).

are scraped off the plant surface using highly specialized basitarsal brushes on the fore-, mid-, and hindlegs of the females. Based on examination of females in observation nests, these plant secretions are visible as gooey droplets on the outside of the wool brood cell. Eltz and colleagues (2015b) demonstrated that these glandular exudates help reduce the incidence of chalcidoid wasp parasitism.

Unlined brood cells

Not all bees line their brood cells. Many wood- and pith-nesting species (e.g., Ceratinini [Xylocopinae: Apidae] and Lithurginae [Megachilidae]) and some cavity-nesting Megachilidae (Osmiini and Megachilini) produce no visible brood-cell lining. Perhaps wood has some antimicrobial properties or is simply dry enough to prevent significant microbial growth on their pollen/nectar provisions. Several ground-nesting bee groups also forgo a brood-cell lining. Females in the subfamily Dasypodainae (Melittidae) produce entirely unlined brood cells (Fig. 6-18; Rozen 1987a, 2016a [*Hesperapis*]; Celary 2002; Yang et al. 2010 [*Dasypoda*]), and brood cells of the subfamily Meganomiinae (Melittidae) appear to be lined with nectar but not with glandular secretions. In this latter group, the nectar consolidates the soil in the cell wall and adds to its strength but does not render it waterproof (Rozen 1977c). Three "primitive" subfamilies of Megachilidae (Fideliinae, Pararhophitinae, and Lithurginae) all make unlined cells (Litman et al. 2011), as does an unrelated apid bee, *Eremapis parvula*, from the deserts of Argentina (Neff 1984, Rozen 2011). Lack of a cell lining might be a trait retained in some basal bee lineages (such as Dasypodainae and basal Megachilidae) from their wasp ancestors, or it could be an adaptation to nesting in dry, arid environments where waterproofing is no longer required. Many

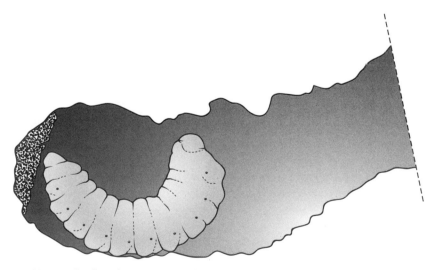

Figure 6-18. Brood-cell and overwintering last-instar larva of *Hesperapis trochanterata* (Dasypodainae: Melittidae), illustrating the unlined brood cell of members of the subfamily Dasypodainae (redrawn from Rozen 1987a, Fig. 4).

species in these groups nest in dry, sandy soils, which supports the idea that a cell lining may not be critical for bees that nest in well-drained soils in which cell provisions have a low probability of fungal infection (Litman et al. 2011).

Finally, species of the North American bee genus *Perdita* (Andrenidae), which occur almost entirely in the desert, make unlined cells but apparently coat the pollen ball with Dufour's gland secretions (Danforth 1989a, Rozen 1967a). Why female *Perdita* do not line their brood cells is unclear, but with more than 650 species, unlined cells can be a successful strategy in arid environments.

SOLITARY BEES AS "PESTS"

The nesting activities of bees can sometimes cause problems for humans. The mason bee *Megachile parietina* normally builds it nests on vertical surfaces, and its mortar nests have been found to severely damage ancient inscriptions on the walls of 2,000-year-old Egyptian temples (Comstock 1909). Ground-nesting bees can also be archaeological pests. *Centris muralis*, a large, gregarious bee that excavates nests in vertical banks, was found to severely damage historic adobe ruins in Argentina when large numbers used them as nest sites (Rolón and Cilla 2012). Carpenter bees can cause significant damage to trim and sometimes structural elements of wooden houses (Barrows 1980). Nests of an *Ashmeadiella* (Megachilidae) were once blamed for an airplane crash. The case made it all the way to trial before the bees were ultimately found to be innocent. It turned out the nests in the plane's fuel lines, the supposed cause of the crash, had been built long after the crash while the wrecked plane was in storage (Eickwort and Rozen 1997). Compared to the many benefits we receive in our interactions with solitary bees, this short list of negative ones only emphasizes the positive nature of our interactions with solitary bees.

BOX 6-4: FACIAL FOVEAE—INTRA-NEST COMMUNICATION IN COMMUNAL BEES?

Female ground-, stem-, and wood-nesting bees spend much of their adult lives in the dark. When in the nest, these bees must navigate by sensory modalities other than vision. Tactile cues are likely important; ground- and stem-nesting bees are likely to navigate within the nest by simply remembering the touch and feel of their burrow walls. Gravitational cues are also likely to be important. Insects can detect gravity, and for a bee nesting in a deep, vertical, below-ground nest, knowing which way is down and which way is up would provide important information as they move around the network of below-ground tunnels. Bees are also able to detect changes in humidity, which could provide additional cues as to where they are located within a subterranean burrow.

Chemical cues may also be important. There is considerable evidence that both ground- and stem-nesting bees mark their nest entrance with individual-specific odors that allow them to recognize their own nest entrance among many nests in an aggregation (Guédot et al. 2006, Hefetz 1992, Rajotte 1979, Raw 1992). Solitary bees may also have a means of chemical communication among individuals within the nest. In the families Andrenidae and Colletidae, females and males of many genera have well-developed facial foveae. These paired, depressed patches of shiny or velvety cuticle are located just mesad of the compound eyes. They come in various shapes and sizes, from narrow, slit-like lines to more broadly expanded, quadrate structures. The shape and texture of facial foveae vary so much among species that they are often used in species-level identification (e.g., in *Andrena*). Facial foveae are generally larger and more clearly developed in females than in males. Much more weakly developed facial foveae can also be seen in pollen-collecting Apidae and some basal Halictidae (Rophitinae).

Schönitzer and Schuberth (1993) and Schuberth and Schönitzer (1993) showed, with a combination of scanning electron microscopy and histological studies, that the facial foveae are connected to secretory glands located beneath the surface of the cuticle. These glands are connected to small pores that open at the lower margin of the individual facial fovea. We suspect that these glands play a key role in intra-nest communication in bees. When two bees encounter each other in the nest (e.g., in communal species), the first thing their antennae are likely to contact will be the face of the other individual. Recognizing whether the individual is a nest mate or not or whether it is a potential brood parasite or nest usurper is extremely important. Communal nesting is common in Andrenidae and, to a lesser extent, Colletidae. Female *Andrena* have some of the largest and most well-developed facial foveae of all bees, and they often form communal nests (Paxton et al. 1996, Paxton and Tengö 1996, Paxton et al. 1999a,b). Panurgine bees, in which the facial foveae are well developed, also form communal associations (Danforth 1989a, Rozen 1967a).

Additional evidence that facial foveae are important in intra-nest interactions comes from observations of macrocephalic males in the genus *Macrotera* (Andrenidae). In large-headed males, the facial foveae are enormously enlarged and protuberant, with a clear layer of glandular tissue underlying the cuticle (Danforth 1991b). Male-male interactions within the nest are important in *M. portalis* (Box 4-3). Males fight to the death within nests, and male-male combat begins with a face-to-face standoff in the darkness of the below-ground burrow. Perhaps the expanded facial foveae in these bees provides a signal to other males: "Watch out, I am a bad-ass, big-headed male." While there is little hard evidence for the function of facial foveae, we suspect that these conspicuous glandular structures are key to some of the unexplored, below-ground interactions that take place among male and female solitary bees.

CHAPTER 7

The Tools of the Trade: Floral Rewards and How Bees Harvest Them

Think of a foraging female solitary bee as a flying Swiss Army knife (Buchmann 1987). Her tools include a multi-functional set of mouthparts for accessing nectar that can be folded neatly beneath her head or, when needed, extended, in some species, to a length greater than her own body. She has some combination of combs, mops, sponges, rakes, and brushes on her six legs for gathering and manipulating floral rewards, such as pollen and floral oils. She also has, in most bees, a well-developed structure (or structures) somewhere on her body for transporting these floral rewards back to the nest. Finally, like all flying insects, she is positively charged, which causes negatively charged pollen to stick to her body surface while foraging near or on her host plant. In short, a female bee is a flying, electrostatically charged tool kit equipped for gathering, manipulating, and transporting a diversity of plant-derived products for nest construction and larval provisioning.

The floral rewards that bees gather vary in energetic value to the bee and energetic cost to the plant that produces them (Table 7-1). Nectar is a relatively cheap, easily synthesized, universal floral reward that attracts a wide array of floral visitors, including birds, bats, and a wide variety of insects. Pollen, because it is rich in protein, is a more costly floral reward. Some pollens have twice the protein content of raw meat and therefore provide a valuable floral reward to any pollinator that can harvest them. Many animals feed on pollen as adults, but bees and masarid wasps are the two main lineages of insects that actively gather pollen as food for larvae. Finally, floral oils, used by some bees as a brood-cell lining and as a source of nutrition for developing brood, are even more energy-rich than pollen but require specialized structures for effective harvesting.

The floral rewards that plants produce are not given away for free, and the more costly a floral reward, the more likely plants are to protect it. As we shall see later in this chapter, plants have evolved a variety of mechanisms for making these floral rewards difficult to harvest. Nectar can be presented at the bottom

Table 7-1. Comparison of the energetic value of various floral rewards.
Data from Simpson, B. B., and Neff, J. L. (1983). Evolution and diversity of floral rewards. In *Handbook of Experimental Pollination Biology*, C.E. Jones, and R. J. Little, eds. (New York: Van Nostrand Reinhold), pp. 142–159 (Table 6-1).

	Reward	Energetic value*
Sugar	Glucose	15.69
	Sucrose	16.53
	Nectar (20% sucrose)	3.31
	Nectar (40% sucrose)	6.61
	Nectar (60% sucrose)	9.92
Pollen	Wind-pollinated monocots	21.76
	Wind-pollinated dicots	24.60
	Animal-pollinated dicots	24.14
	Animal-pollinated dicots	23.84
Oil	Stearic acid	40.17
	Flaiophore lipids**	33.07

* Measured as heat of combustion (kilojoules/g). Data come from various sources, as described in Simpson and Neff (1983).

** Average based on analysis of elaiophore oils from *Aspicarpa*, *Krameria*, *Malpighia*, and *Callaeum* (from Buchmann 1987).

of a long, tubular corolla, pollen can be released in small doses to encourage frequent revisitation, and floral oils can be difficult to extract except by the most well-equipped floral visitors. Because the interests of plants (pollination) and bees (collecting floral resources for brood-cell provisioning) are not the same, the interaction between bees and flowering plants is best viewed as a form of "balanced mutual exploitation" (Westerkamp 1996).

In this chapter, we describe the morphologies and behaviors associated with collecting nectar, pollen, and floral oils—the three primary floral resources used in brood-cell provisioning. Bees collect other rewards, such as resins for nest construction and fragrances for attracting females. How bees gather and transport nesting materials is covered in Chapter 5, and how male orchid bees collect floral fragrances to make their attractive pheromonal signals is covered in Chapter 4. Bees are also known to collect non-floral resources, such as the sugary honeydew secretions from scale insects (Meiners et al. 2017). Such non-floral rewards may comprise an important source of sugar when floral resources are limited.

NECTAR

Eighty-seven percent of approximately 300,000 extant flowering plant species rely on animal pollinators (Ollerton et al. 2011), and for many floral visitors, nectar is the primary reward.

Biological properties of nectar

Flowering plants produce nectar in specialized floral and extra-floral structures called nectaries. In general, floral nectaries function to attract pollinators, whereas extra-floral nectaries function to attract predators and parasitoids that aid in plant defense. This functional distinction between floral and extra-floral nectaries is sometimes blurred. In *Euphorbia* (Euphorbiaceae), the true flowers are simple and nectar-less, but extra-floral nectaries are situated so close to the flowers that they play a role in attracting pollinators.

Nectar composition varies widely among angiosperm species, and some of this variation is likely associated with attracting particular pollinators with varying nectar preferences. The major components of floral nectar include water, carbohydrates, ions, amino acids, and low-molecular-weight proteins (Baker and Baker 1983, Nicolson and Thornburg 2007). Nectar can also be scented to attract and train pollinators (Raguso 2004), and it has recently been shown, in a few cases, to contain stimulants (caffeine) to further aid pollinator memory (Wright et al. 2013). Some nectars may also contain toxic compounds to discourage unwanted floral visitors (Adler 2000).

The carbohydrate portion of floral nectars primarily consists of the disaccharide sucrose and its component monosaccharides, glucose and fructose. Nectar sugar concentration can vary from 7% to 70% (on a weight:weight basis), although most nectar falls within a range of 25–55%. By comparison, Coca Cola is approximately 10% sugar on a weight:weight basis (35 g sugar/366 g). Other monosaccharides, disaccharides, and oligosaccharides may be present in minor amounts. Nectar concentration tends to be relatively constant in flowers with nectar hidden in long tubes, but the concentration can vary radically through the day when nectar is more exposed to evaporation (Corbet et al. 1979, Neff and Simpson 1990). Variation in the mechanisms by which flower visitors lap, lick, or aspirate floral nectars may determine the optimal nectar concentration from the perspective of the pollinator (Borrell 2004, Heyneman 1983).

Nectars also contain essential and non-essential amino acids and proteins, typically in very low concentration, although in some cases these may be high enough to be of nutritional value to certain insects (Baker and Baker 1973, Mevi-Schütz and Erhardt 2005). The protein component of nectar includes enzymes and preservatives. For bees, which collect and ingest pollen, the protein component of nectar is likely of negligible nutritional significance, although it may have a role in nectar taste (Baker and Baker 1973). Minor components of floral nectar include antioxidants (such as ascorbate), which are involved in nectar homeostasis, trace amounts of lipids, terpenoids (Baker and Baker 1973), and secondary compounds such as phenols and alkaloids, which attract some visitors but deter others (Adler 2000).

Floral nectar may also contain an assortment of viruses, bacteria, and fungi. Some bacteria (*Erwinia amylovora*, the causative agent of fire blight) enter plants via the floral nectaries. Increasing evidence suggests that floral visitors, principally bees, may inoculate nectar with fungi, bacteria, and viruses that are then transmitted to subsequent visitors (Graystock et al. 2015). We are just beginning to understand the extent to which flowers may be "hubs" in a network of bee pathogen transmission (Graystock et al. 2016). Intra- and inter-specific pathogen transmission via flowers are potentially a serious threat to pollinator health (Chapter 14).

Bee mouthpart morphology—how bees collect nectar

Bee mouthparts are built on the basic ground-plan elements that all insects possess: mandibles for chewing, maxillae for tasting and manipulating food, and the labium for various functions, including tasting, lapping, licking, sucking, and even capturing prey (Krenn et al. 2005). Bee mandibles are fairly simple structures located on either side of the head, just below the lower margin of the compound eye (Fig. 7-1). In female bees, they are primarily used for chewing either food or nesting materials. Males often use their mandibles for grasping females or sometimes fighting other males. Mandibles are powered by two large, opposing muscles (the mandibular adductor and abductor) that occupy much of

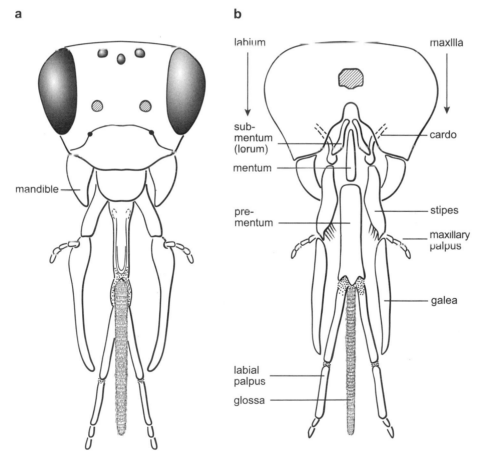

Figure 7-1. Bee mouthparts in (a) frontal view and (b) posterior view. Original drawing based on female *Peponapis pruinosa* (Apidae).

the head capsule of bees. The apex of the mandibles can be simple (untoothed), bidentate (with one large outer and one small inner tooth), multidentate (with many teeth; e.g., *Lithurgus* and some *Megachile*), and even serrated (e.g., the wool-carder bees). Variation in bee mandibular morphology was reviewed by Michener and Fraser (1978).

Unlike the mandibles, the maxilla and labium of bees—in fact, all Hymenoptera—are highly modified relative to that found in other insects. The major modification involves fusion of the paired maxillae and labium to form a "labiomaxillary complex" (Fig. 7-1). The formation of the labiomaxillary complex is a shared, derived feature of all Hymenoptera (including sawflies, ants, wasps, and bees), but it has become especially elaborated, elongated, and modified in the bees. The entire labiomaxillary complex consists of no less than 17 interlocking, movable elements that allow bees to lap and even aspirate nectar from flowers. As we shall see below, it is also involved in pollen harvesting (Müller 1996a). The labiomaxillary complex can be extended, in some bees, to more than the adult body length and then folded neatly beneath the head capsule when not in use.

The maxillary elements of the labiomaxillary complex include two slender rods, the cardines (singular, cardo) that attach to the underside of the head on either side of the proboscidial fossa. At the apex of the cardines are the stipites (singular, stipes) and, more distally, the blade-like galeae (singular, galea). The bee maxilla also possesses a maxillary palpus arising near the apex of the stipes. In most bees, the maxillary palpus has six segments, but it may be greatly reduced (down to just one segment).

The labial elements of the labiomaxillary complex are located between the paired maxillae, at the distal ends of the cardines (Fig. 7-2). The labium in bees does not bear a direct attachment to the head capsule. Instead, it is attached to the distal ends of the maxillary cardines. The labium consists, at its base, of two small, variably shaped sclerites, the mentum and submentum (or lorum), and a much larger, elongate prementum, which bears apically the labial palpi, paraglossae, and the flexible glossa. Most bees have only four (or fewer) labial palpi, but in a few cases, there may be as many as nine (Urban and Graf 2000). The mentum and prementum allow the labium to move outward (distally) away from the cardines through a complex pivoting of the Y-shaped lorum on the base of the cardines. This movement involves rotation of the lorum downward, which swings the mentum and prementum away from the base of the cardines and outward away from the head of the bee. Depending on how long the arms of the lorum are, this movement can give rise to more or less pronounced elongation of the labium while the bee is feeding.

The entire labiomaxillary complex can be folded up under the head (in the proboscidial fossa) and extended downward and forward when feeding (Fig. 7-2). When folded, the labiomaxillary complex can be thought of as forming a Z. The Z shape consists of the cardines (the horizontal upper line of the Z), the stipes and prementum (the diagonal line), and the galea and glossa (the horizontal lower line). When folded, the cardines are pointed backward (relative to the long axis of the bee's body), the stipes and prementum is pointed forward, overlapping the cardines, and the galea and glossa is pointing backward,

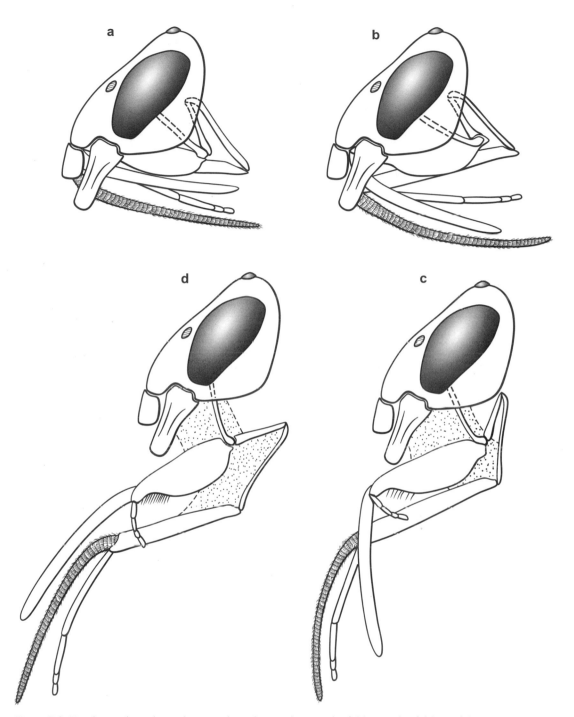

Figure 7-2. Head capsule and mouthparts in lateral view, showing the folding and unfolding of the labiomaxillary complex: (a) mouthparts folded into the proboscidial fossa; (b) mouthparts partially unfolded; (c) mouthparts partially extended; (d) mouthparts fully extended. See Plate 8 for a color version of this figure.

overlapping the stipes and prementum (Fig. 7-2a). This folding and unfolding of the mouthparts involves muscles that are located within the head capsule as well as within the labiomaxillary complex itself.

"Short-tongued" and "long-tongued" bees

The seven families of bees have historically been divided into the "short-tongued" (ST) and "long-tongued" (LT) bees. Unfortunately, these terms are a bit confusing, because the distinction between these two groups is not based on tongue length per se, but on the details of mouthpart structure. Thus there are ST bees with long mouthparts and LT bees with short mouthparts (see below). The ST bee families include Colletidae, Stenotritidae, Halictidae, Andrenidae, and Melittidae, whereas the Apidae and Megachilidae are LT bees. The major feature that distinguishes these two groups is the morphology of the labial palpi (Fig. 7-3a,b). In ST bees (Fig. 7-3a), the four labial palpal segments are morphologically similar and not substantially different from the labial palpi of hunting wasps. However, in the LT bees, the first two segments of the labial palpi are elongated, slender, flattened, and blade-like, whereas the apical two segments are greatly reduced in size (Fig. 7-3b). ST bees exhibit the primitive condition, whereas the LT bee morphology is derived. Hence, the modified labial palpi of LT bees group Apidae and Megachilidae together as a monophyletic group. Recent phylogenetic analyses of bee family-level relationships suggest that ST bees are a paraphyletic group from which the LT bees arose (Fig. 2-2). There are other differences between ST and LT bees in mouthpart morphology. First, in LT bees, the galea and glossa are typically, but not always (Michener and Greenberg 1980), equal to or longer than the stipes and prementum. The galea and glossa of most ST bees are short relative to the prementum. Second, many ST bees possess a comb on the inner surface of the galea (galeal comb) that is important in grooming pollen from the forelegs. In LT bees, this comb is replaced by a comb at the apex of the stipes (stipital comb) that performs the same function (Jander 1976).

The modification of the labial palpi in LT bees is an important adaptation that allows LT bees to form an elongate tube around the glossa when feeding. This tube is formed from the paired galeae (located above the glossa, when the glossa is extended) and the paired labial palpi (located below the glossa, when the glossa is extended) (Fig. 7-2). This tube supports the glossa as nectar is lapped up and withdrawn toward the mouth. There also is a suctorial component to the fluid feeding of bees that is powered by the cibarium, a muscular organ in the preoral cavity of the head (Snodgrass 1985). Most bees use a lapping-sucking method and must load the feeding tube formed by the galea and labial palpi with lapping motions of the glossa in order to ingest fluids (Krenn et al. 2002).

The mouthparts of ST and LT bees work in quite different ways. Harder (1982, 1983) made detailed observations of mouthpart movements in *Andrena carlini* (ST bee; Andrenidae) and *Bombus pensylvanicus* (LT bee; Apidae) while bees were feeding on nectar. He found that when *Andrena* feeds on nectar, the entire labium (prementum and glossa) is repeatedly extended and retracted. This occurs through a rocking motion of the Y-shaped lorum as the bee is lapping

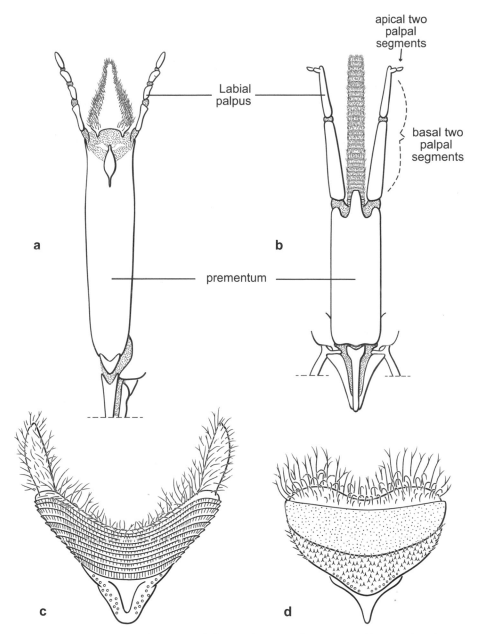

Figure 7-3. Variation in mouthpart morphology in bees: (a) short-tongued bee, *Melitta leporina* (Melittidae) (redrawn from Winston 1979, Fig. 17); (b) generalized long-tongued bee (redrawn from Winston 1979, Fig. 2); (c) bifid glossa of *Cadeguala occidentalis* (Diphaglossinae) (redrawn from McGinley 1980, Fig. 11); (d) bifid glossa of *Scrapter heterodoxus* (Scrapterinae) (redrawn from McGinley 1980, Fig. 9; originally labeled as "*Parapolyglossa paradoxa*," a junior synonym of *Scrapter heterodoxus* [Davies and Brothers 2006]).

up liquid. By contrast, in *Bombus*, once the mouthparts are extended into the nectar, the glossa moves in and out, but the prementum remains fixed in place relative to the apex of the stipes. From these observations, he concluded that "functional tongue length" in ST bees is the length of the prementum and glossa, whereas "functional tongue length" in LT bees is the glossa alone. By measuring nectar-uptake rates of ST and LT bees of varying body size, he found that small ST and LT bees take up nectar at roughly the same rate, but large LT bees can take up far more nectar than large ST bees. This suggests that the LT bee morphology can be more effectively "scaled up" as bee body size increases. Indeed, LT bees are, on average, larger than ST bees.

The bifid glossa of Colletidae

One of the more dramatic and conspicuous modifications of the basic ground-plan morphology described above involves the so-called "bifid" glossa of Colletidae. The colletid glossa, rather than being rounded or acutely pointed apically, as in all other families of bees, is blunt, bilobed, and, in the subfamily Diphaglossinae, deeply forked (Fig. 7-3c,d). This bizarre morphology—present in all female and most male colletid bees—has been historically interpreted as a primitive trait because, superficially, it resembles the glossa of a crabronid or sphecid wasp. However, phylogenetic studies of bee family-level relationships consistently place Colletidae as a highly derived group nested well within the phylogeny of bees (Fig. 2-2), indicating that the bifid glossa arose *from* the pointed glossa in the rest of the bees. McGinley (1980) came to a similar conclusion based on a detailed morphological study of the colletid glossa; this idea can be traced all the way back to Perkins (1912). It now seems that the bifid glossa of Colletidae is a derived morphological trait associated with the application of the cellophane brood-cell lining that is unique to this family (see Almeida 2008 and Chapter 5). The specialized function of the bifid glossa in Colletidae puts considerable evolutionary constraints on the evolution of mouthparts in this family. While elongate mouthparts have evolved in ST bees many times (see below), unusually long mouthparts, when they occur in Colletidae, rarely involve elongation of the glossa (Houston 1983).

"Short-tongued" bees with long tongues

While most short-tongued bees have short mouthparts for lapping up nectar from shallow flowers, elongate mouthparts have repeatedly evolved among these bees, allowing them to access nectar (and sometimes pollen) from longer, tubular flowers. These modifications include elongation of the head, as well as elongation of one or more elements of the labiomaxillary complex (Table 7-2), or both. Diverse members of the families Andrenidae, Halictidae, Colletidae, and Melittidae have evolved—via different routes—the tools necessary to access nectar in flowers with a deep corolla.

Elongate mouthparts have evolved multiple times in two of the three subfamilies of andrenid bees (Andreninae and Panurginae; Table 7-2). At least two

Table 7-2. Short-tongued bees with elongate mouthparts.

Family:Subfamily	Genus/species	Morphology	Host-plant	Reference
Andrenidae: Andreninae	*Andrena (Callandrena) micheneriana*	Elongate galea, glossa, labial palpi (especially segments 1 and 2)	Asteraceae	LaBerge 1978
Andrenidae: Andreninae	*Andrena (Iomelissa) violae*	Elongate glossa, galea, and labial palpi	Violaceae: *Viola*	LaBerge 1986
Andrenidae: Panurginae	*Neffapis longilingua*	Glossa and 3rd labial palp extremely long; galea short; maxillary palpi greatly reduced	Malesherbiaceae: *Malesherbia*	Rozen & Ruz 1995
Andrenidae: Panurginae	*Nolanomelissa toroi*	Greatly elongate glossa; first 3 labial palpi elongate (2nd longest); galea moderately elongate	Solanaceae: *Nolana*	Rozen 2003b
Andrenidae: Panurginae	*Perdita (Glossoperdita) hurdi*	Elongate stipes, prementum, galea, labial palpi, and glossa (glossa much longer than galea and palpi); combined lengths equal to female body length	Martyniaceae: *Proboscidea althaeifolia*	Hurd & Linsley 1963
Andrenidae: Panurginae	*Chaeturginus testaceus*	Elongate glossa; lateral portions of glossal tube formed by rows of flattened setae	Rubiaceae: *Psychotria*	Ruz & Melo 1999
Colletidae: Euryglossinae	*Euryglossa (Euhesma) tubulifera*	Elongate maxillary palpi form tube as long as head and body	Myrtaceae: *Calothamnus* (bird adapted)	Houston 1983
Colletidae: Hyaleinae	*Palaeorhiza (Palaeorhiza) papuana*	Glossa of male as long as head, that of female much shorter	?	Michener 1965
Colletidae: Hylaeinae	*Hylaeus (Pseudhylaeus) spp. (4)*	Maxillary palpi elongate, as long as head	Scrophulariaceae: *Eremophila*	Houston 1983
Colletidae: Neopasiphaeinae	*Leioproctus (Filiglossa) filamentosa*	Apex of galea with 4–12 huge, elongate setae, labial and maxillary palpi elongate, glossa short	Proteaceae: *Persoonia*	Bernhardt & Walker 1996

(continued)

Table 7-2. (*Continued*)

Family:Subfamily	Genus/species	Morphology	Host-plant	Reference
Colletidae: Neopasiphaeinae	*Niltonia virgilii*	Elongate labial palpi	Bignoniaceae: *Jacaranda*	Laroca et al. 1989
Colletidae: Xeromelissinae	*Geodiscelis longiceps*	Elongate head, cardo, stipes, and prementum; maxillary palpi not elongated	Boraginaceae: *Tiquilia*	Packer 2005, 2008
Colletidae: Xeromelissinae	*Xeromelissa rozeni*	Elongate head, stipes, prementum; maxillary palpi extremely long, much longer than glossa	Solanaceae: *Nolana*	Packer 2005
Halictidae: Halictinae	*Ariphanartha palpalis*	Glossa short, but maxillary palpi extraordinarily long, reaching metasoma in repose	Rubiaceae: *Psychotria*	Eickwort 1969, Silva & Vieira 2015
Halictidae: Nomiinae	*Lipotriches testacea*	Elongate glossa	?	Pauly 1984
Halictidae: Rophitinae	*Ceblurgus longipalpus*	Elongate stipes, galea, prementum, glossa, and labial palpi	Boraginaceae: *Cordia leucocephala*	Milet-Pinheiro & Schlindwein 2010
Halictidae: Rophitinae	*Penapis* spp.(*4*)	Elongate glossa, galea, and labial palpi; glossa reaching between forecoxae in repose	Bignoniaceae: *Argylia*	Rozen 1997
Megachilidae: Megachilinae	*Osmia (Orientosmia) maxschwarzi*	Elongate glossa and labial palpi; glossa extends beyond apex of metasoma in repose	Fabaceae: *Astragalus*	Müller 2012
Melittidae: Dasypodainae	*Hesperapis (Hesperapis) palpalis*	Elongate labium and labial palpi	Polemoniaceae: *Eriastrum, Langloisia,* and *Gilia*	Stage 1966
Melittidae; Meganomiinae	*Pseudophilanthus tsavoensis*	Elongate glossa, as long as prementum	?	Michener 1981

independent origins of elongate mouthparts have occurred in the large genus *Andrena*, and multiple panurgine genera have evolved elongate mouthparts, including *Calliopsis, Melitturga, Neffapis, Nolanomelissa, Protandrena,* and *Perdita* (Box 7-1). The aptly named *Neffapis longilingua*, a specialist bee on *Malesherbia* (Malesherbiaceae) from the margins of the Atacama Desert of northern Chile, has evolved an elongate glossa and a labial palpus for accessing the nectaries of its host plant (Fig. 7-4; Rozen and Ruz 1995). *M. humilis* is a low-growing, desert herb with small (4–6 mm long), white flowers. The basal parts of the petals and sepals form a narrow funnel with nectaries at the bottom, while the apical parts of the sepals and petals splay outward to form a flat surface upon which foraging females land (Rozen and Ruz 1995). The stamens and petals are exserted well beyond the corolla, making pollen easily accessible, but nectar is restricted to the bottom of the funnel-shaped corolla. Instead of inserting their heads into the flower, female *Neffapis* perch on the anthers, where they collect pollen while simultaneously using their elongate mouthparts to reach the nectar far below.

BOX 7-1: *PERDITA HURDI*—A SHORT-TONGUED BEE WITH A REALLY LONG TONGUE

Perdita is an enormous genus of over 600 described species of desert-loving, oligolectic bees (Michener 2007). The center of diversity is in the arid southwestern United States and adjacent Mexico. Most species are small to tiny, and the majority of species have mouthparts that are typical of a short-tongued bee. However, in some species, the mouthparts can become greatly elongated or otherwise modified for accessing floral rewards. The most extreme case is *P.* (*Glossoperdita*) *hurdi*, which has an elongate head, enlarged mandibles, and greatly elongated mouthparts. In *P. hurdi*, the prementum, stipes, galea, and especially the glossa have become enlarged so that the mouthparts extend to nearly the length of the female's body (Fig. 1 in Hurd and Linsley 1963). Observations on the behavior of this bee were made by Paul Hurd and Gort Linsley (1963) in Pinal County, Arizona, in the summer of 1962. Female *P. hurdi* are highly specialized visitors to the "unicorn plant" (*Proboscidea althaeifolia*, Martyniaceae). *Proboscidea althaeifolia* has large, yellow, gullet-shaped flowers that, like other Martyniaceae, are primarily visited by large, apid bees such as *Bombus*, *Xylocopa*, *Anthophora*, and *Peponapis*. Female *P. hurdi* have a deviously effective approach to accessing the floral rewards of *Proboscidea*. They use their elongate, acutely pointed mandibles to cut a small hole in the base of *Proboscidea* flowers before the flowers are open and available to larger bees. Once inside the flower, female *P. hurdi* gather pollen from the dehiscent anthers and then exit via the opening they created. Later in the day, once flowers have opened, females visit *Proboscidea* flowers for nectar. This is where their elongate mouthparts come into use. Female *P. hurdi* are too small to regularly contact the stigma of *Proboscidea* flowers, so few visits are likely to result in pollination. However, sometimes they encounter a waiting male, and during their amorous gymnastics, stigmatic contact (and pollination) may occur (Hurd and Linsley 1963). So, what looks like pure floral larceny may, at least occasionally, have a beneficial impact on the host plant.

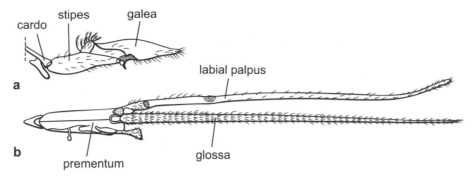

Figure 7-4. Elongate mouthparts of *Neffapis longilingua*: (a) maxilla, showing cardo, stipes, galea, and reduced maxillary palpi and (b) labium, showing mentum, prementum, paraglossae, and greatly elongated glossa and labial palpi (redrawn from Rozen and Ruz 1995, Figs. 12 and 13).

Even the family Melittidae, which is not a group that we generally associate with elongate mouthparts, has some members that break the rules. *Hesperapis* (*Hesperapis*) *palpalis* has an elongate labium and elongate labial palpi associated with accessing nectar from the long, tubular flowers of its host plants: *Eriastrum, Langloisia,* and *Gilia* (Polemoniaceae) (Stage 1966). While most melittids have a short glossa, in *Pseudophilanthus* (Meganomiinae), the glossa can be as long as the prementum (Michener 1981).

Some members of the colletid subfamily Xeromelissinae have evolved absurdly long mouthparts for accessing nectar from long, tubular flowers. In *Geodiscelis longiceps,* the head is greatly elongated, and a combination of elongation in the cardo, stipes, and prementum gives this bee a proboscis that is longer than the adult female. These females gather nectar from the flowers of *Tiquilia* (Boraginaceae) while resting comfortably on the flower petals scanning for predators (Packer 2005). Other colletids have achieved long mouthparts through modifications of the maxillary (e.g., *Euhesma tubulifera*) or labial (e.g., *Niltonia virgilii*) palpi (Table 7-2).

A common path to creating relatively long mouthparts in the Colletidae is simply to elongate the prementum and malar space, as in *Colletes validus* or *Chilicola rostrata.* This also occurs in various Halictidae, particularly in certain Augochlorini, such as *Chlerogelloides, Megaloptidia, Megommation,* or *Micrommation.* These bees have short palpi and a short glossa, but the prementum is remarkably narrow and elongate, typically 10 to 20 times as long as broad. A related genus, *Ariphanarthra,* has a similarly elongate prementum but differs in having extremely elongate maxillary palpi that trail behind it as it flies (Eickwort 1969). It is known to visit the distylous flowers of several species of *Psychotria* (Rubiaceae; Silva and Vieira 2015).

Perhaps the most remarkable adaptation for creating long mouthparts in short-tongued bees occurs in the genus *Leioproctus* (Colletidae: Neopasiphaeinae). In *Leioproctus* (*Filiglossa*) *filamentosa* (and two other species in the same subgenus), the apex of galea has 4–12 huge, elongate setae, and the labial and maxillary palpi are also elongated (Fig. 7-5). The subgenus *Filiglossa* was erected by (Rayment 1959) to accommodate three, small, closely related

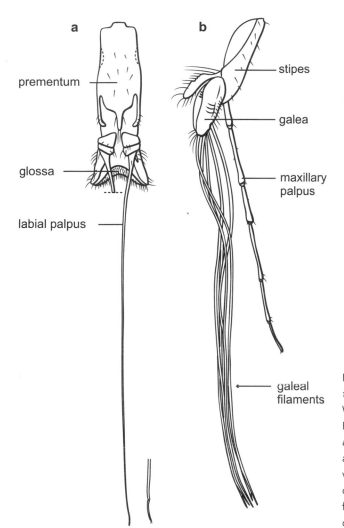

Figure 7-5. Elongate mouthparts of *Leioproctus* (*Filiglossa*) *filamentosa* (Bernhardt and Walker 1996) (redrawn from Michener 2007, Fig. 19-6): (a) labium, with mentum, lorum, and most of one palpus omitted (enlarged apex of palpus shown below) and (b) outer view of maxilla, with cardo omitted. The combined filamentous labial palpi plus galeal filaments draw nectar up to the mouth by capillary action.

species with these highly modified, elongate, filamentous mouthparts. At the time of their discovery, the host plant was unknown, but Rayment predicted that these bees would be associated with a "remarkable flower" (Rayment 1959). It turns out, he was wrong. Bernhardt and Walker (1996) determined, based on collections and analysis of pollen loads, that *Filiglossa* bees are all associated with *Persoonia* (Proteaceae), one of the most common genera of shrubby Proteaceae in Australia, and that they are nectar thieves, not pollinators. These bees insert their filamentous mouthparts into the base of the inverted flower (between the tepals) and extract nectar without coming into contact with the reproductive parts of the flower. Females also collect pollen from the anthers of *Persoonia* flowers but, due to their small size, rarely contact the receptive stigma. The extraordinary mouthparts of these bees are clearly adaptations for extracting floral resources without actually doing any effective pollination.

"Long-tongued" bees with short tongues

While there are many paths to making a long tongue with an ST morphology, making a short tongue with an LT morphology is much easier, as all one has to do is decrease the length of the glossa and galea. This occurs in various tiny *Anthophorula* species (Apidae: Exomalopsini) associated with small shallow flowers of *Chamaesyce* (Euphorbiaceae) or *Tidestroemia* (Amaranthaceae). In *Ctenoplectra* (Apidae), the mouthparts are foreshortened, and the labial palpi resemble those of an ST bee (i.e., the labial palpal segments are all similar in morphology). This has historically created a great deal of controversy over whether *Ctenoplectra* are ST bees (closely related to Melittidae; Michener and Greenberg 1980) or whether they are LT bees (as we now believe, based on a variety of evidence; Bossert et al. 2018). The unique mouthpart morphology of *Ctenoplectra* appears to be an adaptation allowing the labial palpi and glossa to make an abrupt turn when probing for nectar in flowers of Cucurbitaceae (Plant and Paulus 2016, Vogel 1990).

Euglossini—bees that drink through soda straws

All short-tongued and most long-tongued bees rely primarily on lapping to bring nectar to the mouth, where it can be ingested. However, one group of bees—the Euglossini—have evolved truly suctorial feeding (Borrell 2004). In euglossines, the elongate galeae and the long labial palpi form a tightly coupled tube through which nectar can be aspirated (Fig. 7-6). Euglossine mouthparts are exceptionally long (>40 mm) and sometimes extend well beyond the tip of ·

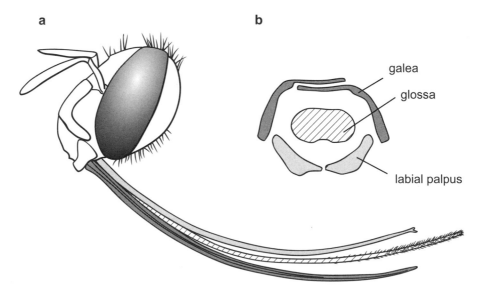

a **b**

galea

glossa

labial palpus

Figure 7-6. Straw-like mouthparts used for aspirating nectar in Euglossini: (a) lateral view of head with mouthparts extended and (b) cross section of suctorial apparatus, showing paired galeae (above) and paired labial palpi (below) (redrawn from Borrell 2004, Fig. 1).

the metasoma when folded. Because nectar viscosity increases exponentially with nectar concentration, suctorial feeding puts an upper limit on the nectar-sugar concentration that euglossines can effectively aspirate. Borrell (2004) analyzed the functional morphology of euglossine mouthparts and found that intermediate nectar concentrations (~35% sugar) maximized energy intake in euglossine bees. A variety of neotropical flowers exploit euglossines as pollinators by producing optimal sugar concentrations for their unique mode of feeding.

POLLEN

Pollen is by far the most important floral resource harvested by solitary bees. Pollen provides the protein needed for larval development (Dobson and Peng 1997) as well as the protein needed for adult female egg production (Cane et al. 2017). Pollen, the haploid, male gametophytic stage of flowering plants, is highly variable in size, shape, chemistry, and nutritional value.

Biological properties of pollen

Pollen can contain more than 60% protein, but it is also a challenging resource to utilize. Recent studies of pollen digestibility (discussed below) indicate that pollen may be both mechanically (Lunau et al. 2015) and chemically (Müller and Kuhlmann 2008, Praz et al. 2008b) difficult for generalist pollenivores to digest. Because pollen is both energetically costly to produce—certainly more costly than nectar—and because pollen is essential to plant reproductive success, it seems logical that plants should not give it away for free (Box 7-2; Schlindwein and Wittmann 1997, Schlindwein et al. 2005). As we shall see below, flowering plants have evolved a variety of mechanisms for restricting access to their valuable pollen reserves.

Pollen grains vary widely in size, shape, and morphology. Typical pollen grains are between 10 and 100 micrometers but can range from extremely fine (<5 μm) to enormous (<210 μm). Pollen varies widely in shape, from typically round or ovoid grains, to triangular grains, to linear grains, as well as some truly bizarre morphologies. Some pollen grains are held together in viscin threads (Onagraceae). Most pollen is released in discrete grains, but sometimes, as in many Ericaceae, Onagraceae, and most mimosoid legumes, pollen is released as polyads, closely adhering clusters of four or more pollen grains. Pollen polyads are generally digestible by bees, but pollinaria masses of hundreds or even thousands of grains found in the Asclepiadaceae and Orchidaceae are not.

Pollen grains consist of multiple, discrete layers, some of which are indigestible to virtually all pollenivores. The outermost layer of pollen is the pollenkitt, a thin, semi-liquid coating that consists of lipids, hydrocarbons, terpenoids, and carotenoid pigments (Pacini and Hesse 2005). Pollenkitt has an important role in facilitating pollen clumping and adherence to the bodies of bees and other flower visitors (Amador et al. 2017, Roulston and Cane 2000). It may also play a role as an odor attractant for floral visitors (Dobson 1988). In a few cases,

BOX 7-2: *CHELOSTOMA RAPUNCULI*—AN OLIGOLECTIC POLLINATOR OF *CAMPANULA* (CAMPANULACEAE)

Chelostoma is a Holarctic genus of approximately 60 species of very slender stem- and cavity-nesting bees. The majority of species occur in Western and Central Europe (Michener 2007). Most species are host-plant specialists on a variety of plant families—most importantly, Campanulaceae, Hydrophyllaceae, and Ranunculaceae, but also Asteraceae, Dipsacaceae, and Brassicaceae (Sedivy et al. 2008). *C. rapunculi* is a particularly well-studied species native to Europe. It is a narrow host-plant specialist on *Campanula* (Campanulaceae). Experiments by Christophe Praz and colleagues (2008b) have demonstrated that *C. rapunculi* larvae are incapable of developing on pollen other than that of their preferred host plant, indicating a clear physiological adaptation to the chemistry, morphology, or nutritional composition of *Campanula* pollen. Campanulaceae pollen is extremely high in protein content (~50% protein; Roulston et al. 2000), suggesting that they are a valuable floral reward for solitary bee specialists that can utilize them.

The interaction between *C. rapunculi* and its host plant (*Campanula rapunculus*) was examined in detail by Schlindwein and colleagues (2005). They asked a simple question: How much of the pollen that is produced by the host plant ends up as food for the bee, and how much contributes to plant reproduction? They quantified the total number of pollen grains produced by an average *Campanula* flower, the number of pollen grains collected by foraging bees (and used for brood-cell provisioning), and the number of pollen grains that are deposited on receptive stigmas. *Campanula* flowers produce on average 83,000 pollen grains, and female *C. rapunculi* gather 95.5% of these pollen grains for brood-cell provisioning. Just 3.7% of the pollen grains produced by *Campanula* ended up on the stigma of a conspecific plant. Nevertheless, the pollen deposited per stigma was 4–10 times the amount needed to achieve full seed set. So while the bees were capturing the vast majority of the pollen for rearing of more bees, their visits were still highly effective in contributing to fertilization of the host plant. This study demonstrates beautifully what is probably a fairly typical dynamic between oligolectic bees and their host plants. Many oligolectic bees are highly effective at extracting pollen from the host-plant flowers, but they are not always floral larcenists (Hargreaves et al. 2009). They can, and do, contribute significantly to the seed set of their host-plant partners.

copious pollenkitt can also be a floral reward for bees (Houston et al. 1993). Pollenkitt is greatly reduced in wind- and buzz-pollinated plants. The structural elements of pollen are formed by the exine, a matrix of the carbohydrate sporopollenin, and the intine, an inner layer of cellulose and pectin that encloses the nutrient-rich cytoplasm. These outer layers, particularly the exine, give pollen its shape and texture. The exine is highly resistant to decay and digestion, which means pollen grains are exceptionally well-preserved in fossil deposits dating to the mid-Cretaceous (Cenomanian, 94–100 mya; Hu et al. 2008). Virtually all pollen grains have one to three apertures (pores or slits) that allow the germinating pollen tube to extrude from the interior of the pollen grain.

We know surprisingly little about how bees digest pollen. The major challenge for a pollenivore is how to liberate the nutrient-rich cytoplasm from the mostly indigestible, rigid exine. Insects and other animals use a variety of methods, including cracking open the pollen wall mechanically, piercing the pollen wall with sharp mouthparts, dissolving the pollen wall with digestive enzymes, inducing germination (allowing the cytoplasm to extrude through the germination pores; also called pseudogermination), or bursting the pollen grain with "osmotic shock" (Roulston and Cane 2000). The literature on pollen digestion in bees is confusing and sometimes contradictory, and the methods used by adult bees to digest pollen may not be the same as the methods used by larval bees (Roulston and Cane 2000). How honey bees digest pollen is also likely to be a poor indicator of how solitary bees digest pollen, because honey bees store pollen for long periods of time and process the pollen before progressively feeding it to their larvae.

The most detailed study of pollen digestion by solitary bees was conducted by Dobson and Peng (1997). Their studies focused on *Chelostoma florisomne* (Megachilidae), a solitary, stem-nesting bee in Europe that is oligolectic on *Ranunculus*. They examined how lipids, carbohydrates, and proteins were assimilated as the pollen moved through the larval midgut. During digestion, the cytoplasm of the pollen grains was observed to gradually extrude through the germination pores. What caused this extrusion was not clear, but it may have been made possible through digestion of the intine layer early in the digestive process. Once the intine layer is breached, the cytoplasm is free to move out through the germination pores, and digestive enzymes are free to move in. Nutrients were gradually assimilated as the pollen moved along the length of the midgut lumen, with lipids, proteins, and carbohydrates all gradually disappearing from stained pollen grains. In the posterior midgut, the digested pollen grains consisted of both intact and shattered exine husks, suggesting that some enzymatic degradation of the exine takes place during larval digestion. The authors concluded that neither pseudogermination nor "osmotic shock" was involved in pollen digestion.

The nutritional value of pollen ranges widely among angiosperm taxa. The primary nutritional components of pollen include protein, amino acids, starches, sterols, and lipids. Broad surveys of pollen nutritional value for bees have focused primarily on protein content. Roulston and colleagues (2000) conducted an extensive analysis of protein content across 377 plant species and 93 plant families, including gymnosperms. Protein content varied from low (2.5%) to extremely high (61%). For comparison, beef is approximately 26% protein, meaning that some pollen has 2.3 times the protein concentration of a sirloin steak.

Roulston and colleagues (2000) examined factors that drive protein content across gymnosperms and angiosperms, including whether a plant is primarily wind- or animal-pollinated, whether a plant is primarily pollinated by bees or other (non-bee) animals, and, finally, the distance from the stigma to the ovule (i.e., the distance a pollen tube has to grow to achieve fertilization). While many anemophilous (wind-pollinated) plants have extremely low protein content, when comparisons were made between closely related wind- and animal-pollinated plants, there was no significant elevation in protein content among the animal-pollinated species or lineages. Likewise, bee-visited plants did not differ significantly in pollen protein content from plants visited by other pollinators. The single best predictor of pollen protein content was the distance from stigma to ovule, suggesting that pollen protein content may be largely driven by the energetic needs of pollen-tube growth.

One clear pattern that emerged from the data was that plants with poricidal anthers that deliver their pollen to bees that can "buzz" the flowers (reviewed by De Luca and Vallejo-Marin 2013; see below) had extremely protein-rich pollen. *Solanum*, a classic buzz-pollinated plant, has pollen that ranges from 34% to 55% protein. There was a clear phylogenetic signal in the data as well—members of the same family, tribe, or genus had similar levels of protein content. The authors found no evidence that the flowering-plant families that host a large number of specialist pollinators (e.g., Asteraceae, Cactaceae, Malvaceae, Convolvulaceae, Boraginaceae, Zygophyllaceae) had higher protein content than those flowering-plant families that are primarily visited by generalists. In summary, pollen is an incredibly protein-rich food, but why flowering-plant families, tribes, and genera produce pollen of variable protein content, and how variation in protein content impacts bee foraging and host-plant choice, is still not clear (see Chapter 12 for an evolutionary perspective on these topics).

Tools for harvesting and transporting pollen

Pollen is a very different resource from the arthropod prey that is captured, transported, and consumed by the ancestors of bees, the crabronid wasps. Pollen comes in the form of a powder that must be swept, groomed, brushed, combed, raked, or (in some cases) vibrated off of the reproductive parts of flowers. Bees have evolved a diversity of structures for extracting pollen from flowers, moving pollen around on their bodies through grooming movements and, ultimately, storing pollen internally or externally for efficient transport back to the nest. The finely branched, plumose hairs that characterize bees are likely an adaptation to improve the uptake and retention of pollen on their bodies. Bees are also thought to use electrostatic charges to facilitate pollen uptake from flowers, especially when pollen grains are small (Erickson and Buchmann 1983). Thorp (1979, 2000) provides an excellent overview of the behaviors and morphologies associated with pollen collection and transport in bees. We cover the morphological structures associated with pollen manipulation and transport below.

The movements associated with grooming and pollen manipulation in bees are likely derived from the widespread grooming movements exhibited by all insects (Jander 1976). Grooming movements are used both to transport pollen to the mouth of adult bees (when they consume pollen) and to transport pollen backward to the regions of the body used for pollen transport back to the nest (the scopa; see below). Bees are highly effective at grooming pollen off the vast majority of their body. They can remove pollen from their mouthparts, head, legs, and most regions of the meso- and metasoma. However, there are surfaces that bees cannot reach, including the posterior surface of the head, the pronotum (just posterior to the head), and the dorsal surfaces of the propodeum and metasoma (in Kimsey 1984, see Fig. 8). Many plants exploit these "safe sites" for pollen deposition as a strategy to avoid losing all of their pollen to the digestive tracts of bees (Beattie 1971, 1972; Cane and Payne 1988; Green and Bohart 1975).

The morphological structures used by bees for gathering, grooming, and manipulating pollen are located in the mouthparts (Figs. 7-1, 7-2) and all three pairs of legs (Fig. 7-7). The galeal comb is used to groom pollen from the forelegs in ST bees, and the stipital comb is used for the same purpose in LT bees (although the development of both combs is highly variable across taxa; Jander 1976). These combs are drawn across the forelegs during and after flower visits and allow bees to move pollen collected on the forelegs to the mouth. These structures are used when adult female bees consume pollen, which is essential for oocyte development (Cane et al. 2017). All bees (even brood parasites) possess an antennal cleaner on the foreleg for grooming pollen off the antenna (Fig. 7-7a). All pollen-collecting bees also possess a foreleg basitarsal brush for grooming pollen from the mouthparts, head, and anterior regions of the mesosoma. The midlegs possess a diversity of brushes and combs for moving pollen that collects on the forelegs and mesosoma backward toward the scopa (Fig. 7-7b). An inner basitarsal brush and outer (apical) tibial brush are widespread in pollen-collecting bees. Bees groom pollen off the foreleg by passing the foreleg between the appressed midfemoral and midtibial combs (Jander 1976, Fig. 3). Additional combs and brushes are located on the undersurface of the femur (near the base of the leg) and the underside of the trochanter. Structures for pollen grooming on the hindleg include an inner hind basitarsal brush and an inner (apical) tibial brush (Thorp 1979) (Fig. 7-7c).

The morphological structures for transporting pollen back to the nest (scopa) are variable in location and ultrastructure. Fig. 7-8 provides a very rough overview of the distribution of scopal structures across bee groups. The scopa in most bees is located on the hind trochanter, femur, tibia, and basitarsus. Many Colletidae, Andrenidae, and Halictidae (excluding Hylaeinae and Euryglossinae) carry pollen widely distributed on the trochanter, femur, and tibia. Some Andrenidae and Melittidae possess a well-developed trochanteral scopa at the base of the leg (referred to as the flocculus in *Andrena*). In Melittidae and Apidae, the scopa is generally concentrated on the tibia and basitarsus. Many bees incorporate scopal hairs on the lateral surfaces of the propodeum and on the underside of the metasoma. Megachilidae and some Halictidae (*Lasioglossum* [*Homalictus*] and *Systropha*) have the scopa largely restricted to the underside of the metasoma. Euryglossinae and Hylaeinae are the only bee subfamilies that carry pollen exclusively within the crop, but there are other bee groups that have convergently evolved crop transport of pollen (see below).

There are many exceptions to these general rules about scopal morphology. Within Apidae, for example, many Xylocopinae sensu stricto have a reduced tibial scopa, because they carry substantial amounts of pollen internally, within the crop. In the apid tribe Exomalopsini, the genus *Eremapis* has what appears to be a sternal scopa (Neff 1984). Most pollen is transported on the hind tibia and basitarsus, but there is significant pollen in the sparse sternal arrays of erect

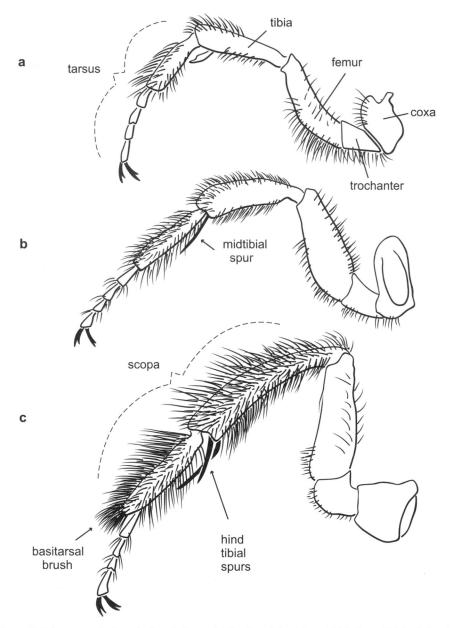

Figure 7-7. Leg morphology in female bees: (a) foreleg; (b) midleg; (c) hindleg. Original drawing based on female *Peponapis pruinosa* (Apidae).

arched hairs on the underside of the metasoma. These hairs may play a role in pollen harvesting, but they seem to be used for pollen transport as well. While most pollen-collecting megachilids transport pollen exclusively in the metasomal scopa, there are some groups that have evolved other modes of pollen transport. The Pararhophitinae, a relatively basal branch of Megachilidae, carry pollen primarily on the hind femur and tibia (Michener 2007). The metasomal scopa is greatly reduced or absent. In Lithurginae, the genus *Lithurgopsis* has a weakly

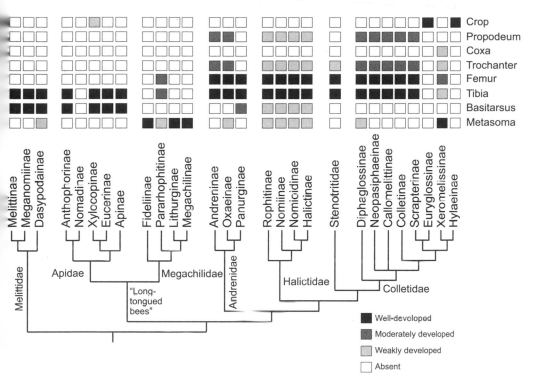

Figure 7-8. Distribution of scopal location across the subfamilies of bees (modified from Table 1 in Thorp 1979).

developed femoral and tibial scopa, in addition to the metasomal scopa. And in the largest subfamily (Megachilinae), the bizarre and relictual genus *Aspidosmia* has a clearly developed tibial scopa (Gonzalez et al. 2012). All South American Neopasiphaeinae (Colletidae) transport pollen on their hindlegs, but some genera, especially those that specialize on flowers with large pollen grains, appear to have a well-developed metasomal scopa. *Brachyglossula* (associated with Cactaceae) and *Tetraglossula* (associated with Onagraceae) are two such examples.

Internal pollen transport has evolved repeatedly in bees and is often associated with stem-, wood-, or cavity-nesting bees (e.g., some Xylocopinae and Hylaeinae). However, a small group of species within two ground-nesting genera of Colletidae have also evolved internal pollen transport: the *fasciatus* group of *Colletes* (Kuhlmann 2006) and *Leioproctus* (*Euryglossidia*) (Houston 1981). In both cases, the transition from external to internal pollen transport appears to have occurred as a consequence of a host-plant shift from plants with large, coarse pollen grains to host plants with small, fine pollen grains. Widely spaced scopal hairs, ideally suited for carrying large pollen grains, are poorly suited for fine, powdery pollen. In both the *fasciatus* group of *Colletes* and *Leioproctus* (*Euryglossidia*) *cyanescens*, female bees appear to have given up on external pollen collection in order to efficiently harvest pollen from their newly acquired host plants.

Scopal structure varies widely among bees and seems largely driven by the size and texture of the pollen collected. Variation in scopal morphology among bees was reviewed by Thorp (1979, 2000) and Roberts and Vallespir (1978), and the ultrastructure of scopal hairs across bee families was documented using scanning electron microscopy by Pasteels and Pasteels (1974, 1975, 1976, 1979). The Pasteels and Pasteels studies demonstrate that scopal hairs are far more complex than light-microscopic analysis reveals. There are scopal hairs with linear, parallel furrows, scopal hairs with ridges that spiral apically in a sort of corkscrew pattern, and scopal hairs with enlarged, spatulate apices, to name just a few. We are far from understanding the adaptive significance of these modified hairs. There are some general patterns in scopal morphology, however. Bees that collect small (<40 μm) pollen grains have dense, closely spaced, finely branched scopal hairs (e.g., Oxaeinae, Diphaglossinae), and pollen is carried largely dry. These bees are often specialists on plants with poricidal anthers, and they extract the pollen by "buzzing" the flowers (see below). In Africa, species in the genus *Lipotriches* (subgenus *Lipotriches*) are narrowly oligolectic grass specialists that fly early in the morning before wind disperses the grass pollen (Tchuenguem Fohouo et al. 2002, 2004; Immelman and Eardley 2000). These bees have a finely branched scopa for capturing the fine, normally wind-dispersed pollen of their grass hosts.

Bees that carry large (>100 μm) pollen grains, such as those from plants in the family Cucurbitaceae, Malvaceae, and Cactaceae, usually have widely spaced, stout, weakly branched scopal hairs (Gaglianone 2000, Linsley 1958). Bees that specialize on the family Onagraceae, which have large pollen grains connected by "viscin threads," have a highly modified scopa for raking up huge quantities of onagraceous pollen. In the *Andrena* subgenus *Onagrandrena*, all species of which are specialists on Onagraceae, the scopal hairs are long, sparse, and unbranched. Similar scopal morphology occurs in other, distantly related bees that specialize on Onagraceae (e.g., *Oenothera* specialists in the *Lasioglossum* subgenus *Sphecodogastra*, some *Andrena* in the subgenus *Diandrena*, some *Melissodes*, *Diadasia*, *Megachile*, *Synhalonia*, *Anthophora*, *Perdita*, and *Hesperapis*). Generalist bees have a generalized, multipurpose scopa that works reasonably well with a range of pollen morphologies, but there is clear evidence that some pollen is avoided by these generalist visitors (e.g., Malvaceae [Azo'o et al. 2011], and Convolvulaceae [Pick and Schlindwein 2011]), either because they are toxic to bee larvae or because their texture makes them difficult to manipulate for a foraging adult worker.

How bees collect pollen

Scrabbling

The most widely used method of pollen collection in bees is called "scrabbling" (Thorp 2000). Scrabbling describes a variety of movements that bees make as they move across a flat plane or column of anthers and gather pollen on their head, mouthparts, legs, or mesosoma. A scrabbling female occasionally stops to groom pollen—first onto her midlegs and then onto her hindlegs, mesosoma, and metasoma (depending on the location of her scopa). MacSwain and colleagues (1973) described these movements rather poetically as "pollen dances" when observing several oligolectic species of bee on *Clarkia* (Onagraceae). Bees also use their mandibles to extract pollen from the anther (as described by Neff and Rozen [1995] for *Anthemurgus* visiting *Passiflora* flowers). Other modes of pollen collection involve abdominal "drumming," "tapping," or "patting" (Cane 2017). Most megachilids have a metasomal scopa that is brushed across the surface of the flower head during pollen collection. Pollen accumulates directly onto the metasomal scopa without the need to groom the pollen from fore- and midlegs. This kind of abdominal pollen collection even occurs in some bees in which the scopa is located on the hindlegs. *Macropis* (Cane et al. 1983a) and *Dieunomia* (Minckley et al. 1994) have also been observed tapping or patting the metasoma across the anther column of their respective host plants.

Scrabbling, drumming, and tapping work effectively when anthers are exposed and pollen is presented in a relatively flat plane (e.g., Asteraceae, Apiaceae) or on a centralized anther column (e.g., Rosaceae, Cactaceae, Malvaceae). However, many plants hide their pollen within a deep corolla tube or within poricidal anthers, and these plants require more specialized modes of pollen extraction.

Bees that visit "nototribic" flowers

Some flowers, especially in the families Lamiaceae and Scrophulariaceae, hide their anthers deep within an elongate corolla. Bees (and masarine wasps) that visit these flowers must insert their heads into the corolla to extract pollen. As they do so, pollen is deposited on the face or dorsum of the bee, where it can be difficult to reach—a mode of pollen deposition called "nototribic" pollen deposition. Müller (1996a) described a group of bee species in the Apidae, Megachilidae, and Halictidae (Fig. 7-9) that have morphological adaptations for collecting pollen from these types of flowers. In this group of mostly oligolectic bees, the face bears erect, stout, recurved, or wavy setae (Fig. 7-10), distinctly different from the facial setae of close relatives that do not visit these plant families. In cases where the flower visitation has been observed, these bees appear to insert their head into the corolla and then vibrate their flight muscles to dislodge the pollen. Pollen is collected on the stout hairs of the face and then groomed off by the forelegs. The fact that this trait has evolved in multiple lineages of Lamiaceae and Scrophulariaceae specialists suggests that it is an important morphological adaptation to effectively extract pollen from these two host-plant families. We have expanded Müller's (1996a) list based on additional taxa known from the New World and more recent publications (Table 7-3). The bee families that exhibit this type of modified facial morphology for harvesting pollen from these nototribic flowers now include Apidae, Megachilidae, Andrenidae, Colletidae, and Halictidae (Table 7-3).

Grappling hooks for extracting pollen

A diverse group of mostly oligolectic bees from five of the seven bee families have evolved striking morphological adaptations for extracting pollen from the flowers of Boraginaceae, Pontederiaceae, and a number of other plant families with comparable floral morphology (Table 7-4). In these plants, the anthers are hidden deep within a narrow, constricted, tubular corolla. Bees that specialize upon these flowers have highly modified, recurved setae on their mouthparts or

Figure 7-9. Habitus drawing of female *Rophites algirus* (Halictidae: Rophitinae). Original artwork by Frances Fawcett.

a

b

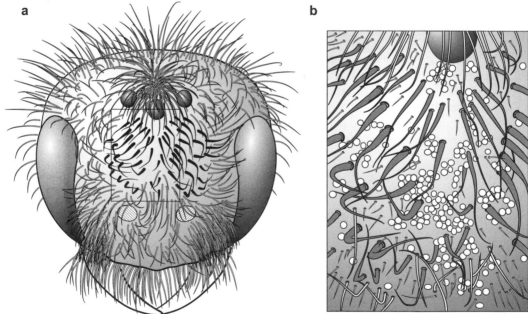

Figure 7-10. Modifications of facial hairs associated with accessing pollen from Lamiaceae. *Rophites algirus*: (a) frontal view of head and (b) close-up of frons (redrawn from Müller 1996a, Figs. 4 and 5).

Table 7-3. Lamiaceae specialists—bees with modified facial hairs associated with collecting pollen from Lamiaceae and Scrophulariaceae (nototribic pollen collection).

See Müller et al. (1996).

Family:Subfamily	Genus/species	Modified hairs	Host plant	Reference
Andrenidae: Panurginae: Panurgini	*Panurginus* n. sp.	Downward-oriented simple hairs on clypeus	Campanulaceae: *Downingia*	Thorp 2000
Apidae: Anthophorinae: Anthophorini	*Amegilla paradoxa*	Short erect bristles on flattened clypeus	host plant unknown	Brooks 1988
Apidae: Anthophorinae: Anthophorini	*Anthophora (Mystacanthophora)* spp. (>20 spp.)	Most species with clypeus and supraclypeal area flat, surface with erect hooked hairs	Mostly polylectic but frequently on Lamiaceae	Brooks 1988, Müller 1996a
Apidae: Anthophorinae: Anthophorini	*Deltoptila elefas*	Erect apically hooked hair on flattened clypeus and supraclypeal area	Lamiaceae?	JLN pers. obs., G. Dierenger pers. comm.
Apidae: Anthophorinae: Anthophorini	*Habropoda salviarum*	Hooked hair or clypeus, erect stout hair on supraclypeal area	Lamiaceae: *Salvia*	JLN pers. obs.
Apidae: Anthophorinae: Anthophorini	*Anthophora (Clisodon) furcata*	Erect, apically curved hair on clypeus	Lamiaceae, Scrophulariaceae	Müller 1996a
Apidae: Eucerinae: Eucerini	*Tetraloniella (Pectinapis)* spp. (4 spp.)	Clypeus and supraclypeal area flat; erect hooked or wavy hair on flattened clypeus and supraclypeal area; a pecten basally of clypeus of *T. fasciata*	Lamiaceae	LaBerge 1989, LaBerge 1970
Apidae: Eucerinae: Exomalopsini	*Exomalopsis (Stilbomalopsis) binotata*	Hooked hair on clypeus, supraclypeal area, and frons	Lamiaceae (?)	JLN pers. obs., G. Dierenger pers. comm.
Colletidae: Colletinae: Colletini	*Colletes* n. sp. MX-Oaxaca	Hooked hair on clypeus, supraclypeal area, and frons	Lamiaceae	JLN pers. obs., G. Dierenger pers. comm.
Colletidae: Diphaglossinae: Caupolicanini	*Zikanapis* spp. (4+ spp.)	Hooked hairs on flattened clypeus	Lamiaceae	Compagnucci 2006, JLN pers. obs., Thorp 2000
Halictidae: Halictinae: Halictini	*Lasioglossum (Lasioglossum) tropidonotum*	Hooked hair on frons	host plant unknown	McGinley 1986

(continued)

Table 7-3. (Continued)

Family:Subfamily	Genus/species	Modified hairs	Host plant	Reference
Halictidae: Rophitinae	*Rophites* spp. (3 spp.)	Hooked or wavy hair on frons	Lamiaceae	Müller 1996a
Megachilidae: Megachilinae: Anthidiini	*Anthidium* spp. (at least 24 spp.)	All with hooked or wavy hair on clypeus; some with flattened clypeus and specialized hair on frons or supraclypeal area	various, Lamiaceae in part	Müller 1996a, Gonzalez & Griswold 2013
Megachilidae: Megachilinae: Anthidiini	*Trachusa (Ulanthidium)*— Mitchelli group (2 spp.)	Hooked hair on clypeus, supraclypeal area, and frons	Lamiaceae	Thorp & Brooks 1994, Michener 1948
Megachilidae: Megachilinae: Megachilini	*Megachile (Chalicodoma) albocristata*	Hooked hair on clypeus	host plant unknown	Müller 1996a, Alqarni et al. 2012
Megachilidae: Megachilinae: Megachilini	*Megachile (Pseudomegachile)* spp. (9 spp.)	Hooked hair on clypeus and often supraclypeal area	Acanthaceae: *Blepharis*	Alqarni et al. 2012, Müller 1996a
Megachilidae: Megachilinae: Megachilini	*Matangapis alticola*	Hooked hair on clypeus and supraclypeal area	host plant unknown	Baker & Engel 2006
Megachilidae: Megachilinae: Megachilini	*Megachile (Dasymegachile) mitchelli*	Hooked hair on flattened clypeus and supraclypeal area	host plant unknown	Mitchell 1980
Megachilidae: Megachilinae: Osmiini	*Osmia (Diceratosmia)* spp. (2 spp.)	Short erect hair, often hooked, on clypeus, supraclypeal area, and frons	Lamiaceae: *Salvia*	Ayala & Griswold 2005, Rightmyer et al. 2011
Megachilidae: Megachilinae: Osmiini	*Osmia (Erythosmia)* spp. (4 spp.)	Erect, curved hair on frons and clypeus	Lamiaceae, Scrophulariaceae	Müller 1996a, 2018
Megachilidae: Megachilinae: Osmiini	*Osmia (Helicosmia)* spp. (12 spp.)	Hooked hairs on clypeus	Polylectic	Müller 1996a, 2018
Megachilidae: Megachilinae: Osmiini	*Osmia (Melanosmia) calaminthae*	Short, erect hair on clypeus and frons	Lamiaceae: *Calamintha*	Rightmyer et al. 2011

Table 7-4. Boraginaceae specialists—bees with highly modified leg and mouthparts for visiting Boraginaceae, Pontederiaceae, and Loasaceae.

From Müller (1995).

Family:Subfamily	Genus/species	Morphology	Host plant	Reference
Andrenidae: Andreninae	*Andrena (Hamandrena) nasuta*	Mouthparts: prementum, stipes, galea	Boraginaceae: *Anchusa*	Müller 1995
Andrenidae: Andreninae	*Andrena (Scoliandrena)* (2 spp.)	Mouthparts: prementum and galea with short hooked hair (stipes bare)	Boraginaceae: *Cryptantha*	Thorp 1979, LaBerge 1986
Andrenidae: Panurginae: Calliopsini	*Calliopsis (Verbenapis)* (4 spp.)	Forelegs: hooked hair on foretarsi	Verbenaceae: *Verbena, Phyla*	Michener 1944, Shinn 1967
Andrenidae: Panurginae: Perditini	*Perdita (Heteroperdita)* (22 spp.)	Mouthparts: long, wavy hair on gena and forecoxa	Boraginaceae: *Tiquilia*	Portman et al. 2016, Thorp 1979, Timberlake 1954
Apidae: Anthophorinae: Anthophorini	*Anthophora (Anthophora) plumipes*	Mouthparts: galea (with specialized hairs)	polylectic with preference for Boraginaceae	Müller 1995
Apidae: Eucerinae: Emphorini	*Ancyloscelis gigas*	Mouthparts: erect hair on prementum and stipes; labial palp 1 has hooked hair	Pontederiaceae: *Eichhornia*	Alves-dos-Santos & Wittmann 1999
Apidae: Eucerinae: Emphorini	*Ancyloscelis turmalis* and *A. ursinus*	Mouthparts: hooked hair on prementum, stipes, and labial palpi 1 and 2	Pontederiaceae: *Pontederia*	Alves-dos-Santos & Wittmann 1999
Apidae: Eucerinae: Eucerini	*Cubitalia (Pseudeucera) parvicornis*	Mouthparts: erect hair on galea and labial palpi 1 and 2	Boraginaceae	Müller 1995
Apidae: Eucerinae: Eucerini	*Florilegus (Euflorilegus)* (5 spp.)	Mouthparts: hooked hairs on prementum, stipes, galea and labial palpi 1 and 2	Pontederiaceae: *Pontederia, Eichhornia*	Alves-dos-Santos 2003, Urban 1970
Apidae: Eucerinae: Eucerini	*Melissodes (Apomelissodes) apicatus*	Mouthparts: long hooked hairs on galea	Pontederiaceae: *Pontederia*	LaBerge 1956, Alves-dos-Santos 2003
Apidae: Eucerinae: Eucerini	*Melissodes (Callimelissodes) stearnsi*	Mouthparts: hooked hair on galea	apparently polylectic	LaBerge 1961
Apidae: Eucerinae: Tapinotaspidini	*Paratetrapedia (Paratetrapedia)* spp. (2 spp.)	Mouthparts: prementum, stipes and labial palp 1	Pontederiaceae: *Eichhornia*	Alves-dos-Santos 2003
Apidae: Apinae: Euglossini	*Euglossa (Euglossa) cordata, E. ignita*	Mouthparts: hair on basal part of proboscis	Rubiaceae: *Sabicea*	Michener & Fraser 1978

(continued)

Table 7-4. (*Continued*)

Family:Subfamily	Genus/species	Morphology	Host plant	Reference
Colletidae: Colletinae	*Colletes anchusae* & *C. wolfi*	Forelegs: short, flattened recurved hairs on basitarsi	Boraginaceae: *Cynoglottis*	Müller & Kuhlmann 2003
Colletidae: Colletinae	*Colletes bumeliae* & *C. inuncantipedis*	Forelegs: hooked hair on basitarsi	Sapotaceae: *Sideroxylon*	Neff 2004
Colletidae: Colletinae	*Colletes nasutus*	Foreleg: hooked hair on foretarsi	Boraginaceae: *Anchusa*	Müller 1995
Colletidae: Neopasiphaeinae	*Leioproctus* (*Leioproctus*) *macmillani*	Forelegs: hooked hairs on tarsi	Ericaceae: *Austroloma xerophyllum*	Houston 1991b
Halictidae: Rophitinae	*Ceblurgus longipalpus*	Mouthparts: long plumose hairs on labial palpi [1 and 2]	Boraginaceae: *Cordia leucocephala*	Milet-Pinheiro & Schlindwein 2010
Halictidae: Rophitinae	*Dufourea novaeangliae*	Mouthparts: prementum, galea, labial palpi	Pontederiaceae: *Pontederia*	Eickwort et al. 1986, Alves-dos-Santos 2003
Megachilidae: Megachilinae: Osmiini	*Haetosmia* spp. (3 spp.)	Mouthparts: labial palp 2 with long, apically curved, capitate hair	Boraginaceae: *Heliotropium*	Peters 1974, Gotlieb et al. 2014
Megachilidae: Megachilinae: Osmiini	*Hoplitis* (*Hoplitis*) *pici*	Mouthparts: stiff, erect or right-angled bristles on galea	Hyacinthaceae: *Muscari*	Müller 2006
Megachilidae: Megachilinae: Osmiini	*Hoplitis* (*Proteriades*) spp. (24 spp.)	Mouthparts: labial palpi [1 and 2]	Boraginaceae: *Cryptantha*	Hurd & Michener 1955, Thorp 1979, Müller 1995
Megachilidae: Megachilinae: Osmiini	*Osmia* (*Melanosmia*) *pilicornis*	Mouthparts: hooked hair on galea	polylectic with preference for Boraginaceae	Müller 1995
Megachilidae: Megachilinae: Osmiini	*Osmia* (*Melanosmia*) *sculleni*	Mouthparts: hooked hair on galea, labial palpi 1 and 2, and maxillary palpi; glossa with long bristles	Boraginaceae: *Cryptantha*	Parker & Tepedino 1982a
Melittidae: Dasypodainae	*Hesperapis* (*Hesperapis*) *trochanterata*	Mouthparts: wavy hair on gena	Hydrophyllaceae: *Nama*	Rozen 1987a
Melittidae: Dasypodainae	*Hesperapis* (*Hesperapis*) *elegantula* group (3 spp.)	Mouthparts: mandibles, stipes, and gena near fossa	Boraginaceae: *Tiquilia*	Stage 1966, Thorp 1979

forelegs that are used to probe into the narrow, constricted corolla and scrape out pollen (Fig. 7-11a,b,c). In cases where the bees have been observed visiting flowers (Alves-dos-Santos and Wittmann 1999, Müller 1995, Müller and Kuhlmann 2003, Müller 2006), the bees probe into the flowers with legs or mouthparts and repeatedly move these structures up and down within the

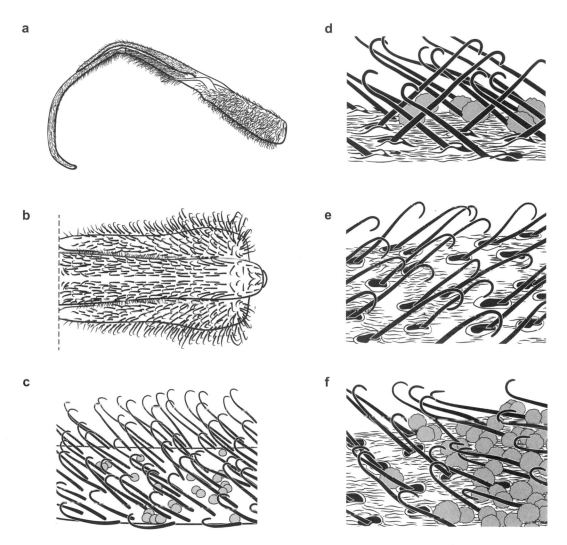

Figure 7-11. Modifications of mouthparts for accessing floral rewards in Boraginaceae. *Ancyloscelis turmalis*: (a) mouthparts; (b) underside of galea and prementum showing recurved hairs; (c) first segment of labial palpus showing recurved hairs and numerous pollen grains; (d) "switch sockets" that allow the hairs to switch abruptly from pointing proximally (as the mouthparts are inserted into the flower) to apically (as the mouthparts are withdrawn from the flower); (e) detail of recurved hairs as mouthparts are inserted into the flower; (f) detail of recurved hairs as mouthparts are withdrawn from the flower with adhering pollen grains (redrawn from Alves-dos-Santos and Wittmann 1999, Figs. 7–12).

corolla. Pollen that accumulates on the legs or mouthparts is then groomed into the hindleg or metasomal scopa. Modified hairs are typically located on the foretarsi, as well as the prementum, stipes, galea, labial palpi, and even the mandibles and genal region of the head (Table 7-4). In some *Ancyloscelis*, there are "switch sockets" (Fig. 7-11d) that allow the hairs to switch abruptly from pointing basally (as the mouthparts are inserted into the flower; Fig. 7-11e) to apically (as the mouthparts are withdrawn from the flower; Fig. 7-11f). The family Boraginaceae hosts an enormous number of oligolectic bees, and many have evolved very similar morphological tools for extracting these hard-to-reach floral rewards.

Poricidal flowers and "buzz pollination"

One effective way flowers can limit access to pollen is to enclose their pollen grains within a tubular (poricidal) anther or corolla (Corbet and Huang 2014, Harder and Barclay 1994, Houston and Ladd 2002). Pollen is released by poricidal flowers only through a narrow pore or slit at the apex of the flower. While pollen can be removed from these poricidal flowers by a variety of methods (including chewing, milking, and "buzz-milking"; Cane and Buchmann 1989), the most widely used method involves sonication ("buzzing") of the flowers. A total of at least 80 angiosperm families include species with poricidal flowers, and about 22,000 species of flowering plants use this mode of pollen delivery (Buchmann 1983, De Luca and Vallejo-Marin 2013). Buzz-pollinated flowers share a number of traits in common. Pollen is typically small (10–30 μm), with a smooth, largely unsculptured exine, and it lacks a thick coating of sticky pollen-kitt typical of non-poricidal flowers (Faegri 1986, Vogel 1978). Buzz-pollinated flowers typically lack nectar (except in the case of blueberries, cranberries, and their relatives); pollen is the sole floral reward for bees. Many species in more than 17 families also share a common floral morphology, termed the solanoid flower form (Faegri 1986, Russell et al. 2016). Such flowers resemble a "bull's-eye," with a yellow (to the human eye) cone of poricidal anthers centered on the flattened blue or purple corolla.

The pollen of buzz-pollinated flowers is also extremely rich in protein. For example, shooting star (*Dodecatheon*) pollen is composed of up to 61.7% protein (by dry weight; Buchmann 1983, Roulston et al. 2000), possibly because these flowers lack other nutritional rewards for pollinators. A common and widely distributed buzz-pollinated plant family is the Solanaceae, which includes the enormous genus *Solanum* (nightshades, including tomatoes and potatoes), with over 2,000 described species. In tropical regions, melastomes (Melastomataceae), dillenias (Dilleniaceae), and sennas (Fabaceae) are common buzz-pollinated flowers. Many bee groups can sonicate ("buzz") flowers. Floral-sonication behavior has been reported from all seven families of bees, and as many as 74 genera have been observed buzzing flowers (Cardinal et al. 2018). Floral sonication behavior has evolved at least 45 times independently in bees, and approximately 58% of bee species are inferred to be capable of buzz pollination (Cardinal et al. 2018). Honey bees are, notably, incapable of buzz pollination, which limits their utility as greenhouse pollinators of tomatoes. That certain bees do not use floral sonication is surprising, since all bees have the same thoracic musculature and many use vibration during nest construction or for defense.

Many genera in the families Apidae, Colletidae, and Halictidae have been observed using floral-sonication behavior, but relatively few genera in the Melittidae, Andrenidae, and Megachilidae have been observed buzzing flowers. Large bees are more likely to buzz than small bees, perhaps because large bees can more easily generate the energy necessary to expel the pollen from poricidal flowers.

The behavior of bees while buzzing is stereotyped to a large degree (Russell et al. 2016). Females grasp the anther column with their legs and then, curling the metasoma below the flower, vigorously vibrate their indirect flight muscles with wings decoupled such that the mesosoma vibrates, but the wings are held stationary over the back of the body.[1] This floral sonication motor routine is fully functional, even at first expression (on the first buzz by a flower-naïve bee), suggesting a strong innate component (Russell et al. 2016). The resulting vibrations are transmitted to the flower, and pollen is rapidly expelled onto the venter of the bee. Individual bouts of buzzing can last from 0.01 to several seconds, and buzzing bouts typically consist of multiple, discrete pulses of vibration. The fundamental frequency of each buzz is largely determined by the biophysical properties of the bee (including body size), but it typically ranges from 200 to 400 Hz (cycles/second), depending on the species (De Luca and Vallejo-Marin 2013). Bees assess the amount of pollen expelled onto their bodies after buzzing poricidal anthers and reduce the amount of time spent on flowers that have been depleted or experimentally glued shut (Buchmann and Cane 1989, Burkart et al. 2013, Cardinal et al. 2018).

FLORAL OILS

Of the three floral rewards covered in this chapter, floral oils were the most recently discovered. Pollen and nectar have been recognized as floral rewards at least since Kölreuter (1761–1766) and Sprengel (1793) discovered animal-mediated pollination, but floral oils were discovered only in the late 1960s. In a series of ground-breaking studies, Vogel (reviewed in Neff and Simpson 2017) documented the use of fatty oils as a floral reward by flowering plants and their collection by a diverse group of oil-collecting bees.

Floral oil production

We now know that oil production occurs in at least 11 families and 1,500–1,800 species of flowering plants in both the Old World and the New World. Oil production occurs in the families Orchidaceae, Iridaceae, Primulaceae, Solanaceae, Calceolariaceae, Plantaginaceae, Scrophulariaceae, Stilbaceae, Krameriaceae, Cucurbitaceae, and Malpighiaceae (Neff and Simpson 2017, Renner and Schaefer 2010). Because of multiple independent origins in families like the Orchidaceae and Iridaceae, oil production is estimated to have as many as 28 separate origins and to have been lost more than 40 times (Renner and Schaefer 2010). Oil production likely first arose during the upper Cretaceous (more than 65 mya) in New World Malpighiaceae, followed by separate Eocene origins in Old World Cucurbitaceae and Primulaceae. Oil production evolved more recently in other lineages, perhaps as recently as 13 million years ago in

---

[1] There is one known exception to this stereotypic floral sonication behavior. *Amegilla murrayensis* (Apidae) in Australia uses a completely different approach to extracting pollen from poricidal flowers. Female *A. murrayensis* tap the anther cone with their heads at extremely high frequencies (~350 Hz) by rapidly vibrating their indirect flight musculature (Switzer et al. 2016).

the Calceolariaceae (Nylinder et al. 2012), a group that includes 220 species with oil-producing flowers (Sérsic 2004).

While the oil-producing flowers of *Monttea* (Plantaginaceae) (Sérsic and Cocucci 1999) and some male cucurbits (Vogel 1990) produce nectar, all other oil flowers are nectar-less, so floral oil is their primary reward. Floral oils are produced in specialized glands, termed elaiophores, that are located on the sepals, petals, tepals, and stamens of oil-producing plants. Elaiophores occur in two forms: trichome elaiophores and epithelial elaiophores. Trichome elaiophores consist of fields of hair-like glands that secrete lipids that typically pool between the glands. Trichome elaiophores are the most widespread type of elaiophore, occurring in all the oil-flower-bearing families except the Malpighiaceae and Krameriaceae. Epithelial elaiophores consist of areas of secretory epithelial cells that secrete oils beneath a thin, protective cuticle, forming a blister-like structure that must be ruptured to release the floral oils. Epithelial elaiophores are present in all the oil flowers of the Krameriaceae, Malpighiaceae, many species of Orchidaceae, and one species of Iridaceae. Floral oils have the consistency of olive oil and are typically colorless or slightly yellowish. At least some are scented. Oils of the Malpighiaceae are reported to have a mustard-like smell (Gaglianone 2005), while *Krameria* oils have a rose-like scent. Floral oils are primarily composed of long-chain (C16-C20) acetoxy-substituted free fatty acids, but considerable variability exists in the chemical composition (Neff and Simpson 2017). While the energetic value has been quantified for few floral oils, they are likely to be among the most energetically valuable floral rewards. On a per-gram basis, lipids are expected to yield over twice the energy of an equal amount of carbohydrate (Buchmann 1987). Simpson and Neff (1983) used stearic acid as a proxy for floral oil and found that stearic acid had nearly twice the energy content of an equivalent amount of pollen and nearly six times the energy content of a 40% sucrose solution (Table 7-1). Buchmann (1987) measured the energetic value of elaiophore lipids and found a slightly lower value, on average, than stearic acid. Nevertheless, the average energy content of elaiophore lipids was more than either nectar or pollen. Floral oils are energetically costly to produce but are attractive, highly valuable floral rewards for bees that can harvest them.

The "oil bees"

A total of 440 bee species in two families (Apidae and Melittidae) have become morphologically and behaviorally specialized upon oil-producing host plants (Table 7-5). These "oil bees" fall into seven tribes. Depending on how the phylogenetic evidence is interpreted, there have been between three and six separate origins of oil-collecting in bees (Danforth et al. 2013, Martins et al. 2014, Neff and Simpson 2017). These bees occur in the tropical and subtropical parts of the New World (Centridini, Tapinotaspidini, and Tetrapediini), the paleotropics (*Ctenoplectra*), and arid southern Africa (*Rediviva*), and there is one Holarctic group (*Macropis*; Fig. 7-12). "Oil bees" are found in a wide variety of habitats, including high montane regions, warm deserts, and temperate forests and grasslands, but they seem to be most common in humid tropical habitats. The floral oils are used both as a brood-cell lining and as a larval food, largely replacing

Table 7-5. Oil bees.

Family	Subfamily	Tribe	Genus	Total no. of species	Total no. of oil-collecting species	Biogeographic distribution	Oil host family
Melittidae	Melittinae	Macropidini	*Macropis*	16	16	NW/OW (Holarctic)	Primulaceae: *Lysimachia*
		Melittini	*Rediviva*	29	22	OW (Africa)	Scrophulariaceae, Orchidaceae, Iridaceae, Stilbaceae
Apidae	Apinae	Centridini	*Centris*	220	210	NW (South America)	Krameriaceae, Malpighiaceae, Calceolariaceae, Solanaceae, Plantaginaceae, Iridaceae
			Epicharis	34	34	NW (South America)	Malpighiaceae; rarely Krameriaceae
	Eucerinae	Tapinotaspidini	*Arhysoceble*	3	3	NW (South America)	Plantaginaceae, Solanaceae
			Caenonomada	3	3	NW (South America)	Plantaginaceae
			Chalepogenus	21	21	NW (South America)	Calceolariaceae, Iridaceae, Solanaceae, Plantaginaceae
			Lanthanomelissa	5	5	NW (South America)	Iridaceae
			Monoeca	14	14	NW (South America)	Malpighiaceae, Ochidaceae
			Paratetrapedia	79	79	NW (South America)	Vainly Malpighiaceae and Orchidaceae; also Iridaceae and Plantaginaceae
			Tapinotaspis	4	4	NW (South America)	Plantaginaceae, Solanaceae, Iridaceae
			Trigonopedia	6	6	NW (South America)	No records; likely on Malpighiaceae
	Xylocopinae	Ctenoplectrini	*Ctenoplectra*	19	16	OW (Africa, Asia)	Cucurbitaceae
		Tetrapediini	*Tetrapedia*	25	25	NW (South America)	Malpighiaceae, Orchidaceae; also Plantaginaceae

Figure 7-12. Habitus drawing of female *Macropis nuda* (Melittidae: Melittinae). Original artwork by Frances Fawcett.

floral nectar (Neff and Simpson 2017). There is no evidence that adult bees consume floral oils. It is notable that none of the "oil bee" taxa so far reported are social. Stingless bees occasionally visit oil flowers and collect oils, but they lack the morphological specializations (described below) for harvesting, manipulating, and transporting floral oils.

The tools that solitary bees have evolved for harvesting floral oils are highly variable, depending on the type and location of the elaiophores exploited. All oil-collecting bees have arrays of highly modified hairs for mopping up floral oils and/or scraping open epithelial elaiophores using blade-like, lamellate setae. In most cases, these hairs are located on the fore- and midbasitarsis, but sometimes they are located only on the forelegs or, more rarely, only the midlegs or even the metasomal sterna. Once collected, the floral oils are groomed onto the well-developed scopa of the hind tibia and basitarsis with the aid of the usually enlarged, strongly pectinate hind tibial spur. The scopa of "oil bees" often consists of a mix of long erect, simple, stiff hairs, and shorter, finer, more flexible, densely branched hairs. In some cases, there is only a single hair type that consists of a basal, densely branched region with a long, stiff, apical tip (Roberts and Vallespir 1978).

Macropidini

The melittid tribe Macropidini consists of three extant (*Macropis*, *Afrodasypoda*, and *Promelitta*) and two extinct (*Eomacropis* and *Palaeomacropis*; Michez et al. 2009b) genera. The 16 extant species of *Macropis* are all oligolectic bees that obtain both pollen and floral oils from *Lysimachia* (Primulaceae; Michez and Patiny 2005). Since *Lysimachia* flowers, like almost all oil flowers, produce no nectar, female *Macropis* must forage elsewhere for their energetic needs. The geographic distribution of *Macropis* matches nearly perfectly the geographic distribution of the 75 oil-producing species of *Lysimachia* (Michez and Patiny 2005), with a center of diversity in southern China. In *Macropis*, the ventral surfaces of the fore- and midbasitarsi are densely covered with odd, flattened setae with fingerlike projections that harvest the floral oils produced by the trichome elaiophores located on the lower portion of the stamens within the floral corolla (Cane et al. 1983a). Floral oils are mixed into the brood-cell provisions and are also used in the construction of the brood-cell lining (Box 7-3). A waterproof brood-cell lining is essential in a bee that nests in moist, humid soils along the edge of watercourses. The Eocene fossil *Palaeomacropis eocenicus*, discovered in 52-million-year-old amber deposits from the Oise region of France, has been proposed as the oldest oil-collecting bee based on the presence of dense plumose setae on its midtarsi (Michez et al. 2007). However, it lacks the specialized setae of modern *Macropis* as well as all the other features of an "oil bee" (broadened hindtarsal spurs, dense scopal hair), rendering its status as an "oil bee" unclear.

Melittini

The genus *Rediviva* is an iconic oil-collecting bee lineage restricted to the winter and summer rainfall regions of southern Africa. This area is both a biodiversity hotspot for flowering plants (the Greater Cape Floristic Region), as well as an area rich in enigmatic, relictual bee lineages (e.g., Fideliinae, *Patellapis*, *Samba*). The 29 species of *Rediviva* are responsible for pollinating 140 species of oil-producing flowers in four families (Scrophulariaceae, Orchidaceae, Iridaceae, Stilbaceae). The importance of *Rediviva* as a pollinator for rare plants is beautifully illustrated by work by Anton Pauw on southern African orchids. He has found that a guild of six rare orchid species in three genera is entirely dependent on a single *Rediviva* species (*R. peringueyi*) for pollination (Pauw 2006, 2007). Extinction of this one "oil bee" would likely lead to extinction of the entire guild of oil-producing orchids.

Most of the *Rediviva* host plants have trichome elaiophores, but some Iridaceae and Orchidaceae have epithelial elaiophores. Kuhlmann and Hollens (2015) described variation in the morphology of foretarsal setae across the genus. Foretarsal setae range from dense, finely branched absorptive brushes (*R. nitida*) to a mixture of finely branched hairs and lamellar, blade-like hairs for rupturing

BOX 7-3: *MACROPIS FULVIPES*—A PARTICULARLY WELL-STUDIED "OIL BEE"

Few "oil bees" have been studied in as much detail as members of the genus *Macropis* (Cane et al. 1983a, Michez and Patiny 2005, Rozen and Jacobson 1980, Schäffler and Dötterl 2011). *Macropis* has a Holarctic distribution with 4 rare species known from North America and 12 more common species known from the Palearctic region (Michez and Patiny 2005). All *Macropis* specialize on the genus *Lysimachia* for both floral oils and pollen (Michez et al. 2008). A recent study of *M. fulvipes* in a greenhouse setting (Schäffler and Dötterl 2011) provides a detailed look at "a day in the life of an oil bee."

Schäffler and Dötterl placed soil containing nests of *M. fulvipes* in a greenhouse on the campus of Bayreuth University. They provided the bees with both *Lysimachia* flowers and a variety of nectar plants and studied annual emergence and brood-cell provisioning over a four-year period. Foraging behavior of individually marked females was monitored, and the behavior of bees within their nests was observed using a flexible endoscope. For each brood cell, females make a very predictable set of (1) oil-collecting trips and (2) oil- and pollen-collecting trips. A typical brood cell requires roughly four, relatively short (four- minute) oil-collecting trips followed by roughly seven, relatively long (12-minute), oil and pollen trips. After each oil trip, females brushed the oils onto the walls of the brood cell using their hind tibial and basitarsal scopa. Females also licked the brood-cell walls with their mouthparts, potentially adding glandular secretions. Females spent slightly more than one hour provisioning a single brood cell, and each brood cell required a total of 460 floral visits (70 oil and 390 oil and pollen). Interestingly, females provisioned one brood cell in the morning and one brood cell in the afternoon, and the provisioning behavior of females in the greenhouse appeared synchronous, in that females were all making the same kind of foraging trip (oil or oil and pollen) at roughly the same time of day! This synchrony in female foraging behavior might be due to temporal variation in floral oil or pollen production by the *Lysimachia* flowers, but this was not confirmed by the authors. This detailed study provides the first in-depth view of how "oil bees" partition foraging for floral oils and lining the brood cell, and foraging for oil plus pollen for brood-cell provisioning. While previous studies had established that oils were used to line the brood cells via chemical analyses of brood-cell lining (Cane et al. 1983a), this is the first study to document exactly how and when the oils for brood-cell lining are collected.

epithelial elaiophores (*R. macgregori, R. intermixta*). The majority of *Rediviva* (19 species) specialize on the genus *Diascia* (Scrophulariaceae) for floral oils and pollen. *Diascia* flowers produce their oil at the base of paired, variously elongate floral spurs. Female *Rediviva* access the floral oils by inserting their forelegs into the spur and mopping up the oils (Fig. 7-13). Nests of *R. intermixta* that were

Figure 7-13. Female *Rediviva* mopping up oils from the elongate floral spurs of *Diascia* (Scrophulariaceae) (redrawn from Steiner and Whitehead 1990, Fig. 1).

excavated by Michael Kuhlmann (2014) confirmed that the oils are used both as larval food and to line the brood cell. *Rediviva* and *Diascia* appear to be locked in an evolutionary arms race whereby elongation of the floral spurs requires a concomitant elongation of the *Rediviva* forelegs (Steiner and Whitehead 1990). Seven of the 29 *Rediviva* species have absurdly long legs in which the forelegs exceed the adult body length (e.g., *R. longimanus*, Fig. 1C in Kuhlmann and Hollens 2015). Interestingly, extremely long legs have evolved at least twice in *Rediviva* (Pauw et al. 2017).

Oil collection has also been lost in *Rediviva*. The subgenus *Redivivoides* (previously considered a distinct genus) is a group of seven species (Kuhlmann 2012) that are nested phylogenetically within the oil-collecting *Rediviva* but no longer collect floral oils (Kahnt et al. 2017). They do have vestiges of the oil-collecting hairs on their foretarsi, supporting the hypothesis that they lost the ability to collect floral oils (Kuhlmann 2012). All *Redivivoides* occur in the driest part of the range of *Rediviva*, so as in other cases, loss of oil collecting in *Redivivoides* may be associated with living in an arid environment where the need to water-proof the brood cell may have been reduced.

Ctenoplectrini

The paleotropic tribe Ctenoplectrini consists of 2 genera: *Ctenoplectra* (with 17 oil-collecting species) and *Ctenoplectrina* (with 3 species believed to be clepto-parasites of *Ctenoplectra*; Michener 2007). *Ctenoplectra* are oligolectic, special-izing upon a small group of oil-producing cucurbits (*Momordica, Thladiantha,* and close relatives) for pollen, nectar, and oil. Like most cucurbits, the flowers are unisexual, with both sexual forms producing oil, but only staminate flow-ers secreting nectar. Females of *Ctenoplectra* differ from all other oil-collecting bees in that they use their metasomal venters to collect oils from the exposed trichome elaiophores at the base of the staminal filaments and the petals of their cucurbit hosts (Michener and Greenberg 1980, Vogel 1981). Females have stiff combs of erect, recurrent setae on the underside of the metasomal sterna (Fig. 7-14; Michener and Greenberg 1980, Vogel 1981, 1990). As they move

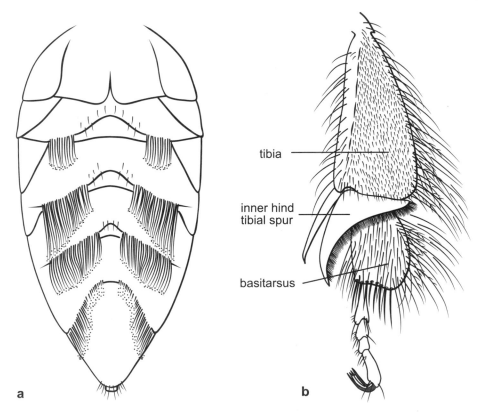

Figure 7-14. Oil-collecting structures in *Ctenoplectra*: (a) ventral view of metasoma, showing the rows of erect, comb-like setae along the posterior margins of the metasomal sterna (*C. bequaerti*; redrawn from Michener and Greenberg 1980, Fig. 25) and (b) enlarged, comb-like inner hind tibial spur used to groom oils off the metasoma (*C. albolimbata*, redrawn from Michener and Greenberg 1980, Fig. 14).

across the elaiophores, they waggle the metasoma from side to side, and oil accumulates on the metasomal combs. Oil is then groomed into the hindleg scopa using the massively enlarged, comb-like inner hind tibial spur (in Michener and Greenberg 1980, see Fig. 14). Little is known of the nesting biology of *Ctenoplectra* other than that they use preexisting cavities in wood and apparently use floral oils in nest construction.

Neotropical Apidae

The vast majority of oil-collecting bees are in the New World apid tribes Tapinotaspidini, Centridini, and Tetrapediini (Table 7-5). The host-plant associations and morphologies associated with oil collecting have been extensively documented by Neff and Simpson (1981, 2017) and Cocucci and colleagues (2000). The most common approach utilized by oil-collecting Centridini involves the use of four legs. In most species, the basitarsi of both fore- and midlegs have well-developed combs (elaiospathes) of specialized overlapping, lamellate setae. Centridines usually have an apical series of opposing giant setae on their forelegs, but these are lacking in *Epicharis*. These combs are ideally suited for harvesting oils from the epithelial elaiophores of the Malpighiaceae, their primary hosts. In fact, the stereotypic oil-collecting activities of these bees are the likely explanation for the strong floral conservatism seen in the New World Malpighiaceae (Anderson 1979, Davis et al. 2014), a group of more than 1,100 species. Interestingly, the Malpighiaceae have migrated to the Old World at least six times but without their oil-bee pollinators (Davis et al. 2014). There, in the absence of the stabilizing selection of the "oil bees," they have undergone a surprising radiation in floral form (Zhang et al. 2010). Some centridines have switched to specializing on flowers with trichome elaiophores (Neff and Simpson 1981, 2017; Vogel 1974), particularly the odd slipper flowers of *Calceolaria*. Here, the elaiophore is hidden on the lower lip of the corolla, and the elaiospathes are restricted to the forelegs, where, due to reduction of the primary elaiospathe and enlargement of the opposing setae, they have a scoop-like form (Vogel 1974). Although not as extreme as the long-legged species of *Rediviva*, one *Centris* associated with long-spurred species of *Angelonia* (Plantaginaceae) has evolved elongate forelegs in order to exploit the elaiophores within the spurs (Machado et al. 2002). Additional variants on the four-legged elaiospathe arrangement are discussed in Neff and Simpson (2017). All *Epicharis* are oil collectors, but oil collection has been lost at least three times in *Centris*, all in lineages in xeric habitats (Neff and Simpson 2017). Although data are sparse, most *Centris* and *Epicharis* are believed to use the floral oils both for nest construction and in the provisions, although it is known that in at least one oil-collecting clade (the *Hemisiella/Heterocentris* group), nest provisions are a mix of pollen and nectar, but not oil (Vinson et al. 2006). While many *Centris* and *Epicharis* are known to collect pollen from their oil-producing hosts, particularly the Malpighiaceae, often using vibratile methods, none appear to be truly oligolectic.

While the centridines are generally medium to large bees, all tapinotaspidines are on the small side. Despite their modest stature, the Tapinotaspidini display a greater diversity of oil-collecting structures than any other oil-collecting group. Most of the tapinotaspidines are associated with flowers with trichome elaiophores. Accordingly, they have various combinations of specialized setae, often involving absorbent pads of finely branched setae, on their foretarsi (Cocucci et al. 2000, Neff and Simpson 2017). *Tapinotaspis* takes a different tack and has arrays of specialized hairs only on the midlegs. Most species are primarily associated with *Neirembergia* (Solanaceae), but *T. nordestina*, a species with elongate midlegs, extracts oils from the spurs of *Angelonia* (Plantaginaceae). Other taspinotaspidines, such as *Paratetrapedia* and *Arhysoceble*, exploit epithelial elaiophores. These bees combine tarsal pads with combs formed of multiple rows of erect simple setae on their flattened forebasitarsi. In addition, *Paratetrapedia* have a giant hooked hair on the first tarsomere that probably plays a role in ripping epithelial elaiophores open. The other tapinotaspidine genus specializing on epithelial elaiophores is *Monoeca*, a genus that has essentially reinvented the four-legged elaiospathe arrangement, although in this case the combs are compound arrays of simple setae, not the giant flattened setae of the centridines. One tapinotaspidine genus, *Tapinotaspoides*, even has sternal brushes similar to those of *Ctenoplectra*, but in this case, they are used for collecting vegetative secretions, not floral oils (Melo and Gaglianone 2005). The pollen-collecting habits of the tapinotaspidines are not well known, but at least some, like *Lanthanomelissa* (Rozen et al. 2006) and some *Monoeca* (Torretta and Roig-Alsina 2016), are known to be oligolectic on the pollen of their oil hosts.

As in the tapinotaspidines, the oil-collecting structures of *Tetrapedia*, the sole genus of the Tetrapediini, are restricted to the forelegs. *Tetrapedia* are small bees that are known to collect oils from both epithelial and trichome elaiophores. Like *Paratetrapedia*, *Tetrapedia* have a compound comb on the posterior margin of the forebasitarsi, but unlike in *Paratetrapedia*, the basitarsi are strangely contorted, so that when harvesting oils from malpigh flowers, their most frequent floral hosts, the tarsal setal comb is oriented so that it works best when the bee is beneath the flower, far from the anthers and stigma. As a result, *Tetrapedia* do not pollinate malpigh flowers when collecting oils. *Tetrapedia* appear to be polylectic, sometimes collecting malpigh pollen, in which cases they may serve as a pollinator.

Male *Tetrapedia* and *Paratetrapedia* have morphological structures for harvesting floral oils that are essentially identical to those of their females (Neff and Simpson 1981) and have been observed collecting floral oils from flowers in the Malpighiaceae (Cappellari et al. 2012). Males in both genera have specialized patches of hair on their hindlegs for storing oils, and *Tetrapedia* have additional pads of wooly setae on their distal metasomal terga. The role of these floral oils in male biology is not known. Presumably they play some role in mating behavior, but exactly what is currently a mystery. This behavior is discussed in more detail in Chapter 5.

CHAPTER 8

Foraging and Provisioning Behavior

Female solitary bees are like hardworking single mothers of our own species. Just like a single human mother struggling to make ends meet, a solitary bee must juggle a number of time-consuming and exhausting tasks, such as protecting the home (nest) from predators and parasites, shopping for food (floral resources), and rearing a brood of healthy offspring (larvae). As in humans, gathering food is a dangerous business involving travel to and from the grocery store (flower patches). When the grocery store is located far from home and rush-hour traffic makes the trip unusually long, the house and children are left unprotected and vulnerable to attack. Solitary bees, like humans, are central place foragers faced with the same conundrum—how to gather food away from home while at the same time keeping your offspring safe at home. Unlike human mothers, bees cannot take their offspring along on foraging trips.

As we shall see in this chapter, solitary bees are highly specialized central place foragers. They have advanced landmark learning and navigational skills for returning to their nest, even after foraging trips over distances of several kilometers. They can facultatively vary their offspring sex ratio (daughters are produced from fertilized eggs and sons from unfertilized eggs), and they adjust the amount of pollen and nectar provided to each offspring accordingly. They are highly responsive to local environmental conditions and alter the body size and the sex ratio of their brood in response to varying floral resource abundance, foraging flight distances, and even parasitism rates. Just as in humans, age impacts athletic prowess (and foraging efficiency) such that older females become less-effective foragers over time. Solitary bees have evolved effective strategies for overcoming the challenges imposed by central place foraging, as we shall see below.

FORAGING AND THE CELL-PROVISIONING CYCLE

All non-parasitic solitary bees are mass-provisioning, which means that all the floral resources needed for complete larval development are placed within a brood cell and, once the cell is closed, the female has no further contact with

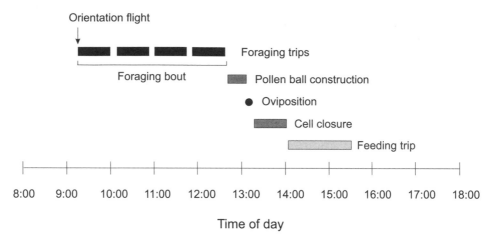

Figure 8-1. Generalized solitary bee provisioning cycle. This pattern would be typical of solitary bees in most families, including Andrenidae, Melittidae, Apidae, and Megachilidae. In some bees (e.g., Megachilinae, *Andrena*, some halictids, and species that make very liquid provisions, such as colletids, anthophorines, eucerines, and centridines), the discrete stage of pollen ball construction is eliminated because the pollen/nectar provisions are constructed gradually over the course of the provisioning cycle.

her developing offspring. Many social bees, including honey bees, bumble bees, and allodapine bees, are progressive-provisioning, meaning that adults repeatedly visit their developing larvae and provide them with food, as needed, over at least part of their larval development. Mass provisioning imposes certain requirements on solitary bees. First, they must insure that they have provided sufficient resources for the completion of larval development before closing the brood cell. This must be particularly challenging for polylectic bees, in which pollen nutritional quality varies widely depending on the type of pollen collected. Second, female solitary bees cannot directly defend the developing larvae against the diverse pathogens, parasites, and predators that can attack while she is out foraging or after her death. The only way mass-provisioning bees can provide long-term defense against such attackers is by meticulous preparation of the brood cell prior to oviposition and cell closure (as described in Chapter 6).

The typical solitary bee provisioning cycle involves a series of well-defined stages (Fig. 8-1), including (a) foraging for pollen, nectar, and, in some bees, floral oils, (b) construction of the pollen-provision mass, (c) oviposition, (d) closure of the brood cell, and then, in some solitary bees, (e) a feeding trip during which the adult female gathers nectar and pollen for her own nutritional needs.

Foraging

The provisioning of a brood cell typically involves a series of closely spaced foraging trips during which a female gathers pollen and nectar for larval consumption. These foraging trips vary in length from short (two minutes or less in *Lasioglossum* [*Sphecodogastra*] *lusorium*; Bohart and Youssef (1976) [as

Evylaeus galpinsiae]) to long (over an hour in several *Andrena* species; Gebhardt and Rohr 1987) to extremely long (nearly two hours in *Eulaema meriana*; Cameron and Ramírez 2001). Some host-plant specialists provision their brood cells with remarkable speed and intensity (see Box 8-1).

BOX 8-1: *PTILOTHRIX PLUMATA*—A NARROW OLIGOLEGE IN A RACE AGAINST A WEEVIL

The bee genus *Ptilothrix* is a member of the tribe Emphorini (sensu Michener 2007)—a New World group of narrowly oligolectic genera that specialize mostly on host plants with extremely large pollen grains, such as Cactaceae, Malvaceae, Convolvulaceae, and Onagraceae. *P. plumata* has been studied extensively in eastern Brazil, and we know a great deal about its nesting biology, foraging behavior, and life cycle (Martins et al. 1996, 1999, 2001; Schlindwein and Martins 2000; Schlindwein et al. 2009). *P. plumata* occurs from northern Brazil (Pará) to northern Argentina. This bee, like other species of *Ptilothrix* (Rust 1980), constructs shallow nests consisting of just one to two brood cells in hard-packed clay soil. *Ptilothrix* females have a trick for excavating their nests. They visit standing water (puddles, small ponds) in the vicinity of the nest site and, with their mouthparts, fill their crops with water as they float on the water surface. They then take the water back to the nest site and wet the soil surface to render it more malleable. These trips are extremely rapid (taking less than one minute). The bees use the water-collection trips both to excavate nests and to close them after provisioning.

P. plumata is a narrow host-plant specialist on plants in the family Malvaceae (Schlindwein et al. 2009). Based on microscopic analysis of pollen loads and provisions, these bees specialize on four species of *Pavonia* and, to a lesser extent, on the mallow genera *Sida* and *Herissantia*. Pollen grains of their preferred host plant (*Pavonia*) are extremely large (140–180 µm) and spiny, and they possess a thick, oily pollenkitt layer. The scopa of *Ptilothrix* consists of widely spaced hairs that facilitate the collection of these large pollen grains.

What makes this bee so remarkable is the speed and intensity with which females provision brood cells. Starting at 6:00 a.m., females make between 30 and 40 rapid foraging trips over a roughly two-hour period to provision a single brood cell. The majority of foraging trips take less than three minutes, and on each foraging trip females collect 2,000–3,000 pollen grains—equivalent to the pollen in a single, virgin *Pavonia* flower. Between foraging trips, females spend less than one minute depositing the pollen in the brood cell. Based on studies of Clemens Schlindwein and colleagues (2000, 2009), the reason this bee seems to be in such a rush is that she is in a race against another floral visitor—the weevil *Pristemerus calcaratus*. These weevils are frequent visitors to *Pavonia* flowers, but they are not effective pollinators. Female weevils damage the flower petals by frequently puncturing them with their chewing mouthparts and causing the flower to wilt prematurely. Once the petals begin to wilt, the weevil pulls the flower petals inward to create a place to sleep. Weevils can cause 50% of the flowers to close prematurely by 10:00 a.m., and nearly all flowers are closed, and inaccessible to bees, by noon. *Ptilothrix plumata* and *Pristemerus calcaratus* co-occur over much of northeastern Brazil, and the presence of the weevil appears to drive this solitary bee to forage early and extremely fast.

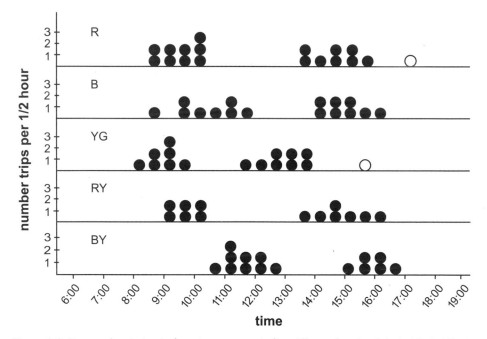

Figure 8-2. Temporal variation in foraging patterns in five different females (labeled R, B, YG, RY, and BY) sharing a common communal nest in *Perdita coreopsidis*, an oligolectic desert bee studied in southern Arizona (Danforth 1989a). Closed circles indicate pollen-collecting trips. Open circles indicate adult "feeding trips." Note that each female provisioned two brood cells in two, clearly discrete foraging bouts, constructing pollen balls, laying eggs, and closing the first cell in the gap between the first and second foraging bouts. Note also that two bees (R and YG) made feeding trips at the end of the day. Females in this species specialize on *Gaillardia* (Asteraceae).

The time spent in the nest between foraging trips is usually short relative to the length of the foraging trip. Typically, a female enters her nest, deposits her load of pollen, nectar, and sometimes oil, and quickly leaves on her next foraging trip.[1] When one observes actively provisioning solitary bees in the field or laboratory, these closely spaced foraging trips are evident as a defined series or "bout" of foraging (Figs. 8-2, 8-3). The number of foraging trips within a bout

[1] In most bees, the pollen is simply deposited in the brood cell, but in some bees, females spend additional time molding or sculpting the pollen mass before leaving on the next foraging trip.

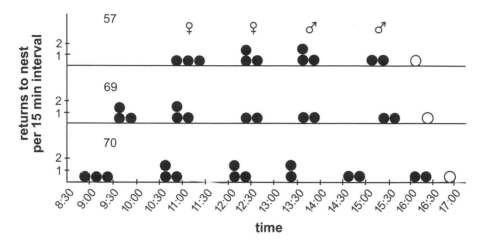

Figure 8-3. Temporal variation in foraging patterns in three individual females of *Calliopsis persimilis* over the course of a day in southern Arizona (Danforth 1990). Females in this species specialize on *Physalis* (Solanaceae) pollen. Note that each female made four to six discrete foraging bouts over the course of the day, with breaks between foraging bouts for pollen-ball construction, oviposition, and cell closure. The sexual identities of the offspring produced by the first female (#57) were determined by excavating the nests later in the season, after the larvae had completed development. In this species, foraging bouts of three trips generally corresponded to the production of a female offspring, whereas foraging bouts of two trips corresponded to the production of one male offspring. Each female made a late-afternoon feeding trip.

of foraging varies for many reasons. Females often make more foraging trips when provisioning female cells than when provisioning male cells (Danforth 1990) (Fig. 8-3). Foraging trips may take longer when floral resources are in short supply or are located far from the nest (Minckley et al. 1994). Female solitary bees also make more foraging trips per cell as they age, and their ability to carry full pollen loads decreases (Sugiura and Maeta 1989).

One conspicuous aspect of the first trip of the day is the orientation flight (Fig. 8-4) that females make on the first departure from the nest. That solitary, ground-nesting wasps perform orientation flights when they leave the nest was first demonstrated by Tinbergen and Kruyt in the 1930s in a series of classic experiments using the beewolf (*Philanthus triangulum*; Tinbergen 1932, Tinbergen and Kruyt 1938). Tinbergen noticed that wasps leaving their nests for the first flight of the day flew a distinctive, looping, zigzag flight. The initial loops are short and flown at low elevation (just above the soil surface), but the loops become larger as the wasp gains altitude. The wasp flies back and forth in an arc of increasing radius as she backs away from the nest entrance, with each successive loop giving her a larger and larger view of the surrounding landscape. In a series of simple (but elegant) experiments, Tinbergen demonstrated that landmark learning was occurring by placing a ring of pine cones around the nest entrance before the orientation flight and then displacing the ring of pine cones nearby (but away from the nest entrance). Wasps invariably returned to the center of the ring of pine cones, but not to the actual nest entrance.

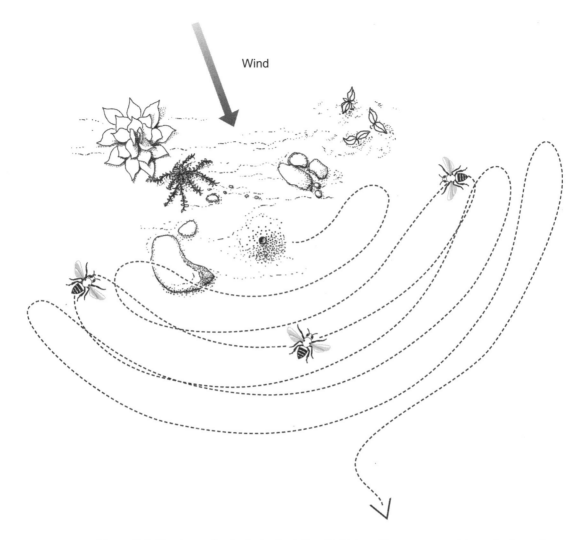

Wind

Figure 8-4. Structure of a typical orientation flight in solitary wasps and bees (redrawn from Tinbergen [1932]).

Orientation flights of this kind have since been described in a variety of solitary and social wasps and bees, including *Odynerus* and *Vespula* (Vespidae), *Cerceris* and *Philanthus* (Crabronidae), *Dasypoda* (Melittidae), *Andrena* (Andrenidae), *Lasioglossum* (Halictidae), and *Apis* and *Bombus* (Apidae) (Zeil and Kelber 1991, Zeil et al. 1996). The structure of these orientation flights is remarkably similar among both wasps and bees (Zeil and Kelber 1991). That orientation flights are so widespread suggests that they are likely present in all nest-making Apoidea. Even cleptoparasites may be able to learn landmarks and therefore relocate potential host nests (Rosenheim 1987).

Since the original description by Tinbergen, research by Zeil and colleagues (Brünnert et al. 1994, Stürzl et al. 2016, Zeil and Kelber 1991, Zeil 1993, Zeil et al. 1996, Zeil 2012) has provided a more detailed picture of the structure and function of these orientation flights in both wasps and bees (Zeil and Kelber

1991). High-speed video recordings show that as the wasp or bee moves along each looping arc of the orientation flight, she maintains the nest entrance at an angle of approximately 30–70° relative to the long axis of the body and between 30° and 60° below the horizon. Only at the end of each looping arc, when the bee changes direction, does she actually face directly at the nest entrance. By maintaining the constant angle and elevation as she moves back and forth across the nest entrance, she is effectively fixing the image of the nest entrance in a single region of the compound eye. Zeil and colleagues have also shown that, on the return to the nest, *Cerceris* wasps follow a very similar looping flight, apparently matching the configuration of landmarks to the original images stored on departure (Stürzl et al. 2016). Matching images obtained on the return flight to those stored in memory from the original orientation flight allows wasps (and presumably bees) to reliably relocate their nests.

Once a female has successfully returned to her nest after a foraging trip, she no longer repeats the orientation flight. It seems that once the landmarks are learned and the nest successfully relocated, females do not waste their time relearning this information. However, if a female has trouble finding her nest, she will repeat the orientation flight on a subsequent departure, indicating that there is some flexibility in the timing of the orientation flight. The first departure of each new day involves an orientation flight, presumably because landmarks can change overnight and nest relocation is an essential aspect of central place foraging. The ability of solitary bees (and wasps) to relocate their nests is remarkable. Keep in mind that the nest entrance is often filled with soil upon departure, that nests are often located in a relatively homogeneous, sparsely vegetated landscape of sand and rocks, and that there may be hundreds of other nests in the neighborhood. Interestingly, we have observed some male bees conducting orientation flights as they leave nests (Danforth and Neff 1992, Neff and Danforth 1991). This behavior occurs in species of *Macrotera* (*M. texana*), in which males commonly enter the nests of females late in the day, where they remain until the following morning. We believe this behavior reflects a male mating strategy in *M. texana*. Males entering nests late in the day have the possibility of mating with females immediately prior to oviposition, which can be a strategy for increased paternity insurance (Danforth and Neff 1992). This strategy can work most effectively if males can retain information on the location of nests, especially nests that have a large number of reproductively active females. Apparently, males in this species, and possibly others in which mating takes place within the nest, have evolved the ability to learn landmarks to help them relocate nests. Males that reuse communal sleeping sites are presumably also using landmarks to relocate these sites (Chapter 5).

The number of foraging trips needed to provision a brood cell varies widely across solitary bees. Some species carry large pollen loads and provision cells with just a few trips, while other species carry smaller pollen loads and provision cells with many short trips. Neff (2008) reviewed data on the number of trips required to provision a cell and found a range of just two trips per cell (in some small species of *Perdita*; Danforth 1989a) to up to 40 trips per cell (in *Megachile pugnata*; Frohlich and Parker 1983). As pointed out by Neff (2008), caution needs to be taken when interpreting published studies because it is not always clear whether authors are reporting the number of trips per cell or the number of trips per day. These numbers are not equivalent, because some female solitary bees can (and do) provision multiple cells per day and others sometimes require more than one day to provision a brood cell.

One factor that determines the number of foraging trips needed to provision a cell is the quantity of pollen carried per foraging trip. As described in Chapter 7, bees carry pollen in a variety of ways. Pollen can be carried externally in a scopa located on the legs, mesosoma, or metasoma or internally in the crop. Some bees carry pollen in a loose, dry form or agglutinated with nectar or, sometimes, floral oils. Neff (2008) analyzed load weights in solitary bees as a fraction of adult dry or wet body weight (Table 8-1). When load weights are measured as a percentage of adult dry body weight, the loads carried by solitary bees varied from 18% (*Andrena rudbeckiae*; Neff and Simpson 1997) to 124% (*Perdita difficilis*; Danforth 1989a). When load weights are measured as a percentage

Table 8-1. Components of cell provisioning in solitary bees (Table II from Neff 2008).

Species are arranged by the efficiency of provisioning (provisioning rate), which is measured as percent of adult body weight (BW) gathered per hour of foraging.

Species	Family	Mode	Pollen load wt (% BW dry)	Pollen load wt (% BW wet)	Trips per hour	Provisioning rate (% BW/hr)
Melissodes rustica	Apidae	dry		16.7	1.4	23.3
Andrena rudbeckiae	Andrenidae	dry	18.3	10.5	1.46	26.7
Osmia lignaria	Megachilidae	dry		7.1	6.7	47.3
Exomalopsis solani	Apidae	agglutinated	60		0.94	56.4
Panurginus polytrichus	Andrenidae	agglutinated	52	25.1	1.12	58.2
Andrena sitiliae	Andrenidae	dry	36.3	19.2	2.12	77
Dieunomia triangulifera	Halictidae	dry	55.5		1.74	96.6
Andrena senticulosa	Andrenidae	dry	39.7	16.8	2.68	106.4
Macrotera portalis	Andrenidae	agglutinated	68		2.13	144.8
Diadasia afflicta	Apidae	dry	21.7	64.1	9	195.3
Pseudopanurgus aethiops	Andrenidae	agglutinated	38.9	17.6	5.1	198.4
Diadasia australis	Apidae	dry	32.7	13.2	6.86	224.3
Pseudopanurgus rugosus	Andrenidae	agglutinated	37	23.3	6.7	247.9
Macrotera texana	Andrenidae	agglutinated	38.1	14.3	6.57	250.3
Perdita difficilis	Andrenidae	agglutinated	124		2.06	255.4
Hesperapis sp. A	Melittidae	agglutinated	116.9	50.3	2.61	305.1
Diadasia rinconis	Apidae	dry	34.2	12.4	9	307.8
Hesperapis sp. B	Melittidae	agglutinated	89.3		5.28	471.5
Calliopsis persimilis	Andrenidae	agglutinated	105		4.8	504

of adult wet body weight, the loads carried by solitary bees ranged from 7% (*Osmia lignaria*) to 64% (*Hesperapis* sp. A[2]). Smaller bees might be expected to carry proportionally more pollen per trip, because wing loading decreases with overall body weight. Neff (2008) found this relationship supported by the dry-weight data available to him (the data he considered most reliable) but not wet weights. Given the small sample sizes, more data are needed to confirm or refute this relationship.

Stone (1994), in a study of *Anthophora plumipes*, showed that load weights increased with air temperature; hotter bees can carry heavier loads. This makes sense given that thoracic temperature is correlated with the power output of the flight musculature. The difference was quite striking: a 15°C increase in ambient temperature resulted in a doubling of the total weight of pollen and nectar carried by foraging females (Stone 1994). Bees in the family Megachilidae, which carry pollen in the metasomal scopa, tend to make many short foraging trips with small pollen loads, presumably because the quantity of pollen that can be carried in the metasomal scopa is small relative to the quantity that can be carried on the legs and adjoining regions of the mesosoma. There is some evidence that oligolectic bees that collect extremely protein-rich pollen may be able to provision cells more quickly than those that collect protein-poor pollen, simply because less pollen needs to be collected for successful larval development when the protein content is high (Danforth 1990).

Construction of the pollen-provision mass

The provisions that solitary bees construct for their offspring vary widely in shape and consistency, ranging from amorphous soupy concoctions to tidy symmetrical spheres or loaves (Linsley 1958, Michener 1964). Soupy mixes of pollen and nectar are characteristic of the Colletidae (Figs. 6-7b, 6-8a, 8-5a, 9-3); extreme cases are found in the Diphaglossinae, in which provisions are nearly completely liquid (Roberts 1971). As these provisions contain yeasts and have been observed to ferment, it has been suggested that for these bees, yeasts are taking the place of pollen as a protein source for developing larvae (Chapter 9). Other groups that produce fairly liquid provisions include the Oxaeinae (Andrenidae) and some Eucerini, Anthophorini (Fig. 8-5h), and Centridini (Apidae). There are relatively few direct observations of how females construct these soup-like provisions. Norden (1984) observed female *Anthophora abrupta* depositing loose pollen on the cell floor after their first few provisioning trips, then switching to tamping in pollen they mixed with nectar and Dufour's gland secretions before finally adding just nectar on the final provisioning trips. This pattern of adding most of the nectar late in the provisioning bout seems to be typical of solitary bees, whether the pollen provisions are solid or soupy.

The impervious, cellophane-like coating of the colletid brood cell is almost certainly an adaptation for containing highly liquid provisions, while the other taxa mentioned above all have strong, waterproof cell linings. Some females have been observed adding glandular secretions to the soupy pollen/nectar provisions. Torchio (1984a) described how female *Hylaeus bisinuatus* add a volatile

[2] *Hesperapis* sp. A was incorrectly labeled in the Neff (2008) Table II as *Diadasia afflicta*.

substance from the apex of the metasoma that gives the pollen provisions a "slightly rancid" odor, while Batra and Norden (1996) provide a detailed description of how female *Anthophora abrupta* and *A. villosula* add glandular secretions from the Dufour's gland by "churning" the Dufour's gland products into the pollen/nectar provisions. It is not clear how widespread the addition of such glandular secretions is or what role these secretions play. Perhaps they prevent microbial attack or are a nutritional supplement for developing larvae.

Soupy provisions are the exceptions among solitary bees. In most bees, the pollen/nectar provisions are either loosely constructed masses or more compact loaves or balls. In Lithurginae (Megachilidae), female bees fill their elongate brood cells with loose pollen; the only structured portions are slightly moistened pollen disks that form one side of a small cavity in which the egg is laid (Fig. 8-5f) (Rozen and Hall 2014). Similarly, female Fideliinae construct largely amorphous provisions consisting of large, relatively loose pollen masses slightly moistened with nectar. In some cases, this mass is a cuplike structure with a concave face placed in the rear of the cell (*Neofidelia profuga*; Rozen 1973), while in others, the cup is closed, forming a central cavity (*Fidelia villosa, F. pallidula*; Rozen 1970, 1977a) (Fig. 8-5e). Such loosely constructed pollen masses, constructed in unlined nests in the soil, are probably possible only in extremely arid environments, such as the Atacama Desert for *Neofidelia* and African deserts for *Fidelia*.

Other bees that construct minimally structured provision masses include some emphorines, like *Diadasia*, which typically pack dry pollen into their waterproofed, urn-shaped cells (Neff et al. 1982). A variant on this pattern is seen in *Pararhophites*, whose provisions consist of a moist pollen and nectar mass placed in a rough, cuplike receptacle formed from sand and nectar in the unlined cell (Fig. 8-5d) (McGinley and Rozen 1987). Most members of the Megachilinae pack a gooey mix of pollen and nectar into the rear of their cells. The faces of such provision masses are typically flat (Fig. 6-16b) or a simple diagonal slant (Fig. 8-5g). Nectar is added to the enlarging provision mass after each pollen load and is mixed into the provision mass by working with the mandibles. In *Megachile rotundata*, the percentage of nectar as a proportion of the total amount of nectar and pollen added each trip increases through a provisioning cycle, so that while the first trips are nearly all pollen, the last trips are nearly all nectar (Klostermeyer and Gerber 1969). Most megachilid brood cells lack a waterproof lining and are thus ill suited for truly soupy provisions. Oil-collecting bees, such as *Tetrapedia, Epicharis* (Fig. 6-17), *Monoeca, Ctenoplectra* (Sung et al. 2009), and *Rediviva* (Kuhlmann 2014), produce amorphous, pasty provision masses consisting of pollen mixed with floral oils. Other oil-collecting bees, like *Chalepogenus caerulea* (Vogel 1974), fill the bottom of the cell with a pasty provision mix and then form additional provisions into a rough ball or projection that does not touch the cell walls.

In most S-T bees (excluding Colletidae and Oxaeinae; see above) and some L-T bees (e.g., Xylocopinae *sensu sticto* and many Eucerinae), pollen is molded into a discrete loaf or ball. Most Halictidae, Andrenidae, and many Melittidae produce a more or less spherical pollen ball (Figs. 3-3b, 3-6a, 8-5b, 9-1). However, there are many variants. Some *Andrena* produce kettle-shaped provisions (Johnson 1981, Stephen 1966), and various *Dieunomia* (Halictidae) produce

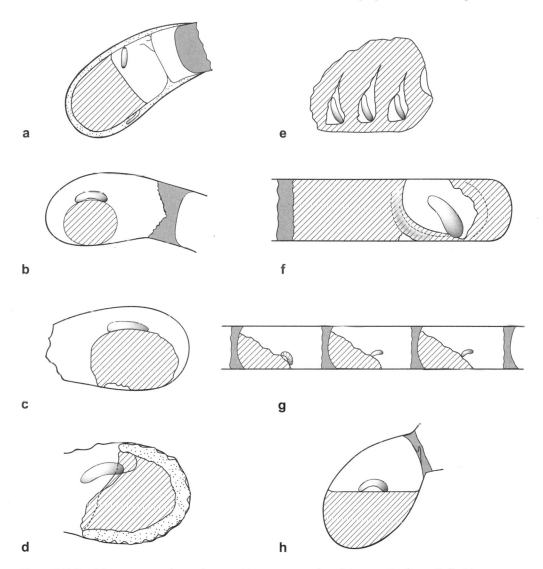

Figure 8-5. Provision masses and egg placement in some ground- and stem-nesting bees. Colletidae: (a) *Colletes compactus* (redrawn from Rozen and Favreau 1968, Fig. 1); Andrenidae: (b) *Meliturgula braunsi* (redrawn from Rozen 1968, Fig. 2); Melittidae: (c) *Macropis nuda* (redrawn from Rozen and Jacobson 1980, Fig. 4); Megachilidae: (d) *Pararhophites orobinus* (redrawn from McGinley and Rozen 1987, Fig. 13); (e) *Fidelia pallidula* (redrawn from Rozen 1977a, Fig. 13); (f) *Lithurgus chrysurus* (redrawn from Rozen and Wyman 2014, Fig. 7); (g) *Osmia lignaria* (redrawn from Torchio 1989a, Fig. 9); Apidae: (h) *Anthophora plumipes* (redrawn from Loonstra 2012, Fig. 6).

rimmed, cupcake-like masses (Stephen et al. 1969). In most *Anthophorula* and some *Exomalopsis* (Apidae: Exomalopsini), the loaf-like provisions have an anterior foot (Rozen 2011), as do the very similar provisions of *Macropis nuda* (Fig. 8-5c) (Rozen and Jacobson 1980). Provision masses in *Dasypoda thoracica* (Melittidae) are spheroidal in shape, with three small feet (Celary 2002). The loaf-like provision masses of many *Xylocopa* species are attached to the rear of the cell, with a small foot supporting the anterior end (Gerling et al. 1983). These spherical shapes and the various legs and protuberances likely serve to minimize contact of the provisions with the brood-cell wall. Some bees, including many Calliopsini (*Acamptopoeum, Arhysosage, Calliopsis*, and *Spinoliella*; Rozen 1958, Rozen and Yanega 1999, Rozen 2008b, 2013) as well as some Rophitinae (*Dufourea, Conanthalictus*; Rozen 1993, Torchio et al. 1967), provide additional protection for the provision mass with a glandular coating. In some lineages, one can see a trade-off between lined cells and unlined provisions. All *Perdita*, a panurgine group with unlined cell walls, have lined spherical pollen masses (Danforth 1989a, Rozen 1967a), while in its sister group *Macrotera*, the cell walls are lined but the pollen balls are unlined (Danforth 1991b). While interesting, this trade-off is hardly universal. All the other taxa with lined provision balls have lined cells, while in *Hesperapis* (Melittidae), both the spherical pollen balls and the cell walls are unlined.

Molded provision masses may be constructed progressively as the cell is being provisioned (e.g., in many *Andrena*, most Calliopsini, some Rophitinae, and *Hesperapis*), at the end of the provisioning cycle when all the pollen has been stored in the brood cell (e.g., in *Macrotera portalis*; Danforth 1991b) or in some hybrid manner. In *Agapostemon texanus*, for example, females deposit the first three or four pollen loads as a loose mass before forming a spherical mass to which the female adds additional pollen and nectar (Roberts 1969). In species where the pollen ball is constructed progressively, the pollen ball in its early stages is usually much drier than the final product (Rozen 1958).

Oviposition

The bee egg is a gently curved, sausage-shaped structure of varying size (described in more detail in Chapter 3). In most pollen-collecting bees, eggs are deposited on the surface of the provisions (Fig. 8-5b,c,d). In spherical or near-spherical provisions, the egg is on top of the mass, while in more loaf-like provisions, it is usually on the top or the anterior face or, more rarely, as in some *Ceratina*, the posterior face of the provision mass. In species with liquid provisions, the egg usually floats on the surface (Fig. 8-5h). However, eggs can also be glued to the roof of the brood cell (in *Colletes*; Fig. 8-5a) (Rozen and Favreau 1968, Torchio 1965, Torchio et al. 1988). In Emphorini (Apidae), the egg is placed in a groove beneath the provision mass (Martins et al. 1996, Rust 1980). In the megachilid subfamily Lithurginae, one or more eggs may be placed in cavities within a massive pollen-provision mass (Fig. 8-5f) (Rozen and Hall 2012, Rozen and Wyman 2014, Sarzetti et al. 2012). In the Fideliinae, the egg is usually placed in a concavity on the face of the provision mass, but in *Fidelia pallidula*, a series of such masses are constructed (Fig. 8-5e), so that the inner eggs end up inside a cavity (Rozen 1977a). *Tetrapedia* place their eggs in a cavity within the pollen-oil provision mass that largely fills their cells. According to Camillo (2005), oviposition occurs before provisioning is completed, but Alvez-dos-Santos and colleagues (2002) observed females ovipositing in completed provisions. In *Osmia*, most species place the egg on the front of the slanted provision mass (Fig. 8-5g), but some place the egg in a cavity within the pollen mass, so this trait can be labile, even among closely related species (Torchio 1989a). Egg position is not necessarily a fixed trait within a species. A female *Andrena perplexa* usually places the egg in a cuplike depression on top of the provision mass, but if the provisions are unusually moist, she may attach the egg to the cell wall (Stephen 1966).

Closure of the brood cell

Once the pollen-provision mass has been prepared and the egg (or eggs) laid, solitary females typically close the brood cell. In wood excavators and cavity-renting bees, the closures also serve as the partition separating cells and are discussed below. In most ground-burrowing bees, a cell is closed with a thin soil plug. The closure is built in a spiral manner, and the resulting spiral structure in usually evident on the inner surface of the closure (Fig. 8-6). These closures are typically flat or slightly concave, but in some *Centris*, they have a distinct medial, spout-like protuberance that apparently aids in air exchange (Rozen and Buchmann 1990). Cells of the basal megachilids (e.g., the fideliines [Rozen 1970, 1977a] and pararhophitines [McGinley and Rozen 1987]) lack special closures, loose soil in the closed burrow filling that function. Well-defined cell closures are also absent in the nests of various *Hesperapis* species (Rozen 1987a), although other melittids construct spiral cell closures (Celary 2002, Rozen 1977c, Rozen and Jacobson 1980). In most colletids, the cell closure is formed with the same

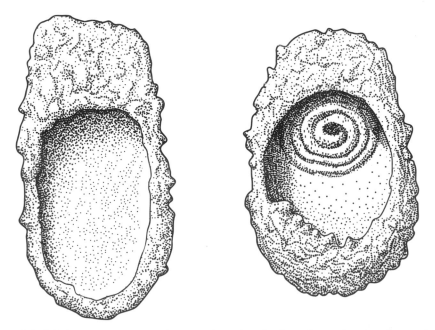

Figure 8-6. Spiral closure in a ground-nesting bee: *Ancylandrena larreae* (Andrenidae) (redrawn from Rozen 1994, Fig. 4).

cellophane-like material used to line the cell (Almeida 2008, Torchio 1965), but the odd, goose-necked cells of the diphaglossines have thin, circular closures formed by a disk of cotton-like material (Rozen 1984b).

In some bee groups, the first stage of cell closure is started well before the cell is fully provisioned. Some *Tetrapedia* and *Centris* construct a "collar" at the cell entrance during the provisioning phase (Alves-dos-Santos et al. 2002, Jesus and Garófalo 2000). Many cavity-renting megachilids construct something called "Fabre's threshold." Fabre (1914) noted that many of the "bramble bees" he was observing constructed a small rim at the front of the cells they were working on. Fabre wondered if these rims were a measuring device so that when provisioning, the bees would know how long the provision mass should be. Although some subsequent authors have accepted that explanation, Fabre pointed out that it is incorrect since when burrow diameter is very close to the width of the bee, the bees often omit the rim entirely, yet they have no problem constructing a provision mass of the proper size. Rather, the threshold serves as the start of the cell closure. In wider burrows, it may be a complete disk, with a central hole just large enough for the female to pass through. With this design, the time to close the cell following oviposition is minimized, regardless of cell diameter, reducing exposure of the brood cell to pests and parasites.

Adult feeding trips

Some solitary bees end the day with what has been described as a "feeding trip." This behavior was first reported in ground-nesting *Dasypoda plumipes* (Lind 1968) but has since been observed in many species. Feeding trips differ from provisioning trips in that the females return to the nest without external pollen loads. Bees that do not provision every day, such as many *Andrena*, typically make a long feeding trip on non-provisioning days (Neff and Simpson 1997). In other cases, feeding trips are made at the end of the day after oviposition and cell closure. Feeding trips tend to be longer than normal provisioning trips, typically lasting 30 minutes or more, although in *Calliopsis persimilis*, they averaged only 9.4 minutes (Danforth 1990).

Dissections of females returning from feeding trips typically reveal that they are carrying large quantities of pollen internally (in the crop). Females in the process of cell provisioning often have a nearly empty crop or a crop filled with nectar and only trace amounts of pollen. Feeding trips are likely essential for egg production in solitary bees. Eggs are costly to produce in terms of protein, and females must be able to replenish their protein stores on a regular basis in order to produce eggs. This would be especially true in bees that produce multiple cells per day (e.g., *C. persimilis*; Box 8-2) or those that have especially large eggs (e.g., *Xylocopa*). A detailed study of *Nomia melanderi* (Cane et al. 2017) indicates that the volume of the pollen consumed by adult females on a daily basis is a small fraction (10%) of the pollen collected for a single brood cell. The solitary, carnivorous wasps from which bees arose also need protein as adults in order to produce eggs. However, they obtain their protein in a different way—hunting females will occasionally feed on the blood and bodies of their arthropod prey (Evans and West-Eberhard 1970).

BOX 8-2: *CALLIOPSIS PERSIMILIS*—THE RECORD HOLDER FOR DAILY OFFSPRING PRODUCTION

Calliopsis persimilis is a hardworking bee. This species holds the record for the number of brood cells provisioned per day by a solitary species under natural conditions. *C. persimilis* is a ground-nesting, solitary, andrenid bee that occurs from southern California and Arizona eastward to southwestern New Mexico and southward to the Baja Peninsula and northwestern Mexico (Danforth 1994). Females are active in the late summer (August–October) in association with the monsoonal rains in the Chihuahuan Desert. Females are narrowly host-plant specific on members of the genus *Physalis* (Solanaceae). *C. persimilis* constructs shallow nests (5–12 cm deep) in sparsely vegetated, silty soil. The foraging behavior of *C. persimilis* was studied in detail by Danforth (1990) in the vicinity of Animas, New Mexico.

Female *C. persimilis* can provision up to six cells per day—a remarkable number compared to many solitary bees. During the night, females construct a single brood cell at the end of a short lateral tunnel. They begin foraging by 8:00 a.m. the following day and provision this first brood cell with three foraging trips. Foraging trips are short (~10 minutes in length), and foraging trip times become shorter as the day warms up. A female constructs the pollen mass, lays an egg, and closes the brood cell. She then constructs the next brood cell immediately in front of the first by enlarging the lateral tunnel—a trick that allows females to construct all the subsequent cells in a very short period of time. It takes females just 27 minutes, on average, to construct these additional brood cells within the confines of the original lateral tunnel. Subsequent cells are all provisioned with two or three foraging trips. The occurrence of foraging trips over the course of the day reveals a clear pattern of discrete foraging "bouts," each with either two or three trips per bout (Fig. 8-3). However, the sequence of these two- and three-trip provisioning bouts is not random; the earliest cells of the day are invariably provisioned with three foraging trips, whereas the later cells are provisioned with just two trips (Fig. 8-3). By excavating nests at particular times during this process, Danforth (1990) determined that cells provisioned with three trips were mostly female cells and those provisioned with two trips were always male cells (Fig. 8-3).

Females clearly shift from investment in females in the morning to males in the afternoon. Why they do this is not clear. It is possible that floral-resource availability is driving this shift. If floral resources are most abundant in the morning, it might make sense for females to invest in the more costly sex (females) when resources are easy to collect and the less costly sex (males) when resources are in short supply. While females can produce six cells per day, the median number of cells produced per day is actually one. This is because many early-season cells (shallow cells containing the oldest larvae) were often located singly, not in series. Females likely produce just one brood cell per day early in their foraging career before "scaling up" to a larger number of cells per day as they become more experienced foragers. Females in this study produced a maximum of 17 brood cells over the course of their lifetime.

How can *C. persimilis* produce so many offspring per day? It is possible that their use of *Physalis*, a highly protein-rich pollen (Roulston and Cane 2000), allows females to produce an unusually large number of eggs per day. Another possibility is that by provisioning their brood cells with protein-rich pollen, they do not need to collect as much pollen per larva as a bee that collects protein-poor pollen. The fact that *C. persimilis* has the highest rate of cell provisioning of any solitary bee, when measured as a percentage of total adult body weight per hour (Table 8-1), supports this hypothesis.

SEXUAL DIMORPHISM AND THE PROVISIONING CYCLE

In species of bees with clear sexual size dimorphism—which is to say the vast majority of bees—females alter both the brood-cell size and the amount of pollen/nectar provisions collected according to whether they are producing a male or female offspring. In most, but not all, solitary bees, females are larger than males, and provisioning mothers make larger brood cells and collect more pollen for their daughters than for sons. In *Calliopsis persimilis* (Danforth 1990) (Fig. 8-3; Box 8-2), females typically make just two foraging trips when they are provisioning sons and three when they are provisioning daughters, and they spend, overall, less time provisioning male cells than females cells. The fact that a female constructs an appropriately sized brood cell prior to cell provisioning indicates that she is making the decision on whether to provision a son or a daughter *before* she has even started foraging for pollen and nectar (Rooijakkers and Sommeijer 2009). Once a female is committed to making a male or a female, she typically provisions the cell with the appropriate amount of food and lays an egg of the appropriate gender. However, females do appear to either make mistakes or change their minds. Rooijakkers and Sommeijer (2009) found that female *Colletes halophilus* produced male offspring in female-sized cells on occasion. This decision to switch from producing a daughter to producing a son even though the cell had been constructed for a daughter is likely driven by resource availability. When resources are limited, females may be forced to alter their preliminary sex-allocation decisions.

STUDIES OF INTRA-NEST BEHAVIOR IN SOLITARY BEES

While nest excavation and careful examination of pollen provisions can reveal an enormous amount about the process of cell provisioning in solitary bees, a more complete understanding of this process can be obtained only by establishing bees in observation nests (Table 8-2). For many cavity-renting bees, this can be done with glass tubes (Torchio 1984a), split canes, or grooves in wood with one side covered with a transparent material (Matthews and Fischer 1964).

Table 8-2. Studies of intra-nest behavior in solitary bees.

Family/species	Reference
Colletidae	
Colletes kincaidii	Torchio et al. 1988
Hylaeus bisinuatus	Torchio 1984a
Halictidae	
Lasioglossum erythrurum	Knerer & Schwarz 1976, Kukuk & Schwarz 1987
Neocorynura fumipennis	Batra 1968
Agapostemon radiatus	Roberts 1969
Agapostemon splendens	Roberts 1969
Agapostemon texanus	Roberts 1969
Augochlora pura	Stockhammer 1966
Nomia melanderi	Batra 1970
Andrenidae	
Macrotera portalis	Danforth 1991a,b
Melittidae	
Macropis fulvipes	Schäffler & Dötterl 2011
Megachilidae	
Dianthidium ulkei	Frohlich & Parker 1985
Megachile pugnata	Frohlich & Parker 1983
Megachile rotundata	Klostermeyer & Gerber 1969, Gerber & Klostermeyer 1970
Osmia bruneri	Frohlich 1983
Osmia californica	Torchio 1989a
Osmia lignaria	Phillips & Klostermeyer 1978, Torchio 1989a
Osmia marginata	Parker & Tepedino 1982b
Osmia montana	Torchio 1989a
Apidae	
Anthophora abrupta	Batra & Norden 1996
Anthophora villosula	Batra & Norden 1996
Xylocopa pubescens	Gerling et al. 1981, 1983
Xylocopa sulcatipes	Gerling et al. 1983, Stark et al. 1990
Ceratina flavipes	Sakagami & Maeta 1987
Braunsapis sauteriella	Maeta et al. 1985
Euglossa townsendi	Augusto & Garófalo 2004
Euglossa fimbriata	Augusto & Garófalo 2009
Euglossa cordatra	Garófalo 1985
Euglossa atroveneta	Ramírez-Arriaga et al. 1996
Euglossa viridissima	Cocom Pech et al. 2008
Euglossa carolina	Augusto & Garófalo 2011

Ground-nesting species are more difficult to deal with, but with some prodding, a number of species have been coaxed to establish nests in ant-farm-like observation structures consisting of soil sandwiched between two transparent panels (Danforth 1991a). This technique was developed for social halictines (Michener and Brothers 1971) but has proven amenable for other taxa. In bees in which the cells are only a short distance from the entrance, much can be learned from direct observations, often aided with an ophthalmoscope for better illumination. Bees that construct their nests in large cavities, like many euglossines, can be observed in larger observation boxes, often connected to the outside to permit foraging. Since bees have very low sensitivity to red light, red lights are often used to illuminate the interior of the nest. An alternative technique for studying the in-nest behavior of wood-nesting bees is the use of X-rays (Gerling et al. 1981). These studies have revealed important details about the provisioning behavior of solitary bees, including how they prepare and waterproof their brood cells, how pollen and nectar are deposited in the brood cell, how and when glandular secretions are added to the provision mass, how the pollen/nectar provision mass is constructed, how females lay eggs, and how the cell is closed at the end of the provisioning cycle. Most such studies have been conducted in flight cages or other controlled environments (Torchio 1984a, Torchio et al. 1988, Torchio 1989a), but some studies have been done entirely under field conditions (Danforth 1991a).

CELL-PROVISIONING RATES

The number of cells provisioned per day, and hence the number of offspring produced per day, varies widely among solitary bees (Neff 2008). Most solitary bees probably provision, on average, about 1 cell per day; however, there is substantial variation among species. Under ideal, greenhouse conditions with essentially unlimited floral and nesting resources, a female *Osmia lignaria* was able to complete 9 cells in a 24-hour period (Torchio 1989a). This is an amazingly high number for a species usually requiring 35 trips for a female cell and 31 for a male cell (Phillips and Klostermeyer 1978). Under field conditions, the highest rate of cell provisioning reported is by *Calliopsis persimilis*, which occasionally completes 6 cells per day and regularly completes 4 cells per day (Danforth 1990). While these rates are impressive, it is important to note that they are not sustained for long periods of time. *O. lignaria* normally average a little over 1 cell per day (Torchio 1989a), while *C. persimilis* averages only 1 cell per day over its lifetime (Danforth 1990). In all these cases, egg limitation seems to constrain the number of cells that can be completed.

For short-term rates, the current champions are individuals of *Lasioglossum* (*Sphecodogastra*) *lusorium*, which can provision 4 cells in under 2 hours (Bohart and Youssef 1976). *L. lusorium* is a specialist on *Oenothera* and normally forages for an hour or so early in the morning and again for a brief period at dusk. Several factors contribute to its extremely rapid provisioning rates. First, unlike the other taxa mentioned above, all the cells to be provisioned in a day are constructed before the start of provisioning. Second, only 3 or 4 pollen trips are required per cell. Third, pollen trips are extremely rapid: 81% of morning trips take less than 5 minutes, and 64% of evening trips less than 2 minutes. In the Bohart and Youssef study, *L. lusorium* was nesting amid its host plants, so foraging distances were very short. The females apparently are very efficient at handling the very large *Oenothera* pollen grains, so floral handling times are brief. Finally, females do not always oviposit in all provisioned cells immediately after completing a provision mass. Some provisioned cells are often left open for extended periods before oviposition, potentially exposing the cell to predators and parasites. Parasite pressure may explain why few bees are known to construct multiple cells before starting provisioning.

In contrast, many bees construct fewer than 1 cell per day, some averaging a cell every 4 days or more (Field 1996, Gebhardt and Rohr 1987, Giovanetti and Lasso 2005). Besides inclement weather, the prime suspects for low rates of cell completion are resource availability, reflected in low provisioning rates, and egg limitation (the lack of mature eggs for oviposition). Resource limitation is seen in the many species that simply do not harvest enough resources per day to provision a cell. In some cases, bees are able to provision a cell of the smaller sex (usually a male) in one day, but not a female (Neff and Simpson 1997, Neff 2003). Egg limitation also seems to be widespread among solitary bees. It is probably the major reason for cell-completion rates of only 1 cell per day for species with high provisioning rates, such as *Macrotera texana* (Neff and Danforth 1991). It is also the likely explanation for the common phenomenon of pollen-foraging days followed by one or more days of nectar-only foraging, seen in many *Andrena* (Bischoff et al. 2003, Neff and Simpson 1997) and other taxa (Field 1996, Minckley et al. 1994). Early-spring *Andrena* have particularly low rates of cell provisioning (Giovanetti and Lasso 2005), perhaps due to low ambient temperatures and unpredictable weather that exists in the spring. However, even when temperatures were not limiting, several *Andrena* in Texas were found to average 2 cells in 4 days (1 day for a male, 2 days for a female, and then 1 or 2 nectar-only days; JLN pers. obs.). While egg limitation seems to be widespread among solitary bees, it is not obvious why this should be so, since brood parasitic bees and social bees can ramp up egg production far beyond 1 egg per day (Alexander and Rozen 1987). At the extreme end of the spectrum, honey bee queens may lay over a thousand eggs per day (Winston 1987). The low egg production of solitary bees probably represents one of the trade-offs involved in their multitasking lifestyles. Their time and energy must be spread between nest construction, provisioning, and oviposition rather than just egg production.

Neff (2008) combined data on adult dry body weight, pollen load weight (both wet and dry), and the number of foraging trips per hour when females are actively provisioning to calculate a provisioning rate measure across 19 species of solitary bees for which sufficient data are available (Table 8-1). Provisioning rate, as measured by the percentage of adult dry or wet body weight gathered per hour by foraging female solitary bees, is a standardized measure that allows provision rates to be compared across species and studies. This value ranged from 23% to just over 500% (Table 8-1). At the low end were female *Melissodes rustica*, which provisioned at the rate of one adult female body weight equivalent every four hours. In contrast, at the high end, a female *Calliopsis persimilis* can provision five times her own body weight per hour. This is an enormous range of variation in provisioning efficiency across solitary bee species.

CONVERSION RATIO

The efficiency with which bee larvae convert pollen and nectar provisions into bee biomass varies among bee species and even between males and females. The efficiency of conversion can be represented as a "conversion ratio"—the average weight of the provision mass divided by the average weight of an adult male or female bee. Conversion ratio can be expressed as either dry or wet weight. Note that higher values represent *less* efficient conversion of pollen and nectar to bee biomass. Neff (2008) summarized data on conversion ratio among solitary bee species and found a range of 2.75–8.33, with an average of 4.66 based on adult female dry body weight. In other words, a foraging solitary bee must gather slightly less than five times her own *dry* body weight in pollen and nectar to produce a female offspring. For a much smaller sample of wet weights, Neff found a range of 1.90–3.67, with an average conversion ratio of 2.87. On a wet-weight basis, a foraging solitary bee must gather between two and four times her own wet weight in pollen and nectar to produce a female offspring.

Within the few species of solitary bees (and wasps) that have been examined, there is a consistent difference in conversion ratio between male and female offspring (Danforth 1990). Female offspring appear to convert pollen/nectar provisions to bee biomass more efficiently than male offspring. In *Calliopsis persimilis*, for example, males are approximately 90% as efficient as a female in converting food to bee biomass. This difference in gender appears to be due to differences between males and females in larval metabolic rate (Boomsma and Isaaks 1985). On a per-gram basis, males are therefore slightly more expensive to produce than females.

What factors explain the variation among bee species in conversion ratio? Pollen nutritional value is an obvious factor that might explain the nearly four-fold difference in dry weight conversion across the 17 bee species analyzed by Neff (2008). Protein content is widely viewed as the primary determinant of pollen nutritional value (Roulston and Cane 2000, Roulston et al. 2000), although some authors have emphasized sterol and polypeptide content (Vanderplanck et al. 2017). Pollen protein content varies from 12% to 61% for pollen commonly collected by both specialist and generalist bees (Roulston et al. 2000). We might expect that higher pollen protein content would be correlated with higher conversion ratios and lower pollen requirements. Unfortunately, given the extremely limited amount of data on conversion ratios among solitary bees (Neff 2008), it is difficult to draw any clear connections between pollen protein content and conversion ratios. *Calliopsis persimilis* (Andrenidae) specializes on plants in the family Solanaceae, a family known for high pollen content (Roulston et al. 2000), and has a conversion ratio of 3.12, one of the lowest conversion ratios recorded. However, *C. persimilis* is less efficient than a related polylectic species (*Panurginus polytrichus*, Andrenidae; conversion ratio 2.75), making the correlation between efficiency and pollen protein content less clear. Species of bees in the Megachilidae have some of the highest conversion ratios recorded by Neff (2008), presumably because of the high energetic cost of producing a cocoon. Cocoon weight can average nearly 50% of adult body weight in species of *Osmia* (Bosch and Vicens 2002).

FORAGING DISTANCES

Given that all non-parasitic solitary bees are central place foragers, it is worth considering what we know about foraging distances in these bees. Some social bees, like honey bees, can forage over enormous geographic areas. In one study, worker honey bees were estimated to forage at sites more than 14 kilometers from the hive (Beekman and Ratnieks 2000). In contrast, solitary bees tend to forage much more locally. Maximum foraging distances (meaning the maximum distance between the nest site and floral resources) recorded for solitary bees range from as little as 100 meters to as much as 6 kilometers (Zurbuchen et al. 2010b). For most solitary bees, foraging range is less than 500 meters, and it is rare for solitary bees to forage over 1 kilometer. Some studies report extremely small foraging ranges for ground-nesting bees. Celary (2006) found that female *Melitta leporina*, which is approximately the size of a honey bee, foraged only 25–30 meters from their nests when nesting in an alfalfa field, although distances were somewhat longer (61 m) when nesting in natural habitats. A number of studies have documented a clear positive correlation between body size and foraging distance (Gathmann and Tscharntke 2002, Greenleaf et al. 2007, Guédot et al. 2009). For truly tiny bees (3 mm or less), foraging distances may be 10 meters or less, although even this would be 3,000 to 5,000 times their body length, the equivalent of a human considering a supermarket 2 miles away an easy stroll.

Measures of maximum foraging distance in solitary bees have been obtained in a variety of ways, and different methods often yield different results. Maximum foraging ranges have been obtained using mark recapture, trap nesting, pollen analysis, radio transmitters, and, most commonly, identifying the closest patch of potential host plants to the nest site (Zurbuchen et al. 2010b). But is maximum foraging distance really what we should be measuring? Zurbuchen and colleagues (2010b) argued that the *maximum* foraging range is not as relevant as some measure of *average* foraging range. They defined average foraging range as the maximum distance over which 50% of the females foraged.

Zurbuchen and colleagues used a clever experimental approach to determine average and maximum foraging distances They studied trap-nesting bees belonging to three genera and two families: *Hylaeus punctulatissimus* (Colletidae), *Chelostoma rapunculi*, and *Hoplitis adunca* (both Megachilidae) by establishing trap nests of each species in the center of a relatively homogeneous agricultural landscape dominated by field crops and devoid of potential host plants for the three experimental bees. Arrays of potted host plants were moved on a set of rails over increasing distances from the trap nests. With each move, a trail of potted plants was placed along the rails at 10-meter intervals to facilitate location of the moving experimental arrays. Distances ranged from immediately adjacent to the trap nests (<1 m) to over 1,600 meters (for the largest species, *Hoplitis adunca*). By moving the host-plant patch progressively farther from the nest site and recording paint-marked bee visits, the researchers were able to establish both the maximum and average (as defined above) foraging distance for two of the three species. For *Hylaeus* and *Hoplitis*, maximum foraging distances (1,100 and 1,400 m, respectively) were considerably greater than the average foraging distances (200 and 300 m, respectively). For *Chelostoma*, 90% of foraging females were able to reach the flower patch at 1,000 meters from the nest, and many individuals were foraging at non-experimental plants more than 1,275 meters from the nests, meaning that the average foraging range of this species was beyond the maximum distance used in the experiment. The Zurbuchen and colleagues (2010b) study shows that the maximum foraging distances obtained from earlier studies may greatly overestimate the foraging distances that could successfully maintain a viable bee population. In summary, foraging distances in many solitary bees appear to be quite small, which has important implications for the long-term viability of solitary bee populations in fragmented habitats (see Chapter 14).

INTRINSIC AND EXTRINSIC IMPACTS ON FORAGING BEHAVIOR AND FITNESS

Foraging behavior in solitary bees is remarkably sensitive to both extrinsic and intrinsic factors. Extrinsic factors include ecological conditions, such as floral resource levels, distance to floral resources, and even levels of parasitism. Intrinsic factors include physiological traits, such as body size, age, and the availability of mature oocytes. How solitary bees respond to these changes is sometimes obvious (reduced offspring production when floral resources are limited) but can also be remarkably subtle, including adaptive changes in sex allocation or offspring body size (see below). Figure 8-7 provides a conceptual framework for visualizing the interacting effects of extrinsic and intrinsic factors on individual offspring production, offspring body size, and the sex ratios of individual bees.

Offspring production

The number of offspring produced by female solitary bees is highly dependent on a variety of factors, including floral resource abundance (Bosch 2008, Buchmann and Cane 1989, Klein et al. 2004, Minckley et al. 1994, Neff and Simpson 1991, Strickler 1982, Williams and Kremen 2007), foraging distance (Peterson and Roitberg 2006a, Zurbuchen et al. 2010a), the presence of brood parasites (Goodell 2003), female age (Seidelmann 2006), female body size (Kim 1997, Seidelmann et al. 2010), and the availability of mature oocytes (Bosch 2008, Kim 1999). These studies have been based on both field observations of foraging rates in relation to floral resource availability (e.g., Minckley et al. 1994) and experimental manipulation of floral reward levels in greenhouse settings (e.g., Kim 1999).

Foraging trip times, provisioning rates, and overall offspring production are intimately related in solitary bees. In a three-year study of foraging and provisioning behavior of *Dieunomia triangulifera*, a species oligolectic on sunflowers, Minckley and colleagues (1994) found that foraging trip times over the course of the day, over the season, and even between years tracked floral resource availability. *D. triangulifera* foraging trip times were short in the morning, long around midday, and short again in the afternoon. Pollen loads were likewise large in the morning, smaller at midday, and large again late in the day. This pattern reflects the bimodal pattern of pollen presentation by the host plant, *Helianthus annuus* (Neff and Simpson 1990). Pollen was most abundant between 9:00 a.m. and 10:30 a.m., dropped at midday, and then increased again later in the afternoon. Thus, solitary bees respond to floral resource availability over the course of the day. What about over the course of the season and among years?

D. triangulifera foraging trip times, pollen load weights, and foraging rates all tracked floral resource availability over the course of the season as well. Early in the season, when few flowers were open, trip times were longer than later in

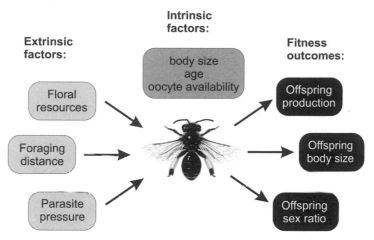

Extrinsic factors:

Intrinsic factors:

Fitness outcomes:

Floral resources

Foraging distance

Parasite pressure

body size
age
oocyte availability

Offspring production

Offspring body size

Offspring sex ratio

Figure 8-7. Interacting effects of extrinsic and intrinsic factors on foraging and offspring production in solitary bees. Extrinsic factors include the level of floral resources available in the surrounding habitat, the distance that bees need to forage to gather enough pollen and nectar for brood provisioning, and the level of parasite pressure (most importantly, open-celled parasites). Intrinsic factors affecting provisioning and reproduction include body size, age, and overall health of the foraging female bee. The behavioral responses to changes in these intrinsic and extrinsic factors include changes in the total number of offspring produced, the body size of male and female offspring produced, and the sex ratio of the resulting brood. The bee at the center of the figure is a female *Macropis europaea* (Melittidae) (original artwork by Birgitte Rubaek, *https://www.behance.net/brubaek*).

the season. Minckley and colleagues (1994) estimated the area over which bees were likely foraging across the season and over the three years of their study and found that early-season bees would have had to forage over a radius of >8 kilometers, whereas females foraging at peak flowering would have only needed to forage over a radius of just 3 kilometers. Variation among years in floral resources also translated into changes in foraging trip times. In 1988, the best year in terms of floral resource availability, foraging trip duration was half that recorded during the same time of day in 1989 and 1990. This translated directly into reproductive success because females produced nearly six cells per nest in 1988 and only between two and four in 1989 and 1990.

The clearest demonstration that offspring production is linked to resource availability comes from studies in which resource levels were manipulated experimentally under greenhouse conditions. Kim (1999) reared *Megachile apicalis* in screened cages under high and low resource levels. He found that with doubled resource levels, females provisioned 1.3 times more cells per day and made a higher investment in females (60% vs 35%) than those under the low-resource levels. While not too surprising, this study clearly demonstrates the impact of floral resource levels on solitary bee provisioning rates.

Foraging distance also impacts foraging activity and provisioning rates. Zurbuchen and colleagues (2010a) experimentally manipulated foraging distances in two host-plant-specific bees (*Hoplitis adunca* and *Chelostoma rapunculi*) under field conditions and demonstrated that increased distance to floral rewards decreased the number of brood cells completed per unit time in both species. For

H. adunca, an increase in foraging distance from 100 to 300 meters resulted in a 31% decrease in the number of cells per unit time, whereas for *C. rapunculi*, an increase from 500 to 1,000 meters led to a 46% decrease.

The potential threat of brood parasitism is one factor that likely interacts with floral resource levels to shape foraging decisions in solitary bees. The logic goes as follows. If floral resource levels are low and females are required to spend more time away from the nest when provisioning cells, the brood will be more exposed to brood parasites (Chapter 10). Under these conditions, females might be expected to make shorter foraging trips, to spend more time in the nest inspecting the cell and provision mass for potential brood-parasitic eggs, or to produce a more male-biased brood (in order to reduce the time spent away from the nest). Based on this logic, one might expect increased parasitism under conditions of low floral resource availability. Goodell (2003) examined the impact of floral resource availability on parasitism rates in *Osmia pumila* by a brood-parasitic wasp (*Sapyga centrata*, Sapygidae). *O. pumila* were established in cages at low and high resource levels and in the presence and absence of the brood-parasitic wasp. Goodell monitored foraging rates, reproductive success, and parasitism rates. Her results provide a fascinating glimpse into how solitary bees manage the trade-offs between foraging and nest defense. First, as expected, females in rich floral resource cages provisioned more cells per day than the females in cages with sparse floral resources. Second, females in cages with brood parasites provisioned slightly fewer cells per day than those in cages without brood parasites—consistent with the notion that females alter their behavior in order to reduce exposure to brood-parasite attack. Females also produced more nests with fewer cells per nest in the presence of the brood parasites—consistent with the hypothesis that they may be dispersing their brood cells among nests to avoid attack by brood parasites. Overall, there was a significant impact of floral resource availability on parasitism rate. Females in the sparse-resource cages had a 15-fold higher rate of parasitism than those in the rich-resource cages. While this study was based on relatively small numbers of bees and experimental replicates, it does suggest that important trade-offs exist between the need to forage and the need to protect the brood from attack by brood parasites.

Intrinsic factors that impact offspring production include size, age, and oocyte availability (Fig. 8-7). Larger females can typically produce more offspring than smaller females (Kim 1997, Seidelmann et al. 2010), and younger females gather more pollen and make fewer trips per cell than older females (Seidelmann 2006, Sugiura and Maeta 1989). Offspring production may also be limited by the availability of eggs rather than floral resources to provision cells. In Kim's studies of *Megachile apicalis* under high and low resource levels, females in the high-resource-level treatment occasionally completed cell provisioning but failed to lay an egg (Kim 1999). At least two studies under field conditions have documented the fact that female bees sometimes do not provision cells, even when there are abundant floral resources and other females are provisioning cells at the same site (Field 1996, Minckley et al. 1994). The explanation for these observations is that egg limitation, rather than floral resource abundance, puts limits on foraging and reproduction in solitary bees. Oocytes are energetically costly and require a substantial protein investment on the part of the female to produce.

Offspring body size

Several studies have shown that female solitary bees, when faced with diminished floral rewards or increased distance to flower patches, respond by producing smaller male and/or female offspring (Bosch 2008, Kim 1999, Peterson and Roitberg 2006a,b, Seidelmann 2006, Seidelmann et al. 2010). Reduced offspring body size can be viewed as a "making the best of a bad lot" approach, because while producing some offspring is better than producing no offspring, smaller offspring have been shown to have higher overwintering mortality and lower reproductive success than larger offspring (Bosch 2008). Kim (1999) found that female *Megachile apicalis* produced smaller daughters under low resource levels but that reduced resource levels had a non-significant impact on the size of sons. Peterson and Roitberg (2006a,b) manipulated both floral resource levels and distance to floral resources in *M. rotundata* and found that under decreased floral-reward levels, females produced smaller daughters but did not produce smaller sons (Peterson and Roitberg 2006a). When they experimentally manipulated foraging distance, they found that females responded by producing both smaller sons and smaller daughters (Peterson and Roitberg 2006b). In studies of *Osmia bicornis* (the red mason bee), Seidelmann and colleagues found that body size (Seidelmann et al. 2010) and age (Seidelmann 2006) impacted individual offspring body size. Smaller females produced undersized daughters (but not sons), and older females produced both sons and daughters that were undersized.

However, there are limits to how much body size can be reduced in the face of diminished floral rewards. Bosch (2008) studied foraging and provisioning behavior in *Osmia cornuta* in almond orchards in Spain over a three-year period. He found that in years when resources were limited, females produced significantly smaller female offspring, whereas male offspring body size was more or less constant across years. However, he also found that in all years, there were some female offspring produced that were dramatically smaller than the average; in fact, these females were smaller than the average male in body weight. These tiny females are unlikely to be able to overwinter, found new nests, and successfully rear offspring. Tiny females were particularly common in 1991, the year of significantly reduced floral resource levels. As we discuss below, females are expected to shift their sex ratio to a more male-biased brood under conditions of limited floral resource levels. Why would females produce undersized daughters when they could have shifted to producing "cheap" sons? Bosch (2008) discusses various possibilities, but the one that seems to make the most sense is that females are locked into their decision about whether to produce a son or a daughter prior to actually provisioning the cell. Once a female has committed to fertilizing the egg (and producing a daughter), she cannot reverse that decision. If floral resource levels are low and she cannot gather enough pollen for a female offspring, it is too late to switch to producing a male. Hence, females produce significantly undersized daughters as a last-ditch effort to produce an offspring (any offspring) given the current levels of floral resources.

Sex ratio

Female solitary bees appear to make remarkably "intelligent" decisions when it comes to their offspring sex ratio. Varying the sex ratio of their offspring is one adaptive strategy for dealing with variation in resource availability, foraging distance, maternal size, and even age. In most bees, males are smaller than females and therefore "cheaper" to produce; a son typically requires significantly less pollen and nectar than a daughter. When faced with low floral-resource levels, nesting solitary bees have consistently been shown to skew their sex ratio toward males.[3] Kim (1999) provided a particularly elegant demonstration of the impact of floral-resource levels on solitary bee sex allocation. Under low resource levels, female *Megachile apicalis* produced a highly male-biased brood (30% female), whereas under high resource availability, females produced a more female-biased brood (50% female). Likewise, females of *Amegilla dawsoni*, a large apid bee from Australia (Box 4-1), shifted from a mix of females and large males early in the season to exclusively producing small males later as resources became limiting (Alcock et al. 2005).

Foraging distance, and hence the energy needed to completely provision a brood cell, also impacts sex allocation in solitary bees. Peterson and Roitberg (2006b) studied nesting aggregations of *M. rotundata* foraging on alfalfa patches located (1) immediately in front of the nest site and (2) 150 meters from the nest site. They found that the nests located near the alfalfa patch produced sex ratios of 50% females, whereas the nests located 150 meters from the alfalfa patch produced only 31% female offspring. The change was largely due to a *decrease* in female production at the nests located far from the floral resources; the number of males produced per nest at each site was the same. This study shows how a small change in flight distance can impact foraging and reproductive success in solitary bees.

Body size, age, and other intrinsic factors can also impact sex-ratio decisions of solitary bees. Seidelmann and colleagues (2010) showed that smaller females produce a more male-biased brood than larger females. The relationship between maternal body mass and relative investment in daughters shows an almost perfect linear relationship. Females at the low end of the body-size range produced 20% female offspring, whereas those at the high end produced nearly 80% female offspring. Likewise, Seidelmann (2006) showed that as females age they become less-efficient foragers. Provisioning efficiency, measured

---

[3] Trivers and Willard (1973) developed what has been termed Conditional Sex Allocation Theory (CSAT). CSAT makes specific predictions about how resource levels, foraging distance, female health, body size, and age should impact sex allocation. The theory is based on the idea that the relationship between size and fitness is not identical for males and females. If one sex benefits more (in terms of fitness) from large body size than the other, females should selectively alter their sex-ratio decisions accordingly. When resources are limited, CSAT predicts that mothers should invest in the sex that benefits least from increased body size— i.e, the "cheaper" sex. When resources are abundant, CSAT predicts that mothers should invest in the sex that benefits most from increased body size—i.e., the more "costly" sex. Female Hymenoptera, by choosing whether or not to fertilize the egg, can facultatively alter their brood sex ratio.

as milligrams of pollen collected per hour, in older females was one-quarter that of younger females. Older females accommodate for the decline in provisioning efficiency by producing smaller offspring and a more male-biased brood. Sex-ratio and provisioning efficiency were tightly correlated.

A common theme that runs throughout the section above is that female off-spring production is more sensitive to reduced floral resources, increased foraging distances, or intrinsic factors like body size than is male offspring production. This has relevance for conservation of wild bee populations in small, isolated habitat fragments (Franzén et al. 2009, Larsson and Franzén 2007). If females are having a harder time producing robust female offspring and their offspring sex ratio is shifting toward a more male-biased brood, the overall reproductive output of the population will decline over time. One can imagine this reaching a point of no return at a local scale: females are too small and resources too limited to allow for any female reproduction. Males are great, but they don't build nests, provision brood cells, and lay eggs. This phenomenon could establish a kind of downward spiral or "extinction vortex" that drives small solitary bee populations to extinction (Chapter 14).

TEMPORAL FORAGING SPECIALIZATION

One conspicuous aspect of bee foraging and provisioning behavior is that not all bee species forage at the same time of day. The majority of bee species work "banker's hours"—from 9:00 a.m. until 4:00 p.m., with a unimodal distribution of activity that typically peaks during the warmest part of the day. However, many bee species show activity patterns that differ significantly from this typical unimodal pattern. Even among diurnal (day-flying) bees, foraging activity can show specific temporal patterns. Some bees forage early in the day and again late in the afternoon, with a pronounced break in the middle. Some bees complete their foraging activities well before breakfast (e.g., several species of *Lipotriches* that specialize on grasses), while others might not start foraging until dusk (e.g., several species of *Perdita* that are oligolectic on *Mentzelia* [Loasaceae]; Griswold and Parker 1988). And, amazingly, there are even highly specialized nocturnal and crepuscular bees that forage exclusively under low-light conditions. Aside from the intrinsic (innate) preferences of different bee species for particular times of day, there are clearly three main drivers of variation in foraging patterns in solitary bees: (1) temperature, (2) light intensity, and (3) floral resource availability (Willmer and Stone 2004).

Temperature

Temperature is an obvious driver of foraging behavior in bees, because, like most insects, bees are essentially ectothermic organisms whose body temperature varies more or less with ambient conditions. However, unlike many insects, bees are also facultative endotherms (Willmer and Stone 2004) that can regulate body temperature via a variety of mechanisms. This combination of ectothermy and facultative endothermy is termed heterothermy. Heterothermic vertebrates include shrews and hummingbirds. They are endothermic most of the time but can allow body temperature to drop for parts of the body or parts of a day or season when it becomes too expensive to maintain. Not all bees are facultative endotherms, as we shall see below.

The minimum thoracic temperature necessary for flight in bees varies from species to species but seems to be in the range of 20–30°C (Willmer and Stone 2004). Hence, bees living in cool, temperate habitats where overnight temperatures in the spring can drop to as low as 1°C must increase thoracic temperature just to leave the nest in the morning. Likewise, for bees living in hot desert habitats, it may be necessary to cool down the thoracic flight muscles in order to avoid reaching the upper critical temperature (the temperature that can kill a bee; ~45–50°C).

We know from extensive work on the flight physiology of bees that they can both warm up when ambient temperatures are below the minimum thoracic temperature for flight and cool down when ambient temperatures are approaching the upper critical temperature. Warming in solitary bees is achieved through both behavioral and physiological means. On cool, spring mornings in eastern North America, solitary females will bask in sunny spots in order to warm up their thoracic flight motor. Some early-spring bees (e.g., *Colletes inaequalis*; López-Uribe et al. 2015) will construct their nest sites on south-facing slopes, where the rising sun can warm the soil at the nest entrance. Females rest at the nest entrance with the head and mesosoma exposed in order to absorb the heat needed for powered flight. Some bees (e.g., *Andrena*) can even use "intra-floral basking" to warm up. Herrera (1995) studied the pollination of an early-blooming daffodil, *Narcissus longispathus* (Amaryllidaceae) by a solitary, oligolectic, early-spring bee, *Andrena bicolor* (Andrenidae). *Narcissus* blooms in the early spring (February–April) in southeastern Spain. Temperatures ranged from 12°C to 13°C during the flowering period, which is well below the lower thoracic temperature needed for *Andrena* flight. Herrera (1995) found that female *Andrena* were able to increase their body temperature by basking in the long, tubular flowers of *Narcissus*, which are as much as 8°C warmer than the surrounding air due to warming by the sun. At the start of foraging, female *A. bicolor* spend as much as 66% of their time within flowers. Female *Andrena* were unable to raise their thoracic temperature in the laboratory, indicating that they rely on the host plant to help them passively absorb heat during foraging. This has obvious benefits for the bee as well as the flower.

Many bees, especially large bees, can raise their body temperature by purely physiological means. Facultative endothermy in (mostly) large bees is achieved by "shivering thermogenesis"—rapid, opposing contractions of the large indirect flight muscles that can generate significant increases in thoracic temperature (Heinrich 1993). Facultative endothermy appears possible in bees above about 35–50 milligrams of adult body mass (Willmer and Stone 2004). Bees that are active at low temperatures can show warm-up rates as high as 10–15°C per minute (Stone 1993), meaning that they can go from 1°C to 20°C in less than two minutes. This exceeds the warm-up rate of most vertebrate animals! Once in flight, larger bees can maintain stable thoracic temperature of 35°C even at ambient temperatures at or below 20°C. Endothermy allows many bees, especially large-bodied species, to access floral rewards earlier in the day and at lower ambient temperatures than would be possible in a purely endothermic insect.

Solitary bees living in warm desert habitats also need ways to stay cool when ambient temperatures exceed 30°C. Behavioral mechanisms include remaining within the nest or seeking shade. In ground-nesting bees, the nest can provide an essential shelter from broiling temperatures. In some larger bees (e.g., *Amegilla*, Apidae), radiative cooling can also provide a mechanism for eliminating excess heat. This mechanism was first discovered in bumble bees but has since been shown to function in *Amegilla* and, most likely, other anthophorines. This mechanism involves careful control over how blood flows from the mesosoma, which generates heat, to the metasoma, where heat can be dissipated. When bees need to heat up, they establish a counter-current heat exchanger between the mesosoma and metasoma such that the hot blood flowing from mesosoma to metasoma comes in contact with the cool blood flowing from metasoma to mesosoma. This allows heat transfer to take place so that heat remains in the mesosoma and is not dissipated to the metasoma. When it is beneficial to reduce mesosomal body temperature, the counter-current heat exchanger is interrupted, so that the hot mesosomal blood flows directly into the metasoma without the transfer of heat to the incoming cooler metasomal blood. This allows mesosomal heat to be dissipated more quickly and thus leads to a reduction in body temperatures.

That cooling is an important component of thermoregulation in desert bees is demonstrated by studies of *Centris pallida* in the Colorado Desert of southern California, where ambient temperatures range from 25°C to 40°C (Chappell 1984). Thoracic temperature in flying males was 44–46°C, and in flying females it was slightly lower (43–45°C). Males establish territories by hovering over the nest site and pouncing on emerging females. Hovering males, which have the highest metabolic demands, approached thoracic temperatures of 47°C, which is 2–3° below the upper critical temperature for these bees, and they do this for nearly the entire day. Males that fly too low to the ground, where temperatures were at their maximum, can die of heat exhaustion.

It is clear that body size is an important determinant of thermoregulatory ability in bees. Larger bees, because they have lower surface area per unit volume, are better able to warm up under cool conditions, but will be more prone to overheating when temperatures exceed 30°C. In contrast, smaller bees, because they have a relatively large surface area per unit volume, take longer to warm up in the early morning, but are likely to be able to tolerate higher temperatures at midday. A number of studies have demonstrated that in warm, temperate habitats, larger bees will restrict foraging to the early morning and late afternoon, whereas small and medium-sized bees will forage at midday. Willmer (1988), in a study of co-occurring *Xylocopa* species visiting Sodom apple (*Calotropis procera*) in Israel, found that the larger species (*Xylocopa pubescens*) foraged mostly in the morning and late afternoon, whereas the smaller species (*X. sulcatipes*) foraged throughout the middle of the day. These trends are evident within species as well. Sihag (1993) found that *X. fenestra* in India shifted from unimodal (midday) foraging in the early season (February) to bimodal foraging later in the year as ambient temperatures increased. Larger females of *Anthophora plumipes*, a species that has been the subject of extensive study by Graham Stone, are able to warm up at lower temperatures than smaller females and thus are able to forage earlier in the morning. In addition, this species shifts from unimodal foraging at midday on cool days and bimodal foraging on warm days in the United Kingdom (Stone 1994). Finally, *Amegilla sapiens* (Stone 1993), which occurs from sea level to 2,000 meters of elevation in Papua New Guinea, shows intraspecific variation in body size such that higher-elevation populations have, on average, larger body size than lower-elevation populations. This allows the high-elevation bees to forage at lower ambient temperature than their low-elevation relatives, but the high-elevation bees must cease activity at lower temperatures than the low-elevation bees because of their increased risk of overheating.

Light intensity (nocturnal bees)

While most bees are diurnal (daytime) foragers active during the brightest and hottest part of the day, a remarkable group of bees have evolved the ability to forage at extremely low light levels (reviewed by Warrant 2008 and Wcislo and Tierney 2009). These obligate dim-light bees occur in four of the seven families of bees (Table 8-3). They show both behavioral as well as morphological adaptations to dim-light foraging. They can be matinal (active in the early morning), vespertine (active in the early evening), crepuscular (active in both early morning and early evening), and fully nocturnal. Dim-light foraging bees occur in both deserts and tropical rain forests. Desert taxa include the *Andrena* subgenera *Onagrandrena* and *Diandrena* (both specialists on evening primrose, *Oenothera*), the *Perdita* subgenus *Xerophasma* (specialists on *Camissonia* and *Oenothera*; Griswold and Miller 2010), many species of the *Lasioglossum* subgenus *Sphecodogastra* (specialists on *Oenothera*; McGinley 2003), large, fast-flying colletid bees in the genera *Ptiloglossa* and *Caupolicana*, and one species of dune-inhabiting *Colletes* (*C. stepheni*). Tropical species include the *Xylocopa* subgenus *Nyctomelitta*, halictid bees in the genera *Megalopta*, *Megaloptidia*, and *Megommation*, and two Australasian genera of Nomiinae (*Melittidia* and *Reepenia*). Noctural pollen-collecting bees even have nocturnal cleptoparasites. *Megalopta* (*Noctoraptor*) is a presumed cleptoparasite of pollen-collecting *Megalopta* sensu stricto (Engel 2001, Biani and Wcislo 2007), and *Odyneropsis apicalis* (Apidae: Nomadinae) is a cleptoparasite of *Ptiloglossa* (Roberts 1971). This group includes ground nesters (*Andrena, Perdita, Protoxaea, Lasioglossum, Rhinetula, Caupolicana, Ptiloglossa, Colletes, Tetralonia*), stem and cavity nesters (*Megalopta* and relatives), and wood nesters (*Xylocopa*). All are exclusively solitary except for *Megalopta*, which includes species that are facultatively social (i.e., some nests are solitary, while others in the same population are social).

Nocturnal and crepuscular bees share a number of morphological features that appear to be related to their dim-light foraging habits. Many are relatively large-bodied, at least in comparison to closely related diurnal species. This may allow nocturnal bees to warm up more quickly or stay warm in the cool evening hours. Nocturnal and crepuscular bees are often paler in coloration, presumably because they are not exposed to the intense UV radiation that exists in bright sunlight. Dim-light species show modifications of the wing morphology

Table 8-3. Nocturnal bees.

Family	Subfamily	Genus	Subgenus	Number of species	Temporal foraging pattern
Andrenidae	Andreninae	Andrena	Onagrandrena	24	matinal, crepuscular
			Diandrena	2	matinal
	Panurginae	Perdita	Xerophasma	6	? (collected only at light traps)
	Oxaeinae	Protoxaea gloriosa		1	matinal
Halictidae	Halictinae	Lasioglossum	Sphecodogastra	8	crepuscular
			Hemihalictus lustrans	1	matinal
		Rhinetula denticrus		1	? (collected only at light traps)
		Megalopta	Megalopta	27	crepuscular, nocturnal?
			Noctoraptor	3	crepuscular; cleptoparasite of Megalopta s.s.
		Megaloptidia		3	crepuscular
		Megommation insigne		1	nocturnal or crepuscular
	Nomiinae	Melittidia		19	? (enlarged ocelli suggest nocturnality; foraging patterns not known)
		Reepenia		8	? (enlarged ocelli suggest nocturnality; foraging patterns not known)
Colletidae	Diphaglossinae	Caupolicana		42	crepuscular or matinal; some diurnal
		Ptiloglossa		40	matinal, crepuscular, nocturnal
	Colletinae	Colletes stepheni		1	matinal
Apidae	Eucerinae	Eucera speciosa		1	matinal
	Xylocopinae	Xylocopa	Nyctomelitta	3	nocturnal and crepuscular
		Xylocopa tabaniformis	Notoxylocopa	1	matinal and crepuscular
	Nomadinae	Odyneropsis apicalis		1	crepuscular or nocturnal; cleptoparasite of Ptiloglossa (see Roberts 1971)

(Danforth 1989b), including an enlarged stigma, low aspect ratio (relatively short, fat wings), and low wing loading (a relatively large wing area for their body weight). Specialized aspects of the wing morphology may be related to a slower flight speed in nocturnal bees, which could be required under the low light levels at which they forage. Finally, and most conspicuously, dim-light bees have

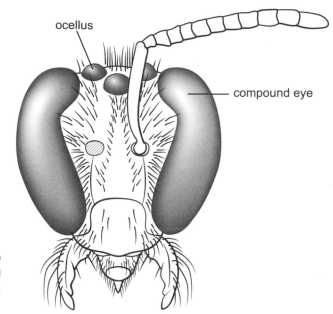

ocellus

compound eye

Figure 8-8. Head of female *Megaloptidia nocturnalis*, showing the enlarged compound eyes and ocelli of nocturnal bees (redrawn from Engel and Brooks 1998, Fig. 1).

greatly enlarged compound eyes and enormous ocelli (Kerfoot 1967a) (Fig. 8-8). These features are obviously related to increased light gathering needed by dim-foraging bees (see below).

Among the best-studied of the dim-light-foraging bees are *Megalopta genalis* and *Xylocopa* (*Nyctomelitta*) *tranquebarica* (Burgett and Sukumalanand 2000, Somanathan and Borges 2001, Somanathan et al. 2008, 2009). *M. genalis* is cre-puscular, foraging in the hour after sunset and the hour before sunrise in dense, neotropical rain-forest habitats. *X. tranquebarica* occurs in tropical southeast Asia and is crepuscular and sometimes fully nocturnal when moonlight allows for an extended foraging period. To put the visual capacity of these bees in perspective, *M. genalis* forages at light levels between 10^{-3} and 10^{-4} cd/m^2 (cd = candela, approximately the amount of light emitted by one candle), and *X. tranquebarica* can forage at light levels as low as 10^{-5} cd/m^2. For comparison, 10^{-4} cd/m^2 is equivalent to the light available on a starry night (with no moon). Light intensity on a sunny day can approach 10^6 cd/m^2, meaning that dim-light bees are able to forage at light levels 10 orders of magnitude lower than in full sunlight. Humans cannot see at 10^{-4} cd/m^2 without the aid of night-vision goggles. Just like their diurnal relatives, crepuscular bees fly an orientation flight when they leave the nest in order to learn visual landmarks. Warrant and colleagues (2004) showed experimentally that *M. genalis* can learn landmarks even under dim-light conditions. Even more remarkably, Somanathan and colleagues (2008) showed that *X. tranquebarica* can distinguish among colors when learn-ing visual landmarks near the nest, even at very reduced light levels. Crepuscular and nocturnal bees are apparently foraging at the limits of their visual capacity. Using video recordings of bees returning to the nest, Theobald and colleagues (2007) found that female *M. genalis* could easily locate the nest entrance a few minutes before sunrise or a few minutes after sunset (light levels around 10^{-3}

cd/m^2). However, at lower light levels (10^{-4} cd/m^2), the bees flew a much more circuitous route when returning to the nest.

How do these nocturnal and crepuscular bees manage to learn landmarks, navigate to and from their nests, and gather floral resources under such dim-light conditions? Two studies have examined the anatomical differences between nocturnal and diurnal bee compound eyes and ocelli in order to answer this question. Greiner and colleagues (2004) examined the anatomy of the *M. genalis* compound eye and ocellus and compared it to the eyes of a closely related, diurnal halictid (*Lasioglossum leucozonium*) and the honey bee (*Apis mellifera*). Somanathan and colleagues (2009) examined the compound eye and ocellus of *X. tranquebarica* and compared it to those of diurnal and crepuscular carpenter bees, as well as *A. mellifera*, *L. leucozonium*, and *M. genalis*. All bees, wasps, and ants have an apposition compound eye that is well suited for bright light conditions but poorly suited for dim-light conditions. In an apposition compound eye, each ommatidium (the individual, light-gathering unit of the eye) consists of a corneal lens that transmits light via a crystalline cone to just one photoreceptive rhabdom (equivalent to our retina). In the apposition eye, light passing through the corneal lens can be detected only by the photoreceptive rhabdom of that ommatidium, which puts limits on how effective these eyes are at dim light levels. (Other insects, including night-flying moths and some beetles, utilize a superposition eye in which each photoreceptive rhabdom receives light coming from hundreds and even thousands of corneal lenses.)

Careful histological examination of the compound eye in nocturnal bees reveals several anatomical modifications of the lens and rhabdom that give nocturnal bees increased light sensitivity. First, nocturnal bees have increased numbers of light-gathering ommatidia compared to their diurnal relatives. Each ommatidium has a greater corneal facet diameter and a thicker corneal lens. Both rhabdom diameter and rhabdom length are increased in nocturnal bees, presumably giving increased light sensitivity. Based on these anatomical changes, nocturnal bees are estimated to have a nearly 30-fold increase in light sensitivity relative to diurnal bees. However, increased light sensitivity has a cost. The larger, thicker corneal lens and the thicker rhabdom give a wider field of view for each individual ommatidium. So, while sensitivity is increased in nocturnal bees, their visual resolution is considerably less than that of a diurnal bee. Nocturnal bees are able to capture more light, but their view of the world is likely far less precise than that of a closely related diurnal species. While a 30-fold increase in light sensitivity may sound impressive, the differences in light levels experienced by nocturnal and diurnal bees is on the order of 100 million-fold (Greiner et al. 2004). It seems likely that nocturnal bees have some other tricks for dealing with low light levels. One possibility is that they carry out some kind of higher-order summation of the ommatidial inputs deep in the optic lobe of the compound eye. Such summation could occur over time (analogous to increased exposure time in a camera) or over space (summing the images gathered by neighboring ommatidia). Because temporal summation would seem to be a poor choice in a fast-flying insect, Greiner and colleagues (2004) speculate that spatial summation within the optic lobe of the brain might provide a more plausible mechanism for enhanced light sensitivity. Clearly, nocturnal and crepuscular bees are living at the edge of what an apposition compound eye can achieve.

BOX 8-3: NEST ROBBING—ANOTHER WAY TO OBTAIN POLLEN AND NECTAR

Almost all pollen and nectar foraging by bees takes place at the obvious place: flowers. However, some bees skip this tedious business and rob the pollen/nectar provisions from the nests of other bees. This behavior is found almost exclusively among the highly social honey bees (Free 1955) and stingless bees (Sakagami et al. 1993), with their large stores of pollen and nectar. Beyond occasional opportunistic nest robbing, some of these bees are professional robbers. Workers of two genera of meliponines, *Lestrimelitta*, a neotropical genus of 23 species, and *Cleptotrigona*, a monotypic African genus, are rarely (if ever) seen at flowers. Rather they obtain all their pollen and nectar from the nests of other stingless bees. These aren't sneak thieves. *Lestrimelitta* attacks en masse and chemically overwhelms the defenses of its victims (Sakagami et al. 1993), while *Cleptotrigona* raiders attempt to kill all the defending workers in the colonies they attack (Portugal-Araújo 1958). Although aggressive interactions, such as attempts at nest usurpation, are not unusual among solitaries, nest robbing seems to be quite rare. The only reasonably well-documented examples seem to be of a few *Xylocopa* species whose females occasionally rob nest provisions from other carpenter bee nests (Lucia et al. 2017, Mordechai et al. 1978, Watmough 1974). Since the nests of solitary bees are necessarily left unguarded for extended periods, it is a mystery why nest robbing is not more common. A possible explanation is that unlike the discrete prey items of predatory wasps, which are often stolen by other wasps, the gummy or even liquid nest provisions of bees are more difficult for the thief to carry (Field 1992). Even rarer are instances of bees collecting pollen directly off the bodies of foragers of other bee species. These acts of thievery, which Thorp and Briggs (1980) called cleptolecty, are performed almost exclusively by social bee workers, although their victims are sometimes solitaries.

Ocelli, the light-gathering structures located on the top of the bee head, also show dramatic differences between diurnal and dim-light foraging bees. In *X. tranquebarica*, the ocelli are nearly 1 millimeter in diameter, which is several times larger than in diurnal relatives (Somanathan et al. 2009). In addition, there is a tracheal tapetum below the retina, presumably to reflect light back to the photoreceptive retina for increased light sensitivity.

Nocturnal bees can sometimes get help from the moon. A number of studies have shown that moonlight can extend foraging behavior in nocturnal and crepuscular bees. A particularly clear example is shown in a study by William Kerfoot (Kerfoot 1967b) of the crepuscular and sometimes nocturnal *Lasioglossum (Sphecodogastra) texana*. *L. texana* belongs within a clade of North American, mostly nocturnal bees that are specialists on plants in the family Onagraceae (evening primroses) (McGinley 2003). The pollen grains of Onagraceae are large and associated with "viscin" threads that make them difficult to manipulate by most bees. *L. texana* (and relatives) have a highly modified femoral scopa consisting of a single row of recurved hairs for raking up large quantities of onagraceous pollen. Because of the viscin threads, pollen loads can be huge, and a female *L. texana* with a full pollen load resembles a bee carrying a large wad of yellowish-white cotton back to her nest. Flowering of *Oenothera* (and other Onagraceae) occurs during the early-morning or evening hours.

Kerfoot (Kerfoot 1967b,c) studied *L. texana* over a three-year period at Kingman County State Park in Kansas. Foraging by female *L. texana* normally occurs just within a one-hour period following sunset. Kerfoot referred to this as the crepuscular foraging period. During this time, females typically made one foraging trip, which was not sufficient to fully provision a cell, and then closed the nest entrance with soil. However, on nights when the moon rose within the one-hour window after sunset, female *L. texana* would continue foraging and could fully provision multiple cells. This occurred on multiple nights, and foraging could be extended to as late as midnight. There was a clear alternation between crepuscular and fully nocturnal foraging that followed the lunar cycle. Excavation of nests also revealed distinct clusters of closely spaced brood cells, presumably all constructed and provisioned over the several days when nocturnal foraging was possible. Oddly enough, if the moon rose after the bees had reentered the nest during the crepuscular foraging period, they would not reemerge to take advantage of the elevated light levels. It seems that once a nest is closed, females will no longer leave the nest. Lunar periodicity may be a common feature of nocturnal bee foraging. However, it is likely that this will be most pronounced in desert bees rather than those living in dense tropical rain forests.

Given that nocturnal foraging has evolved repeatedly in diverse bee groups, one might ask what ecological factors would favor the evolution of nocturnal foraging in bees. Wcislo and Tierney (2009) identified two main hypotheses for the evolution of dim-light foraging: escape from competition and escape from natural enemies. The "escape from competition" hypothesis posits that dim-light foraging allows crepuscular and nocturnal bees to exploit floral resources unavailable to diurnal relatives. Pollen and nectar are limiting resources for solitary bees, and they are often exploited on a first come, first served basis. Early-morning, dim-light foraging would allow bees to exploit flowers of early-blooming plants just as they open and when pollen and nectar resources are at their peak. In the Sonoran Desert of Arizona, *Ptiloglossa arizonensis* (Colletidae) begins foraging on *Solanum* (Solanaceae) one to two hours prior to the arrival of the first diurnal visitor, *Bombus sonorus* (Apidae) (Shelly et al. 1993). Pollen availability drops dramatically over the course of just a few hours, making early-morning foraging clearly advantageous for female *Ptiloglossa*.

A common assumption has been that the evolution of bat-pollinated plants also provides a niche for dim-light foraging bees (Wcislo and Tierney 2009). Flowers visited by bats often produce copious amounts of nectar and protein-rich pollen, making them valuable resources for any bees that can exploit them (Howell 1974, Roulston 1997). If so, one would expect to see bee-visited clades of nocturnal flowering plants nested within bat-visited lineages (i.e., bee visitation evolved *after* bat visitation). A phylogenetic study of *Parkia* (Fabaceae), which includes both bat- and insect-pollinated species, indicated that the bat-pollinated clade is derived from within the bee-pollinated clade (Luckow and

Hopkins 1995). Perhaps nocturnal bees, and their role as pollinators of night-blooming flowering plants, paved the way for the evolution of nectivorous bats, rather than the other way around.

The alternative hypothesis—that noctural foraging is favored because night-flying bees "escape" from diurnal parasites and predators—would seem to be contradicted by the fact that even some nocturnal bees have nocturnal clep-toparasites (e.g., *Megalopta* [*Noctoraptor*] and *Odyneropsis apicalis*). However, Wcislo and colleagues (2004) compared parasitism rates in two nocturnal *Megalopta* species with rates in diurnal stem-nesting bees and found that, indeed, *Megalopta* does show relatively low levels of cell parasitism: 5–6% in *Megalopta*, as compared to 24–29% for a sample of diurnal, twig-nesting bee species. One caveat of this comparison is that nocturnal bees forage during a relatively short window of time compared to diurnal bees. This alone, not nocturnality per se, could explain the low levels of parasitism in *Megalopta* (Wcislo et al. 2004). Nocturnal foraging could certainly be an effective escape from generalist parasites and predators, such as bombyliid and phorid flies, which are nearly exclusively active during the warm, sunny part of the day (Chapter 11).

Floral resource availability

A number of studies have demonstrated that floral resource availability is an important driver of bee foraging activity (Willmer and Stone 2004). Bees are highly sensitive to the floral-resource levels (including nectar, floral oils, and especially pollen) and can adjust their foraging activity in order to maximize floral-resource collection. Stone and colleagues (1999) documented a close correlation between bee foraging and floral-resource presentation in *Anthophora pauperata*, a desert bee that nests in dry river valleys (wadis) in Egypt. *A. pauperata* is a narrow host-plant specialist on *Alkanna orientalis* (Boraginaceae). *Anthophora pauperata* showed clear morning and afternoon peaks of foraging, with a two- to three-hour midday period spent within the nest. Stone and colleagues (1999) found that bimodal foraging was not related to difficulty thermoregulating at midday (bees were capable of maintaining a stable body temperature relatively independent of ambient temperature) but to pollen presentation by the host plant. *Alkanna orientalis* produces protandrous, hermaphroditic flowers that open for the first time in the early afternoon. This produces a peak of pollen availability in the early afternoon that coincides with a peak of foraging by the bee. Flowers open a second time the following morning, providing another early-morning peak of pollen availability that again attracts actively foraging female *Anthophora pauperata*. While nectar presentation is fairly constant over the course of the day, pollen presentation is distinctly bimodal, giving rise to the bimodal pattern of bee foraging.

Other examples exist as well. Neff and Simpson (1991) documented bimodal foraging in *Megachile fortis*, a specialist on common sunflowers (*Helianthus annuus*). Females foraged throughout the day but spent more time in the nest between trips and took longer trips during midday, when *Helianthus* pollen presentation was low. Likewise, bimodal foraging has been documented in *Andrena raveni* (Linsley et al. 1963), *Lasioglossum* [*Sphecodogastra*] *lusorium* [as

Evylaeus galpinsiae] (Bohart and Youssef 1976), and *Ptiloglossa jonesi* (Linsley and Cazier 1970). Neff and Simpson (1991) provide a discussion of bimodal foraging patterns in solitary bees. A clade of desert *Perdita* (Andrenidae) that are specialists on *Mentzelia* forage only late in the day, when the *Mentzelia* flowers are in bloom (Griswold and Parker 1988). The foraging of the bees is highly correlated with the opening of the host-plant flowers at approximately 5:00 p.m. on hot days in the Chihuahuan Desert of southern Arizona and New Mexico.

CHAPTER 9

The Microcosm of the Brood Cell: A Bestiary of In-Nest Mutualists

Despite the efforts of solitary bees to disinfect, waterproof, and seal their brood cells, there is a remarkable diversity of organisms, including bacteria, yeasts, nematodes, annelids, mites, and other insects that manage to gain entry. Some of these organisms are pathogenic, some predatory, and some commensal (with no apparent impact on the bees), and some are clearly beneficial to the developing bee brood (i.e., they are mutualists). We cover groups that have a clear negative impact on solitary bees in Chapter 10 (on brood-parasitic bees) and Chapter 11 (on non-bee parasites and predators). In this chapter, we focus on the microbes (bacteria and fungi), worms, and mites that have been reported from the brood cells of bees. The routes that these organisms take to gain access to the brood cell are diverse. Some arrive in pollen or nectar, some in the digestive or even the reproductive tract of the bee, and some on the surface of the bee's body or in specialized structures designed to carry them. Some in-nest mutualists take a more direct route—they walk into the open cell while it is being provisioned— while others are able to move through cell partitions or through the soil itself. The life cycles of these in-nest mutualists are often highly synchronized with that of the bee host, suggesting that these are ancient, coevolved relationships.

The solitary bee brood cell can be thought of as a "microcosm" or miniature ecosystem (Biani et al. 2009) with a whole host of bacteria, fungi, nematodes, annelids, mites, and other creatures interacting in surprisingly complex ways. Mites recorded from bee brood cells may be parasitic (killing the egg or larva), cleptoparasitic (consuming stored provisions), or mutualistic (beneficial to the bee host) (Biani et al. 2009, Cordeiro et al. 2011). Mites in the family Chaetodactylidae have a close association with many bees in the families Megachilidae and Apidae. Chaetodactylid mites often aid their bee associates by consuming pathogenic fungi within the host's cells, and female bees in both Megachilidae and Apidae have structures (acarinaria) for transporting their mite associates. In certain groups of Apidae—in particular, Xylocopini, Ceratinini, and Tetrapediini (subfamily Xylocopinae)—there are unique mite lineages associated with each bee group, suggesting coevolution of mites and their bee hosts (Klimov et al. 2007a). Commensal nematodes are also known to inhabit the brood cells of many solitary bees. Nematodes feed on the bacteria and fungi and undergo multiple generations during the development of the host bee, apparently without causing any damage to the bee. Some bee provision masses appear to host beneficial yeasts that cause the provisions to ferment and smell like cheese or Vegemite. These yeasts may be playing a beneficial role as a source of food for developing larvae (Steffan et al. 2019) or as a source of in-nest disinfection (Kaltenpoth 2009). Where these yeasts come from and whether they are specific to certain bee groups is not entirely clear.

FUNGI, BACTERIA, AND OTHER MICROBES

More and more evidence suggests that bee provisions are inhabited by a diverse microbial fauna, including bacteria, yeasts, and other microbes. Much of the work on the microbial associates of bees has been done in social bees, especially honey bees (Anderson et al. 2011, Evans and Schwarz 2011), bumble bees (Koch and Schmid-Hempel 2011), and stingless bees (Menezes et al. 2013). However, we are finding intriguing hints that microbial activity may be a normal, and perhaps essential, part of the pollen and nectar provisions of solitary bees as well.

Bacteria

The earliest studies of bacterial associates of solitary bee provision masses (Gilliam et al. 1984, 1990) were conducted using culture-based techniques. These studies documented the presence of spore-forming bacteria in the genus *Bacillus* from provisions of *Centris*, *Anthophora*, *Xylocopa* (Apidae), and *Crawfordapis* (Colletidae). These bacteria produced a number of metabolically active enzymes, including esterases, proteases, amylases, and glycosidases, and Gilliam and colleagues (1990) concluded that "*Bacillus* species are common associates of Apoidea and could participate both in metabolic conversion of food and in the control of competing and/or spoilage microorganisms."

New methods of microbial detection, namely high-throughput metagenomic sequencing, have opened up enormous potential for characterizing the bacterial faunas of solitary and social bees. Martinson and colleagues (2011) screened a broad sample of both solitary and social bees from six of the seven families of bees, plus apoid wasp relatives, and found that the most common bacteria isolated from the midgut of adult solitary bees is a widespread phylotype of *Burkholderia* and the pervasive insect associate *Wolbachia*. This study did not analyze provision masses or larval guts, but does suggest that *Burkholderia* may be an important and ubiquitous associate in the digestive tracts of solitary bees. Whether it occurs in provision masses has not been investigated.

Honey bees (Olofsson and Vásquez 2009, Vásquez and Olofsson 2009) and bumble bees (Olofsson and Vásquez 2009) have been shown to host a variety of bacterial strains (*Lactobacillus* and *Bifidobacterium*) that appear to be involved in protection of stored honey and defense against other pathogenic bacteria, such as American foulbrood (Forsgren and colleagues 2010). Olofsson and Vasquez (2008) hypothesized that *Lactobacillus* contribute to the conversion of nectar to honey through fermentation. *Lactobacillus* have also been detected in solitary (*Augochlora pura*), eusocial (*Halictus ligatus*), and socially polymorphic (*Megalopta centralis* and *M. genalis*) halictid bees (McFrederick et al. 2012). In the two species of *Megalopta*, *Lactobacillus* were abundant in pollen provisions and larval digestive tracts (McFrederick et al. 2014). While the honey bee and bumble bee *Lactobacillus* strains appear to be transmitted vertically (from mother to daughter), the *Lactobacillus* associated with most other bee groups are commonly acquired from the environment, presumably while foraging for nectar (McFrederick et al. 2012, 2013a, 2014). *Lactobacillus* strains identified in both solitary and social bees are likely beneficial to their bee associates either by causing fermentation of the provisions or through the production of antimicrobial products that prevent provisions from spoiling (McFrederick et al. 2013a). The fact that these bacterial associates are regularly acquired from the environment in the few bees that have been examined (*Megalopta*, *Halictus*, *Augochlora*) suggests that this may be more widespread across solitary bee lineages.

Of course, bacteria can also be harmful to bees. Two of the worst honey bee diseases are bacterial in origin: American foulbrood (*Paenibacillus larvae*) and European foulbrood (*Melissococcus plutonius*). The impact of these pathogens on honey bees (as well as bumble bees) is well known (Ellis and Munn 2005). We know much less about the bacterial pathogens of solitary bees.

Fungi and yeasts

A number of studies have documented, both directly and indirectly, the presence of fungi and yeasts in solitary bee provision masses. Yeasts and other fungi can be a major source of mortality in ground-nesting bees, whereas in some bees (Diphaglossinae, Oxaeinae, and Eucerini), yeasts are believed to play an important role in larval nutrition (see below). Batra and colleagues (1973) provide an excellent review of solitary bee mycoflora, especially the pathogenic fungi attacking cells of ground-nesting *Nomia melanderi* (the alkali bee). Larval mortality in *N. melanderi* is often due to a progression of soil microbes initiated by yeasts (primarily species of *Saccharomyces*, *Pichia*, *Hansenula*, *Candida*, *Kloeckera*, and *Rhodotorula*) and bacteria (*Lactobacillus*, *Fusarium*, and *Rhizopus*), which initiate fermentation of the provision mass (Fig. 9-1). The eggs or larvae often die at this stage, and the cell contents are ultimately overgrown by secondary fungal invaders, including *Aspergillus*, *Penicillium*, *Ascosphaera*, *Eurolium*, *Carpenteles*, and *Mucor* (Fig. 9-1). Yeasts and other fungi are common in floral nectar (Golonka and Vilgalys 2013), and this may be the likely site of acquisition for solitary bees. *Aspergillus* fungi are a common cause of larval mortality in other ground-nesting bees. In *Diadasia bituberculata*, a solitary, ground-nesting apid bee, Linsley and MacSwain (1952) reported 34% mortality of larvae due to infection with *Aspergillus flavus* and an undescribed species of *Rhizopus*. Butler (1967) also reported extremely high rates of mortality due to fungal infection in nesting aggregations of *Ptilothrix* in southern Arizona. Cane

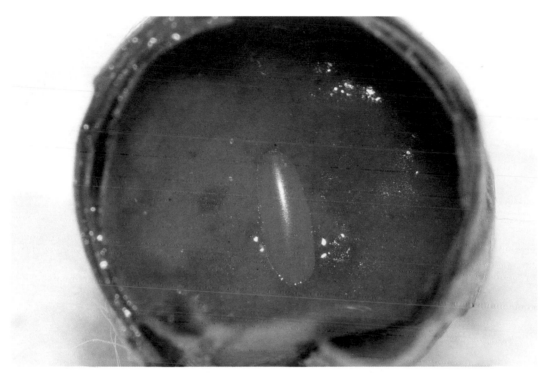

Plate 1. *Megachile circumcincta* (Megachilidae) brood cell with egg in semi-liquid pollen provisions. Photo courtesy of Nico Vereecken.

Plate 2. *Osmia cornuta* (Megachilidae) egg on pollen/nectar provisions. Mud partition visible at left. Photo courtesy of Nico Vereecken.

Plate 3. *Osmia cornuta* (Megachilidae) feeding larva on pollen/nectar provisions. Photo courtesy of Nico Vereecken.

Plate 4. *Colletes cunicularius* (Colletidae), male and female *in copula*. Photo courtesy of Nico Vereecken.

Plate 5. Male *Andrena nigroaenea* (Andrenidae) attempting to copulate with flower of *Ophrys lupercalis* (Orchidaceae) flower. Photo courtesy of Nico Vereecken.

Plate 6. Female *Colletes inaequalis* (Colletidae) emerging from below-ground burrow in early spring. Photo courtesy of Alberto Lopez.

Plate 7. *Rhodanthidium sticticum* (Megachilidae) closing a snail shell nest. Photo courtesy of Nico Vereecken.

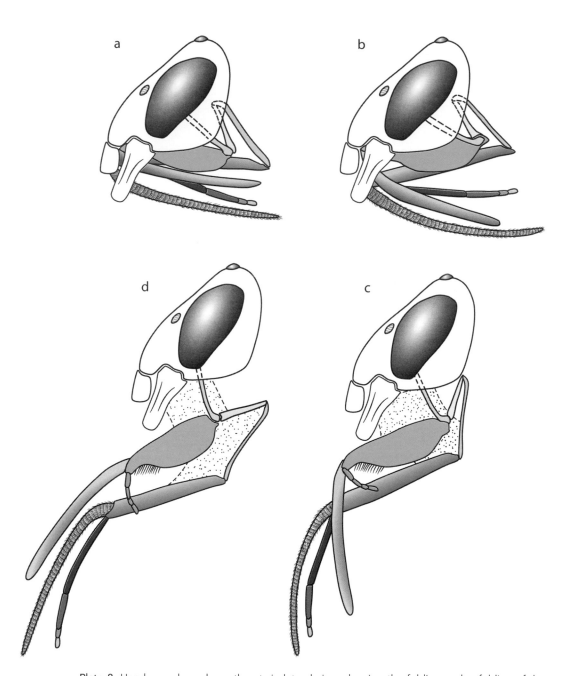

Plate 8. Head capsule and mouthparts in lateral view, showing the folding and unfolding of the labiomaxillary complex: (a) mouthparts folded into the proboscidial fossa; (b) mouthparts partially unfolded; (c) mouthparts partially extended; (d) mouthparts fully extended. Color codes: lorum, mentum, and prementum in green; cardo, stipes, and maxillary palpus in blue; galea in purple; labial palpus in red; glossa in yellow. Original drawing based on female *Peponapis pruinosa* (Apidae). Original artwork by Frances Fawcett.

Plate 9. *Anthophora atriceps* (Apidae) approaching flower of *Trifolium stellatum* (Fabaceae) with mouthparts extended. Photo courtesy of Nico Vereecken

Plate 10. *Dasypoda hirtipes* (Melittidae) visiting *Leontodon* (Asteraceae) flower with fully loaded scopa. Photo courtesy of Nico Vereecken.

Plate 11. *Thyreus splendidulus* (Apidae) sleeping on a twig. Note mandibles grasping the twig as the bee remains in a sleep-like state. Photo courtesy of Nico Vereecken.

Plate 12. Bombyliid fly (*Bombylius* sp.: Bombyliidae) on *Myosotis* (Boraginaceae) flower. Photo courtesy of Nico Vereecken.

Plate 13. Chrysidid (jewel) wasp (*Chrysis* sp.: Chrysididae). Photo courtesy of Nico Vereecken.

Plate 14. Crab spider (*Xysticus cristatus*) with immobilized honey bee (*Apis mellifera*). Photo courtesy of Nico Vereecken.

Plate 15. *Dasypoda crassicornis* (Melittidae) visiting a flower of *Cistus salviifolius* (Cistaceae). Photo courtesy of Nico Vereecken.

Plate 16. *Hoplitis perezi* (Megachilidae) on flower of *Convolvulus althaeoides* (Convolvulacae). Photo courtesy of Nico Vereecken

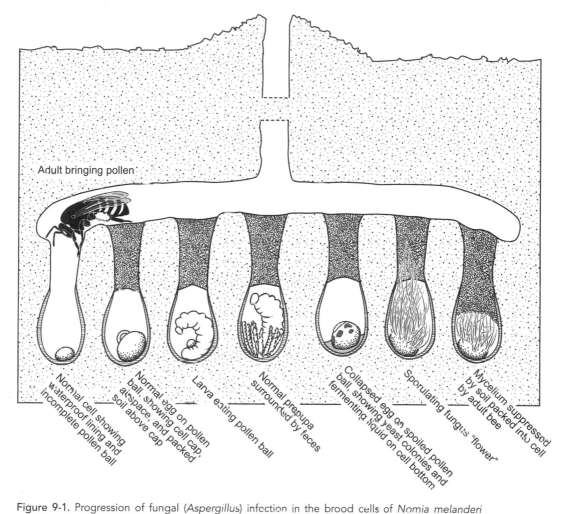

Figure 9-1. Progression of fungal (*Aspergillus*) infection in the brood cells of *Nomia melanderi* (Halictidae) (redrawn from Batra et al. 1973, Fig. 1).

and colleagues (1983b, Table 1) provide a summary of data on larval mortality related to fungal infection across a diverse array of solitary bee species.

Among the most damaging of the fungal pathogens of bees are members of the genus *Ascosphaera*; it includes 29 described species, all of which have an obligate association with solitary and social bees (Wynns et al. 2013). Approximately half of the *Ascosphaera* species are saprophytic in the brood chambers of bees, feeding on pollen provisions, nesting materials, and larval feces. However, some species are clearly bee pathogens. Two species in particular, *A. aggregata* and *A. apis*, cause a disease called chalkbrood in both solitary and social bees. Spores are transferred by adults to larvae, where they germinate in the larval midgut and ultimately lead to larval death. *Megachile rotundata*, an important managed pollinator of alfalfa, is heavily impacted by *Ascosphaera* (Bissett 1988).

Nevertheless, in some bee species, yeasts appear to be an important component of the provision mass and are most likely beneficial to larval development. Bees with very liquid provisions, including colletid bees in the subfamilies Hylaeinae (Michener 1960) and Diphaglossinae (Roberts 1971), bees in the andrenid subfamily Oxaeinae (Fig. 9-2; Rozen 2018), and apid bees in the tribe Eucerini (Miliczky 1985) have been reported to have provisions that smell of fermenting yeasts. Roberts (1971) provides a graphic description of active fermentation in the brood cells of *Ptiloglossa guinnae* (Diphaglossinae) studied in Costa Rica:

> The contents of each of the more than one hundred cells examined was actively fermenting. The odor of fermentation was unmistakable and bubbles of gas could be seen rising from the ropey slime at the bottom of each cell. Some fermentation of provisions has been reported for other bees, but in no case is it so obvious and vigorous a process as with *Ptiloglossa*. Cultures taken from the cells of *Pt. guinnae* were found to contain the yeasts *Saccharomyces* sp. and *Candida pulcherrima* (Lindner) Windische. The cells were found to contain so much nectar and so little pollen that yeast, rather than pollen, would seem to be the principal source of amino acids, vitamins, and sterols for the developing larva.

Figure 9-2. Habitus drawing of female *Protoxaea gloriosa* (Andrenidae: Oxaeinae). Original artwork by Frances Fawcett.

Interestingly, the cells of many diphaglossine bees (*Diphaglossa*, *Ptiloglossa*, *Crawfordapis*, *Zikanapis*, and *Cadeguala*) lack a clear cell closure, and this has been hypothesized to allow gas to escape from the active fermentation going on in the brood chambers (Fig. 9-3; Roberts 1971, Sarzetti et al. 2013). Whether these bees have actually switched from pollen as their primary protein source to yeast-derived proteins remains to be seen.

Similar reports have been published for the apid tribe Eucerini. Miliczky (1985), in a study of *Eucera hamata* (as *Tetralonia*), reported a "faint, sourish

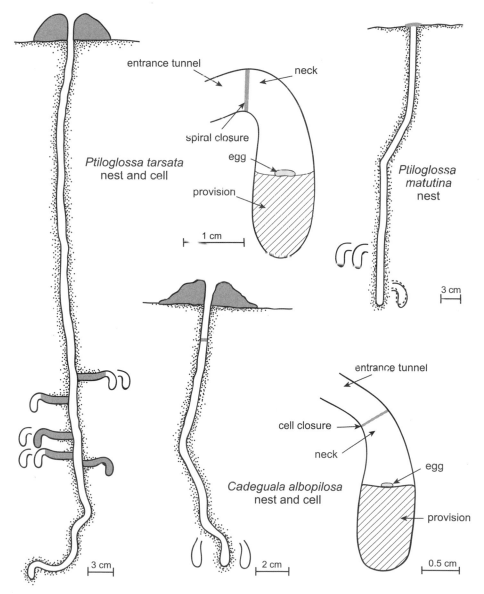

Figure 9-3. Nests of bees in the subfamily Diphaglossinae (Colletidae), showing liquid provision mass and vertical cell orientation (redrawn from Sarzetti et al. 2013, Figs. 19–23).

odor detectable from a short distance away" from the liquid provisions as well as bubbles arising from the liquid provisions, suggestive of fermentation. Similar observations have been made in other eucerine bees, including *Eucera*, *Svastra*, *Melissodes*, and *Peponapis* (see references in Miliczky 1985), although there has been no definitive confirmation of the yeast species involved.

These intriguing reports of fermentation have not yet been investigated using new methods of metagenomic sequencing. We do not know how many yeasts may be involved, how the yeasts are introduced into the brood chamber, or how infection by other pathogenic fungi (e.g., *Aspergillus*) is prevented. There is indeed an interesting story here that needs to be investigated in more detail.

Several studies have actually characterized the yeast associates of solitary bee brood cells using a combination of culture-based and molecular methods (Inglis et al. 1993, Pimentel et al. 2005, Rosa et al. 1999). Rosa and colleagues (1999) cultured and sequenced a new species of asexual ascomycetous yeast (*Candida batistae*) from larval provision masses, larvae, and pupae of two ground-nesting bees, *Diadasina distincta* and *Ptilothrix plumata* in Minas Gerais, Brazil. Another species of yeast, an undescribed *Mucor* species, was also isolated from larval provisions and larvae. Together, these two yeasts were the dominant components of the brood-provision mass, producing between 10^4 and 10^7 colony-forming units per provision mass. *Candida batistae* are nutritionally specialized. Of the 45 carbon sources tested, the various strains of *C. batistae* were able to assimilate only glycerol, mannitol, glucitol, D-gluconate, and glucono-Δ-lactone. *C. batistae* is able to ferment glucose and to produce lipase. The *Mucor* species was found to produce proteases and pectinases, as well as toxins that are active against other yeasts (but not *C. batistae*) that might be instrumental in keeping the provisions free of non-beneficial yeasts and/or molds.

Sequencing of the D1/D2 domain of the 28S ribosomal gene revealed that *C. batistae* is related to a group of nutritionally specialized yeast species associated with bees and flowers (the "*Starmerella*" group). Other bee-associated *Candida* species have been described in the *Starmerella* group: *C. bombicola*, a yeast originally isolated from honey bees (Barnett et al. 1990), was identified in nectar, pollen, and provision masses of solitary *Megachile rotundata*, the alfalfa leaf-cutter bee (Inglis et al. 1993), *C. riodocensis* was isolated from pollen provisions, larvae, and fecal pellets of a species of *Megachile* (Pimentel et al. 2005), and *C. cellae* was isolated from pollen-nectar provisions of *Centris tarsata* (Pimentel et al. 2005). Rosa and colleagues (2003) sequenced additional *Candida* species associated with the pollen provisions of stingless bees. It is likely that these yeasts are obtained from either pollen or nectar while bees are foraging (Inglis et al. 1993). This clade of bee-associated yeasts is extremely interesting given the reports of fermentation in the provisions of many bee species (see above) and may suggest that yeasts are a far more widespread element of bee larval nutrition than we have previously realized (Rosa et al. 2003). Unfortunately, the exact nutritional role of these yeasts is not clear. Are they "defensive mutualists" (Kaltenpoth 2009, Kaltenpoth and Engl 2014), producing compounds that keep the provisions clear of infecting microbes, or are they playing a key role in the breakdown of the pollen/nectar provisions?

NEMATODES

Among the most interesting, but poorly understood, in-nest associates are commensal nematodes that inhabit the reproductive tracts and brood cells of solitary and social bees (Giblin-Davis et al. 1990). Commensal nematodes have been reported from all the families of bees except the tiny Australian family Stenotritidae (and no one has really looked for them, we suspect). Commensal nematodes seem to be most commonly associated with Halictidae—social halictids in particular—although they also occur in solitary bees such as *Anthophora*, *Colletes*, and *Andrena*. Genera of bee-inhabiting nematodes include *Acrostichus* and *Aduncospiculum* (associated exclusively with halictid bees), *Bursaphelenchus* (associated with halictid, melittid, and apid bees), *Aphelenchoides* (associated with the nests of large carpenter bees, Xylocopini), and *Koerneria* (associated with andrenid and colletid bees) (see references in Hazir et al. 2010). In all cases, these nematodes have a life history that is similar to that described below for *Acrostichus*: a resting (dauer) stage of the nematode inhabits the reproductive tract of adult bees, while the adult and larval forms of the nematode consume fungi and bacteria within the brood cells of their host bee.

The genus *Acrostichus* (family Diplogastridae) has been studied in detail by Quinn McFrederick and colleagues (McFrederick et al. 2013b, McFrederick and Taylor 2013) and provides a good model for understanding the life history (and impact) of these bee-associated nematodes. *Acrostichus* inhabit the reproductive tracts of solitary and social halictid bees. Hosts include species in the genera *Augochlora*, *Augochlorella*, *Megalopta* (Augochlorini), and *Halictus* (Halictini). The life cycle of *Acrostichus* is closely linked to that of the host bee. Transmission of nematodes is largely vertical (from mother to daughter or son) but can also occur via mating (since males host nematodes in their genitalia!). The dauer (non-feeding, resting stage) of the nematode inhabits the Dufour's glands and male genitalia of the host bee. When a female coats a newly constructed brood cell with Dufour's gland products, the nematodes are transmitted to her brood chamber (Fig. 9-4), where they feed on yeast and bacteria. Nematodes develop to adulthood and mate within the brood cell, and multiple generations of nematodes occur before the host bee completes its own development. Once the host bee molts to adulthood, juvenile nematodes molt to the non-feeding (dauer) stage and enter the Dufour's gland (if the host bee is female) or the male genitalia (if the host bee is a male).

What impact does *Acrostichus* have on bee growth and development? McFrederick and colleagues (2013b) compared the number of offspring produced and the number of cells provisioned in nests of solitary (*Augochlora pura*) and social (*Halictus ligatus*) bees with and without *Acrostichus* nematodes and found no detectable differences. *Acrostichus* appear to be in-nest commensals with no significant impact on host reproduction and survival.

An extensive survey of bees in Virginia by McFrederick and Taylor (2013) indicated that bee-associated nematodes are highly host-specific. The researchers surveyed 106 bee species for the presence of nematodes, and only 6 bee species in 3 genera were found to host them: *Halictus*, *Augochlora*, and *Augochlorella*. Some species showed extremely high levels of infection (100% of

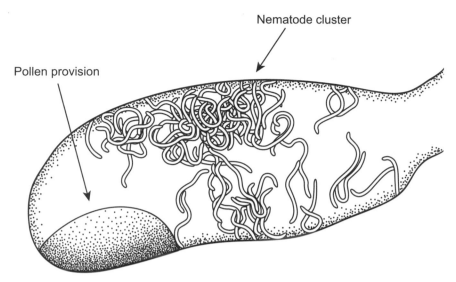

Figure 9-4. Nematodes (*Acrostichus halicti*) present in the brood cells of *Halictus ligatus* (Halictidae) (redrawn from McFrederick et al. 2013b, Fig. 1).

female *H. parallelus* were infected), while infection was below 20% for male and female *H. ligatus*. Other species of these same genera tested negative for nematodes. Three species of bee-inhabiting *Acrostichus* have been described, but a molecular phylogeny suggests that there may be multiple cryptic species, each one associated with a different bee species.

What factors drive the host associations between *Acrostichus* and their bee hosts? Clearly, *Acrostichus* are specific to the family Halictidae. Beyond that, they seem to show a preference for social species: five of the six species that tested positive are social, and *Augochlora pura*, the sole solitary species, is actually a secondarily solitary species (they are a solitary species descended from a group of mostly eusocial species). Why do these nematodes prefer social hosts when they are vertically transmitted and thus unlikely to benefit from the increased bee-to-bee contact within a social nest? It seems we have only scratched the surface when it comes to an understanding of the nematode associates of bees.

MITES

Mites are among the most common mutualists—and also some of the worst predators—within the nests of bees. Eickwort (1994) provided a comprehensive review of bee-mite associations, and we have relied heavily on his paper as well as more recent papers by Pavel Klimov and Barry O'Connor. There are over 30 genera of astigmatid, prostigmatid, and mesostigmatid mites from 12 families that show associations with solitary bees (Table 9-1). The vast majority of these mites are fungal feeders that seem to have either no impact on bee development or, in some cases, a beneficial effect (by controlling fungal growth within the brood cells; Biani et al. 2009, Cordeiro et al. 2011). Some bee mites feed on pollen and/or nectar, but they consume such small amounts that their impact on bee larval development is likely minimal. A small group of bee-associated mites are predatory on other mites. *Cheletophyes*, for example, have been shown to prey on mites that attack carpenter bee larvae and are thus clearly beneficial to the bee host (O'Connor 1993). However, some mite associates are devastating predators, attacking bee eggs or early-instar larvae so that they can feed without competition on either the pollen provisions or fungi that develop on the provisions.

All bee-associated mites have a phoretic stage that attaches to the body of the adult bee prior to its departure from the brood cell. These "hitchhikers" attach to specific regions of the bee's body, including regions that are difficult for the bee to reach—the underside of the head; the junction between the head and mesosoma; the propodeum (the last mesosomal segment) or the first metasomal segment, beneath the metasomal sterna and terga; the wings; and even the genital chamber. Mite life histories are also tightly synchronized with the host's life history. Mites produce one generation within a brood cell, and the phoretic stage is produced in the late-larval or pupal stage of the bee; the mites remain in the brood cell until the bee emerges as an adult and then attach to the adult bee, sometimes in large numbers (>300 mites/bee!). Mite dispersal appears to be primarily based on the phoretic stage, but some mites can burrow through the substrate and enter neighboring cells or catch a ride on another bee exiting through the cell (in stem nesters).

Mites are associated with every bee family except Melittidae (Table 9-1). This is interesting in light of the new view of bee phylogeny that suggests Melittidae are the basal branch of bee phylogeny. Perhaps mite associations arose only in the common ancestor of the remaining bee families, after the early split between Melittidae and the other bee families. Bee mites are particularly common on Halictidae, Xylocopinae, and Megachilinae (Table 9-1). Cavity-nesting (Megachilinae) and wood-nesting (Xylocopinae) appear to provide greater opportunities for mite associates. Neither Megachilinae nor Xylocopinae sensu stricto line their brood cells, and cells are often constructed in linear series, perhaps allowing mites to move easily from brood cell to brood cell. Xylocopinae sensu stricto (small and large carpenter bees) are long-lived, and parents and offspring coexist within the nest for long periods of time, allowing transfer of mites among nest mates. The high frequency of mites associated with Halictinae is not simply due to the relatively large number of eusocial Halictini, because mites inhabit the nests of both solitary and social halictines.

Among the best studied of the mite lineages associated with bees are the Chaetodactylidae (Fig. 9-5). Chaetodactylids are obligate associates of solitary and facultatively social bees in the families Apidae and Megachilidae. There are more than 200 described chaetodactylid species, and they include parasitic/predatory species that kill bee eggs and larvae, mutualists that feed on fungi and nest wastes (like fecal material), and cleptoparasites that consume pollen/nectar provisions. There are highly host-specific chaetodactylid lineages, such as *Sennertia* (170 species), which have a commensal relationship with carpenter bees (Xylocopini) (Fig. 9-5). Other bee-mite commensals exist, such as *Tetrapedia* (bee)–*Roubikia* (mite); *Ceratina* (bee)–*Achaetodactylus* (mite); and *Centris*

Table 9-1. Summary of mite lineages associated with bees, their host associations, geographic distributions, and presumed ecological interactions with bees.

Modified from Tables 9.1 and 9.2 in Eickwort (1994).

Mite lineages exclusively associated with social bees are excluded.

Distribution codes: AU = Australia, ET = Ethiopian, NE = Nearctic, NT = Neotropical, OR = Oriental, PA = Palearctic.

	Genus	No. of species	Host (bee family or subfamily)	Distribution	Ecology
ASTIGMATA					
Histiostomatidae	*Anoetus*	10	Halictinae	NE, NT, PA, AU	scavenger/saprophyte
	Glyphanoetus	1	Nomiinae	NE	scavenger/saprophyte
	Histiostoma	2	Apidae, Diphaglossinae	NE, NT, ET	scavenger/saprophyte
Acaridae	*Cerophagopsis*	2	Megachilinae	OR, PA, NE	scavenger/saprophyte?
	Ctenocolletacarus	3	Stenotritidae	AU	cleptoparasite/scavenger
	Diadasiopus	1	Apidae (*Diadasia opuntiae*)	NE	scavenger/saprophyte?
	Halictacarus	1	Halictinae	ET	scavenger/saprophyte?
	Horstia	14	Xylocopinae, Megachilinae	NE, NT, PA, ET, OR, AU	cleptoparasite/scavenger
	Horstiella	2	Euglossini	NT	scavenger/saprophyte?
	Konoglyphus	1	Halictinae	NT	scavenger/saprophyte?
	Medeus	3	Apidae (*Anthophora bomboides*)	NE	scavenger/saprophyte?
	Megachilopus	1	Megachilinae	ET	scavenger/saprophyte?
	Neohorstia	1	Megachilinae	PA	scavenger/saprophyte?
	Sancassania	1	Halictinae, Nomiinae, Megachilinae	NE	cleptoparasite/scavenger
	Schulzea	2	Halictinae, Hylaeinae, Megachilinae	NT, PA, OR	scavenger/saprophyte?
	Sennertionyx	1	Megachilinae	PA, NE	scavenger/saprophyte?
Winterschmidtiidae	*Vidia*	7	Megachilinae	PA, NE, OR, ET	scavenger/saprophyte
Suisiidae	*Tortonia*	3	Megachilinae, Xylocopinae	PA, OR, NE, NT	cleptoparasite/scavenger

Chaetodactylidae	*Achaetodactylus*	3	Xyloccpinae (*Ceratina*)	ET	scavenger/saprophyte
	Centriacarus	2	Apidae (Centridini)	NT	scavenger/saprophyte
	Chaetodactylus	15	Megachilinae, Lithurginae, Apidae	NE, NT, PA, ET	pollen-feeder
	Roubikia	1	Apidae	NT	scavenger/saprophyte
	Sennertia	59	Xylocopinae, Megachilinae	NT, AU, PA, NE, OR, ET	scavenger/saprophyte
PROSTIGMATA					
Trochometridiidae	*Trochometridium*	1	Halictinae, Nomiinae, Panurginae, Apidae	NE, NT	scavenger/saprophyte
Pygmephoridae	*Parapygmephorus*	6	Halictinae, Nomiinae	ET, OR	scavenger/saprophyte
Scutacaridae	*Imparipes*	11	Halictidae, Andrenidae	NE, NT, PA	scavenger/saprophyte
	Nasutiscutacarus	2	Nomiinae	OR	scavenger/saprophyte
	Scutacarus	1	Halictinae	NE	scavenger/saprophyte
Cheyletidae	*Cheletophyes*	13	Xylocopinae	ET, OR, NT	predator of mites
Tarsonemidae	*Tarsonemus*	2	Xylocopinae	NT, OR	scavenger/saprophyte?
MESOSTIGMATIDAE					
Ameroseiidae	*Afrocypholaelaps*	2	Megachilinae, Xylocopinae	ET	cleptoparasite
	Neocypholaelaps	7	Apidae, Megachilinae, Hylaeinae	OR, ET	cleptoparasite
Laelapidae	*Dinogamasus*	36	Xylocopinae	PA, ET, OR	scavenger/saprophyte?
	Hypoaspis	3	Diphaglossinae, Xylocopinae, Megachilinae	NT	predator
	Laelaspoides	1	Halictinae	NE, AU	cleptoparasite
	Raymentia	3?	Halictinae	AU	scavenger/saprophyte?

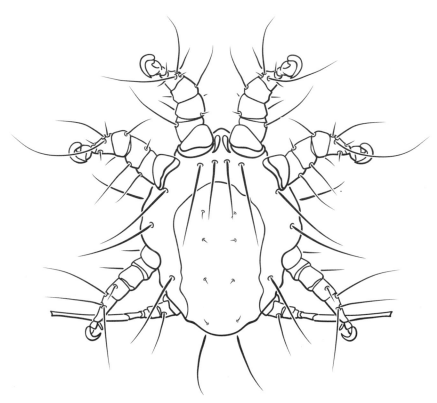

Figure 9-5. *Sennertia lauta* (Chaetodactylidae) heteromorphic deutonymph, dorsal view. This mite is reported from Asian *Xylocopa*: *Xylocopa* (*Zonohirsuta*) *fuliginata* and *X. (Z.) dejeanii* (redrawn from Klimov et al. 2007b, Fig. 11).

(bee)–*Centriacarus* (mite) (Table 9-2). Mites in the genus *Chaetodactylus* attack a wide variety of bee hosts and seem largely detrimental to bee larval development (Qu et al. 2002).

When bees benefit from these mite associates, one finds highly specialized "acarinaria" (mite homes) located on the body of adult female bees. Acarinaria have been identified on the metasoma, mesosoma, and even within the genital tract (in *Xylocopa fimbriata*; Klimov et al. 2006). The acarinaria provide safe passage of the non-feeding stage of the mite, the so-called heteromorphic deuteromorph (or hypopi). The hypopi are highly adapted for dispersal: they are a non-feeding stage that lacks mouthparts and a digestive tract; they have a heavily sclerotized carapace, which prevents dehydration; and they have robust pretarsal claws and caudoventral suckers for attaching to the host (Houck and O'Connor 1991). Studies of the coevolutionary history of chaetodactylid mites and their long-tongued bee hosts by Klimov and colleagues (2007a) failed to show a clear pattern of co-cladogenesis. However, our understanding of the phylogeny of Apidae has changed dramatically since 2007 (*Tetrapedia* is now considered part of Xylocopinae; Bossert et al. 2018), so it may be necessary to reexamine these findings in light of this new understanding of apid phylogeny.

Table 9-2. Summary of chaetodactylid mites associated with bees in the families Megachilidae and Apidae.

Modified from Klimov and O'Connor (2007).

Note that the genus *Chaetodactylus* has an extremely broad host range.

Bee family	Bee subfamily	Bee tribe	Bee genus	Mite genus
Megachilidae	Lithurginae	Lithurgini	*Lithurgus*	*Chaetodactylus*
			Trichothurgus	*Chaetodactylus*
			Microthurge	*Chaetodactylus*
	Megachilinae	Osmiini	*Osmia*	*Chaetodactylus*
			Hoplitis	*Chaetodactylus*
			Chelostoma	*Chaetodactylus*
		Anthidiini	*Rhodanthidium*	*Chaetodactylus*
			Anthidium	*Chaetodactylus*
		Megachilini	*Megachile*	*Chaetodactylus*
Apidae	Apinae	Centridini	*Centris*	*Centriacarus*
	Eucerinae	Emphorini	*Ancyloscelis*	*Chaetodactylus*
			Diadasia	*Chaetodactylus*
			Melitoma	*Chaetodactylus*
			Ptilothrix	*Chaetodactylus*
		Tapinotaspidini	*Chalepogenus*	*Chaetodactylus*
	Xylocopinae	Ceratinini	*Ceratina*	*Sennertia*
			Ceratina	*Achaetodactylus*
		Tetrapediini	*Tetrapedia*	*Roubikia*
		Xylocopini	*Xylocopa*	*Sennertia*

Two studies have demonstrated a clear benefit of mites to bees. Cordeiro and colleagues (2011) studied the biology of *Roubikia* mites in *Tetrapedia diversipes* nests in Brazil. Based on a massive sample of over 900 nests, they found that roughly 18% of nests were inhabited by mites. In nests inhabited by mites, 81% of the emerging adult bees bore mites on their bodies. The average mite load was 78 mites per bee, and the maximum number of mites on a single bee was 385. Mites were slightly more common on male bees than on female bees. *Roubikia* was clearly beneficial to *Tetrapedia*, based on a number of observations. First, nests infected with mites showed zero mortality from fungi, whereas nests lacking mites showed 15% mortality due to fungal infection. Second, across the four localities studied, there was a significant negative correlation between the percentage of nests infested with mites and the mortality rate due to fungal infection.

In a second study, Biani and colleagues (2009) observed mesostigmatid mites (*Laelaspoides*) that occur in the twig nests of two nocturnal halictid bees (*Megalopta genalis* and *M. ecuadoria*) in Panama. They demonstrated a clear benefit to the bees of having mites. Naturally occurring mite frequencies varied between the two hosts, with 63% of *M. genalis* cells having mites and 17% of *M. ecuadoria* cells having mites. Mite numbers ranged up to 25 per cell. When mites were present in cells, there were lower rates of fungal infection than when they were absent, suggesting that they benefit the bees by decreasing the proliferation of potentially pathogenic fungi.

Chaetodactylid mites can also be harmful to their hosts, especially in the case of *Chaetodactylus* (Table 9-2). Furthermore, the bees associated with *Chaetodactylus* generally lack acarinaria, suggesting that, from the perspective of the bee, these associations are not beneficial. *Chaetodactylus* mites have dimorphic deuteronymphs. Besides the phoretic form, there is a cyst-like, non-dispersing form. These immobile mites remain in the nest, ready to attack any bee that makes the mistake of reusing the nest. These cyst-like, non-dispersing forms are probably key to the survival of predatory *Chaetodactylus* species, because once the hosts are all killed, there is little opportunity for phoretic dispersal.

Cross and Bohart (1992) documented a surprisingly intimate (and temporally synchronized) association between the alkali bee (*Nomia melanderi*, Halictidae) and a mite (*Imparipes apicola*, Scutacaridae). *I. apicola* appears to feed exclusively on hyphae of several undescribed species of *Ascosphaera*. The most common species of *Ascosphaera* develops in the midgut of the bee larva and is passed in the feces. Cross and Bohart (1992) describe the fungal hyphae as "cordlike . . . connecting the fecal pellets together necklace-like" (p. 161). Defecation of the fully fed bee larva appears to trigger ovogenesis in *I. apicola* so that eggs are deposited coincident with larval defecation. The juvenile and adult stages of *I. apicola* feed on the fungal hyphae growing in the bee's fecal material. *N. melanderi* is a multivoltine bee, and the life cycle of *I. apicola* is perfectly synchronized with that of the bee. While some species of *Ascosphaera* are pathogenic to bees (e.g., *A. apis*, the causative agent of chalkbrood in honey bees), the species found in *N. melanderi* cells did not cause bee mortality and appeared only in nests with healthy larvae. It appears that *I. apicola* is a highly adapted, in-nest commensal of *N. melanderi*. It causes no harm to *N. melanderi* but does not seem to provide any benefits either. In other halictid bees, especially in the genus *Lasioglossum*, mites appear to have beneficial effects. Many *Lasioglossum* in North America, Europe, and Australia have mite associates that attach to the propodeum or the first abdominal segment. Walter and colleagues (2002) described massive, seemingly beneficial mites associated with the endemic Australian *Lasioglossum* subgenus *Parasphecodes*.

In an even more bizarre twist, some mites associated with bees carry their own fungal spores within specialized pouches on the underside of their bodies. *Trochometridium tribulatum*, a mite associated with a number of genera, including *Calliopsis* and *Anthemurgus* (Andrenidae), *Halictus* and *Nomia* (Halictidae), and *Exomalopsis* (Apidae), carries ascospores in a specialized structure called a sporotheca. When it invades the nest of its host, the mite inoculates the brood chamber (and provision mass) with the fungus. It is not clear if it is the mite or the fungus that kills the host egg or early-instar larva, but the mites have a much more developed pharyngeal pump and cheliceral stylets than in strictly fungivorous mites, so hospicidal (host-killing) behavior by the mites seems likely (Lindquist 1985). With the bee host out of the way, the fungus grows rapidly on the nest provisions, and the fungivorous mites multiply. This is an interesting form of agricultural behavior by the mites at the expense of the bees. Sporothecae have been reported in other bee-associated mites (Ebermann et al. 2013, Ebermann and Hall 2003).

A different type of bee-mite interaction is found among some Old World carpenter bees (*Xylocopa* spp.). These bees have acarinaria that transport *Cheletophyes* mites (Prostigmata: Cheyletidae). *Cheletophyes* mites are predaceous and feed on stigmatid mites that attack bee larvae (O'Connor 1993). Rather than transporting a cleanup crew that consumes harmful microorganisms, these bees are transporting a crew of bodyguards!

For commensal mites, an obvious problem is ending up in a cell that will produce a male bee. Since males do not go on to produce nests and provision cells, these mites must find a way to get themselves onto the body of a reproductive, provisioning female. In *Horstiella*, which are acarid mites associated with *Epicharis* (Apidae), while mites on female bees usually are found under the lateral edges of the metasomal terga and sterna, mites on males tend to position themselves on the undersurface of the mesosoma, since this is likely to facilitate transfer to a female during copulation (Ochoa and O'Connor 2000). Transfer from male to female during copulation is believed to be common among bee-associated mites, whether they are beneficial to the bee or not.

As an interesting aside, there is good evidence that cleptoparasitic bees can serve as vectors for mites that occur within the nests of their hosts. Thus *Stelis* can serve as a vector of mites among nests of its host, *Osmia*, and *Coelioxoides* can serve as a vector for mites among nests of its host, *Tetrapedia* (Klimov et al. 2007b). Such a cleptoparasite-mediated mite infection may be a mechanism for mites moving from one host bee species to another, when hosts share the same cleptoparasitic species.

CHAPTER 10

Brood-Parasitic (Cuckoo) Bees

As we have seen in previous chapters of this book, solitary bees are hard workers. They must build a nest (Chapter 6), prepare a waterproof, hermetically sealed brood cell (Chapter 6), and provision the brood cell with pollen, nectar, and, sometimes, floral oils (Chapters 7 and 8). They must then lay an egg and close the brood cell with soil, wood, mud, or plant materials. This whole process can take from one to several days of work, and brood cell preparation can take place only when weather conditions and floral resources permit. What if there was an easier way? What if female bees could produce offspring without carrying out the hard work described above? What if bees could heed the advice of Jean-Henri Fabre (1914) and "Throw aside your exhausting labours, follow the evolutionists' advice and, as you have the means at your disposal, become a parasite!"? In fact, many bee lineages have heeded Fabre's advice. There are more brood parasitic than social bee species, and brood parasitism has arisen more frequently in bees than sociality. Brood parasitism is a common feature of animals that build nests, including birds, fish, dung beetles, wasps, and bees (Litman et al. 2013). As we shall see below, certain bee groups have evolved remarkably effective (and devious) modes of brood parasitism.

Brood-parasitic[1] bees are remarkable creatures (Fig. 10-1). Female brood parasites enter the nests of their pollen-collecting hosts, sometimes a closely related species, and lay one or more eggs in a fully or partially completed cell. Either the adult brood parasite or her larva kills the host egg or larva, and the brood parasite then consumes the pollen/nectar provisions so meticulously collected by the host. Some brood parasites are so discrete and subtle that the host female may not even know she is being attacked. Others wage all-out war on their hosts and kill adult residents in the process. Adult brood-parasitic bees are beautiful, ominous, hunchbacked, wasp-like creatures that skulk around the periphery of nest sites looking for opportunities to enter nests and wreak havoc. Larvae

---

[1] In this chapter and throughout the book, we use the terms *brood parasite* and *cleptoparasite* interchangeably. The term *cleptoparasite* is widely used in the solitary bee literature, but these terms can have different meanings in the vertebrate literature, where *cleptoparasitism* can refer to the physical removal of nesting materials and food from a nest. *Brood parasitism* (the laying of an egg in the nest of another organism) is a more precise term for the behavior we see in the bees described in this chapter.

Figure 10-1. Habitus drawing of female *Holcopasites calliopsidis* (Apidae: Nomadinae). Original artwork by Frances Fawcett.

in many bee brood parasites are highly adapted host-killers with ridiculously elongate mandibles for puncturing host eggs or first-instar larvae, as well as for doing battle with other brood-parasitic larvae within the same brood chamber. Brood parasitism has arisen repeatedly in the families Halictidae, Megachilidae, and Apidae, whereas in some families (Melittidae, Andrenidae, Stenotritidae), brood parasitism has never been reported (Fig. 2-2). Brood parasites comprise a large proportion of bee diversity: we estimate that 13% of all bees and 20% of apid bees are brood parasites (Chapter 2). There are more brood-parasitic species than eusocial species of bees. In Europe, one-quarter of all bee species are cleptoparasitic (Bogusch et al. 2006). Nearly every species of solitary bee seems to have its own brood parasite (there are exceptions), and there seem to be very few ways for solitary bees to avoid brood parasitism (although we will see some examples below).

Brood parasitism is just one of two modes of parasitism we described in Chapter 2. Another mode of parasitism is social parasitism. In socially parasitic species, females enter the nest of a social species and replace the queen as the primary egg layer in the colony. She can either physically evict the host queen, kill her, or simply coexist with her but suppress her egg laying. Social parasites occur in Allodapini, Bombini (both Apidae), and Halictinae (Halictidae) (see Chapter 2 for more details on social parasitism).

ADAPTATIONS FOR BROOD PARASITISM

Brood-parasitic bees often differ dramatically in morphology and behavior from their pollen-collecting relatives. Brood parasites lack pollen-collecting structures (like the scopa), and they no longer need the highly specialized hairs, combs, and brushes used by pollen-collecting bees to gather, manipulate, and transport pollen (Chapter 7). In fact, some brood-parasitic groups have been identified based solely on the fact that they lack pollen-gathering structures. We still do not have detailed behavioral observations on the mode of parasitism in many brood-parasitic bees, and host identity is entirely unknown for some groups. Freed from having to construct nests, brood parasites lack the structures needed for digging and packing soil (in ground-nesting bees) or manipulating nesting materials (in stem- and cavity-nesting species). They are generally heavily armored relative to pollen-collecting bees, and they have additional ridges, crests, spines, and carinae for protecting their vulnerable (membranous) body regions from attack by host females.

Brood parasites also show modifications to their internal anatomy, especially their reproductive system (Alexander and Rozen 1987). Female bees have two ovaries, and each ovary typically consists of three (most bee families) or four (Apidae) ovarioles per ovary (Fig. 10-2). A female pollen-collecting bee that is actively provisioning a nest would normally have one fully mature oocyte (egg) present in one of her ovarioles, with additional oocytes in various stages of development (Fig. 10-2a,b,c). This makes sense because female solitary bees generally lay just one egg per day. However, brood-parasitic females may lay many eggs per day if the opportunity arises. A large nesting aggregation of the appropriate host species or a communal nest with many actively provisioning host adult females offers a potential reproductive bonanza for a brood-parasitic bee ready to exploit the host provisions. It is not surprising that many brood parasites have a highly modified reproductive tract, with more mature oocytes per ovariole or more ovarioles per ovary—two strategies that allow more rapid and sustained oviposition (Fig. 10-2d). In Nomadinae, a diverse clade of apid brood parasites, the number of ovarioles per ovary is always more than the typical number (4) for pollen-collecting apids and can range as high as 17 in *Rhopalolemma*, a parasite of *Protodufourea* (Rozen et al. 1997). This pattern was documented in an important study by Alexander and Rozen (1987), and a similar pattern has been observed in apoid wasp brood parasites (Ohl and Linde 2003).

There is some evidence that brood-parasitic bees may use chemical mimicry to gain entry into the host nest. *Nomada* (Apidae) is a large genus of brood-parasitic bees that exclusively attack nests of bees in the genus *Andrena* (Andrenidae). In studies of four species of *Nomada* and their *Andrena* hosts in Europe, Tengö and Bergström (1977) documented a remarkable level of similarity in the

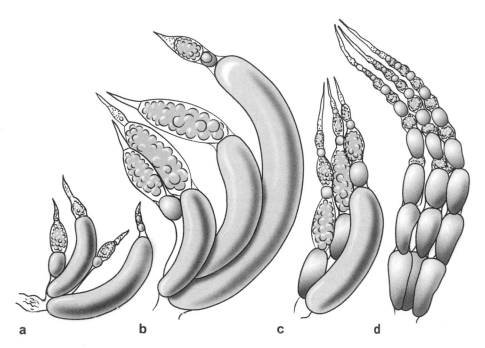

a   b   c   d

Figure 10-2. Variation in egg size and ovarian morphology among solitary and brood-parasitic bees: (a) *Anthophora florea* (Apidae) (redrawn from Iwata and Sakagami 1966, Fig. B); (b) *Xylocopa appendiculata* (Apidae) (redrawn from Iwata and Sakagami 1966, Fig. C); (c) *Megachile sculpturalis* (Megachilidae) (redrawn from Iwata and Sakagami 1966, Fig. D); (d) *Coelioxys fenestratus* (Apidae) (redrawn from Iwata and Sakagami 1966, Fig. E) Figures a–c represent pollen-collecting solitary bees. Figure d indicates the morphology for a brood-parasitic bee.

chemicals produced by the Dufour's gland of female *Andrena*, which are used to mark nests and line the brood cells, and the mandibular gland secretions of males in species of *Nomada* that attack these same species of *Andrena*. Similarity between female *Andrena* and male *Nomada* chemical profiles was found in multiple host-parasite pairs. The authors speculated that the cephalic secretions of the male *Nomada* are applied to the external surface of the females during mating, thus giving female *Nomada* the same chemical odors as the Dufour's gland secretions of their hosts. Such chemical mimicry of the host odors might give *Nomada* females the ability to slip into host nests undetected. This fascinating study has never been corroborated and raises more questions than it answers (e.g., why are male, and not female, *Nomada* the source of the chemical mimicry?). Clearly more work needs to be done on the fascinating interactions between *Nomada* and their *Andrena* hosts.

Torchio (1989b) described a remarkable behavior in female *Stelis montana* that suggests that they actively acquire host nest odors in order to gain entry to *Osmia* nests. Two hosts of *S. montana*, *O. californica* and *O. montana*, mix chewed leaf tissue into their nest partitions and closures. Leaves selected for this purpose were relatively hairy, and Torchio reasoned that he should provide plants of this type to his greenhouse bee populations. He therefore placed *Malva* and *Oenothera* plants in the greenhouse—both species with hairy leaves. On several occasions, he observed female *Stelis* chewing these leaves and spreading the masticated leaf tissue (and presumably odors) onto the surface of their bodies. He also saw female *Stelis* robbing pollen/nectar provisions from partially provisioned cells and then spreading this pollen/nectar mix on their bodies. Both observations strongly suggest that the brood parasites are actively masking their odors in order to gain entry into host nests. The fact that host females seemed to behave the same, whether their nests had been visited by a parasite or not, supports the idea that *Stelis* females can mask any evidence of their visits to host nests. Whether other modes of chemical mimicry exist within brood-parasitic bees remains to be seen. We have only glimpses of the complexity of brood-parasite behavior to suggest that mimicry may be an essential tool in the brood parasite's arsenal of weapons.

When watching brood-parasitic female bees near host nests, one sometimes has the impression that the brood parasites are watching for host females entering and leaving the nest and that they may be able to time their attacks for the moment when the host nest is unoccupied. Female brood parasites sometimes perch on small stems and vegetation around or within nesting aggregations facing toward nests and apparently scanning for host activity. When host females leave their nests, brood-parasitic females have been observed to immediately approach and enter the host nest. Such anecdotal observations have been made in a wide variety of brood-parasitic bee species.[2]

Brood-parasitic females may also be capable of learning the locations of host nests and returning to them repeatedly. Such nest-location learning has been reported in *Epeolus minimus* (Graenicher 1906), *Nomada opacella* (Linsley and MacSwain 1955), *Leiopodus singularis* (Rozen et al. 1978), *Melecta separata callura* (Thorp 1969a), and *Holcopasites ruthae* (Danforth and Visscher 1993); according to Rozen and colleagues (1978), it may be widespread among parasitic bees. By returning to previously parasitized nests, the brood parasite may be better able to assess when cells are at the appropriate stage for attack.

MODES OF BROOD PARASITISM

Long-tongued bees (Megachilidae and Apidae)

The long-tongued bee families (Apidae and Megachilidae) include a large proportion of the brood-parasitic bee species and include up to 10 independent origins of parasitism (Litman et al. 2013). Of the nearly 6,000 described species of Apidae, more than 25% are brood-parasitic. Evolution of brood parasitism in Apidae and Megachilidae was recently analyzed based on extensive molecular phylogenetic studies of each family (Cardinal et al. 2010, Litman et al. 2013).

[2] Examples include: *Coelioxys sodalis* (as *ribis*) (Graenicher 1927), *C. octodentata* (Michener 1953b), *C. flagrata* (Bohart and Youssef 1972), *Stelis lateralis* (Michener 1955) (all Megachilidae), *Leiopodus singularis* (Rozen et al. 1978), *Exaerete smaragdina* (Garófalo and Rozen 2001), *Coelioxoides waltheriae* (Alves-dos-Santos et al. 2002), and *Holcopasites ruthae* (Danforth and Visscher 1993) (all Apidae).

Modes of brood parasitism in the long-tongued bees are enormously variable but can be grouped into three distinct categories, as described in Litman and colleagues (2013).

Adult, closed-cell parasitism (AC)

In adult, closed-cell parasitism, females enter the nest of a host bee and locate a closed, fully provisioned cell. These females gain access to the host cells and then kill the host eggs or larvae using either their mandibles or their sting. This strategy is typified by parasites such as *Euaspis* in the *Stelis* clade (Megachilidae; Iwata 1976) and *Exaerete* (Apidae; Garófalo and Rozen 2001) (Table 10-1). Females that attack closed brood cells typically produce much larger eggs than those that attack open brood cells (Rozen 2003a; Fig. 10-3).

Garófalo and Rozen (2001) observed six nests of an orchid bee, *Eulaema nigrita*, on the campus of the University of Sao Paulo, Ribeirão Preto, Brazil, and documented this mode of parasitism in *Exaerete smaragdina*, a closely related brood parasite in the same tribe (Euglossini). Some nests were in glass observation containers, which provided an unusually detailed view of the behavior and mode of parasitism. Nests of the host are made of resin mixed with fecal material, and females of the brood parasite (*E. smaragdina*) were observed from the time they entered the nests to their departure after oviposition. Female *Exaerete*, upon entering the nest, moved rapidly over the cluster of brood cells, vibrating their wings and touching their antennae to the cell surfaces. Female *Exaerete* preferred cells that had recently been completed because they appeared to be easier to open but would attack cells up to six days old. Some cells were parasitized by multiple female *Exaerete*.

A female *Exaerete* opens the closed cells of the host by cutting a small hole in the top of the cell above the level of pollen and nectar provisions. She then introduces her metasoma into the hole and stings the host egg or first-instar larva before ovipositing in the cell. After laying her egg, the female *Exaerete* closes the cell, using resin collected from a nearby cell under construction or from nest-building materials stored on the floor of the nest. *Exaerete* females attacked up to three cells per visit. Host females, when they encountered female parasites within their nests, did not show any signs of aggression, but the parasitic females usually immediately left the nest. Female *E. smaragdina* made repeated visits to the same nest over successive days, suggesting that females learn the locations of potential nests. This may explain the high rate of parasitism reported in the Garófalo and Rozen (2001) study: 11–100% of cells per nest were parasitized. Similarly high rates of parasitism have been observed in other studies of *Exaerete* (see references in Garófalo and Rozen 2001).

It appears that female *E. smaragdina* sting both host eggs as well as co-occurring *Exaerete* eggs prior to oviposition within cells (Garófalo and Rozen 2001). Other *Exaerete* species (e.g., *E. dentata*; Bennett 1972) use their mandibles to kill or remove the host egg. Surprisingly, the second-instar larva of *E. smaragdina* is equipped with sharply pointed, fang-like mandibles to kill host or conspecific eggs and larvae, perhaps as a backup in the event the adult female brood parasite does not eliminate all available eggs.

Table 10-1. Multiple origins of brood parasitism in bees.

Each entry in the table represents a unique, monophyletic lineage of brood-cell parasites.

Hosts are known for most, but not all, of these brood parasites. In Halictidae, the distinction between brood and social parasite is blurred, and we have included several lineages of halictids that might better be described as social parasites (in *Lasioglossum* and *Halictus*).

AC = adult, closed-cell parasitism; LC = larval, closed-cell parasitism; LO = larval, open-cell parasitism.

Cleptoparasite (genus, tribe, or clade)	Cleptoparasite family	Host	Host family	Mode of parasitism	Reference
Nomadinae	Apidae	diverse hosts [see Table 10-2]	Apidae, Colletidae, Melittidae, Andrenidae, Halictidae	LC, LO	Cardinal et al. 2010, Litman et al. 2013, Rozen 2000
Euglossini (*Aglae* + *Exaerete*)	Apidae	*Eulaema, Eufriesea*	Apidae	AC	Bennett 1972, Garófalo & Rozen 2001, Michener 2007
Ctenoplectrina	Apidae	*Ctenoplectra*	Apidae	unknown	Schaefer & Renner 2008
Hylaeus (*Nesoprosopis*)	Colletidae	*Hylaeus* (*Nesoprosopis*)	Colletidae	unknown	Daly & Magnacca 2003
Halictus (*Paraseladonia*)	Halictidae	*Halictus* (*Seladonia*)	Halictidae	social parasite	Pauly 1997
Lasioglossum (*Dialictus*)	Halictidae	unknown	Halictidae?	social parasite?	Gibbs 2009
Lasioglossum (*Echthralictus*)	Halictidae	*Lasioglossum* (*Homalictus*)	Halictidae	unknown	Michener 1978
Lasioglossum (*Paradialictus*)	Halictidae	unknown	Halictidae?	social parasite	Pauly 1984, Arduser & Michener 1987
Lasioglossum (*Paralictus*; now placed in *Dialictus*)	Halictidae	*Lasioglossum* (*Dialictus*)	Halictidae	social parasite	Wcislo 1997, Gibbs et al. 2012a
Megalopta (*Noctoraptor*)	Halictidae	*Megalopta*	Halictidae	social parasite?	Biani & Wcislo 2007
Megommation (*Cleptommation*)	Halictidae	*Megommation*	Halictidae	unknown	Engel 2000
Parathrincostoma	Halictidae	*Thrinchostoma*	Halictidae	unknown	Danforth et al. 2008

Sphecodes clade	Halictidae	diverse hosts	Halictidae, Colletidae, Andrenidae	AC?	Sick et al. 1994, Knerer & Atwood 1966, Ordway 1964, Eickwort & Eickwort 1972, Bogusch et al. 2006
Temnosoma	Halictidae	Augochloropsis	Halictidae	unknown	Engel 2000
Austrostelis + Hoplostelis	Megachilidae	Anthidiini	Megachilidae	AC	Zanella & Ferreira 2005, Bennett 1966
Coelioxys	Megachilidae	mostly Megachile also Centris, Anthophora, and Euglossa	Megachilidae, Apidae	LO	Gonzalez et al. 2012, Rozen et al. 2010b and references therein
Dioxyini	Megachilidae	Osmiini, Anthidiini, Megachilini	Megachilidae	LC, LO	Rozen & Favreau 1967, Westrich 1989, Rozen & Özbek 2005
Hoplitis (erythrogaster group)	Megachilidae	Hoplitis	Megachilidae	unknown	Sedivy et al. 2013a,b, Eickwort 1975
Radoszkowskiana	Megachilidae	Megachile	Megachilidae	LO (likely)	Rozen 2007
Stelis clade	Megachilidae	Anthidiini	Megachilidae	AC, LO	Fabre 1914, Iwata 1976, Parker et al. 1987, Rozen 1987b, Rozen & Kamel 2009, Rust & Thorp 1973, Taylor 1962, Westrich 1989

Larval, closed-cell parasitism (LC)

In larval, closed-cell parasitism, female parasites locate closed cells of the host and introduce their egg via a small opening in the cell. Eggs are deposited within the nest, and the cell is closed by the host female. Because the host female does not enter the cell to kill the host egg or larva, one or more of the parasite's larval instars is modified for killing the host. Such host-killing larvae are referred to as "hospicidal" and generally have modified mandibles for killing both the host and co-occurring parasitic larvae (Rozen 1989a). Eggs in the larval, closed-cell mode of parasitism are generally large (as large as the eggs of pollen-collecting hosts) and are not modified, as in the larval, open-cell parasites (see below). This mode of parasitism is exemplified by the megachilid tribe Dioxyini and a variety of apid tribes, including Ericrocidini (Rozen 1969a), Melectini (Rozen 1969b [*Thyreus*], Torchio and Trostle 1986 [*Xeromelecta*]), Rhathymini (Rozen 1969a), Osirini (Michener 2007), and the genus *Coelioxoides* (Alves-dos-Santos et al. 2002) (Tables 10-1 and 10-2).

Rozen and colleagues (2011b) provide a detailed description of the biology and mode of parasitism of *Mesoplia sapphirina*, a brood parasite of *Centris flavofasciata* (Apidae). Nests of the host were widely scattered along a curved, sandy beach with overhanging trees at Playa Grande, Guanacaste Province, on the Pacific coast of Costa Rica. Each nest contained a single cell, which was located at a depth of 12 to 17 centimeters. Female *M. sapphirina* were observed flying rapidly over the nest site, occasionally inspecting potential host nests. A total of 10 *Centris* cells were excavated in 2010, and a whopping 9 of them were parasitized by *M. sapphirina*. In some cases, cells contained as many as 6 *Mesoplia* eggs or larvae, indicating an extremely high rate of parasitism. Parasitism rates were much lower (30%) in 2011. In recently parasitized cells, *M. sapphirina* eggs were attached to the inner surface of the cell closure, and a small hole was evident in the cell closure itself, indicating that females oviposit into closed cells. The first-instar *M. sapphirina* larva has large, sickle-shaped mandibles (Figs. 46, 47 in Rozen et al. 2011b) and is an aggressive assassin of both host and co-occurring parasite eggs or larvae. *Mesoplia* first and second instars were described by the authors as "agile and feisty," in spite of the fact that the first instars remain attached to the chorion of the egg while they move around the inside of the host cell! *Mesoplia* larvae develop more rapidly than *Centris* larvae, presumably an adaptation that allows the larval brood parasite to locate and kill the defenseless host egg.

Larval, open-cell parasitism (LO)

The largest group of brood-parasitic long-tongued bees fall into this category of larval, open-cell parasites. Females in these species lay their eggs discretely in open, partially provisioned cells of their hosts. The brood-parasitic egg is often small, asymmetrical, and cryptically sculptured, and the egg is laid directly into the wall of the host cell in some groups (in Fig. 10-3, cells above the horizontal line). Larvae are "hospicidal" (host-killing) and modified for attacking host eggs and larvae (and perhaps, most importantly, other conspecific brood parasites). These larvae bear sharply pointed, sometimes sickle-shaped mandibles,

Table 10-2. Tribes of subfamily Nomadinae (sensu Bossert et al. 2018), with their host associations and modes of parasitism.

Note that several tribes (Melectini, Rhathymini, Ericrocidini, Isepeolini, Proepeolini, Coelioxoidini, and Osirini) were previously classified in the subfamily Apinae. The remaining tribes have traditionally been placed in the subfamily Nomadinae.

Modified from Litman et al. (2013). Genera of Osirini are listed because th s tribe is potentially paraphyletic.

Tribe	Genus	Host	Host family	Number of described species	Mode of parasitism
Ammobatini		Ancyla, Tetralcniella, Anthophora, Panurginae, Nomioides, Nomiinae, Scrapter	Apidae, Andrenidae, Halictidae, Colletidae	127	LO
Ammobatoidini		Melitturga, Melitturgula, Calliopsis Camptopoeum, Pseudopanurgus, Protandrena	Andrenidae	33	LO
Biastini		Protodufourea, Systrcpha, Rophites, Dufourea	Halictidae	13	LO
Brachynomadini		Exomalopsis, Anthophoruia, Liphanthus, Protandrena, Psaenythia, Leioproctus	Apidae, Andrenidae, Colletidae	26	LO
Caenoprosopidini		Arhysosage, Callonychium	Andrenidae	2	?
Coelioxoidini		Tetrapedia	Apidae	4	LC
Epeolini		Eucerini, Arthophorini, Centridini, Emphorini, Colletes, Ptiloglossa, Oxaea, Protoxaea, Dieunomia	Apidae, Colletidae, Andrenidae, Halictidae	297	LO
Ericrocidini		Centris, Ep charis	Apidae	42	LC
Hexepeolini		Ancylandrena	Andrenidae	1	LO
Isepeolini		Colletes, Mourecotelles, Caneohorula	Colletidae, Apidae	21	LO?
Melectini		Anthopho ini, Eucerini	Apidae	196	LC
Neolarrini		Perdita, Calliopsis	Andrenidae	16	LO

(continued)

Table 10-2. (Continued)

Tribe	Genus	Host	Host family	Number of described species	Mode of parasitism
Nomadini		mostly *Andrena*, other hosts include *Agapostemon*, *Halictus*, *Lasioglossum*, *Lipotriches*, *Panurgus*, *Melitta*, *Colletes*, *Exomalopsis*, *Eucera*	Andrenidae, Halictidae, Melittidae, Colletidae, Apidae	709	LO
Osirini	*Osiris*	*Tapinotaspidini*, *Tetrapedia*, *Macropis*	Apidae, Melittidae	32	LC?
	Epeoloides	*Macropis*	Melittidae	2	LC?
	Parepeolus	*Tapinotaspoides*, *Chalepogenus*, *Lanthanomelissa*	Apidae	5	LC?
	Protosiris	*Monoeca*	Apidae	5	LC?
Protepeolini		*Diadasia*, *Melitoma*, and *Ptilothrix*	Apidae	5	LO
Rhathymini		*Epicharis*	Apidae	19	LC
Townsendiellini		*Hesperapis*	Melittidae	4	LO

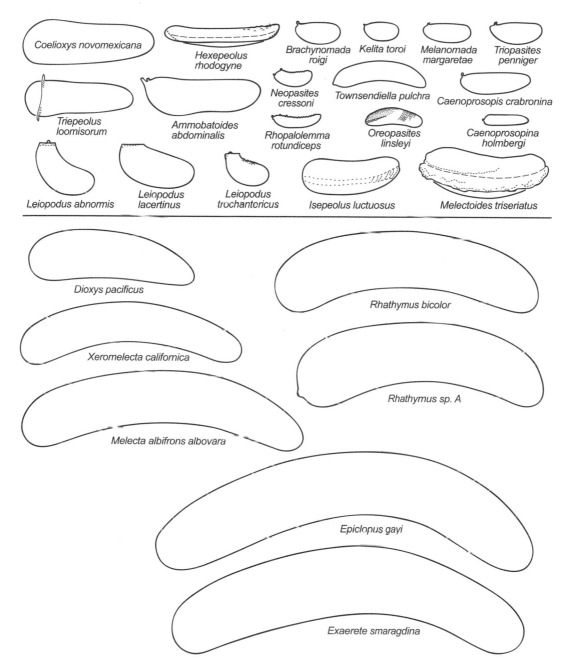

Figure 10-3. Variation in egg morphology in brood-parasitic bees in the family Apidae. Eggs are all drawn to scale. Eggs that are hidden within open cells are above the horizontal line; eggs that are introduced into closed cells are below the horizontal line (redrawn from Rozen 2003a, Fig. 53).

protuberances on the labrum apparently designed for killing other bee larvae within the host cell, heavily sclerotized head capsules packed with mandibular musculature, and lateral appendages on the abdominal segments for movement within the pollen-nectar provisions (Fig. 10-4). Larval, open-cell parasitism characterizes some of the largest and most species-rich brood-parasitic groups, including the *Stelis* clade in the Megachilidae and the majority of tribes in the apid subfamily Nomadinae (Tables 10-1 and 10-2).

Stelis montana, an anthidiine parasite of various *Osmia* species, exhibits this mode of parasitism. Torchio (1989b) conducted a detailed field, greenhouse, and laboratory study of *S. montana* attacking three *Osmia* species: *O. lignaria*, *O. californica*, and *O. montana*. The study spanned a 14-year period and was based on over 65,000 *Osmia* nests and 320,000 opened brood cells. Torchio documented rates of parasitism, parasite sex ratio, details of egg deposition, interactions between parasite and host larvae, and behaviors of the host to combat the parasite. Because host bees were nesting in glass observation tubes, the interactions between host and parasite could be directly observed.

S. montana females, like all female *Stelis*, enter open, partially provisioned cells of the host and deposit their eggs on or within the host's cell provisions some distance away from where the host female will eventually lay her egg. Egg deposition by the parasitic females varied among the three hosts, indicating that the *Stelis* females can detect host species identity or at least modify its behavior based on cues from the host nest or provision mass. Female *Stelis* preferentially parasitized cells that are more than half completed by either measuring (with their antennae) the angle of the provision mass (in *O. lignaria* and *O. montana*) or the distance between the cell threshold and the provision mass (in *O. californica*). Cells nearing completion (more than half provisioned) were parasitized, whereas those that were in an early stage of completion were not. How eggs were hidden within the provision mass varied among host species. When parasitizing *O. lignaria*, female *Stelis* construct a pyramid-like barrier behind which they deposit the egg, effectively preventing host detection. When parasitizing *O. californica* or *O. montana*, female *Stelis* dig into the host provision mass to create a cavity into which the egg is deposited. Following oviposition, the egg chamber was covered with pollen/nectar from the edge of the provision mass.

Host species varied in their response to discovering an adult *Stelis* within their nests and in how they treat their provision masses prior to egg deposition. In *O. lignaria*, females chased the intruding brood parasite from the nest but did not manipulate or investigate the pollen provisions. In *O. californica*, females that detected and physically evicted a female parasite chewed the entire provision mass prior to oviposition, thus destroying parasite eggs. In *O. montana*, females systematically excavated the entire provision mass prior to oviposition, thus killing any parasite eggs hidden in the provision mass. This may explain why *O. montana* had the lowest rate of parasitism in the study (0.8% overall, as opposed to 2.3% for *O. lignaria* and *O. californica*).

The host-killing behavior of *S. montana* and interactions between *Stelis* larvae and co-parasitized cells was observed in detail via glass observation tubes. While all *Stelis* larval instars have acutely pointed mandibles that are seemingly capable of piercing host larvae, Torchio (1989b) reported that only

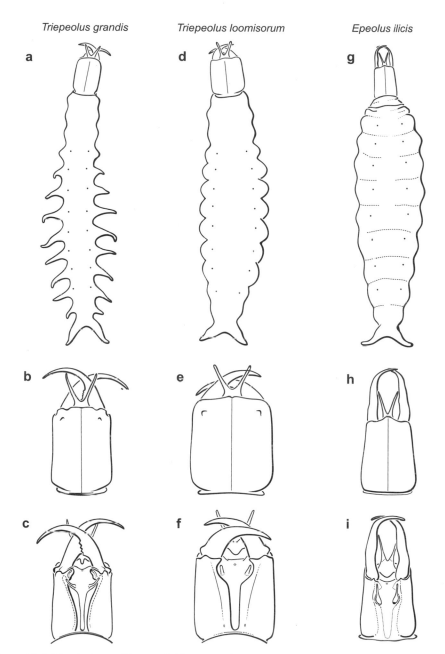

Figure 10-4. Morphology of hospicidal first-instar larvae in the tribe Epeolini (Apidae). First-instar larva (a) and close-up of the dorsal (b) and ventral (c) views of the head capsule in *Triepeolus grandis* (redrawn from Rozen 1989a, Figs. 5, 1, 2); first-instar larva (d) and close-up of the dorsal (e) and ventral (f) views of the head capsule in *T. loomisorum* (redrawn from Rozen 1989a, Figs. 23, 19, 20); first-instar larva (g) and close-up of the dorsal (h) and ventral (i) views of the head capsule in *Epeolus ilicis* (redrawn from Rozen 1989a, Figs. 29, 31, 32).

fifth instars attack and kill host larvae either on or within the pollen/nectar provision mass. Fifth instars, in contrast to earlier instars, are mobile and move aggressively (with mandibles open) toward physical stimuli. They kill host larvae by aggressively biting and puncturing the host cuticle and imbibing the liquid contents of the host larva. *Stelis* larvae are capable of killing up to six host larvae consecutively before they begin to ignore additional larvae experimentally added to the cell. When cells are parasitized by multiple *Stelis* larvae, they aggressively attack one another until one larva remains alive. When feeding on host provisions, larvae burrow through the pollen/nectar mass using peristaltic movements of the body.

S. montana has one additional adaptation to parasitizing stem- or cavity-nesting hosts. After consuming between two-thirds and three-quarters of the provision mass, larvae defecate and then construct a tough, highly resistant cocoon. Construction of the cocoon consumes a whopping 40% of total larval body weight but renders the *Stelis* larva (and subsequently, the adult) entirely resistant to damage when male and female *Osmia* dig their way out of the tunnel the following spring. This would appear to be a good investment in larval resources given the potential for host adults to kill the parasite as they exit the nests.

Colletidae

Brood parasitism is known from only one, geographically isolated, group of colletid bees, but we know almost nothing about their mode of parasitism. The *Hylaeus* subgenus *Nesoprosopis* includes the only native, solitary bees in the Hawaiian Islands. The group consists of a total of 60 species (Daly and Magnacca 2003) that all descend from a single colonization event that occurred less than 1 million years ago (Magnacca and Danforth 2006, 2007). Like many Hawaiian groups (e.g., *Laupala* crickets; Mendelson and Shaw 2005), the Hawaiian *Hylaeus* show an extraordinarily rapid diversification following their arrival in the islands.

Perkins (1899) was the first to comment on the existence of brood parasitism among Hawaiian *Hylaeus*; he based this on the reduced pollen-gathering hairs on the front tarsi of females as well as observations of the brood-parasitic species within the nests of other *Hylaeus*. Molecular phylogenetic studies indicate that the five presumed brood-parasitic species (the "inquiline" group) represent a single origin of brood parasitism (Magnacca and Danforth 2006). Detailed studies of the behavior and host associations of the Hawaiian *Hylaeus* have not been conducted, but such studies would be extremely interesting given the rarity of brood parasitism in colletid bees.

Halictidae

Few details are known about the modes of brood parasitism in halictid bees. For many groups, the existence of brood-parasitic taxa has been determined only on the basis of the lack of pollen-collecting structures or other morphological details suggestive of a brood-parasitic life history. In a few cases, we have solid evidence for associating hosts and parasites, but few studies have documented the details of the mode of parasitism. In comparison to Apidae and Megachilidae, we know very little about brood parasitism in Halictidae, in spite of the fact that there are more origins of brood parasitism in Halictidae than in any other bee family. Based on the few detailed studies that have been conducted of halictid brood parasites (see below), the mode of parasitism in Halictidae would be categorized as adult, closed-cell parasitism, as described above for certain long-tongued bee groups. However, halictid brood parasites behave quite differently from the apid and megachilid brood parasites, as we will see below.

An additional feature that adds confusion to understanding the modes of parasitism in Halictidae is that the boundary between brood parasite and social parasite is blurred. This is largely because in halictids there are both socially polymorphic species and clades that include a mix of solitary and social taxa (Biani and Wcislo 2007, Gibbs et al. 2012b). In the absence of detailed behavioral data, one can infer, based on the host life history and nesting biology, the likely mode of parasitism for many halictid parasites. For example, the following parasitic taxa are known (or inferred) to attack exclusively solitary or communal hosts: *Lasioglossum* (*Echthralictus*), *Megommation* (*Cleptommation*), *Parathrincostoma*, and *Temnosoma* (Table 10-1). Hence, we would refer to these as brood parasites rather than social parasites. For other taxa that attack largely eusocial hosts, we can reasonably infer that they are social (rather than brood) parasites: *Halictus* (*Paraseladonia*), parasitic members of the *Lasioglossum* subgenus *Dialictus* (formerly referred to as *Paralictus*), and *Lasioglossum* (*Paradialictus*) (Table 10-1). Indeed, behavioral observations of *Lasioglossum* (*Dialictus*) *asteris* (Wcislo 1997) confirm that parasitic *Lasioglossum* (*Dialictus*) behave more as social parasites than as brood parasites. The boundary between social and brood parasite becomes even more fluid in taxa that attack socially polymorphic hosts. For example, *Megalopta* (*Noctoraptor*) attacks socially polymorphic species of *Megalopta* (*Megalopta*), and the tribe Sphecodini (Chapter 2; Fig. 2-7) includes members that attack solitary hosts and others that attack social hosts. Given the complexity of parasitism in Halictidae, we have chosen to list all parasitic halictid groups in Table 10-1, even though some may fall into the social parasite category.

The largest group of parasitic halictid bees are in the tribe Sphecodini, which includes diverse brood-parasitic and socially parasitic taxa. The Sphecodini includes several genera (*Sphecodes*, *Microsphecodes*, *Ptilocleptis*, *Eupetersia*, and *Nesosphecodes*); as a whole, they attack a wide variety of solitary, communal, and eusocial hosts in the families Halictidae, Andrenidae, and Colletidae. Unlike the apid brood parasites, *Sphecodes* larvae are relatively unmodified (Fig. 10-5). The genus *Microsphecodes* behaves clearly as a social parasite based on detailed studies of Eickwort and Eickwort (1972) in Costa Rica. *M. kathleenae* attacks nests of *Lasioglossum* (*Dialictus*) *umbripenne*, a eusocial halictine bee. *Microsphecodes* females (sometimes multiple females per nest) coexist with the host queen in large eusocial nests and replace the queen's eggs with their own, thus behaving much like a *Psythirus* parasite in a bumble bee nest (Kearns and Thomson 2001). Other genera, such as *Ptilocleptis* and *Eupetersia*, have never been studied in detail, and a recently described genus (*Nesosphecodes*; Engel 2006) is known only from a few specimens collected in Cuba, Puerto Rico, and

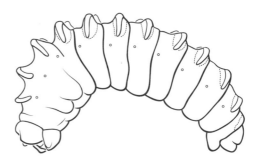

Figure 10-5. Post-defecating (last-instar) larva of *Sphecodes kathleenae* (Halictidae; redrawn from Eickwort and Eickwort 1972, Fig. 1).

the Dominican Republic. Brood-parasitic behavior in these species has been inferred primarily based on female morphology.

Parasitism of solitary, communal, and eusocial hosts has been reported for several species of *Sphecodes* (Bogusch et al. 2006, Danforth 1989a, Knerer and Atwood 1966, Ordway 1964, Sick et al. 1994). Taken together, these studies reveal a diversity of modes of parasitism and methods of host attack. Two main strategies seem to exist, depending on the biology of the host. First, some species use what Eickwort and Eickwort (1972) described as a "sneak attack." In these species, the female *Sphecodes* enters the host nest while the host is absent and lays her eggs in recently closed cells containing either host eggs or early-instar larvae. Thus, *Sphecodes* would be described as following the "adult closed" (AC) mode of attack described in Apidae and Megachilidae. Battles may ensue if the host encounters the parasites within the nest, and some species have been reported to render their hosts "akinetic" (Knerer and Atwood 1966). This mode of attack is used also on eusocial hosts, but during the foundress phase, when nests are functionally solitary. A second mode of attack, as described by Eickwort and Eickwort (1972), is "colony attack." In this case, the hosts are either communal (e.g., *Perdita*; Danforth 1989a) or eusocial (e.g., *Lasioglossum*; Knerer 1993); in both cases, there are large numbers of adult females in the nest that actively battle the intruding *Sphecodes*. In this case, the *Sphecodes* females kill or repel, apparently chemically, large numbers of host bees. Ordway (1964), for example, describes how *S. pimpinellae* cause social *Augochlorella aurata* to leave the nest either in flight or staggering about as if completely disoriented. These kinds of observations strongly suggest that *Sphecodes* may use pheromones to disorient their hosts when the host populations are large enough to put up resistance to the invader. Other studies have reported large-scale slaughter of the resident female population prior to accessing the host brood chambers (Danforth 1989a, Knerer 1993; described in graphic detail in Packer 2010). One is tempted to call this the "scorched earth attack." Overall, the few reports we have on parasitic behavior in *Sphecodes* indicate substantial variability in the mode of attack and intriguing hints that chemical weapons may be involved.

Host-parasite associations have been examined extensively in European *Sphecodes*. Some species appear to be widespread host generalists (attacking diverse host taxa, including various genera of Halictidae and Andrenidae; e.g., *S. monilicornis* and *S. ephippius*), whereas others are more host-specific (attacking just a single genus or subgenus). Habermannová and colleagues (2013) examined the evolution of host range using a molecular phylogeny of 37 *Sphecodes* species and found evidence of repeated host switching and both expansions and contractions of host range. The ancestral state for the genus appears to be a broad host range. Within *Sphecodes* there are clearly defined clades that have evolved more narrow host preferences. *S. majalis* and *S. spinulosus* specialize on members of *Lasioglossum* sensu stricto (Halictidae); *S. pinguiculus* specializes on members of the *Halictus* subgenus *Seladonia*; *S. autumnalis* specializes only on members of the North American genus *Perdita* (Andrenidae); and the clade of *S. rubicundus* and *S. ruficrus* specializes exclusively on members of the genus *Andrena* (Andrenidae). However, even within the host generalists, there are hints of individual preferences for certain host taxa. Bogusch and colleagues (2006) examined the host choice among individual members of two "generalist" species: *S. monilicornis* and *S. ephippius*. They found that individual female *Sphecodes* showed a strong preference for the same host across repeated nest entries. These observations indicate that individual female brood parasites may have clear host preferences when patterns of host choice are examined closely.

EVOLUTION OF BROOD PARASITISM

Brood-parasitic bees have posed a particular evolutionary challenge. While it is not hard to understand *why* brood parasitism is such an appealing strategy (no hard work constructing a nest or gathering pollen and nectar and seemingly limitless possibilities for offspring production), and *how* brood parasitism might originally have evolved (via intra-specific brood parasitism, as we shall see below), it has been extremely difficult to reconstruct evolutionary relationships among brood parasites and their hosts. Two predominant models exist: (1) brood parasites have evolved repeatedly and recently from within the lineage (clade) of their hosts (Fig. 10-6a), and (2) brood parasites evolved rarely in the more distant past and have radiated onto diverse host taxa (Fig. 10-6b). According to the first view, the morphological similarity among brood-parasitic taxa (loss of pollen-collecting and nest-digging structures; heavily sclerotized cuticle; short, appressed, scale-like hairs) have arisen repeatedly from pollen-collecting ancestors. According to the second view, the morphological similarities are due to a shared, ancient brood-parasitic ancestor. Both patterns certainly exist among bees, but knowing which pattern applies to which brood-parasitic bee lineage requires a clear understanding of bee phylogeny.

Bee phylogeny, like the phylogeny of most organisms, was primarily constructed based on anatomical features (morphology) up until the early 1990s, when DNA sequencing and comparison among gene-sequence data became widely available. Excellent morphological studies of bee higher-level relationships were published by Michener and colleagues in the mid-1990s, and re-reading these papers illustrates just how hard it was to analyze the brood-parasitic bees. Roig-Alsina and Michener (1993), for example, analyzed adult and larval morphological features in the long-tongued bees (Apidae and Megachilidae). Because they believed that the morphological similarities among brood parasites across subfamilies of Apidae were due to convergent evolution (model 1 above), they explicitly eliminated morphological traits that were presumed to be related to the brood-parasitic

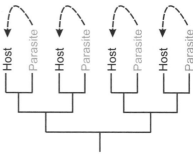

a

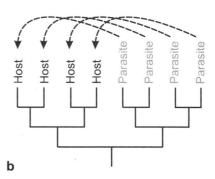

Figure 10-6. Two models for the evolutionary relationships between host and brood parasites in bees:
(a) model 1: brood parasites evolve repeatedly from
within pollen-feeding lineages and attack their close
relatives and (b) model 2: brood parasites evolved
long ago and diversified onto distantly related hosts,
which are not their closest relatives.

b

lifestyle. As such, they effectively eliminated the data that could have supported
model 2. Of course, at the time, they had good reason for this. Each tribe of
pollen-collecting Apidae seemed to have a corresponding brood-parasitic tribe
(Ericrocidini parasitize Centridini, Melectini parasitize Anthophorini), and even
genera show the same (apparent) one-to-one match (*Coelioxoides* parasitize species of *Tetrapedia*). In addition, there were morphological traits (not related to
parasitism) that seemed to support grouping these host-parasite pairs, thus supporting model 1.

DNA data provided a new tool for analyzing these patterns that was (seemingly) independent from parasitism. Cardinal and colleagues (2010) analyzed
relationships among tribes of Apidae based on seven genes and documented a
pattern that previously had not been given much support: most (but not all)
brood-parasitic Apidae share a common brood-parasitic ancestor (model 2
above). Cardinal and colleagues supported the grouping of most brood-parasitic
tribes of Apinae plus the exclusively brood parasitic subfamily Nomadinae into
a large "cleptoparasitic clade" (Table 10-2). Two other (independent) origins of
brood parasitism were identified within the tribes Ctenoplectrini and Euglossini
(Table 10-1). That brood parasites in these tribes had arisen recently and from
within their host clades came as no surprise—*Ctenoplectrina* is morphologically
nearly identical to *Ctenoplectra* except that it lacks the structures to collect pollen
and floral oils, and *Exaerete* and *Aglae* are undoubtedly highly modified brood-
parasitic orchid bees. The existence of an ancient (~95-million-year-old) origin
of brood parasitism raises a very interesting question. Why do brood-parasitic
Apidae seem to converge (in certain morphological features) on their host taxa?

We have seen above that some brood parasites may mimic hosts chemically (*Nomada* and *Andrena*, *Stelis* and *Osmia*). Perhaps there are advantages to brood parasites mimicking host morphology as well when they are attempting to gain access to host nests.[3]

We now know from studies of Halictinae (Gibbs et al. 2012b), Colletidae (Magnacca and Danforth 2006), Apidae (Cardinal et al. 2010), and Megachilidae (Litman et al. 2013) that brood-parasitic bees follow both models 1 and 2. In Halictidae and Megachilidae, there are both ancient origins of brood parasites (Sphecodini in Halictidae and the *Stelis* clade in Megachilidae) and much more recent origins (*Megommation* in Halictidae and *Hoplostelis* in Megachilidae). At present, our best estimate of the number of origins of brood parasitism in bees is 20 (Table 10-1), but this number includes some potential cases of social parasitism in Halictidae. Excluding those cases, we estimate that true brood parasitism has arisen 15 times in bees. An unpublished report of an andrenid brood parasite (Gabriel Melo pers. comm.) may add one additional, but extremely interesting, case.

What are the implications of a shift from pollen collecting to brood parasitism? Litman and colleagues (2013) analyzed the impact of shifts in life history on diversification rates in bees and found a net decrease in diversification with shifts from a solitary, free-living lifestyle to brood parasitism, but a substantial increase in diversification (in both Megachilidae and Apidae) when brood parasites shift from closed-cell to open-celled parasitism. Why would such a pattern exist? Open-celled parasitism appears to be an extremely effective mode of parasitism that allows brood-parasitic females to exploit many cells when bees nest in aggregations (as they often do) and when solitary females share a common nest entrance (i.e., they are communal). Open-cell parasites have tiny, cryptically shaped eggs that can be laid in large numbers when the opportunity arises (Alexander and Rozen 1987). As pointed out by Litman and colleagues (2013), certain nest structures may make this mode of parasitism ineffective, especially when brood cells are widely scattered and lateral tunnels are long (as they are in many Andrenidae, Melittidae, and solitary Halictidae).

Given the frequent origins of brood-parasitic bees and the large proportion of bee species that are brood-parasitic, it is puzzling that brood parasitism has arisen so infrequently in closely related apoid wasps. Brood parasitism has arisen just twice in the apoid wasp family Crabronidae: in the genus *Stizoides* and the tribe Nyssonini (Ohl and Linde 2003). A female *Stizoides* burrows into the closed nests of other apoid wasps, destroys any offspring present, and deposits its egg on the provisions inside the brood cell (O'Neill 2001). In contrast, female Nyssonini hide their eggs on their host's provisions before the host's egg has been laid. The brood-parasitic larva emerges first, kills the host's offspring, and develops while feeding on the provisions (O'Neill 2001). *Stizoides* would be described as an adult, closed-cell parasite, whereas Nyssonini pursues a larval,

---

[3] Subsequent studies, based on much larger molecular data sets, have supported Cardinal's hypothesis of a single, ancient, "cleptoparasitic clade" of Apidae (Hedtke et al. 2013, Peters et al. 2017, Bossert et al. 2018). Further corroborating evidence for the antiquity of the apid brood parasites comes from the fossil record. Dehon et al. (2017) documented a 60-million-year-old nomadine bee (*Paleoepeolus micheneri*, Epeolini) from a compression fossil discovered in Menat (Puy-de-Dôme), France. Bossert et al. (2018) subsequently revised the limits of the subfamily Nomadinae to include all apid brood parasites minus *Aglae* and *Exaerete* (both in Euglossini) and *Ctenoplectrina* (in Ctenoplectrini; see Box 2-9).

closed-cell mode of parasitism. Together, *Stizoides* (30 species) and Nyssonini (227 species) account for just 2.6% of the nearly 10,000 species of apoid wasps (Pulawski 2018). Both the significantly fewer origins and the substantially smaller percentage of brood parasites in apoid wasps as compared to bees suggests that there may be substantial barriers to the evolution of brood parasitism in apoid wasps.

Origins of brood parasitism

How did brood-parasitic behavior arise within bees and their closely related wasp relatives? One hypothesis is that brood parasitism of one species (the host) by another (the parasite) arises through an initial stage in which individuals within the same species opportunistically parasitize each other's brood cells. *Intra-specific brood parasitism* (or *nest usurpation*), as this is called, has been reported widely across bees and hunting wasps (Černá et al. 2013b, Field 1992, Wuellner 1999). Eickwort (1975), in a detailed study of the nesting biology of *Hoplitis anthocopoides* (Megachilidae: Osmiini), documented a remarkably high frequency of intra-specific brood parasitism within a nesting aggregation of these communally nesting bees. Eickwort (1975) studied a nesting aggregation of *H. anthocopoides* in Rensselaerville, New York, in the summers of 1969 and 1970. He marked all females with individual paint marks so he could track individual behaviors and distinguish nest resident from non-resident. *H. anthocopoides* builds freestanding mortar nests on the exposed rock surfaces (Fig. 6-10), making them easy subjects to observe. Females build a cluster of closely spaced (contiguous) cells that they provision with pollen and nectar from a single host plant (*Echium vulgare*: Boraginaceae). When nests were spaced close together, female *H. anthocopoides* were observed frequently entering the partially provisioned cells of their neighbors and attempting to "usurp" open, partially provisioned cells. If a resident female detected such an interloper inspecting her open cell, a fight inevitably erupted, and the resident generally repelled the intruder. However, on two occasions, intruders successfully overcame the resident females and successfully took over the partially provisioned cell (and nest), eventually laying an egg and closing the cell. Such nest usurpation was actually less common than outright brood parasitism of closed cells, which Eickwort describes as follows:

> Cells that have been recently closed, and thus that contain eggs or very young larvae, are also subject to attack by parasitic mason bees. Such a mason bee moistens the cell closure with copious saliva and laboriously bites it, dislodging particles of mortar, until a hole is opened that is large enough to admit her body. The brood parasite then enters the cell head-first and removes the egg, presumably by eating it. She does not remove the host's provisions, although she may then forage and add more provisions to the cell. Either after this brief provisioning or immediately after removing the host's egg, the parasite backs into the cell and lays her egg. She then constructs a closure to the cell. Successful cleptoparasitism of closed cells was more commonly observed (nine times) than the usurpation of open cells (two times).

BOX 10-1: *HOLCOPASITES RUTHAE* AND THE "CUCKOO BEE TASK FORCE"

Few brood-parasitic bee species have been the focus of as much interest as *Holcopasites ruthae*—a small but beautiful apid brood parasite first described from a site on the campus of the University of California, Riverside (UCR). *H. ruthae*, affectionately known as Ruth's *Holcopasites*, was described by a retired chemistry professor turned bee taxonomist—Dr. Kenneth Cooper (Cooper 1993)—and named after his wife, Ruth. Dr. Cooper discovered this bee while collecting in the Biological Control Groves on the edge of the UCR campus. The discovery of a new, brood-parasitic bee species on the UCR campus was quite a surprise because a very famous and productive bee taxonomist—P. H. Timberlake—had regularly collected bees at that precise location for much of his career. How had Timberlake missed this species? Because the bee seemed to be restricted to a very small geographic area just on the edge of campus, Dr. Cooper decided to move forward with a formal listing as an endangered species. His efforts immediately caught the attention of both the UCR administration and the United States Department of Agriculture (USDA), because the site was slated to be developed into a major USDA Salinity Laboratory. Plans for this multimillion-dollar laboratory would be suspended if *H. ruthae* was added to the Red List of endangered plants and animals.

So, in the spring of 1992, the USDA funded an intensive survey effort in the vicinity of UCR in order to assess the conservation status of *H. ruthae*. This effort was led by entomology professor Kirk Visscher, a honey bee expert in the UCR Department of Entomology. Kirk assembled a "rapid response team" that included one of the authors of this book (BND) as well as a team of UCR graduate students, undergraduates, and technical support staff. Our team came to be called the "Cuckoo Bee Task Force" (CBTF). Cooper had already inferred that the likely host of *H. ruthae* was *Calliopsis pugionis* (Andrenidae), a small, gregarious, ground-nesting, early spring panurgine bee. The CBTF set to work finding aggregations of *C. pugionis* in the surrounding area in order to confirm that *H. ruthae* was a brood parasite of *C. pugionis* as well as to assess the geographic range and abundance of *H. ruthae*. In the end, a number of large nesting aggregations of the host bee were discovered both near the UCR campus and at the San Jacinto Wildlife Reserve, 30 miles to the southeast. Adult *H. ruthae* were observed at all sites, and excavations of nearly a thousand brood cells confirmed that *H. ruthae* was a common brood parasite of *C. pugionis*. Parasitism rates varied from site to site but could be as high as 30% at some nest sites. So *H. ruthae* did not seem to be an endangered species, and the USDA could move forward with their planned construction of the Salinity Laboratory. Not only did the CBTF help peacefully resolve a major standoff between the United States government and a small brood-parasitic bee; it also led to three publications on the biology of the host and this very charismatic brood parasite (Danforth and Visscher 1993, Visscher and Danforth 1993, Visscher et al. 1994).

Over half of the females Eickwort observed at one nesting aggregation behaved as brood parasites, but none behaved as obligate brood parasites. Females that were observed behaving as brood parasites also built their own nests and provisioned their own cells. This illustrates a remarkable level of behavioral plasticity for such a tiny organism. Why some females behave as a brood parasite one day and a hardworking, pollen-collecting bee the next is unknown. It is remarkable how closely Eickwort's description above matches the behavior we described earlier as *adult, closed-cell parasitism*—a mode of obligate brood parasitism reported from various lineages of Apidae, Megachilidae, and, possibly, Halictidae (Table 10-1).

The frequent origins of brood parasites from within a clade of their hosts (Table 10-1) supports the view that intra-specific brood parasitism might be a common first step toward obligate parasitism of other bee species. In Halictidae, Megachilidae, Apidae, and Colletidae, the most recent origins of brood parasitism involve brood parasites that attack closely related pollen-collecting hosts. *Parathrincostoma* attack *Thrinchostoma* (Halictidae), *Hoplitis* (erythrogaster group) attack other species of *Hoplitis* (Megachilidae), *Ctenoplectrina* attack closely related species of *Ctenoplectra* (Apidae), and the five brood-parasitic species of *Hylaeus* (*Nesoprosopis*) attack other *Hylaeus* (*Nesoprosopis*) in Hawaii (Table 10-1). In all these cases, phylogenetic studies have placed the brood-parasitic group as either a sister group to or within the host group. These recent and repeated origins of brood parasitism from closely related hosts support the idea that the origins of brood parasitism might involve an initial intra-specific stage. In fact, one might expect closely related species to be the most effective brood parasites because they already possess many of the same preferences as the host in terms of nesting substrate, nest architecture, and host-plant preferences. From the perspective of a bee, your worst enemy might just be your closest relative. Or, in the words of Bob Marley, "Your worst enemy could be your best friend and your best friend your worst enemy."

RARITY, VULNERABILITY, AND IMPORTANCE AS BIOINDICATORS

Brood-parasitic bees are generally rare, and there is good reason to think that they may be more prone to local extinction than their solitary pollen-collecting hosts. Brood parasites are entirely dependent on the presence of one or more host species, they are often host-specific, their population numbers are usually low (sometimes very low) relative to their hosts, and they would seem to be rather poor dispersers. That brood parasites can be extremely rare is illustrated beautifully by *Epeoloides pilosulus*, a brood parasite of *Macropis nuda*.

E. pilosulus is a very rare bee. It is the sole member of the genus *Epeoloides* in the Western Hemisphere (there is another species in Europe), and it has gone uncollected for decades. It was, in fact, thought to be extinct because specimens had not been collected in North America since the 1960s. However, Cory Sheffield collected two male specimens in Nova Scotia, Canada, in 2002 (Sheffield et al. 2004), Sheffield and Heron (2018) reported on a single specimen collected in Alberta, and Achik Dorchin from Cornell University collected a single male in New York State in 2014. We should point out that *Macropis* is not your average bee. The genus *Macropis* includes a handful of species (16) ranging from North America to Europe and eastward to China (Michez and Patiny 2005). *Macropis* is an oil-collecting bee that relies exclusively on *Lysimachia* plants for its floral oils, which are mixed into the provision mass and used to waterproof brood cells. *Macropis* is a rare bee by most standards. They can be found only along the edges of streams, ponds, and lakes, where their oil-producing host plant is present. The rarity and narrow ecological preferences of *Macropis* would certainly explain the rarity of *E. pilosulus* in North America. But is *E. pilosulus* on the verge of extinction? We really don't know. Perhaps this rare tripartite relationship between host plant (*Lysimachia*), pollinator (*Macropis*), and parasite (*E. pilosulus*) has always been rare and managed to hang on due to some remarkable dispersal (and recolonization) abilities of host and parasite. Alternatively, perhaps we are witnessing the steady decline and disappearance of one of North America's most fascinating bees.

Because brood-parasitic bees represent the apex of bee communities, the diversity and abundance of brood parasites relative to other bees might be a good indicator of the status of the entire bee community. This idea was suggested by Cory Sheffield and coauthors (2013) based on studies of native bee communities in apple orchards. Sheffield surveyed bees in apple orchards and adjacent areas in the Annapolis Valley, Nova Scotia, in 2001 and 2002 using "yellow pan traps," a common method for surveying bees. He sampled 19 sites over a range of habitat types from relatively undisturbed sites (unmanaged meadows and abandoned orchards) to heavily disturbed sites (commercially managed apple orchards). Overall, he detected 146 native bee species. He found that brood-parasitic bees were diverse and abundant in old fields and abandoned orchards, but were nearly absent in heavily managed commercial orchards. Brood-parasitic species richness and abundance declined dramatically in the more disturbed habitats and showed a more predictable relationship to disturbance than did the entire bee community. Sheffield and colleagues (2013) argued that brood parasites, like predators and other kinds of parasites, may have a stabilizing influence on bee communities because common host species are likely to be hit more heavily by brood parasites than rare species.

CHAPTER 11

Non-Bee Parasites and Predators

In 1918, John Lovell wrote, "Bees are frequently described as roaming among flowers leading a joyous, care-free existence . . ." Indeed, solitary bees do seem to be blessed with the good life. They often live in hot, sunny, Mediterranean environments surrounded by an abundance of flowers in full bloom—the same kinds of places to which people like to retire, cultivate olives, or grow fine wine. But Lowell's sentence continues: ". . . but they often meet a terrible fate and are seized by a monster as remorseless as the fabled Scylla of ancient mythology." The remorseless monster that Lowell was referring to was a crab spider, one of the many parasites, predators, and brood parasites that attack the adults and larvae of solitary bees (Box 11-1). These skillful attackers include a bewildering diversity of animals, spanning mammals, birds, spiders, mites, true bugs, wasps, flies, beetles, and an obscure but fascinating group called the twisted-wing parasites (Strepsiptera). Some of these natural enemies attack adult bees, some attack larval bees, and others feed on the pollen and nectar stored in the brood cell. Still others prey upon or parasitize the bee attackers, rather than the bees themselves (Mihajlovic et al. 1989). The kinds of non-bee predators, parasitoids, and brood parasites and their impacts on solitary bee populations have received considerable study by bee biologists (Clausen 1940, 1976; Danks 1971; Jayasingh and Freeman 1980; Linsley 1958; Wcislo and Cane 1996; Westrich 1989). In this chapter, we focus on the life history, modes of parasitism, and impact on solitary bees of Lovell's "remorseless monsters" (Table 11-1).

PREDATION ON FORAGING ADULT BEES

Perhaps because many female bees sting, the risk that adult bees face as they forage on flowers was thought to be insignificant (e.g., Danks 1971). However, recent studies have demonstrated that predation on foraging bees is common (Dukas 2001, Gonzálvez et al. 2013). In most cases, the arthropod predators of adult bees have stylet-like mouthparts for injecting venoms that quickly disable the bee and, shortly thereafter, digestive enzymes that liquefy the internal organs. The predators then drink the liquefied internal organs of the prey through their straw-like mouthparts. Once the internal organs are consumed, the remaining

BOX 11-1: A GLOSSARY OF PREDATORY AND PARASITIC LIFESTYLES

Predators consume the adult or larval bee. Death is more or less instantaneous, and the predators are generally, but not always, larger than the bee itself. Predators like spiders and true bugs (Reduviidae) kill the adult bee using venoms and then liquefy the internal organs prior to consumption (extra-oral digestion). Others, such as ants, tear the prey into pieces.

Parasitoids consume the adult or larval bee gradually over time. Larvae feed either internally (*endoparasitoids*) or externally (*ectoparasitoids*), and parasitoids are always smaller than their hosts. Parasitoids can be further divided into *idiobiont* parasitoids, which prevent further development of the host following parasitism, and *koinobiont* parasitoids, which allow the host to continue its normal development. Idiobionts are generally ectoparasites, whereas koinobionts are generally endoparasitoids.

Brood parasites (cleptoparasites) consume larval provisions (pollen, nectar, floral oils) rather than the larval bee itself. Includes Sapygidae, some Meloidae, Phoridae, and *Grotea* (Ichneumonidae).

Parasites are organisms that live for extended periods of time within a host, obtaining resources from it without returning any benefits. Generally, they are much smaller than the host (think nematodes), but for bees, Strepsiptera would fall into this category.

Depredators are similar to brood parasites in that they consume the pollen/nectar provisions of the host. However, they do not kill the host larva directly, as do brood parasites. Rather, nest depredators compete with the host for pollen and nectar and, thus, eventually starve the host of food. The end result is generally death of the host.

Scavengers consume remains of the provisions left behind by bee larvae that have completed development and departed the nest; there is no direct mortality impact on bees.

exoskeleton is discarded, and the predator moves on to their next victim. This mode of prey consumption has been called *extra-oral digestion* and is common in arthropods that prey on fast-flying insects (e.g., assassin bugs, crab spiders, and robber flies; Cohen 1995, 1998). This gruesome mode of feeding allows these predators to consume prey several times their own body size. Most flower-hunting predators fall into two groups: those that "sit and wait," and others that actively hunt and capture foraging adult bees.

Ambush, sit-and-wait predators

The most common sit-and-wait predators of bees are ambush bugs (Reduviidae: Phymatinae), assassin bugs (Reduviidae: Harpactorinae: Apiomerini), crab spiders (Thomisidae), and ants (Formicidae). Many ambush predators rest

Table 11-1. Common arthropod predators of solitary bees and their modes of attack.

Class	Order	Family/superfamily	Subfamily	Mode of attack	Life stage attacked	Reference
Arachnida	Acariformes	Chaetodactylidae		predators	larva	Eickwort 1994
	Araneae	Thomisidae		attack foraging adults	adult	Dukas & Morse 2003
Hexapoda	Hemiptera	Reduviidae	Phymatinae	attack foraging adults	adult	Masonick et al. 2017,
			Harpactorinae (Apiomerini)	attack foraging adults	adult	Zhang et al. 2016
	Coleoptera	Meloidae		brood parasites/ectoparasites	provisions/ larva	Askew 1971
		Ripiphoridae		endoparasites	larva	Linsley et al. 1952a
		Cleridae		predators	larva	Linsley & MacSwain 1943
	Diptera	Asilidae		attack foraging adults	adult	Linsley 1960
		Anthomyiidae		parasitoids	larva	Clausen 1940
		Sarcophagidae	Miltogramminae	parasitoids	larva	Clausen 1940
		Phoridae		brood parasites	provisions	Clausen 1940
		Conopidae		endoparasitoids	adult	Clausen 1940
		Bombyliidae		brood parasites/ectoparasites	larva	Yeates & Greathead 1997
		Drosophilidae		nest depredators	provisions	Krunić et al. 2005
	Hymenoptera	Chalcididae		endoparasitoids	larva	Grissell 2007
		Encyrtidae		endoparasitoids	larva	Grissell 2007
		Eulophidae		endoparasitoids	larva	Grissell 2007
		Eupelmidae		endoparasitoids	larva	Grissell 2007
		Eurytomidae		endoparasitoids	larva	Grissell 2007
		Leucospidae		endoparasitoids	larva	Grissell 2007
		Mymaridae		endoparasitoids	larva	Grissell 2007

	Perilampidae		endoparasitoids	larva	Grissell 2007
	Pteromalidae		endoparasitoids	larva	Grissell 2007
	Torymidae		endoparasitoids	larva	Grissell 2007
	Gasteruptiidae		brood parasites/ ectoparasites	larva	Askew 1971
	Ichneumonidae		parasitoids/brood parasites	provisions	Slobodchikoff 1967
	Chrysididae		ectoparasitoids	larva	Askew 1971
	Mutillidae		ectoparasitoids	larva	Askew 1971
	Sapygidae		brood parasites	provisions	Askew 1971
	Formicidae		attack provisions and foraging adults	adult	Hölldobler & Wilson 1990
	Crabronidae	Philanthinae	attack foraging adults	adult	O'Neill 2001
	Crabronidae	Larrinae	attack foraging acults	adult	O'Neill 2001
	Crabronidae	Nyssoninae	attack foraging adults	adult	O'Neill 2001
Strepsiptera	Stylopidae		endoparasites	larva/adult	Kathirithamby 2009

cryptically on flowers, where they hunt foraging bees, but others can attack at the nest entrance (Schatz and Wcislo 1999, Wcislo and Schatz 2003). Mantids (Caron 1978) and web-spinning spiders are other groups of sedentary predators that opportunistically capture bees, as are many toads, frogs, and lizards. All the sit-and-wait predators that have been studied so far appear to capture any appropriately sized floral visitor. There are no known sit-and-wait predators that specialize exclusively on bees (Pekar et al. 2011).

Ambush bugs are rather small, squat, cryptic flower inhabitants with multi-toothed raptorial forelegs that they use to capture prey many times their own body size (Masonick et al. 2017). While females regularly take large flower-visiting prey such as bees, males are more opportunistic, rarely capturing large prey. Sometimes males feed while *in copula* on prey captured by their female partner (Mason 1986).

The assassin bugs (Apiomerini) are among the most highly specialized predators of adult solitary bees. In contrast to the cryptic ambush bugs and crab spiders, these are brightly colored, aposematic hunters that find their prey at many sites besides flowers. Instead of relying on purely mechanical means to grasp their prey, they have a trick for prey capture. All Apiomerini have hairy foretarsi to which they add sticky, plant-derived resins that serve as adhesives to more easily capture fast-flying prey, such as bees (Choe and Rust 2007, Forero et al. 2011). Glandular secretions, probably from an abdominal gland, are added to prevent hardening and keep the resins pliable and sticky (Forero et al. 2011).

Crab spiders (Thomisidae) are the other classic group of floral ambush predators; they earn their name from the vaguely crab-like appearance of some species and the ability to move laterally in a crab-like manner. Some crab spiders can change color, typically from yellow to white, or white to yellow, depending on the flowers they are sitting on. Unlike in many color-shifting animals, this is not a rapid process. *Misumena vatia* takes up to seven days to switch from white to yellow and vice versa (Morse 2007).

Ants are common in flowers, mainly foraging for nectar (Lach 2008, Ness 2006) but sometimes hunting floral visitors (Rodríguez-Gironés et al. 2013). Plants are under constant attack by herbivores, and members of more than 90 families attract ants in order to control herbivores (Koptur 1992). This creates an evolutionary conundrum if the presence of ants provides protection against herbivores but at the same time deters floral visitors that play a role in pollination. Not surprisingly, bees avoid flowers when predatory ants are present (Gonzálvez and Rodríguez-Gironés 2013, Ness 2006).

Remarkably, some bees are able to use chemical and visual cues to discriminate predatory ant species from those that pose little threat. *Nomia strigata* (Halictidae) hovered about twice as long around flowers of *Nephilium lappaceum* (Sapindaceae) in which a chemical extract or a dead ant of the predatory species *Oecophylla smaragdina* was experimentally introduced, as they did when the same experiment was repeated with *Polyrachis dives*, an ant that does not prey on bees (Gonzálvez and Rodríguez-Gironés 2013). Flowers of *Melastoma malabathricum* (Melastomataceae) with the same predaceous ant (*O. smaragdina*) were avoided by small and medium-sized bees, which allowed ant-inhabited flowers to accumulate nectar and pollen. These flowers with increased resources were preferentially visited by large carpenter bees (*Xylocopa*: Apidae), which were too big for the ants to attack. This relationship provides benefits to the plant in terms of pollination—*M. malabathricum* flowers had greater success at setting fruit when visited by *Xylocopa* than when visited by small and medium-sized bees (Gonzálvez et al. 2013). The presence of ants, ironically, had a positive effect on pollination.

Active predators

Predators that actively chase and capture solitary bees include robber flies (Asilidae), beewolves (Crabronidae), and birds. These mobile, opportunistic predators are often difficult to study, but they are likely an important source of mortality for solitary bees.

Robber flies (Asilidae) are aggressive hunters with elongate, slender bodies, enormous bulging eyes, long, hairy legs, and piercing mouthparts. They perch attentively on plant stems and dart out to capture their prey in flight. Most robber flies perch upright, but flies in the genus *Diogmites* have the odd habit of hanging from twigs by their forelegs. Many robber flies mimic bees in shape and coloration.[1] Like predaceous bugs, robber flies pierce the body of their prey with their stylet-like mouthparts and inject venom that quickly disables the victim. The internal organs are liquefied and aspirated back through the stylet-like mouthparts. Robber flies are opportunistic hunters that are often found at patches of flowers (Linsley 1960), in solitary bee nesting aggregations (O'Neill and Bjostad 1987), and at the entrances of social bee colonies (Dennis and Lavigne 2007). Although no robber fly is known to hunt only bees, some clearly discriminate among prey in interesting ways when prey are sufficiently plentiful. For example, Linsley (1960) speculated that *Mallophora bromleyi* avoided bees that are banded and "wasp-like" based in part on his observation that all green female *Agapostemon* (Halictidae) were frequently captured but the males of these same bee species, which have a banded metasoma, were not attacked. A species in Wyoming was twice observed to drop a prey item and attack a larger insect that suddenly arrived at a flower (O'Neill and O'Neill 2001). On average, robber flies attack prey up to about half their own body size, so most large bees are probably safe from attack (Dennis and Lavigne 2007, Linsley 1960, O'Neill and O'Neill 2001).

Many crabronid wasps in the subfamily Philanthinae and a few species of Larrinae and Nyssoninae provision their nests with bees (Evans and O'Neill 1988). By far the most famous of these are the so-called beewolves of the genus *Philanthus*. Most *Philanthus* take a mix of bees, but the European *P. triangulum* exclusively hunts honey bees (Fabre 1921). *P. bicincta* has a strong preference for bumble bees, and *P. inversus* specializes on male *Agapostemon* (Halictidae) (Evans and O'Neill 1988). Evans and O'Neill reported 30 different wasps and 51 bee species in the nests of *P. pulcher*. While *P. triangulum* is known to hunt at bee hives, the prey of most beewolves are usually captured at flowers, where they are paralyzed with a quick sting and then transported back

---

[1] This is presumably a form of Müllerian mimicry that provides them with some added protection from vertebrate predators (Brower et al. 1960, Linsley 1960).

to the nest. Curiously, while bees captured on flowers often bear scopal pollen loads, *P. sanborni* removes the pollen of the bees stored in its nests (Kurczewski and Miller 1983). Since multiple prey items are needed per cell, these wasps are likely to take a heavy toll on local bee populations. A population of 200 *P. bicincta* was estimated to have killed at least 7,500 bumble bees in a single season (Gwynn 1981).

A wide array of birds is known to at least occasionally prey on bees, particularly honey bees, but for the most part, their impact on solitary bees has been poorly studied. Birds sometimes take advantage of the dense prey concentration when large numbers of bees nest in a small area (Alcock 1995, O'Neill and Bjostad 1987, Rust 2003). John Alcock examined bird predation on *Centris pallida* in Arizona (Alcock 1995) and *Amegilla dawsoni* in western Australia (Alcock 1996a). Both these species nest in large aggregations and produce males that range widely in body size (Chapter 4). Large-bodied males fly and walk incessantly over the nest site in search of prospective mates, and small-bodied males establish aerial territories along the nest-site perimeter and at flowers. Large-bodied males of both species are prone to bird predation as they dig out emerging virgin females from the ground or wrestle with copulating female-male pairs. In Arizona, four bird species were observed simply walking around the nest aggregation and making quick meals of otherwise distracted *C. pallida* males. Birds rarely attacked flying males. Measurements of head size showed that the victims were disproportionately large-bodied males (Alcock 1995). In Australia, the pied butcherbird (*Cracticus nigrogularis*) attacks *A. dawsoni* at nesting sites and, as with *C. pallida* in Arizona, large-bodied males are most often the victims of predation.

Birds that capture a female bee run the risk of being stung and presumably either learn to distinguish female bees from male bees or cease hunting bees altogether. Rust (2003) recorded seven bird species capturing *Nomia melanderi* in Washington State and examined stomach contents of the horned lark (*Eremophila alpestris*) at two sites before and after emergence of female bees. As in most solitary bees, male *N. melanderi* emerge a few days before females. Prior to female emergence, horned larks consumed large numbers of male *N. melanderi*, but after female emergence, the consumption of *N. melanderi* dropped dramatically, presumably because the birds began to encounter stinging females. Predation on males and avoidance of females greatly reduced the impact these birds had on the *N. melanderi* population.

PREDATION ON ADULTS IN NESTS

Adult bees may be less prone to attack in their nests than when they are foraging outside of the nest, but studies occasionally report mortality of adult bees in nests. Nine-banded armadillos often destroy solitary bee nests in central Texas. Although they are primarily in search of larvae, any female bee in the nest is probably taken as well (JLN pers. obs.). The nests of large carpenter bees (*Xylocopa*: Apidae) are often attacked by vertebrate predators. Watmough (1974, 1983) documented predation by baboons, woodpeckers, small mammals, and man (besides ants and "other invertebrates") in southern Africa. Carpenter bees fall into two nesting categories depending on the persistence of the wood they use for nesting (Hurd 1958). Some use soft woods such as floral stalks that last one year, and others use dead tree branches that remain structurally sound for years. *Xylocopa* species that use short-lived woods rapidly reproduce over one month, before the wood starts to degrade, while species that nest in hard wood reproduce more slowly and asynchronously. Watmough (1983) found that carpenter bee species that reproduce and develop synchronously sometimes suffer heavy predation; more than 60% of nests were attacked by woodpeckers at some sites. Watmough reported in the same study that the ant *Pheidole megacephala* could eliminate entire populations of some carpenter bee species in a local area, but that *X. caffra* could successfully repel ants by "sitting at the entrance and blocking it tightly by pressing the smooth dorsal surface of the metasoma against the rim" of the nest entrance. After several days, the ants moved on. Blocking the entrance in this way is reported in other carpenter bees (Bernardino and Gaglianone 2008) and requires a nest entrance that is concave on the inside, with an opening that has a smaller diameter than the tunnel, an architectural feature found in nests of many bees (Michener 1964).

PARASITOIDS OF ADULT BEES

Compared with the list of parasitoids that attack larval bees (see below), the list of parasitoids of adult solitary bees is relatively short. The main group of parasitoids that attack adult solitary bees are the thick-headed flies (Conopidae). As flies go, these are attractive insects, and many are mimics of bees and wasps. They attack solitary bees in flight but are internal parasitoids rather than predators (Fig. 11-1). Like robber flies, conopids perch on vegetation waiting for a bee to fly by. Returning female bees usually fly slowly as they approach their nests and thus are easier to attack, although both male and female bees have been found with thick-headed fly larvae (Howell 1967). The flies do not directly kill the bees. Instead, a female thick-headed fly grabs the bee in flight and attempts to insert an egg into the body. The terminal segments and genitalia of female conopids are syringe-like and punch a hole through the intersegmental membrane of the adult bee so the conopid can insert her egg. Despite this gruesome mode of attack, the mortality rates among adult bees are quite low. A four-year study of the alkali bee *Nomia melanderi* estimated that attacks by the conopid *Zodion obliquefasciatum* resulted in just 8% mortality overall (Howell 1967). Another thick-headed fly, *Z. fulvifrons*, was responsible for less than 5% mortality at a nest aggregation of *Dieunomia triangulifera* (Halictidae) (Wcislo et al. 1994). Bees attacked by thick-headed flies react "frantically" (Howell 1967) and try to sting their attacker, which unintentionally exposes membranes that would normally be protected by the hard exoskeleton (Fig. 11-1a). Nevertheless, these evasive maneuvers must often work. Howell (1967) estimated that female *Z. obliquefasciatum* successfully oviposited in a host bee on only 7% of their attacks and an average of 4–10 times over their lifetime.

Once inserted into the body of the host, the egg of *Z. obliquefasciatum* hatches in one day and, over the next nine days, matures to a second-instar larva. During this time, the larva imbibes only hemolymph; all soft tissue of the host remains

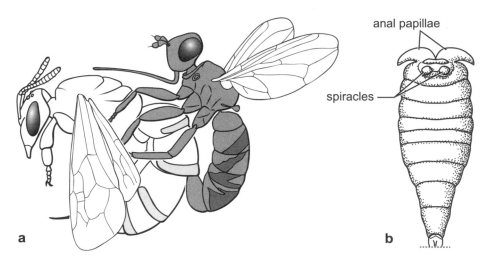

Figure 11-1. Conopidae: (a) *Zodion obliquefasciatum* ovipositing into the abdomen of *Nomia melanderi* and (b) mid-stage third instar of *Z. obliquefasciatum*, showing the position of the posterior spiracles and anal papillae (redrawn from Howell 1967, Figs. 10 and 13).

undamaged. Up to this stage, bee behavior seems little affected; parasitized bees build nests, forage for pollen and nectar, lay eggs, and produce offspring just like any other bee. However, late in the second instar, the conopid attacks from within and kills the bee while she is in her nest. Once the host is dead, the larva consumes all the internal organs of the host in eight hours, escapes the host, and moves into lateral tunnels of the nest, where it pupates. Thick-headed flies have a remarkable adaptation for obtaining oxygen while inside the body of the host. They use specialized papillae near the anus to tap into the host tracheal system (Fig. 11-1b). By avoiding respiratory stress, the larva is able to grow more rapidly and, at the same time, avoid consequences of a strong immune response by the host that would affect an endoparasite that developed more slowly.

PARASITOIDS OF LARVAL BEES

Larval bees, either feeding or resting in closed brood cells, are vulnerable to attack from a variety of wasp, fly, and beetle parasitoids. Adult velvet ants (Hymenoptera: Mutillidae), jewel wasps (Hymenoptera: Chrysididae), and wasps in the superfamily Chalcidoidea enter bee nests or oviposit through the brood cell wall in order to place their eggs in or on developing bee larvae. Bee flies (Bombyliidae) and satellite flies (Anthomyiidae and Miltogramminae) launch the eggs into the open nests of diverse solitary bee hosts. Even beetles have evolved the ability to parasitize bees. The families Meloidae and Ripiphoridae produce "triungulin" larvae that latch onto foraging adult bees as they move from flower to flower. Triungulin larvae are minute, highly mobile first-instar larvae specialized to search out and locate the host. Parasitoid groups that have this unusual stage are said to undergo hypermetamorphosis, because subsequent larval stages are immobile, grub-like creatures. Many triunglulin larvae cling to the adult bees and, if they happen to be on a female, are brought back to the nest, where they can attack the developing larvae. These bee parasitoids have diverse and interesting modes of parasitism, which we describe in detail below.

Attack by adults

Velvet ants (Mutillidae) have winged males and wingless females. The females resemble true ants (Formicidae) and move rapidly across the soil in search of bee nests. Some are nocturnal and dull in coloration, but the species associated with bees typically are diurnal, with strikingly colored black, white, red, and brown patches of erect hairs on the body. Their bright colors are presumably aposematic, as some females have famously painful stings (Schmidt 2016). Despite their conspicuous appearance, we know little about the details of their biology. Host records are known for only 3% of the more than 8,000 mutillid species (Brothers et al. 2000). Some mutillids specialize on hosts that nest in the ground, whereas others specialize on cavity nesters and bees that nest above-ground. Females locate bee nests by odor (kairomones), and if the attack is detected by the host bee, an aggressive encounter may ensue (Fig. 11-2). Adult female velvet ants attack closed brood cells of their hosts (Brothers 1972, 1978). Once inside the nest, they open the completed brood cells and lay an egg on the host larva. Hatching and development is delayed, so the larval mutillid consumes the mature host larva rather than the pollen/nectar provisions. Parasitism rates of solitary bees by mutillid wasps are typically low (Minckley and Danforth 2019).

Jewel wasps (Chrysididae) are brightly colored, heavily sculptured, metallic wasps that are often seen lurking suspiciously around the nests of solitary bees (Evans and West-Eberhard 1970). Adult female jewel wasps enter the nests of solitary bees and oviposit in the closed brood cells using their highly modified telescopic abdomen. When attacked by the host bee, they roll up into a ball by arching their head and mesosoma downward against the underside of the concave metasoma. This behavior appears to be an adaptation to avoid being stung. Solitary wasps have been reported to pick up the chrysidid wasps in their mandibles and discard them outside the nest (Evans and West-Eberhard 1970). Larvae of most jewel wasps do not attack the host until it has consumed its provisions and is ready to enter diapause as a prepupa. In North America, most species and a number of endemic genera are found in the same desert areas where bees are most diverse. Kimsey (2006) speculated that most chrysidid species attack a broad array of hosts.

Melittobia are minute wasps in the family Eulophidae that can cause significant mortality to solitary bees. They have been recorded from nests of almost every genus of cavity-nesting bees (Matthews et al. 2009). *Melittobia* females enter the nest, chew through the brood cell closure, and attempt to sting any immature host they find. Pheromones released during oviposition incite other *Melittobia* to chew into the brood cell (Matthews et al. 2009). Hosts are stung repeatedly, and the venoms induce torpor in the host that prevents further growth and development. *Melittobia* can quickly achieve high population densities because they are able to lay multiple eggs at once (this is called *superparasitism*) and finely control both the sex ratio and the dispersal capabilities

Figure 11-2. Mutillidae: combat between a halictid bee (*Lasioglossum* [*Dialictus*] *pruinosum*) and a mutillid wasp (*Timulla vagans vagans*) that was trying to gain entrance into the nest (redrawn from Melander and Brues 1903, Fig. 4. Original figure caption: "Down the hill they roll, heedless of everything but an inborn desire to annihilate each other").

of their offspring. Those that lay multiple eggs in one host will intentionally skew their offspring toward daughters. If the host is small, few *Melittobia* eggs are laid, and the female offspring are winged forms that can disperse to a new host. If the host is large, some or all female offspring lack the capability to fly and immediately lay eggs in the host alongside their mother. The larvae produced by the two morphs of *Melittobia* also feed differently. Flight-capable females produce larvae with strong mandibles that can consume tissue, while larvae of flight-incapable females have feebly sclerotized mouthparts and feed only on hemolymph. Rapid offspring production and fine-tuned control of the sex ratio allows *Melittobia* to rapidly destroy many brood cells if nests are close together. As a result, *Melittobia* can be a major source of mortality when cavity-nesting bees are raised in large numbers for use as managed, agricultural pollinators (Bosch and Kemp 2002).

Other wasps that attack solitary bee larvae do so by piercing the nest wall or brood cell cap with an elongate ovipositor. Many of these are in the Chalcidoidea, a clade of 22,000 described species and the largest superfamily of parasitoid wasps. Grissell (2007) tabulated the host records of Chalcidoidea and identified 10 families that attack larval bees. Most records are from cavity-nesting species, although this may represent the ease with which trap-nesting bees can be studied, rather than an accurate record of host preference. The family Torymidae had the highest proportion of species associated with bees. Torymid species of the genus *Monodontomerus* sting the host, immobilizing it, and place a number of eggs in the intervening space between the cocoon and the diapausing larva or adult. Larval wasps hatch in one or two days and consume the bee larvae over roughly six days (Eves 1970, Tepedino 1988). Females mate, often with brothers, and then search out new hosts. If the host bees reproduce over an extended period, populations of *Monodontomerus* can increase rapidly and place a substantial burden both on the host population they originally attacked and on other bee species nesting nearby (MacIvor and Salehi 2014).

The Gasteruptiidae are strange-looking, skinny wasps with long necks and long, thread-like ovipositors. They are most famous for attacking the nests of twig-nesting bees such as *Hylaeus* spp. (Colletidae) but are known to attack a wider range of bees and wasps (Jennings and Austin 2004). They oviposit into both open and closed cells, in the latter case using their long ovipositors to penetrate the cell closures. Some larval gasterupiids first consume the egg or early-instar host larva and then consume the cell provisions while others wait and consume the mature bee larva. Since some gasterupiids are significantly larger than their hosts, they may consume the contents of two or more host cells.

Attack by mobile larvae

Adults of two unrelated groups of Diptera, bee flies (Bombyliidae; Fig. 11-3) and satellite flies (Anthomyiidae and Miltogramminae), find the nests of prospective hosts but oviposit near the nest entrance or within a nest tunnel, and rely on the early-instar larvae to find the larval bee host.

Adult female bombyliid flies are agile acrobats that are common around solitary bee nests (Fig. 11-3e). These flies hover above the ground and launch eggs into any hole, crack, or crevice that might have the slightest resemblance to a bee nest. They frequently oviposit toward totally inappropriate objects, such as rocks, stones, roots, and the eyelets of shoes. Their strategy seems to be to lay as many eggs as possible (up to 1,000/day; Yeates and Greathead 1997) in the hopes that at least a few will end up in an active bee nest. Andrietti and colleagues (1997) took high-speed film of *Bombylius fimbriatus* at a nest aggregation of *Andrena agilissima* and determined they oviposited every 4.2 seconds!

First-instar bombyliid larvae (Fig. 11-3a) that find themselves in a bee nest crawl until they reach a brood cell occupied by a developing bee larva. Here, they enter a quiescent stage (Fig. 11-3b) and wait for the bee larva to complete its development. It is only after the bee has reached its maximum size that the bombyliid larva begins to consume the host. For *Heterostylum robustum*, a bee fly that attacks *Nomia melanderi* (Halictidae), the feeding period lasts for

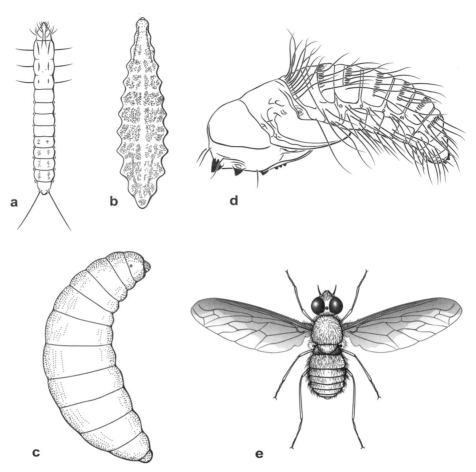

Figure 11-3. Bombyliidae, bee flies: (a) first-instar larva (*Hyperalonia oenomaus*); (b) second-instar larva of *H. oenomaus*; (c) third-instar larva of *H. oenomaus*; (d) pupa of *Thyridanthrax argentifrons*; (e) adult of *H. oenomaus* (redrawn from Clausen 1940; larvae, Fig. 170 A, B, C; pupa, Fig. 171; adult, Fig. 169).

7–10 days (Bohart et al. 1960). Once the host larva is consumed, the fly larvae pupate and remain below-ground throughout the winter. The following spring, the pupae digs itself out of the ground by gyrating the body in a way that pushes dirt behind it, with the aid of an elaborate array of setae and spines (Fig. 11-3d). Once the head and thorax are above-ground, the pupal skin splits open on the dorsal side, allowing the adult to emerge. Adult bee flies are long-lived and sustain themselves on nectar and pollen from some of the same flowers used by the bees they attack. Although the narrow, elongate proboscis of many bombyliids would seem to preclude feeding on pollen, many are able to harvest pollen with modified setae on their foretarsi (Neff et al. 2003). Some bee flies can also be important pollinators (Pellmyr and Thompson 1996).

Satellite flies (Diptera: Anthomyiidae and the subfamily Miltogramminae [Sarcophagidae]) are a biologically diverse, but morphologically homogeneous group of medium-sized brown and black flies. Many feed on roots, dung, or other materials, but a few are known to feed on the cell contents of bees in the families Andrenidae, Halictidae, and Apidae. The name "satellite" fly comes from their habit of closely following female bees as they return to their nests. Once a nest is located, the satellite fly lays an egg at the nest entrance or on the pollen/nectar provisions of the host (Polidori et al. 2005). The anthomyiid satellite fly *Leucophora personata* was studied at a nest aggregation of the communal bee *Andrena agilissima* (Andrenidae), where the flies perched near nest entrances and darted out to follow returning females (Polidori et al. 2005). When followed by a satellite fly, female bees took evasive action, flying in a zigzag pattern away from the nest. This behavior often allowed them to shake off the pursuing satellite fly.

Miltogrammine flies are among the smallest flies that attack solitary bees; they look like tiny house flies to the untrained eye. Most are associated with wasps, but some attack bees. Groups of miltogrammines differ in the ways they gain access to a bee nest, but all lay eggs that hatch within minutes of being laid. Some satellite flies wait at nests until the bee returns, then attempt to lay eggs on the flying host. Others lay eggs at or near the nest entrance (Michener and Ordway 1963). Once inside the nest, the larvae find their way to a brood cell. At several nest aggregations of *Amegilla dawsoni*, Alcock (2000) found *Miltogramma rectangularis* to be present and sometimes abundant. Flies usually attached more than one egg to an adult bee. Excavated brood cells containing fly parasites had an average of 8 and up to 14 fly pupae per brood cell. However, most of the eggs died before reaching a host. On average, less than 4% of the excavated brood cells had fly pupae, and mortality of the host remained low in all years and sites.

Attack by phoretic "triungulin" larvae

"Triungulin" larvae are found in two groups of beetles, the Meloidae and Ripiphoridae. The larvae of these beetle parasites all undergo hypermetamorphosis: the first-instar larva is active, long-lived, and morphologically distinct from the sessile later instars. In meloid and ripiphorid beetles, this first-larval instar (the triungulin) is entirely tasked with the job of gaining access to the bee nest and finding the bee larvae. Ripiphorid beetles are the only beetles that are endoparasitic, and their larvae feed directly on the host. Meloids, in contrast, may feed on the pollen provisions or the late-instar immature bee.

Ripiphorids are small to medium-sized, wedge-shaped beetles with reduced elytra (forewings). The males have enlarged pectinate or flabellate antennae (Fig. 11-4). Although sometimes common at nest sites, most ripiphorids are very rarely observed, and we have only scant knowledge about the scope of hosts they attack, because adults live for just a few days. Some ripiphorids seem to specialize on wasps, but one genus, *Ripiphorus*, is known to mainly attack solitary bees. The biology of *R. smithi* was studied in detail by Linsley and colleagues (1952a) along with its host, *Diadasia consociata* (Apidae), a solitary, ground-nesting pollen specialist on mallows (*Sida*: Malvaceae). Adult *R. smithi* emerge and mate at the nest site of the host. Males flew upwind toward virgin females, indicating that they were attracted by pheromones the females emit. Mated females then fly to a *Sida* plant and pierce through young flower buds with their

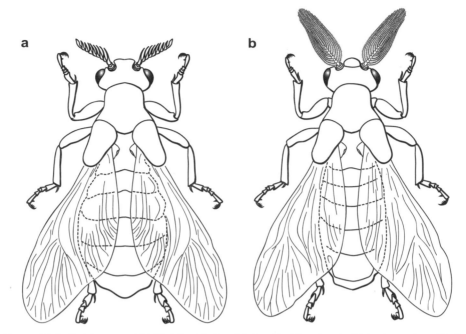

Figure 11-4. *Ripiphorus smithi* (Ripiphoridae): adult female (a) and male (b) (redrawn from Linsley et al. 1952a, Plate 9).

modified abdomen. Eggs are deposited in the developing flower bud and hatch one or two days before the flowers open. The open flowers are where the triungulin larvae wait for a bee to visit and hopefully provide them with transport back to the nest. Hitchhiking on other insects, called *phoresy*, is employed by a variety of insects (Clausen 1976). Ripiphorid triungulins have sucker-like structures on the tarsi and on the end of the abdomen that secure them tightly to their host. Larvae of one species remained attached to wings that were beating at over 150 hertz (cycles/second) (Tomlin and Miller 1989). *R. smithi* larvae that attach to a male *D. consociata* have little chance of finding a host, given that males do not return to a nest to sleep at night. However, male *D. consociata* carried up to 11 triungulins, and females rarely carried more than one. Linsley and colleagues (1952a) speculated that the triungulins fall off quickly when a female returns to the nest and that males accumulate more triungulins because they remain on flowers longer and sleep in flowers at night. Triungulins on males may also transfer to females during mating. Other studies have found evidence that ripiphorid triungulins attach preferentially to females (Tomlin and Miller 1989).

A ripiphorid larva that finds its way to a provisioned brood cell first rests on the pollen mass for a day or more until the immature bee emerges from the egg. It then pierces the bee larva and crawls into its body (Fig. 11-5). The endoparasitic larva remains as a first instar until the host develops and, based on the observation that there was no indication of the mouthparts in motion, acquires nutrition at this stage entirely through the body wall (Linsley et al. 1952a). When the host larva reaches its maximum size, *R. smithi* emerges and immediately molts into a second-instar larva. Feeding then commences on the

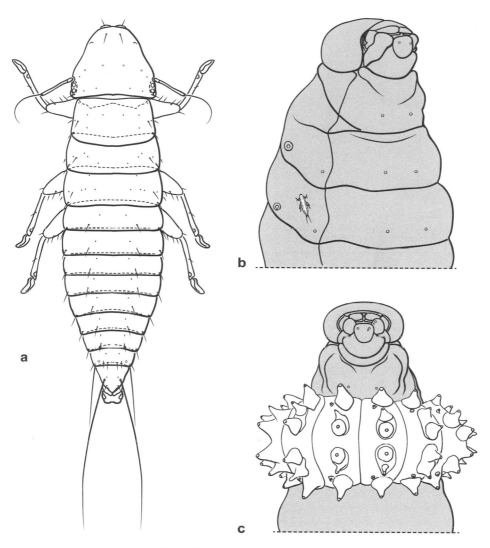

Figure 11-5. *Macrosiagon flabellatum* (Ripiphoridae): (a) first-instar larva of *M. flabellatum* (redrawn from Grandi 1936, Fig. III); (b) host (*Rhynchium oculatum*) larvae with (very small) first-instar larva of *M. flabellatum* on the external surface of the body (redrawn from Grandi 1936, Fig. XVII); (c) the normal external feeding position of the second and following instars upon the larvae of the host (redrawn from Grandi 1936, Fig. XVIII).

outside of the host body and is rapid. Larval *R. smithi* developed from second to sixth instars in about 14 days. The feeding ripiphorid larva consumes the host entirely after repeatedly injecting and reimbibing oral juices; this finally results in only a dry host integument. Host death occurs just before the ripiphorid pupates. To escape from the brood cell, adult *R. smithi* moisten the cell cap with an "oral secretion" and burrow up to the soil surface. Ripiphorids can also fall prey to other non-bee parasites. Linsley and colleagues (1952a) report finding a bombyliid (fly) larva feeding on a larval ripiphorid.

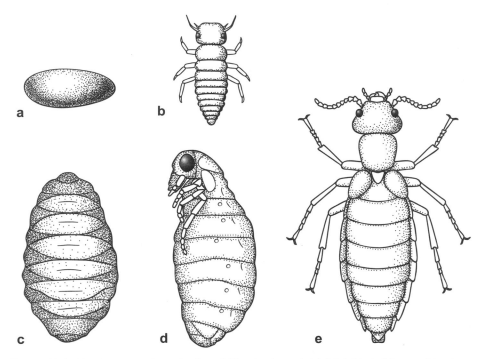

Figure 11-6. *Hornia neomexicana* (a–d) and *H. minutipennis* (e) (Meloidae): (a) egg; (b) primary larva; (c) mature larva; (d) pupa in lateral view; (e) adult. (redrawn from Porter 1951, Figs. 12–16).

Meloid beetles are bizarre creatures. Some females are flightless and have an enormously distended abdomen loaded with eggs and ridiculously small elytra (forewings) that barely cover the first segment of the abdomen (Fig. 11-6). The abdomen of these species is often so enlarged that the beetle seems barely capable of dragging it along the ground. Many adults are brightly colored to advertise the nasty chemicals they harbor. The Spanish fly, *Lytta vesicatoria*, is famous for producing cantharidin, a toxic blistering agent considered to be an aphrodisiac. While the adults are long-lived herbivores, sometimes achieving pest status in agricultural settings, the larvae feed either on grasshopper egg masses or in the cells of solitary bees.

Depending on the species, meloids deposit enormous numbers of minute eggs in flowers or flower buds, on the ground, or in shallow burrows they excavate (Clausen 1940). In some cases, such as species of *Lytta*, the triungulins crawl about and enter the nests of their bee hosts on their own. Once in a cell, they kill the egg or early-instar larva, molt to a more normal feeding form, and then proceed to consume the cell contents (Selander 1960). In most cases, the contents of one cell are enough, but sometimes a late-instar larva needs more food and consumes additional bee larvae. Such a larva thus can behave both as a brood parasite (consuming pollen/nectar provisions) and as a parasitoid of larval bees. Other meloids rely on phoresy to find a host. In the simplest form, the triunglulin attempts to grab onto any adult bee that passes by to get a free ride into the nest. *Hornia* triungulins wait at the nest entrance to catch a ride, but many others catch rides from bees at flowers (Erickson and Werner 1974). Exactly how phoretic triungulins pick the right bee to grab onto is unclear. The wing-beat frequency of the foraging bee may provide a cue that triungulins use to make their choices. Hocking (1949) found that the triungulin larvae of *H. minutipennis* responded to the precise wing-beat frequency (160 Hz) of its host (*Anthophora bomboides*: Apidae).

Some meloids use an unusually complex scheme to get their free ride into a nest. Several species are known to form aggregations of triungulins that use sexual deception to attract male bees. The triungulins hop on the males, which then take them to the females for their final ride into a nest (Vereecken and Mahé 2007; Box 11-2).

BOX 11-2: *MELOE FRANCISCANUS*

Meloe franciscanus, a meloid beetle parasite of *Habropoda pallida* (Apidae), has an extraordinarily devious mode of brood parasitism involving chemical and visual mimicry of host females. *H. pallida* is a solitary, univoltine bee that forms large, persistent nesting aggregations in dune habitats in the Mojave and Lower Sonoran Deserts. Females excavate deep nests (~1.5 m) that end in a single brood cell that they provision with pollen and nectar collected primarily from creosote bush (*Larrea tridentata*) (Bohart et al. 1972, Hurd and Linsley 1975). In the Kelso Dunes of southern California, female *M. franciscanus* emerge in the winter. Mating takes place, and as they are incapable of flight, females walk to nearby plants, where they lay batches of more than 700 eggs in the sand. Triungulins emerge in synchrony and, upon emergence, climb up the nearest plant (Erickson and Werner 1974). However, rather than dispersing and moving to flowers, as larvae of many other meloids do, these larvae aggregate and form a conspicuous ball on a stem while at the same time emitting a blend of compounds that are highly attractive to male *H. pallida* (Saul-Gershenz and Millar 2006). When male bees approach the spherical mass of triungulin larvae, the *M. franciscanus* larvae detect them and extend toward the bee while at the same time "waving" their forelegs in the bee's direction. Particularly amorous male *H. pallida* pounce on the writhing mass of triungulin larvae and end up covered with trianguliuns. Remarkably, groups of triungulins move en masse from branch to branch, and most of the aggregations are successfully transported away by male bees (Hafernik and Saul-Gershenz 2000, Saul-Gershenz and Millar 2006). Detailed chemical analyses and bioassays suggest the compound produced by the triungulins mimic part of the pheromone bouquet produced by mandibular glands in female *H. pallida*. Only male bees were attracted to chemical extracts of triungulins or the live masses of triungulins on plants, and males covered in triungulins were often pounced upon by other males, suggesting they smelled similar to receptive females (Saul-Gershenz and Millar 2006). When male *H. pallida* eventually do pounce on a receptive female, the triungulins rapidly move to the body of the female, where they are carried back to the nest and consume the provisions intended for the bee larva. This is a remarkable system involving chemical mimicry, cooperative behavior among close relatives (the triungulin larvae), and an improbable mode of brood-cell parasitism.

Other meloids take a much more direct route, with females ovipositing directly into the brood cells of their hosts. Members of the genus *Hornia* are specialized nest parasites of gregariously nesting species of *Anthophora*. For the most part, the entire life cycle regularly occurs within a single nest. Adults do not feed and are rarely seen above-ground. Even mating occurs inside the host nest (Linsley and MacSwain 1942a).

Bianchi (1962) observed egg masses of *Cissites auriculata* within 24 centimeters of *Xylocopa fimbriata* and *X. brasilianorum* nest entrances in Guatemala. In southern Africa, meloids in the genus *Synhornia* also lay their eggs near carpenter bee nests. Triungulins were found in brood cells being provisioned and on male *Xylocopa*, where they may be transferred to females during copulation and transferred to nests in other aggregations (Watmough 1974). In both these examples, the triungulin appears to enter the nest and find the immature bee.

INTRA-NEST PREDATORS

Trichodes is a genus of mainly brightly colored beetles in the family Cleridae. The larvae of most clerids are predators of other beetles, although some species attack egg masses of Orthoptera, and others attack wasp and bee larvae. We are fortunate that Linsley and MacSwain (1943) studied one species, *T. ornatus*, in great detail. Their observations and others since (Hurd and Linsley 1950, Parker 1984) confirm that this species attacks bees and wasps in a number of genera that nest in a variety of substrates (above-ground in dead logs and twigs, and below-ground in banks and flat ground) and that span a range of body sizes (Linsley and MacSwain 1942b). The researchers did not observe how the first-instar larva gets to the nest, but it probably does so by phoresy; females oviposited into unopened flower buds, and many nests with *Trichodes* were so far from flowers that it is unlikely the larvae could have found the nest by crawling. *T. ornatus* larvae can develop by feeding on the host larva or pollen; however, in the lab, larvae developed three times faster when fed host larvae than those fed only pollen. This ability to utilize pollen allows them to mature whether or not the host larva dies. *Trichodes* adults in Europe and North America frequent flowers and feed on pollen (Linsley and MacSwain 1942b) and other insects (Linsley and MacSwain 1943). Mawdsley (2004) provides some evidence that they act as pollinators. Other clerids such as species of *Lecontella* also can be pests in bee nests (Mawdsley 2001). Unlike *Trichodes*, these are dull, uninspiring beetles. How they get to bee nests is unclear, but once inside, *Lecontella* larvae are able to chew through and consume the contents of multiple *Osmia* cocoons (JLN pers. obs.)

Mites are often found in bee brood cells, and many are beneficial (Chapter 9), but some genera and species of the mite family Chaetodactylidae can be harmful to their hosts, especially in the case of *Chaetodactylus*, which has a broad host range (Table 9-2). Some *Chaetodactylus* are mere scavengers on residual pollen in a cell, but others are known to kill the host egg or early-instar larva and then consume the cell provisions, resulting in a cell full of mites and pollen husks. As none of the bees associated with *Chaetodactylus* have modifications to carry mites (acarinaria), this strongly indicates that these associations are not beneficial from the perspective of the bee. The mites can find new hosts in different ways. When nests are closely packed, mites can simply walk from one nest to another. More typically they need help. The deuteronymph stage of *Chaetodactylus* mites are dimorphic. A phoretic form is transported externally on the bees, and a cyst-like, non-dispersing form remains in the bee nest, ready to attack any bee that makes the mistake of reusing the nest. These cyst-like, non-dispersing forms are important to the survival of predatory *Chaetodactylus* species if all the hosts in a nest are killed and there is little opportunity for phoretic dispersal. More typically, not all of the cells are infested, and emerging bees are encrusted with mites they have acquired while moving through *Chaetodactylus*-filled cells.

BROOD PARASITES

Brood parasites include those non-bee attackers that consume the pollen/nectar provisions rather than the bee larva itself. Members of the family Sapygidae are mostly black and yellow wasps that superficially resemble hornets or yellow-jackets. But unlike these highly social, predatory wasps, sapygids behave as brood parasites in the nests of solitary bees (O'Neill 2001). In fact, they behave very much like the larval, closed-cell, brood-parasitic bees described in Chapter 10. Adult female sapygids place an egg in a newly provisioned brood cell, and the larvae are hospicidal. Immediately after it hatches, the first-instar larva dispatches the host egg with its elongate, sickle-shaped mandibles (Rozen and Kamel 2009). Once the host is dead, the sapygid larva feeds on the pollen/nectar provision mass. An unusual morphological feature of this group is that females have an ovipositor rather than the sting typical of other aculeate Hymenoptera. The ovipositor is used to pierce the cap of the bee brood cell and place the egg inside (Torchio 1972). Sapygidae attack only bees that nest above-ground. They are known to attack species of Megachilidae and Apidae (*Xylocopa* and *Eufrie-sea*; Rocha-Filho et al. 2016).

An additional group of brood parasites are the phorid or scuttle flies (Diptera: Phoridae). These small, black, agile flies hover around the nest entrances of solitary bees and wasps and, when the opportunity arises, enter the nests to the lay their eggs in the pollen provisions. Wcislo (1990) provided a detailed picture of the life history and behavior of female *Phalacrotophora halictorum* attacking the nests of a solitary halictid bee, *Lasioglossum figueresi* in Costa Rica. *P. halictorum* has a broad host range, including halictid and andrenid bees and even crabronid wasps. Females enter the nest to oviposit in the pollen/nectar provisions in open brood cells. When brood cells were parasitized, there were often multiple phorid larvae per brood cell, and infected brood cells had an odor of fermentation. Once the phorid larva has completed consuming the pollen/nectar provisions, it moves out of the brood cell and pupates along one of the burrows in the nest. Sixteeen percent of brood cells were parasitized in this study.

A final group of brood-parasitic wasps includes the ichneumonid genus *Grotea*, a group of 18 species that attack bee nests and consume the pollen/nectar provisions therein (Herrera-Flórez 2014). Female *Grotea* oviposit into the host brood cells, and immediately upon hatching, the first-instar *Grotea* larva kills the egg or the early-instar host larva and then shifts to feeding on the provision mass (Slobodchikoff 1967). In some cases, provisions in adjacent brood cells are also consumed (Packer 2004). Ichneumonidae is an enormous family of parasitic wasps, and all of the other more than 100,000 species are carnivores, making the obligate pollen feeding by *Grotea* extremely unusual.

ENDOPARASITES

Strepsiptera (the twisted-winged parasites) are endoparasites in a diverse array of hosts, including silverfish (Zygentoma), mantids (Mantodea), roaches (Blattodea), grasshoppers and crickets (Orthoptera), true bugs (Heteroptera), flies (Diptera) and bees, wasps, and ants (Hymenoptera) (Cook 2014, Kathirithamby 2009). Compared to closely related insect orders, the order Strepsiptera is unusually species-poor (700 species total). Strepsiptera are so highly modified for their specialized lifestyle that their relationships to other insect orders has been a source of acrimonious debate among entomologists. They are now thought to be most closely related to beetles (Boussau et al. 2014, Misof et al. 2014). Those that attack bees form a monophyletic group, the Stylopidae, all of which exclusively attack short-tongued bees in the families Andrenidae, Halictidae, Colletidae, and Melittidae. The only short-tongued bee family not known to host stylopids is the Stenotritidae, and no long-tongued bees are known to harbor stylopids. Stylopids seem to be quite host-specific; the different stylopid genera all attack different bee groups, and the various species are all restricted to a single genus or even a single bee species.

Like many meloid and ripiphorid beetles, stylopids begin life as a mobile triungulin larva that rests on flowers waiting for an unsuspecting host bee to stop by for a visit. Little is known of triungulin transport by bees, but in the case of *Stylops pacifica*, the triungulins work their way to the plant nectaries, where they are imbibed along with the nectar and taken up in the crop (Linsley and MacSwain 1956). The larvae are transported back to the nest, where they will be freed when the female bee regurgitates on the provision mass. Larvae picked up by the "wrong" host are probably out of luck.

Once free in the host cell, the mobile triungulin larvae seek out and enter the host egg or larva by dissolving a small hole in the host's cuticle. Once inside, they undergo hypermetamorphosis and become simple legless forms. Developing stylopid larvae remain in the body of the host and go through several molts without shedding their larval skins. Upon pupation of the host, the stylopid larvae migrate to the metasoma, where they take up residence. Up to this point, male and female stylopid larvae have followed a similar developmental trajectory. However, when the adult host bee leaves the safety of her brood cell and ventures out into the world, their paths diverge. Males emerge from the metasoma of the host as winged, mobile adults with enormous bulging compound eyes and multibranched antennae. Departure from the host generally leaves a gaping hole in the metasoma, and bees that give rise to adult male stylopids die shortly thereafter. Female stylopids are morphologically very different from the males. Females are legless, wingless, soft-bodied, bag-like creatures that remain within the abdomen of the host for their entire lives. When fully mature, their cephalothorax (head plus thorax) protrudes out between the overlapping metasomal terga. It is not uncommon to find bees with multiple female stylopids protruding grotesquely from their metasoma. Immobile female stylopids signal their presence to males via pheromones, and mating takes place on the body of the host. When the triungulins are ready for dispersal, they leave through genital openings on the underside of their mother's body. As the host bee forages for pollen and nectar, she deposits mobile triungulin larvae in flowers along the way. In fact, the rate at which she visits flowers accelerates once the triungulins are ready for dispersal (Linsley and MacSwain 1956). A single female strepsipteran can produce thousands of triungulin larvae over her lifetime.

Strepsiptera manipulate their hosts in remarkably devious ways. First, infected female bees do not reproduce like healthy bees. Instead, they construct a below-ground burrow where they safely spend the night, but they never construct or provision brood cells. Second, infected hosts facilitate mating of their parasites. Bees infected with female Strepsiptera climb up on vegetation and sit quietly, usually with their heads facing downward, while the parasite advertises her sexual receptivity to males in the neighborhood (Tolasch et al. 2012).[2] Finally, infected female bees have extended periods of adult activity (Straka et al. 2011). Rather than emerging after males, as would be normal for female *Andrena*, the females infected with Strepsiptera emerge earlier in the season, in synchrony with males. This presumably increases the period over which they can disperse the triungulin larvae prior to the emergence of the majority of the prospective host females. Parasites often manipulate host behavior to enhance the probability they are transmitted to a new victim, and changes in behavior have been noted in other studies of stylopized bees (Linsley 1958, Salt 1927).

Bees and other insects infested with a strepsipteran are called stylopized. Beyond the behavioral changes mentioned above, common side effects of stylopization include changes in morphology so that infected bees often become intersexes, blending male and female characteristics. Females sometimes have a reduced scopa or increased maculation, while infected males often have reduced facial markings or otherwise become more female-like (Salt 1927).

NEST DEPREDATORS

Insects that feed on the pollen/nectar provisions of bees but do not attack the bee larvae are called nest depredators. *Cacoxenus indagator*, a member of the family Drosophilidae, falls into this category. Unlike typical drosophilid flies that deposit their eggs in rotting fruit, mushrooms, or the slime flux of trees (Werner and Jaenike 2017), female *C. indagator* enter open nests of stem-nesting solitary bees (such as *Osmia bicornis*) and deposit their eggs on the pollen/nectar provisions. The fly, which is much smaller than its host, competes with the developing bee larvae for the pollen and nectar provisions. The flies develop more rapidly than the bee larvae, so when many eggs are placed in the brood cell, the bee has little chance of surviving. However, when one or a few eggs are placed in the nest, both flies and bees may mature in the same brood cell (Krunić et al. 2005). Once the

---

[2] Stylopid mating can be prolonged and, as in bed bugs, involves traumatic penetration (Peinert et al. 2016).

fly larvae have consumed their bee hosts, they migrate to the nest entrance, where they wait to emerge the following spring. They have a devious trick for emerging from the nest: Since they cannot penetrate the solid soil plug produced by their host, they always leave a single *Osmia bicornis* alive in the nest. When this bee emerges and excavates its way out of the nest, it liberates the adult *Cacoxenus* flies for another season of brood-cell attack (Ashburner 1981). These flies can be a significant natural enemy of trap-nesting bees in Eurasia. Approximately 50% of the brood of *O. bicornis* (Megachilidae) were lost to *C. indagator* in cane nests in the vicinity of Warsaw, Poland (Madras-Majewska et al. 2011).

THE IMPACT OF NATURAL ENEMIES ON BEES

As the discussion above illustrates, solitary bees are attacked by a bewildering diversity of non-bee parasites and predators. They are also frequently attacked by brood-parasitic bees, as described in Chapter 10. What level of larval mortality do these various organisms impose on solitary bees? Minckley and Danforth (2019) analyzed patterns of larval mortality across 60 published studies in which brood loss was carefully analyzed. They found that larval mortality due to known causes (e.g., identifiable parasites, predators, or brood parasites) ranged from 0% to 50% across the published studies. Mortality in the majority of studies was below 10%, but in 21 studies, overall larval mortality exceeded 10%. The most heavily attacked species included *Diadasia bituberculata* (Linsley and MacSwain 1952) and *D. consociata* (Linsley et al. 1952b), two ground-nesting bees that form large nesting aggregations in arid regions of the western United States.

If one examines the relative impact of the various brood parasites and predators across these studies, it becomes clear that the most significant source of brood mortality in solitary bees is caused by their close relatives, the brood-parasitic bees (Table 11-2). Brood-parasitic bees were reported to cause larval mortality in 43% of the studies. In terms of incidence, meaning the proportion of total larval mortality caused by any one factor,[3] brood-parasitic bees are also the most significant, accounting for an average of nearly 14% of the observed larval mortality (Table 11-2). In terms of overall significance, brood-parasitic bees are followed by meloid beetles, bombyliid flies, clerid beetles, and torymid wasps (Table 11-2). It is intriguing that brood-parasitic bees impose such a dramatic level of mortality on solitary bee nesting aggregations. The last column in Table 11-2 indicates the average brood-cell loss attributed to these various parasites and predators. Despite the large proportion of studies that report parasitism and the incidence of parasitism, the average mortality attributed to each group is low—just under 4% for brood parasitic bees and less than 1% for most of the other taxa listed.

There is another significant source of mortality that emerges from the analysis of these studies—larval mortality due to unknown causes. We include in this

---

[3] We calculated "incidence" in the following way: Assume there are 60 studies, and in those 60 studies there are reports of a total of 300 natural-enemy attacks (because more than one group attacks the brood of any one bee species). Then, if bee flies are responsible for 50 attacks, the incidence (expressed as a percentage) would be 50/300 * 100 = 16.7%.

Table 11-2. Known causes of solitary bee larval mortality ranked by order of significance.

The impact of these parasites, predators, and brood parasites is expressed as (1) the percentage of studies that reported any larval mortality due to the eight most important parasites and predators listed, (2) the "incidence" of attack (meaning the percentage of total larval mortality attributed to each of the eight predators and parasites listed below), and (3) the average percentage of brood-cell loss attributed to each of the predators and parasites listed. Note that the rank order of significance is the same for the first two measures but not for the third.

Order	Family/group	Percent of studies (n=60)	Incidence (n=141)	Percent of brood loss on average (n=60)
Hymenoptera	Brood-parasitic bees	43.3	13.6	3.83
Coleoptera	Meloidae	33.3	10.5	1.83
Diptera	Bombyliidae	31.7	9.9	1.13
Hymenoptera	Torymidae	20.0	6.3	0.73
Coleoptera	Cleridae	18.3	5.8	1.06
Hymenoptera	Mutillidae	15.0	4.7	0.57
Hymenoptera	Chrysididae	13.3	4.2	0.43
Hymenoptera	Leucospidae	8.3	2.6	0.42

category larval death that cannot be attributed to a known parasite, predator, or brood parasite. Brood cells are often found with dead larvae but no clear evidence that an attacker entered the brood cell. Such brood cells often contain partially consumed pollen provisions and a dead larva and are often invaded by fungi. Whether fungi are the cause of death or not is usually impossible to determine. Desiccation is also a likely cause of death in some of these cases. Mortality due to these unknown causes can be exceptionally high. Forty-two percent of *Megachile rufipennis* (Megachilidae) larvae died of unknown causes in studies conducted by Jaysingh and Freeman (1980) in Jamaica. This kind of unexplained larval mortality is possibly an underappreciated limitation on population growth in bees.

DEFENSE AGAINST PARASITES AND PREDATORS

How do solitary bees avoid attack from parasites and predators? We know very little about how bees defend their nests against attackers, but several strategies have been suggested. As described above, some bees (e.g., large carpenter bees) block the nest entrance with their metasoma to impede entry into the nest when under attack. In addition, female large carpenter bees forcibly ejected fecal material to a distance of 2 meters when disturbed (Janzen 1966). The tumulus and various turrets constructed by soil-nesting bees may also function to deter predator attack (Chapter 6). The materials used by cavity nesters and above-ground builders may also deter predator and parasite attack. Resins, for example, are malleable when first applied to the brood-cell lining, but they harden to form a significant physical, and possibly chemical, barrier to attack. More extreme approaches to avoiding parasite attack include moving the nest site from one location to another en masse, which has been proposed as an effective strategy to "leave the parasites behind." Escape from natural enemies may explain why *Centris caesalpiniae* (Apidae), a large bee in the Sonoran Desert of North America, appears to move the entire nest site on an annual basis. A nesting aggregation consisting of more than 400,000 nests had a surprisingly low level (<0.3%) of brood-cell mortality in a study by Rozen and Buchmann (1990). Similar mass-migratory behavior is known for *Diadasia rinconis*, another gregarious ground-nesting bee that is often associated with large numbers of bombyliids (Neff and Simpson 1992). Even nocturnality in bees has been hypothesized to be a strategy to "escape from parasitism" (Wcislo et al. 2004; Chapter 8). Solitary bees are living in a dangerous world. Natural selection has honed the skills of some bees for avoiding parasite attack, but others seem pretty clueless. Otherwise, it is hard to see how so many different animals could specialize on bees as dinner.

CHAPTER 12

Bees and Plants: Love Story, Arms Race, or Something in Between?

Bees are herbivores. We tend not to think of them as such because they don't spend their time chewing leaves like caterpillars, grasshoppers, or gazelles, but their diets are almost exclusively composed of plant products, including pollen, nectar, and occasionally floral oils.[1] While some authors reserve the term *herbivore* for animals that feed on plant vegetative tissues (e.g., Wäckers et al. 2007), a standard dictionary definition is that herbivores are animals that eat only plants. By that definition, bees certainly qualify.

The extraordinary diversity of angiosperms (approx. 300,000 described species; Christenhusz and Byng 2016) and bees (>20,000 described species; Ascher and Pickering 2018) is almost certainly due, at least in part, to the reciprocal evolution observed in insects and plants in general. Bees are under pressure to efficiently exploit floral resources, while plants are "striving" to manipulate insect behavior to their advantage. Every new floral defense (poricidal anthers, for example) leads to radiation in bees that can overcome this floral defense (vibratile, "buzz" pollination, for example). Since bees are significantly more speciose than their closest wasp relatives (the pemphredonine subtribe Ammoplanina, including just 134 described species; Chapter 2), the switch from carnivory to herbivory could have been a significant factor in bee diversification (but see Murray et al. 2018). In this chapter, we explore the bee-plant interaction from an evolutionary perspective. There are elements of both love story and arms race in this ancient coevolved partnership.

---

[1] There are exceptions to this general rule. A small number of social bee species have switched to a diet of animal products. These include the stingless bees that specialize on carrion (Roubik 1982), social wasp larvae (Mateus and Noll 2004), or the tears of large mammals (Bänziger et al. 2009).

BEES AND FLOWERS—A HISTORICAL PERSPECTIVE

Before examining the relationship between modern bees and modern angiosperms, it is worth taking a historical view of bee-flower interactions based on the available, sometimes fragmentary, fossil record for each group.

The time line of bee-angiosperm evolution—angiosperms came first, but bees were not far behind

The interaction between bees and flowers is a coevolved partnership that has shaped terrestrial life on earth. When did this partnership begin, and what impact has this partnership had on each player? In order to understand the timing of bee-angiosperm evolution, we rely on a relatively new method that combines evidence from the fossil record with molecular data to calculate the ages of major lineages of plants and animals. This method has been used widely in the plant and insect literature to calculate clade ages and to put a time line on the evolution of each group (Hedges and Kumar 2009).

For seed plants—the group that includes both angiosperms and gymnosperms—we rely on a study by Magallón and colleagues (2015) that combined 137 fossil calibration points with five genes from the mitochondrial and chloroplast genome to calculate a dated phylogeny (Fig. 12-1). According to this time line, the seed plants arose early in the Paleozoic, over 325 million years ago. Based on the fossil record, gymnosperms dominated terrestrial habitats for over 200 million years. Gymnosperms are largely wind-pollinated. Pollen grains are small and non-sticky (so they travel far in the wind), and the pollen is of relatively low nutritional value for an insect. However, two groups (Gnetaceae and Cycadaceae) have exploited the services of insects for pollination. Gnetaceae rely on flies, whereas Cycadaceae rely primarily on beetles and thrips (Labandeira et al. 2007). As one might expect, these animal-pollinated gymnosperms have relatively protein-rich pollen (Roulston et al. 2000). In the case of these animal-pollinated gymnosperms, the only reward they offer is pollen.

There is an enormous gap—nearly 200 million years—between the origin of seed plants and the first appearance of flowering plants (Fig. 12-1). Darwin famously observed that the sudden appearance of flowering plants in the early Cretaceous was an "abominable mystery" (Friedman 2009). Darwin speculated that there must be a hidden fossil record that would show a gradual transition from gymnosperms to angiosperms. However, no hidden fossil record has been found, and even modern molecular phylogenies document a rapid accumulation of angiosperm lineages in the early Cretaceous.

Molecular studies of basal angiosperms have largely converged on the tree shown in Figure 12-1. The earliest extant angiosperm lineages include the Amborellales, Nymphaeales, and Austrobaileyales (now referred to collectively as the ANITA grade). Other early branches of the angiosperm tree include magnoliids, monocots, and Ceratophyllales (Fig. 12-1). These basal angiosperm lineages are mainly either wind-pollinated or pollinated by beetles, flies, and

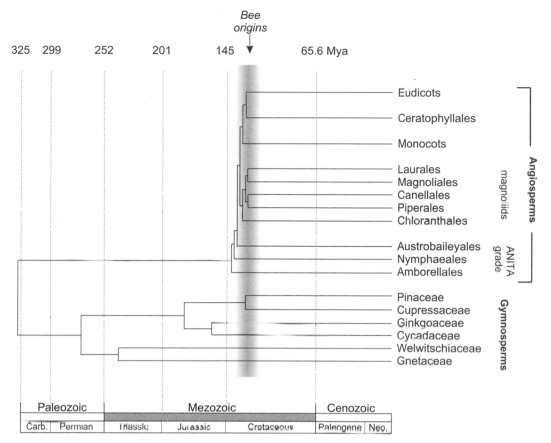

Figure 12-1. Fossil-calibrated phylogeny of the seed plants, showing the major lineages (redrawn from Magallón et al. 2015, Fig. 2). The estimated age of bees is based on Cardinal and Danforth (2013).

thrips. Nectar in these basal angiosperms is uncommon and, if present, is rarely copious (Erbar 2014). Bees are significant visitors to a few modern members of these basal angiosperm lineages, such as some Alismataceae and Araceae (Alismatales), Piperaceae (Piperales), and Nymphaeaceae (Nymphaeales), but in all cases, these seem to be relatively recent developments (Gottsberger 2016). The one group of angiosperms that are largely bee-pollinated are the eudicots. This group includes over 210,000 described species or roughly 73% of extant angiosperms (Fig. 12-2). Eudicots first become evident in the fossil record as distinct tricolporate pollen grains that appeared throughout the world roughly 125 million years ago. Fossil-calibrated studies have largely confirmed this date or pushed eudicots slightly earlier (Magallón et al. 2015).

Where do bees fit into this time line? Estimates of bee antiquity using fossil-calibrated phylogenies have largely converged on a date of roughly 125 million years ago (Cardinal and Danforth 2013). This age coincides almost precisely with the appearance of tricolporate pollen and the estimated age of the eudicots (Fig. 12-1). Eudicots include most of the flowering-plant

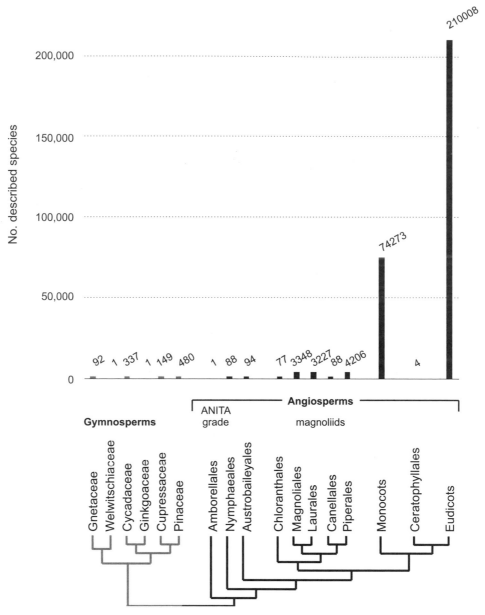

Figure 12-2. Comparison of species richness across seed-plant lineages. Note the dramatic increase in diversity in clade, including monocot and eudicot clades.

lineages with clear affinities to bees, including the rosids and asterids. Bees and eudicots seem to be the real love story we have been looking for. Angiosperms may have gotten their start thanks to pollinators like beetles, flies, moths, and thrips, but the bees are the major players when it comes to eudicot diversification.

The pathway to pollenivory—Malyshev was right, but for the wrong reasons

How did a carnivorous group of solitary wasps give rise to pollen-feeding bees? Put another way, how did the ancestors of bees, which were hunting arthropod prey, switch to using pollen as a protein source for rearing their larvae? A number of hypotheses have been proposed over the years (Evans 1966, Malyshev 1968, Müller 1872), but most previous hypotheses were developed before we had a precise understanding of the higher-level phylogeny of bees or a clear understanding of the origins of bees from crabronid wasps (Chapter 2). A study by Sann and colleagues (2018) provides a fascinating model for how pollen feeding may have arisen. The pathway to pollenivory seems to involve an obscure order of pollen-feeding insects—thrips (Thysanoptera).

Sann and colleagues (2018) provide solid evidence, based on a large molecular data set and thorough sampling of relevant taxa, that bees are the sister group to the crabronid subtribe Ammoplanina (Fig. 12-3). These small, slender,

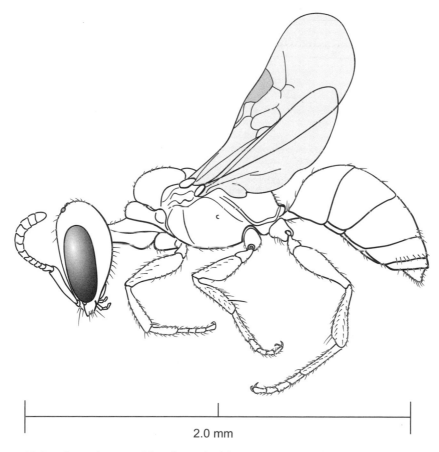

2.0 mm

Figure 12-3. Habitus drawing of female *Timberlakena yucaipa* (Pemphredoninae: Ammoplanina) (redrawn from Bohart and Menke 1976, Fig. 48).

ground-nesting wasps are thrips hunters. Most thrips are tiny insects with bizarre fringed wings and piercing-sucking mouthparts (Fig. 12-4). Thrips feed on a variety of materials, including fungi, leaves, and—most importantly for our story—pollen grains. Many thrips inhabit flowers, where they pierce pollen grains with their stylet-like mouthparts and aspirate the pollen cytoplasm. A thrips can empty a single pollen grain in a few seconds, and a single thrips can consume hundreds of pollen grains per day. Thrips are also known to be common visitors to some very primitive plants, including cycads, and they can play an important role as pollinators (Terry 2001). Thrips were present in the Mesozoic, and fossil evidence indicates that they were common visitors to Mesozoic gymnosperms, including cycads (Peñalver et al. 2012). Individual *Gymnopollisthrips*

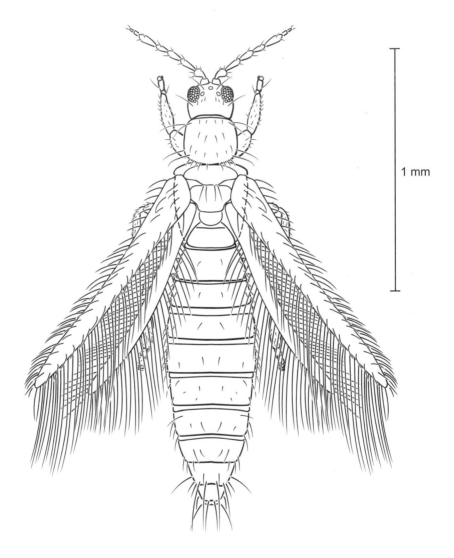

1 mm

Figure 12-4. Habitus drawing of female *Frankliniella occidentalis* (Thysanoptera) (redrawn from Kirk 1996, Plate 4, Fig. 4).

(Melanthripidae) carried up to 140 pollen grains on the external surface of the body, which is a substantial load for a 1-millimeter-long insect. These thrips also bear specialized setae, apparently modified for gathering, and possibly transporting, pollen (Peñalver et al. 2012).

These Mesozoic, pollen-collecting thrips provide a plausible pathway for a Cretaceous pemphredonine wasp to transition from carnivory to pollenivory. By feeding their larvae with pollen-laden thrips, a proto-bee may have been effectively omnivorous before transitioning to a diet of pure pollen. By hunting flower-feeding thrips on Mesozoic gymnosperms, our proto-bee would have come into extensive contact with pollen and perhaps carried some pollen back to the nest externally, thus initiating the transition to external pollen collection and transport. By establishing the sister group to the bees (Ammoplanina), Sann and colleagues (2018) have provided a wonderful glimpse into how pollenivory might have arisen.

The pemphredonine-bee connection is further supported by *Melittosphex burmensis* (Melittosphecidae), a remarkable 100-million-year-old fossil from Burmese amber (Danforth and Poinar 2011, Poinar and Danforth 2006; see also Chapter 2). *M. burmensis* bears a number of very bee-like features; most importantly, it has branched, plumose hairs over much of the body. But *M. burmensis* also shows some wasp-like features. This combination of bee-like and wasp-like features suggests that *M. burmensis* is a transitional form between hunting wasps and modern bees. *M. burmensis* also bears a number of pronounced similarities to pemphredonine wasps that were present in the same amber fossil deposits. Antropov (2000) described several Burmese amber fossil Pemphredoninae. These tiny wasps are similar in size to *M. burmensis* and also share some unusual morphological features with *M. burmensis*, such as acute spines on the propodeum. Perhaps *M. burmensis* exemplifies a lineage of proto bee in the process of transitioning from an omnivorous diet of thrips plus pollen to a diet of pure pollen. The Burmese amber deposits may capture a key point in bee origins and the evolution of pollenivory.

The hypothesis that bees arose from pemphredonine wasps is not a new one. Malyshev (1968) identified two traits that he viewed as evidence of close affinity between "primitive" bees (Colletidae, in his view) and Pemphredoninae. First, he pointed out that some colletid bees carry pollen internally—a trait he considered primitive for bees in general. Second, he noted that pemphredonine wasps line their brood cells with a silken material similar to the cellophane-like material used by Colletidae (Chapter 5). Malyshev speculated that aphid hunting might provide a pathway to pollen feeding. This model is slightly less plausible than a pathway involving thrips, because aphids do not feed on pollen and they do not spend much time among flowers. Based on our current understanding of bee and wasp relationships (Chapter 2), we now know that Colletidae is actually far from being a "primitive" family of bees. Their cellophane brood-cell lining is fundamentally different from (and not homologous to) the silken material used by pemphredonine wasps, and the mode of internal pollen transport observed in Hylaeinae and Euryglossinae is actually derived *from* external pollen transport in the rest of the bees. So Malyshev seems to have come to the right conclusion (bees are derived from pemphredonine wasps) but for the *wrong* reasons (an incorrect view of higher-level bee phylogeny). Nevertheless, his keen natural history observations laid the groundwork for our modern view of bee origins.

BEES AND FLOWERS—A MARRIAGE INVOLVING CONFLICT AND COMPROMISE

One reason we tend not to think of bees as herbivores is that plant-herbivore interactions are typically very one-sided, with one side (the herbivore) benefiting and the other (the plant) losing. In evolutionary terms, plant-herbivore interactions are frequently viewed as arms races, with plants evolving defenses to deter herbivores and herbivores evolving mechanisms to overcome those defenses.[2] Bee-plant interactions share some aspects of that model, but they deviate from it since they often have a mutualistic aspect, with both sides winning: plants winning through the pollination services they receive via bee visits, and bees winning via the floral resources they harvest during flower visits. However, it is not all winning all the time, since bees and plants have different goals. For plants, the goal is a reliable pollination system, with optimal pollen dispersal and receipt at minimal cost (Box 12-1). Feeding bees is not a goal but rather a cost to be minimized. For bees, the goal is optimizing the rate and efficiency of harvesting floral resources, most notably for provisioning nests, but also for individual energetic and nutritional needs. Pollen "wasted" on pollination is something to be minimized for provisioning female bees.

This tension in the bee-plant relationship has been described as *balanced mutual exploitation* by Westerkamp (1996), although sometimes, as in rewardless floral mimicry systems where the bees are losers (Dafni 1984) or cases of floral larceny where the plants are the clear losers (Inouye 1980), the systems are quite unbalanced. Many floral adaptations—like poricidal anthers, nototribic pollen deposition (see Chapter 7), and keel flowers—have been interpreted as defensive mechanisms to minimize pollen harvesting by bees. However, these same features can also serve to increase the efficiency of pollination systems via more precise pollen placement on visitors or increased male function by serving as pollen-parceling mechanisms, so their unequivocal roles as plant defenses against greedy bees are unclear (Harder and Thomson 1989, Harder and Wilson 1994).

Many flower-visiting insects consume pollen, but with the exception of the masarine wasps (Chapter 2), all are minor leaguers compared to bees—professional pollen consumers whose females must typically harvest 3 to 4

---

[2] The arms race analogy was first proposed by Ehrlich and Raven (1964) to describe the relationship between herbivores and their host plants (review in Schoonhoven et al. 2005). The arms race idea can be described as follows. As herbivores cause more and more damage to their host plants, the host plants respond evolutionarily with increased plant defense—toxic chemicals, gooey sap, spiny leaves, and other mechanisms that allow them to deter herbivory. Released from herbivore attack, the plant lineage then diversifies. But herbivores are also evolving. Once an herbivore lineage evolves mechanisms for overcoming the new plant defense, the herbivore lineage can also diversify on the now abundant and species-rich host-plant lineage. This back-and-forth, reciprocal escalation of plant defense and herbivore counter-defense was thought to be the likely explanation for the overwhelming predominance of plants and herbivores on earth (Schoonhoven et al. 2005). The arms race hypothesis is now just one of several plausible ecological and evolutionary processes shaping plant-herbivore interactions.

BOX 12-1: AN ABBREVIATED GUIDE TO POLLINATION BIOLOGY—THE "GOOD," THE "BAD," AND THE "UGLY"

The title of Sergio Leone's classic spaghetti western *The Good, the Bad, and the Ugly* refers to the movie's three main characters, none of whom is particularly good at anything except gunfighting. This has not stopped the phrase from popping up in many different contexts, including pollination biology. Thomson and Thomson (1992) appropriated the phrase to describe pollinator quality. In their formulation, "good" pollinators have both high pollen removal per visit (and hence the possibility of high pollen dispersal) and high pollen deposition; "bad" pollinators have low pollen removal and low deposition; and "ugly" pollinators have high removal but low deposition (Fig. 12-5a). From the plant's viewpoint, visits by the good should be encouraged, the bad tolerated, and the ugly discouraged. Due to their high pollen consumption, pollen-collecting bees will usually fall into either the good or ugly boxes, depending on how much pollination they do. In a study of pollination of *Campanula americana*, ugly pollen-collecting halictines were found to depress floral siring success when the good-nectaring bumble bees were scarce, although not when they were common (Lau and Galloway 2004).

An alternate way to classify good, bad, and ugly pollinators is that, instead of using just pollen removal and deposition per visit as the primary criteria, one might use effectiveness and efficiency (Fig. 12-5b). Here, an effective flower visitor gets the job done in terms of pollen deposition without regard to cost. A visitor might remove high proportions of the available pollen and deliver only a small fraction to receptive stigmas, but if this is sufficient for adequate seed set, this constitutes effective visitation. This is what many authors mean when they talk about pollinator efficiency (Ne'eman et al. 2009). An ineffective pollinator simply doesn't accomplish much pollination per visit. Efficiency adds an element of cost to the formula, so it is not the amount but the proportion of pollen removed that reaches a receptive stigma that counts. If the proportion is high, it is efficient even if it may not actually be effective. One can add other costs, like costs of floral displays, or rewards, such as nectar or floral oil, to move x amount of pollen y distance to get a better estimate of the costs of moving pollen around. In this scheme, a good pollinator is both effective and efficient, a bad pollinator might be effective but inefficient or efficient but ineffective, and an ugly pollinator has no redeeming qualities at all, being both ineffective and inefficient. Many pollen-collecting bees will have trouble staying out of the bad category simply because of their high pollen usage, and small bees collecting pollen on large flowers or plants will often fall into the ugly category. Major parts of floral evolution probably involve fostering visitation by the good, limiting the damage done by the bad, and avoiding the ugly.

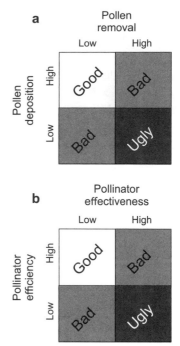

Figure 12-5. The "good," the "bad," and the "ugly": (a) relationship between pollen deposition and pollen removal for four different pollinator scenarios and (b) relationship between pollinator effectiveness and pollinator efficiency for four different pollinator scenarios.

times their dry body weight to provision a single brood cell (Neff 2008). A number of studies have documented the high pollen demands that bees place on local floral resources. Females of *Andrena hattorfiana* (Andrenidae) were estimated to need all the pollen from 72 inflorescences or 11 individual plants of its host, *Knautia arvensis* (Dipsacaceae), to provision a single 6-celled nest (Larsson and Franzén 2007). Similarly, *Chelostoma rapunculi* (Megachilidae) was estimated to require all the pollen from 59 flowers of its host, *Campanula rapunculus* (Campanulaceae), to provision a single brood cell (Schlindwein et al. 2005; Box 7-2). A female of *Dieunomia triangulifera* (Halictidae) should be able to provision 3 or 4 brood cells with pollen from a single capitulum of *Helianthus annuus* (Asteraceae). However, she would not be able to do so quickly, since pollen is made available slowly throughout the day, and florets on a given capitulum open over 8 or more days. Thus, in the real world, she must visit many different flowers to fully provision a brood cell (Minckley et al. 1994). In a study using pollen volume, 6 European solitary bees were estimated to need all the pollen from 4 to 110 flowers, depending on the species. Extrapolating from that data set, different bees in an assemblage of 35 species were estimated to require the pollen from 4 to more than 1,139 flowers to provision a single brood cell (Müller et al. 2006). These estimates are all conservative, as they make the highly dubious assumption that females are able to harvest all available pollen with no loss (which also means no pollination). Real requirements of the numbers of flowers needed to provision a cell are probably often significantly higher. In the case of *Campanula rapunculus*, it has been estimated that as much as 95% of pollen production ended up producing *Chelostoma* offspring (Schlindwein et al. 2005).

DIET BREADTH IN BEES—THE OLIGOLECTY–POLYLECTY CONTINUUM

The vast majority of herbivorous insects are dietary specialists that feed on only a single plant family, genus, or smaller group. Perhaps no more than 10% of herbivorous insect species have diets that include three or more plant families (Bernays 1989). Although the percentage of specialists is undoubtedly lower, the majority of bees have a relatively restricted pollen diet. Some early workers assumed that bees were not particularly picky about the flowers they visited (Müller 1883), but Charles Robertson noted that many bee species restrict their pollen collecting to just a single plant family or a few related genera (Robertson 1899). He called this behavior *oligotropy* and contrasted it with *polytropy*, the tendency to collect pollen from various unrelated plant groups. He used the term *monotropy* for bee species believed to restrict their pollen collection to a single plant species. Later, noting that the oligotropy-polytropy terminology, as originally proposed by Loew (1884), did not refer to specificity of pollen collection, Robertson (1925) proposed the terms we use today: *polylecty* (literally choosing many), *oligolecty* (choosing few), and *monolecty* (choosing one) to reflect the spectrum of pollen host-plant preferences in bees (Fig. 12-6; Box 12-2).

Most subsequent writers discussing the range of pollen used by different bee species have followed that general outline, although often with variants. Some have used a very narrow definition of *oligolecty* ("regularly collect pollen from a single plant species or a group of related plant species") and dropped the term *monolecty*, since they did not consider it sufficiently different from oligolecty, at least in the sense they used the term (Linsley and MacSwain 1957). Others, noting that oligolecty or polylecty are potentially very broad categories, have

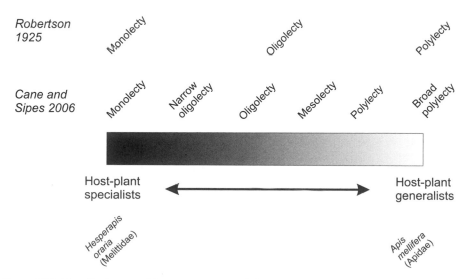

Figure 12-6. The oligolecty–polylecty continuum and alternative terminologies to describe the range of host-plant preferences in bees.

BOX 12-2: *PROTODISCELIS PALPALIS* (COLLETIDAE) AND *HYDROCLEYS MARTII* (ALISMATACEAE)— AN EXTRAORDINARILY INTIMATE BEE-PLANT PARTNERSHIP

The bee *Protodiscelis palpalis* (Colletidae: Neopasiphaeinae) and the plant *Hydrocleys martii* (Limnocharitaceae, the "water poppy" family) are locked in a particularly tight embrace. *H. martii* is entirely dependent on foraging female *P. palpalis* for pollination, and *P. palpalis* obtains all of its pollen and nectar requirements from this single host-plant species. Disappearance of one partner would mean rapid extinction for the other. This remarkable bee-plant love story was documented by Clemens Schlindwein and colleagues in a series of studies conducted in the arid, semi-desert region of northeastern Brazil called the Caatinga (Carvalho et al. 2014, Carvalho and Schlindwein 2011).

The Caatinga is a harsh, tropical, dry forest habitat dominated by succulent and spiny xerophytic woody plants; rainfall is limited and unpredictable. Nine months of the year are exceedingly dry, but rains that typically fall in February, March, and April bring the flowering plants, including *H. martii*, into bloom. *H. martii* is an aquatic monocot that grows in seasonally flooded ponds and rivers. Flower buds develop under water and, when mature, extend briefly above the water surface. Flowers open, and six to eight hours later they close and then recede below the water surface, where fruit development takes place. When open, the flowers of *H. martii* are bright yellow, with a deep orange central disk and three radially arranged petals. The reproductive parts of the flower (the anthers and stigma) are enclosed within a cone of non-reproductive staminodes. For bees to access the pollen, they need to bypass the tightly closed cone of staminodes. *H. martii* are incapable of self-pollination and require floral visitors to set fruit.

Based on Schlindwein's work, *P. palpalis* was by far the most frequent visitor to *H. martii* flowers. Across 25 sites in 5 Brazilian states, *P. palpalis* was the most frequent floral visitor, and at 18 sites it was the only floral visitor. Female *P. palpalis* are capable of entering the staminodal cone by working their bodies between the individual staminodes. Once inside the staminodal cone, they remained there for up to 14 seconds gathering pollen and nectar before departing. Other floral visitors, including stingless and halictid bees and honey bees, were either robbers (extracting pollen and nectar without coming in contact with the stigma) or gleaners (collecting pollen left on the flower petals by previous floral visitors).

P. palpalis are exceptionally effective at harvesting the pollen of *H. martii*. A single flower contains on average 480,000 pollen grains, and female *P. palpalis* harvest 89% of them. Ten percent of the pollen remains attached to the staminodes, anthers, and petals, and 1% is deposited on the stigma. *P. palpalis* appears to be the sole pollinator of *H. martii*, because flowers grown in the absence of *P. palpalis* failed to set fruit. *H. martii* flowers emit an odor (p-methylanisole) that is unique to *H. martii*, that is lacking in a conspecific (*H. nymphoides*), and that is highly attractive to *P. palpalis* (but not to another, closely related *Protodiscelis* species; Carvalho et al. 2014). Thus, there appears to be a private line of chemical communication between *H. martii* and *P. palpalis*. The intimate codependence of *P. palpalis* and *H. martii* is uncommon among bee-plant partnerships, which are typically characterized as more diffuse interrelationships. But this example illustrates beautifully the extent to which specialization can occur in bee-plant associations, especially in arid regions of the world, where host-plant flowering is brief and unpredictable.

added modifiers like "broad" or "narrow": *broad polylecty* for species using the pollen of many plant families, *narrow polylecty* for those using only a few, *broad oligolecty* for species collecting the pollen of many genera in a single large family, and *narrow oligolecty* for those collecting from only a few species in a genus or a few closely related genera. Cane and Sipes (2006) considered broad oligolecty and narrow polylecty to be oxymora and proposed the terms *mesolecty* for species collecting pollen from many species in one, two, or three families, and *eclectic oligolecty* for species collecting pollen from two *unrelated* host-plant families (Fig. 12-6). Many other classifications or definitions have been proposed, but for our purposes, we will stick with the original definitions of Robertson (1899, 1925).

It is important to remember that there is a *continuum* of dietary specialization in bees, ranging from narrow specialists that restrict their pollen diet to a single plant species to broad generalists that will attempt to collect pollen from virtually any flower they encounter; this continuum even includes non-floral sources like fungal spores in their diet. In between will be many intermediate forms of dietary restriction. Terms like oligolectic and polylectic are useful for communication but can give the misleading impression that there are clear distinctions between these categories of pollen use.

Oligolecty is sometimes confused with *floral constancy* (or "foraging constancy"). The two concepts are related but are far from identical. As discussed above, oligolecty concerns the range of pollen sources a given bee species will utilize. In contrast, floral constancy, a phenomenon known from at least the time of Aristotle (c. 350 BC), refers to the tendency of an individual forager to restrict its foraging to flowers of a single species, or even the color morph of a given species, during at least part of a foraging bout. As such, constancy is considered to be an important contributor to the effectiveness of individual foragers as pollinators (Grant 1950). While oligolecty appears to be genetically hardwired, constancy has a strong learned component, with individual foragers learning, via experimentation, what flowers are rewarding and, with repeated visits, how to most effectively forage on them. In many cases, foraging constancy allows an individual forager to forage more efficiently than it would if it was just blundering around, sampling every flower it encountered.

PHYLOGENETIC AND BIOGEOGRAPHIC PATTERNS OF FLORAL HOST USE IN SOLITARY BEES

Phylogenetic patterns of host range at higher levels

For many years, it was assumed that polylecty was primitive among bees and oligolecty was a derived trait (Michener 1954). However, recent phylogenetic analyses, coupled with more extensive data on host-plant utilization patterns, have essentially reversed that assumption. Almost all members of the Melittidae, the sister group to the rest of the bees (Chapter 2), and most of the basal members of the Megachilidae (Fideliinae, Lithurginae), Halictidae (Rophitinae), and Andrenidae (basal Andreninae) are oligolectic, solitary bees.

Patterns are more complex in the Apidae and Colletidae, where the extent of dietary specialization in many groups remains poorly known. In the Colletidae, host-plant relations are poorly known in the Diphaglossinae, the sister group to the rest of the family, while most Neopasiphaeinae and at least the basal Colletinae are mainly oligolectic. We currently lack detailed generic or species-level phylogenies for Hylaeinae and Euryglossinae, and inferences about pollen preferences are complicated by their mode of internal pollen transport. Given our limited understanding of host-plant preferences in basal Colletidae, it is difficult to say whether Colletidae were ancestrally polylectic or oligolectic.

In the Apidae, the subfamilies Anthophorinae, Xylocopinae, and Apinae (Centridini plus corbiculates) are largely host-plant generalists. The subfamily Nomadinae includes exclusively brood-parasitic bees, for which host-plant preferences cannot be determined (the adults do not collect pollen). Only the subfamily Eucerinae includes a large proportion of oligolectic groups, including Ancylini, Emphorini, and Exomalopsini. Given the basal placement of Anthophorinae and Xylocopinae (Bossert et al. 2018), it seems likely that apids were ancestrally polylectic. Most of the "basal" specialists described above are restricted to xeric habitats (e.g., Fideliinae, Rophitinae, basal Andreninae). It is possible that oligolecty is just part of a suite of traits that allows these "relics" to persist in harsh, desert environments.

Phylogenetic patterns of host range at the species level

The use of species-level phylogenies has provided detailed insights into our understanding of host-range evolution among solitary bees. These studies start by generating a phylogeny of the bees at the species level. They then map pollen host preferences onto the phylogeny and examine questions such as whether the group was ancestrally polylectic or oligolectic, how many cases of host-range expansion or contraction have occurred, how frequently host-plant shifts have occurred, whether host switches occur to related or unrelated host-plant genera or families, and whether similar features of the floral-host morphology, pollen chemistry, or scent might explain acquisition of a new host. In the past two decades, these phylogenetic approaches have proven incredibly valuable as road maps to understanding how and why host shifts occur.

Müller (1996b) was the first to use this approach. He mapped pollen host breadth on a phylogeny of Eastern Hemisphere anthidiine bees (Megachilidae) and found that polylectic species were rare and derived from oligolectic ancestors. This same pattern—origins of polylecty from within ancestrally oligolectic lineages—has been documented in other genera, including *Chelostoma* (Megachilidae), *Andrena* (Andrenidae), and *Dasypoda* (Melittidae) (Larkin et al. 2008, Michez et al. 2008, Sedivy et al. 2008).

Reversions from polylecty to oligolecty have occurred in several groups, most notably in the primarily polylectic genus *Lasioglossum* (Danforth 2002). In the subgenus *Sphecodogastra*, there has been a small radiation of bees specialized on the Onagraceae (McGinley 1986), while in the subgenus *Hemihalictus*, one species (*L. lustrans*), has become a specialist on *Pyrrhopappus* (Asteraceae; Michener 1947), while another (*L. nelumbonis*) has specialized on the Nymphaeaceae (Gibbs et al. 2013). Other oligolectic European *Lasioglossum*, such as *L. brevicorne* (on Asteraceae), *L. punctatissimum* (on Lamiaceae), and others probably represent additional independent derivations of oligolecty from polylecty (Pesenko et al. 2000, Scheuchl and Willner 2016). More recently, Haider and colleagues (2014b) documented shifts from polylecty to oligolecty in the osmiine subgenera *Osmia*, *Monosmia*, and *Orientosmia* (Megachilidae). We are finding that host range has been remarkably dynamic through the evolutionary history of bees.

In a textbook example, Sipes and Tepedino (2005) examined floral host use among the 25 North American species of *Diadasia*. In this group, 17 species are specialists of Malvaceae, 6 species use Cactaceae, and 1 species each occur on members of Onagraceae, Convolvulaceae, and Asteraceae. Mapping of host-plant associations onto the phylogeny indicated that the ancestral state for the genus was specialization on Malvaceae (Fig. 12-7). Few host shifts appear to have occurred. In fact, with 5 potential host-plant families, there were only 4 host shifts—the minimum number possible. Once host switches occurred, descendants typically remained associated with the new host-plant family through multiple speciation events. For example, a single switch from Malvaceae to Cactaceae gave rise to a monophyletic group of 6 Cactaceae specialists (Fig. 12-7). Finally, some host shifts seemed to be between unrelated plant families that have similar floral or pollen morphology. The families Cactaceae and Malvaceae are unrelated, but both have open, yellow to orange, cup-shaped flowers with a central column of densely spaced anthers and very large pollen grains. Malvaceae-Cactaceae host switches occur in other bee genera as well (e.g., *Macrotera* [Andrenidae]), suggesting that these two plant families, while unrelated, share a similar set of morphological traits that attract the same lineages of host-plant specialist bees. Similar floral morphology may be one factor that favors host switching among unrelated host-plant families when it does occur.

Another factor that may facilitate host shifts is similarity in pollen chemistry. Sedivy and colleagues (2013c) examined the pattern of host switching in the osmiine genus *Hoplitis*. Species in this group consist of specialists on Boraginaceae or Fabaceae and generalists that visit both. This is puzzling because these plant families are neither closely related nor do they have similar floral morphologies. Fabaceae have bilaterally symmetrical, keel flowers, whereas Boraginaceae have radially symmetrical, cup-shaped flowers. Sedivy and colleagues (2013c)

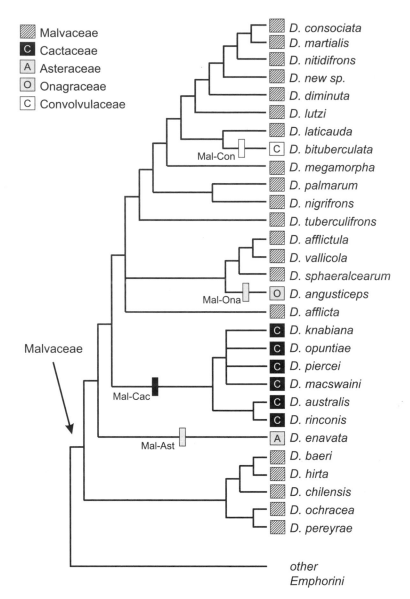

Figure 12-7. Phylogeny and host-plant use in the genus *Diadasia* (Apidae) (redrawn from Sipes and Tepedino 2005, Fig. 2). The phylogeny is derived from nuclear and mitochondrial gene sequences. Host-plant associations are indicated in boxes at the tips of each branch. *Mal-Ast*: host-switch from Malvaceae to Asteraceae; *Mal-Cac*: from Malvaceae to Cactaceae; *Mal-Ona*: from Malvaceae to Onagraceae; *Mal-Con*: from Malvaceae to Convolvulaceae.

constructed a molecular phylogeny at the species level and found that the ancestral state for the *Annosmia-Hoplitis* group was to be oligolectic on Boraginaceae. Multiple switches from Boraginaceae to Fabaceae occurred within the group. They attributed the frequent host switches between Boraginaceae and Fabaceae to either similar secondary plant defenses or possibly similar nutritional components in the pollen. It is certainly not floral morphology that explains the "Boraginaceae-Fabaceae" paradox.

The studies available at present show pollen host range is labile, and transitions from specialization to generalization and back to specialization have occurred repeatedly. In addition, shifts to unrelated host-plant families and orders are not uncommon (Sedivy et al. 2008, Sipes and Tepedino 2005). Major shifts in host-plant preference often involve switches to unrelated plant families with similar floral morphology (Cactaceae-Malvaceae) or pollen chemistry (Boraginaceae-Fabaceae). The North American *Larrea* bees demonstrate just how quickly novel host-plant associations can evolve (Box 12-3).

Intra-specific variation in host-plant range— evidence of a preference hierarchy?

While we sometimes label certain bee species as oligolectic on a particular host plant, several studies have documented considerable potential for intra-specific variation in host-plant preferences. Even oligolectic bees can use alternative host plants under certain situations (Linsley and MacSwain 1958). For example, *Andrena erythronii* (Andrenidae) is an early-spring bee that specializes on collecting the pollen of the dog-toothed violet (*Erythronium americanum*, Liliaceae). However, late in the season, when *Erythronium* is no longer in bloom, female *A. erythronii* will switch to collecting pollen from other flowers (Michener and Rettenmeyer 1956). Similarly, *A. nemophilae*, as the name suggests, normally restricts its pollen collection to *Nemophila* (Hydrophyllaceae) but may collect other pollen late in the season when *Nemophila* is no longer available (Cruden 1972). *Diadasia australis*, normally a cactus specialist, was observed collecting *Phacelia* (Hydrophyllaceae) pollen when the local cactus bloom was exhausted (Linsley and MacSwain 1958).

There is also evidence for intra-specific geographic variation in host use. *Colletes stepheni* was long thought to be a creosote bush (*Larrea tridentata*, Zygophyllaceae) specialist, nesting on sand dunes in the Sonoran Desert (Hurd and Powell 1958). Nelson and Griswold (2015) surveyed sand dunes over a much larger geographic range and found that the range of *C. stepheni* extends into the Mojave and Great Basin deserts. In the Sonoran and Mojave deserts, the bee appears to be a specialist on *L. tridentata*. However, in the Great Basin Desert and in the Eureka Valley (at the junction of the Great Basin and Mojave deserts), females largely specialize on *Psorothamnus* (Fabaceae). *Larrea* is absent from the Great Basin, but both *Larrea* and *Psorothamnus* are present in the Eureka Valley. In the case of *C. stepheni*, the preferred host appears to be *L. tridentata*, but individuals nesting outside the range of *L. tridentata* (and even some nesting within the range of *L. tridentata*) are capable of switching to an alternative (and unrelated) host (*Psorothamnus*).

These switches to alternate hosts when the primary host is not available are what might be expected if diets are *hierarchical*. For any given bee species, there will be some flowers that are always acceptable, some lower on the hierarchy that will be acceptable only after some period during which the highly ranked species are not encountered, and some species that are never acceptable. What

BOX 12-3: THE NORTH AMERICAN CREOSOTE BUSH (*LARREA TRIDENTATA*) BEE FAUNA—RECENT AND RAPID EVOLUTION OF A SUITE OF HOST-PLANT SPECIALISTS

Larrea tridentata (creosote bush, Zygophyllaceae) is a super-abundant resource for bees. In the warm deserts of western North America. *L. tridentata* is the dominant flowering shrub (Turner et al. 2005). *L. tridentata* extends over 862,900 square kilometers and essentially defines the limits of the Sonoran, Chihuahuan, and Mojave deserts. Individual plants can grow to 3.5 meters in height and may live as clonal rings for several thousand years (McAuliffe 1988, Vasek 1980). *L. tridentata* typically blooms in the spring following a sufficient winter rain, but it can also bloom in the late summer in years of heavy monsoon rains (Bowers and Dimmitt 1994). When in bloom, a creosote bush produces hundreds of small, yellow, open flowers with abundant nectar and protein-rich pollen (Minckley et al. 2003).

L. tridentata is a recent arrival in North America. Four closely related *Larrea* species are dominant shrubs in the Monte Desert of southern South America. Of these, only *L. divaricata*, or a close relative, made the journey to North America, giving rise to *L. tridentata*. As there is no evidence for the existence of continuous arid corridors, this almost certainly involved long-distance dispersal, probably via migratory birds. Molecular clock estimates indicate that *L. tridentata* colonized North America between 500,000 and 1.5 million years ago (Laport et al. 2012). In the short time that creosote bush has been in North America, it has attracted a diverse bee fauna of over 120 species of native, mostly solitary bees (Hurd and Linsley 1975, Minckley et al. 2000). The creosote bush bee fauna includes both generalists and a group of 20 essentially monolectic species from 9 genera and 5 families (Table 12-2). These specialists have all arisen in the last few million years, and they are unrelated to the *Larrea* specialists in South America, which include mostly colletids in the subfamily Neopasiphaeinae, species of *Parasarus* (e.g., *P. speculiferus*), a strictly South American panurgine, and species of *Megachile* and *Anthidium* not closely related to any North American species. The arrival of a superabundant, predictable, easily accessible floral resource in North America likely provided an ideal opportunity for endemic bee lineages that could exploit it—and they did, to a remarkable degree.

The rapid evolution of *Larrea* specialization among North American solitary bees demonstrates how labile host associations are in bees. This is consistent with the phylogenetic studies that document host switching between unrelated host-plant families. Rapid host switching might also explain why there is no broad phylogenetic congruence between bees and eudicots: primitive lineages of bees do not appear to be associated with primitive lineages of eudicots. Oligolectic bee lineages can acquire and lose host plants rapidly. In this Cretaceous love story, close relationships are common, but in the long run, infidelity, rather than long-term stable monogamy, seems to be the rule.

makes something acceptable probably rests not on species identity but rather on some combination of cues (color, scent, floral morphology, and so forth). Cane and Sipes (2006) used the term "eclectic oligolecty" (Fig. 12-6) to describe cases such as this, in which members of the same species consistently collect pollen from two *unrelated* host-plant families. Alternatively, this could be viewed as evidence of a preference hierarchy.

Variation in host-plant range and biogeography

As described in Chapter 2, xeric regions of the world host a particularly species-rich solitary bee fauna. Bee species richness peaks in arid, semi-desert regions, such as the circum-Mediterranean region, southern Africa, western Australia, arid regions of South America, and the warm deserts of North America (Michener 1979, 2007). These are also regions of the world where the proportion of host-plant specialists appears to be greatest. Moldenke (1976) analyzed the patterns of oligolecty across North America and found that the highest proportion of oligolectic species was in warm deserts and Mediterranean scrub habits, while the lowest was in high-latitude tundra and muskeg habitats. But does this pattern hold more broadly? We combined Moldenke's original data from North America with more recently published faunal surveys in which the authors provided an estimate of the proportion of species that were oligolectic versus polylectic (Table 12-1). Our survey included 22 studies from sites in North America, Europe, and South America. The proportion of host-plant specialists varied widely, from 0% (in one Brazilian study) to over 60% in arid regions, such as the warm deserts of North America. Our survey supports Moldenke's original observation that oligolecty peaks in arid regions where rainfall is highly seasonal and unpredictable. Chapter 3 provides a more detailed discussion of why solitary, oligolectic bees are so superbly adapted to harsh arid environments.

WHY DO SO MANY BEES SPECIALIZE AT ALL?

On the surface, it seems that bees that fail to visit all the pollen hosts available are foregoing available resources. Wouldn't a bee that visited every available host plant collect more pollen and find the resources it needs more reliably? Below we review some of the physiological, ecological, and evolutionary hypotheses for why bees specialize on a narrow range of host plants.

Specialization reduces resource competition

A common theme among early students of bee biology was that inter-specific competition was the force driving the evolution of oligolecty (Linsley 1958, Michener 1954, Robertson 1899, 1914). Oligolecty, it was argued, was a form of resource partitioning that allowed many species to coexist during periods of pollen scarcity. While this is an attractive hypothesis, a number of authors have noted some flaws with it (Minckley et al. 1999, Sedivy et al. 2008, Williams

Table 12-1. Levels of host-plant specialization (oligolecty) in selected habitats for which data are available and reliable.

Locale	Habitat	Area	Citation	Number of total spp.	% oligolectic
North America	US hot deserts	368,424 km^2	Moldenke 1979	890	66
California, US	Colorado Desert	28,000 km^2	Moldenke 1979	482	62
North America	Mediterranean California	184,000 km^2	Moldenke 1979	830	51
Lower Don Region, Russia	Temperate steppe	?	Pesenko 1975	347	42
Chaparral Wildlife Management Area, TX	Tamaulipan scrub	61 km^2	JLN unpublished	200	37
North America	US Great Plains	2,543,000 km^2	Moldenke 1979	500	37
Travis Co., TX, US	Oak woodland and prairie (now mainly disturbed)	2,564 km^2	JLN unpublished	331	35
San Bernardino Valley, Sonora, MX	Mixed xeric habitats	16 km^2	Minckley 2008	383	30
Mid and NE US[1]	Mixed temperate habitats	740,000 km^2	Fowler 2016	450	30
Brackenridge Field Lab, Travis Co., TX, US	Oak woodland and prairie (habitat island within city)	0.2 km^2	JLN unpublished	234	29
Central-Northern Europe[2]	Mixed temperate habitats	480,000 km^2	Scheuchl & Willner 2016	820	26
Carlinville, IL, US[3]	Mixed temperate agricultural and grassland	811 km^2	Minckley 2008	288	24
North America	US mixed mesophytic forest	921,000 km^2	Moldenke 1979	450	24
North America	US southern mixed forest	829,000 km^2	Moldenke 1979	280	22
North America	NA tundra and muskeg	4,515,000 km^2	Moldenke 1979	84	15
New Jersey, US	Pine barrens	482 km^2	Winfree et al. 2007a	123	15

Black Rock Forest, New York, US	US mixed mesophytic forest	1,520 ha	Giles & Ascher 2006	144	15
New York, New York, US	Urban gardens	0.02 km²	Matteson et al. 2008	54	11
Ukiah, California, US	Urban garden	0.0001 km²	Frankie et al. 2009	68	10
Everglades N. P., Florida, US	Subtropical marshland	4,208 km²	Pascarella et al. 1999	66	6
Chã Grande Pernambuco, Brazil	Agreste: Mixed transitional vegetation, humid tropical forest and caatinga	71 km²	Milet-Pinheiro & Schlindwein 2008	79	5
Bahia, Brazil	Coastal Dunes	0.08 km²	Viana & Kleinert 2006	49	0

1. Combined areas of 13 states plus District of Columbia.

2. Combined areas of Germany, Austria, and Switzerland.

3. 10-mile radius around Carlinville, essentially a day's buggy ride.

Table 12-2. Oligolectic bees with specialization on *Larrea tridentata* in North America.
Modified from Hurd and Linsley (1975).

Species	Family
Ancylandrena larreae	Andrenidae
Calliopsis (Micronomadopsis) foleyi	Andrenidae
Calliopsis (Micronomadopsis) larreae	Andrenidae
Megandrena enceliae	Andrenidae
Perdita (Perdita) semicaerulea	Andrenidae
Perdita (Perdita: Sphaeralceae) flavipes	Andrenidae
Perdita (Perdita: Sphaeralceae) covilleae	Andrenidae
Perdita (Perdita: Sphaeralceae) eremica	Andrenidae
Perdita (Perdita: Sphaeralceae) punctulata	Andrenidae
Perdita (Perdita: Ventralis) lateralis	Andrenidae
Perdita (Perditella) larreae	Andrenidae
Perdita (Pseudomacrotera) turgiceps	Andrenidae
Habropoda pallida	Apidae
Colletes clypeonitens	Colletidae
Colletes covilleae	Colletidae
Colletes larreae	Colletidae
Hoplitis (Alcidamea) biscutellae	Megachilidae
Trachusa (Heteranthidium) larreae	Megachilidae
Hesperapis (Ambylapis) arida	Melittidae
Hesperapis (Ambylapis) larreae	Melittidae

2003). The most obvious flaw is that, rather than the tidy one-bee–one-plant relationships we might expect as the result of resource partitioning, we often find many specialist bees utilizing the same plant at the same place and at the same time along with numerous generalists (Minckley and Roulston 2006). A survey of bees visiting the flowers of creosote bush (*Larrea tridentata*) in the southwestern United States found more than 90 bees species, around 20% of which were *Larrea* specialists (Hurd and Linsley 1975). Although Hurd and Linsley (1975) found considerable seasonal and geographic variation, they commonly encountered five or more co-occurring specialist species, along with various generalists, at a given site. A follow-up survey found, despite methodological differences, that many of these patterns persisted 20 years later (Cane et al. 2005). A similar survey of bees associated with sunflowers (*Helianthus annuus*) reported 284 species collecting *Helianthus* pollen, 131 (46%) of which were considered to be sunflower (or at least Asteraceae) specialists (Hurd et al. 1980). Up to 18 specialists, along with 37 generalists, were collected on a single day at a single site. While these are extreme cases, coexistence of multiple specialists at the same plant at the same time and the same site is not rare (Sedivy et al. 2008). If these are cases of resource partitioning to avoid competition, they must involve some very subtle mechanisms (Williams 2003). More likely, as originally suggested by Lovell (1913), the main factors driving bee host specialization are those that directly benefit the bee rather than the avoidance of competition. In this view, bees specialize because it is good for them.

Specialist bees are more efficient foragers

Increased foraging efficiency has been suggested as a possible advantage of host-plant specialization (Michener 1954, Thorp 1969b). In this view, a specialist should be able to harvest a given amount of pollen more rapidly than a similarly sized generalist foraging on the same plant. While a number of studies have shown that specialists may forage more rapidly than generalists on the same flowers, this is not the same as showing increased efficiency. One can be fast and sloppy or slow and careful and end up with very similar harvesting rates. Strickler (1979) conducted one of the few studies that actually compared foraging efficiency of a specialist bee to a group of generalists on the same host plant. She compared the foraging efficiency of *Hoplitis anthocopoides* (Megachilidae), a specialist on *Echium* (Boraginaceae), to similarly sized generalists visiting the same host plant. She found that female *H. anthocopoides* removed about the same amount of pollen per visit as the generalists. However, *H. anthocopoides* spent less time on each flower and moved more quickly between flowers. Thus, per unit time, female *H. anthocopoides* were more efficient foragers. Many of the morphological specializations mentioned in Chapter 7 almost certainly enhance the foraging ability of specialists by such mechanisms as granting access to otherwise inaccessible pollen (via hooked hairs on the legs or mouthparts) or enhancing the ability to harvest pollen from nototribic flowers (via specialized brushes on the frons or clypeus). Similarly, scopal adaptations, like the very sparse simple scopae of bees associated with giant pollen or pollen with viscin threads, should reduce handling times and increase foraging rates. However, some specialists lack these morphological specializations, and any enhanced foraging ability must come from behavioral adaptations alone. Whether such behavioral adaptations are widespread remains to be seen.

Specialist bees choose host plants with the greatest nutritional value

Another possible mechanism driving specialization is that oligolectic bees are overcoming the "cost" of being a specialist by using especially nutritious pollen. Pollens vary in their protein, carbohydrates, sterols, lipids, and fatty acids (Nicolson 2011, Vanderplanck et al. 2017), and some of these differences affect bee development. Protein is the component of pollen we know most about (Chapter 7). The protein content of pollen ranges from 2% to 60% of total pollen content (Roulston et al. 2000), and the experimental addition of protein to larval bees results in faster development time and heavier adults (Levin and Haydak 1957, Roulston and Cane 2002). Thus, it could be that oligolectic bees specialize on the best pollen rewards and leave the low-quality pollen to the generalists. Roulston and colleagues (2000) noted that while some oligolectic bees specialize on pollen with very high protein content (Hydrophyllaceae), many others specialize on pollen with low protein content (e.g., many Cactaceae,

Malvaceae, and Onagraceae). While the roles of other components of pollen chemistry remain to be explored, it is clear that protein content alone is not a major factor driving the evolution of floral-host specificity.

Specialist bees choose superabundant resources

The *superabundant resource hypothesis*[3] was first proposed by John Lovell (1914), a contemporary of Robertson. Lovell noted that most oligolectic bees had strong preferences for plants that produced flowers in "multitudes"—the same superabundant resources that are also attractive to generalist bees. Lovell doubted Robertson's hypothesis that floral resources were ever in short supply, pointing out that the same hosts that oligolectic bees favored were those that honey bees also used as major nectar and pollen sources. Honey bees are famous for shopping the floral market for conspicuous and abundant flowers from which pollen and nectar can be quickly harvested. Lovell (1914) also argued that there would be particularly strong selection to specialize on superabundant resources in bees that were small-bodied and weak-flying. He cited the genus *Perdita* as an example in support of this phenomenon. In North America, sunflower (*Helianthus annuus*), goldenrod (*Solidago* spp.), creosote bush (*Larrea tridentata*), and willow (*Salix* spp.) host large numbers of oligolectic bees and offer unusually large amounts of pollen and nectar from open, easily accessible flowers (Minckley and Roulston 2006). *H. annuus* growing around a nesting aggregation of *Dieunomia triangulifera* (Halictidae) in eastern Kansas produced an estimated 137.2 kilograms of pollen per square kilometer, and in one year the pollen produced by one plant over the season was enough to provision up to 420 *D. triangulifera* brood cells (Minckley et al. 1994).

Specialist bees benefit by being in temporal synchrony with their host plants

Another possible factor that has been proposed to explain the evolution of oligolecty is the *host-bee synchronization hypothesis*. According to this idea, oligolectic bees gain an advantage over polylectic bees by maintaining close, temporal synchrony with their hosts, especially in harsh, highly seasonal, arid habitats (Linsley 1958, Thorp 1979). Synchronized foraging activity with the daily and seasonal availability of pollen of the floral host can strongly affect the individual fitness of oligolectic bees. Female *Dieunomia triangulifera* are narrow pollen specialists on sunflower (*Helianthus annuus*). Pollen-collecting females made faster foraging trips and more trips, and collected more pollen per trip if they foraged in the morning during peak pollen presentation of the plant. Emergence in three years was also timed so that nests were dug and females were provisioning cells at peak bloom. Minckley and colleagues (1994) estimated that females that initiated nests in synchrony with peak bloom foraged no more than 3 kilometers from the nest and could provision cells for up to 30 days. In contrast, females that were active too early or too late in the bloom would have had to forage up

---

[3] Also called the *predictable plethora hypothesis* (Wcislo and Cane 1996).

to 10 kilometers from their nests over a provisioning season of only 24 days. The cost of being poorly synchronized with host plant bloom was likely substantial, given that an average female in the best year produced only 6 offspring.

There is abundant evidence supporting the idea that the activities of oligolectic species indeed are highly synchronized with the flowering patterns of their hosts on both a seasonal and a diel basis. Many desert bees associated with desert annuals appear to track host-plant bloom using the same cues (soil humidity) used by their host plants (Chapter 3). The ability of bees to predict host-plant bloom when bloom is episodic may be a tremendous advantage in deserts and may underlie the large proportion of oligolectic bees found in xeric areas (Minckley et al. 2000).

A criticism of the synchrony hypothesis is that it is a consequence of oligolecty rather than a selection pressure (Robertson 1925, Wcislo and Cane 1996). A bee that is not active at the same time as the host is in bloom could not be oligolectic. Synchrony is usually determined from captures of bees at host flowers, which may lead to the false impression that phenological overlap is tight. More variation has been reported when synchrony is documented at nest sites or with pan traps (Linsley 1958, Minckley et al. 2013, Wuellner 1999).

Specialist bees have overcome plant defensive chemicals, leaving the generalists behind

Just as plant defensive chemistry is a major factor driving the evolution of many traditional insect herbivores, increasing evidence suggests that *pollen secondary chemistry* is an important factor shaping bee-plant interactions (Box 12-4). The pollen of several plant families, including Asteraceae, Ranunculaceae, Boraginaceae, Fabaceae, and Cucurbitaceae, has been shown experimentally to be toxic, or at least of poor nutritional value, for the larvae of generalist bees (Praz et al. 2008b, Sedivy et al. 2011, Haider et al. 2013, Brochu 2018). Paradoxically, these same plant families host a diverse array of pollen specialists.

The family Asteraceae has been particularly well-studied. While many bee species and even whole lineages of bees specialize on Asteraceae, aster pollen is avoided by generalists. Müller and Kuhlmann (2008) documented host plant preferences in 60 species of European *Colletes* (Colletidae) by analyzing their pollen loads. Fourteen species from four different lineages were found to be aster specialists, whereas closely related generalist species collected only minute quantities of aster pollen. Aster pollen appears to be great if you are an aster specialist, but something to be avoided at all costs if your larvae do not have the physiological mechanisms to digest it.

A number of studies have taken a more direct look at aster pollen and how it impacts larval development in generalist bees. In an early egg-transplant experiment involving artificial provision masses reconstituted from honey bee pellets, Levin and Haydak (1957) found that the generalist *Osmia lignaria* did best on its natural diet of *Hydrophyllum* (Boraginaceae) pollen, less well on several other pollens it did not normally collect, but failed to complete development on various Asteraceae pollen. An egg-transplant study with the generalist *Megachile rotundata* also found Asteraceae pollen to be unsuitable for this species. Larvae developed well, actually better than on the "natural" diet, when reared on mix of either *Medicago, Galega,* or *Melilotus* (all Fabaceae) pollen plus a

---

BOX 12-4: *COLLETES ANCHUSAE* AND *C. WOLFI* (COLLETIDAE; EUROPE)—RARE AND ENDANGERED SPECIALIST VISITORS TO *CYNOGLOTTIS* (BORAGINACEAE)

Plants in the family Boraginaceae seem to do a particularly good job of protecting their pollen from generalist bees. Many genera of Boraginaceae (such as *Anchusa*, *Cynoglossum*, *Cryptantha*, *Heliotropium*, and *Tiquilia*) hide their anthers within a narrow tubular corolla that is inaccessible to many generalist bees. Other genera, like *Cynoglottis*, have overlapping hairy scales that further restrict access to the reproductive parts of the flower. In some genera (such as *Echium*), there is evidence of chemical protection. The pollen of *E. vulgare* is rich in protein (Somerville and Nicol 2006), but it also has high levels of pyrrolizidine alkaloids (Boppré et al. 2005), feeding deterrents for many herbivores that are toxic to some bees during larval development (Praz et al. 2008b). Despite these morphological and chemical barriers, a surprising number of bee lineages have given rise to specialists on Boraginaceae. Many of these borage specialists have morphological modifications for harvesting pollen from these plants.

Colletes anchusae and its closest relative, *C. wolfi*, are examples of extremely narrow host-plant specialists with both morphological and behavioral adaptations for accessing the pollen of a single genus of boraginaceous host plant, *Cynoglottis*. *Cynoglottis* consists of just two species with a combined range extending from Italy eastward to Turkey and Syria and northward to Ukraine and Russia. These species occur on limestone soils in dry, rocky slopes, stony pastures, fallow fields, and open conifer forests; they flower from May through July. *C. wolfi* is known only from Italy, whereas *C. anchusae* is more widespread, occurring from eastern Europe to Turkey. Together, the distributions of these two species occur entirely within the range of *Cynoglottis*. Analysis of scopal loads by Müller and Kuhlmann (2003) at 15 different localities stretching from Italy to Turkey showed that females of these two *Colletes* species exclusively collect the pollen of *Cynoglottis*. Like many specialist visitors to boraginaceous host plants, *C. wolfi* and *C. anchusae* have morphological specializations that facilitate pollen harvesting: their foretarsi are shortened and bear stout, recurved setae. Female bees insert their forelegs into the narrow corolla tube to extract pollen from the anthers hidden within. Specialization does have its drawbacks. *C. wolfi* and *C. anchusae* are both listed as endangered in Europe (Nieto et al. 2014). *C. wolfi* is already in danger because of its very restricted range, and a recent study found that the habitats currently occupied by *C. anchusae* are likely to become greatly reduced and increasingly fragmented as a consequence of a warming climate (Roberts et al. 2011).

30% sucrose solution, but failed to survive past two days on a similar diet with *Haplopappus* (Asteraceae) pollen (Guirguis and Brindley 1974). Another study with two different generalist *Osmia*, *O. bicornis* and *O. cornuta*, found significant differences between the two species in what was suitable or unsuitable. Again, neither species successfully matured on an aster diet (Sedivy et al. 2011).

Exactly why so many bees, particularly generalists, seem to have problems with Asteraceae pollen is unclear. Asteraceae pollen tends to have low protein content (17.5–34.3%), but not radically lower content than that of many other pollen types successfully used by bees (Nicolson and Human 2013, Roulston et al. 2000). Williams (2003) suggested that the copious pollenkitt (Chapter 7) typical of the Asteraceae might be responsible, either directly as a toxin or indirectly by somehow interfering with digestion. Many asteroid taxa have been reported to contain feeding deterrents (pyrrolizidine alkaloids) in leaf tissue, which may also be present in pollen (Müller and Kuhlmann 2008 and references therein).

The observation that some pollen appears to be toxic or indigestible to generalists but perfectly acceptable to diverse lineages of specialists is exactly what one might expect to see in a classic arms race between plant and herbivore. The specialists, with physiological adaptations to overcome plant defensive chemicals, can exploit floral resources unavailable to pollen generalists. Generalists do, in many cases, take small amounts of aster pollen for their brood-cell provisions. But these generalists mix multiple pollen types together, so the detrimental effects of individual chemically defended pollens are likely diluted (Eckhardt et al. 2014).

Specialist bees are self-medicating to deter pathogens and parasites

Perhaps bees specialize on certain pollen types because, by doing so, they reap benefits that are unrelated to larval nutrition. Certain pollens may be more resistant to microbial attack or may act as a deterrent to pathogens and parasites. Bees that specialize on those pollen types may have healthier offspring or lower rates of parasitism (Giacomini et al. 2018). This idea has rarely been explored in solitary bees, but a fascinating study by Spear and colleagues (2016) found evidence that bees that specialize on aster pollen have lower rates of brood parasitism than generalist bees or bees that specialize on other, non-aster pollen.

Sapygid wasps (Sapygidae) are a significant source of brood mortality for stem-nesting bees (Chapter 11). Like many brood-parasitic bees, sapygid wasps enter the nest of a host, kill the host egg or larva, and lay an egg on the pollen provisions. Spear and colleagues (2016) studied six species of *Osmia* in Colorado, all of which were potential hosts for *Sapyga pumila*. Two *Osmia* were generalists, one was a legume specialist, and three were aster specialists. Sapygid parasitism was high (33%) in the generalist and legume-specialist *Osmia*. By contrast, none of the 72 nests of the aster specialists were parasitized by *Sapyga* wasps. *Sapyga* females were either avoiding the nests of the aster specialists, or their larvae were unable to develop on aster pollen. To assess how *Sapyga* larvae do on aster pollen, Spear and colleagues reared *Sapyga* larvae on three types of pollen: pollen collected by the generalist *Osmia*, pollen collected by the legume specialist, and pollen collected by the aster specialists. *Sapyga* larvae reared on aster pollen suffered much higher rates of mortality than those reared on the other pollen types. Aster pollen appears to be either toxic or of poor nutritional value for *Sapyga* larvae. This study demonstrates a clear advantage to aster specialization—lower brood parasitism. It seems entirely possible that specialization on chemically protected pollen may provide a defense against pathogenic microbes and parasites, and this might help explain why many oligolectic bee

BOX 12-5: ARE OLIGOLECTIC (SPECIALIST) BEES GOOD, BAD, OR UGLY POLLINATORS?

One might expect oligolectic (specialist) bees to be highly effective pollinators. Since oligolectic bees restrict their pollen foraging to one or a small set of closely related host-plant species, they come with built-in constancy and thus could potentially be better vectors of conspecific host-plant pollen than a bee that visits many unrelated host plants. However, as we saw in Box 12-1, pollinator effectiveness depends on the relative levels of pollen deposition and pollen removal. High pollen removal coupled with low pollen deposition could put a bee into the "ugly" category of pollinator. Alison Parker and colleagues conducted one of the few studies in which pollen removal and pollen deposition were quantified for specialist and generalist pollinators on the same host plant. Their results suggest that the highly efficient pollen removal that characterizes many specialist bees can be more of a detriment than a benefit to plant reproduction.

Parker and colleagues (2016) studied the pollinators of an early-spring ephemeral host plant in the eastern United States—*Claytonia virginica* (Montiaceae). *C. virginica* produces delicate, white to pink, radially symmetrical flowers that bloom in the forest understory before the deciduous trees leaf out in the spring. Visitors include a specialist bee (*Andrena erigeniae*, Andrenidae), small, polylectic generalist bees (*Hylaeus*, *Lasioglossum*, and *Ceratina*), and bombyliid flies (*Bombylius major*) (Motten et al. 1981). Through careful experiments, the researchers quantified the amount of pollen removed and deposited by each visitor in a single floral visit. A female of *A. erigeniae* (the specialist) removed on average 61% of the available pollen per visit, while bombyliid flies removed just 23.7% and the small generalist bees only 20%. Despite removing most of the available pollen in a single visit, *A. erigeniae* deposited an average of only 39.4 grains on the stigmas following a single visit, a mere 2.3% of the amount it removed. This is scarcely more than the 30 grains deposited by bombyliid flies (4.2% of the amount they removed) or the 15 grains deposited by small generalist bees (2.7% of the amount they removed). Because of its abundance, *A. erigeniae* is the dominant pollinator of *Claytonia* at the site studied by Parker and colleagues, but the whopping differences in pollen removal, coupled with the small differences in pollen deposition, suggest that *A. erigeniae* may often be a mixed blessing. Indeed, the authors conclude that in some contexts, visits by *A. erigeniae* may be "more detrimental than beneficial." Other studies have documented extremely high pollen removal by oligolectic bees (Schlindwein et al. 2005; see Box 7-2). It is entirely possible that in many cases oligolectic bees are more burden than benefit to their preferred host plants—yet another reason to view bee-plant interactions from the perspective of arms race rather than love story.

lineages often converge on the same subset of host-plant families (Asteraceae, Malvaceae, Cactaceae, Boraginaceae, Zygophyllaceae, and Lamiaceae).

Specialists are hardwired to recognize just one host plant

Another hypothesis that addresses why specialization is common in insects, including bees, is the neural limitation hypothesis (Bernays 2001). Any organism must consistently make decisions about where and when to forage, find a mate, build a nest, etc. This hypothesis proposes that processing complex information is costly, and specialists reduce this cost by focusing attention on one or a few hosts. Specialization is an evolutionary mechanism to reduce the complexity of heterogeneous environments. Specialization is further favored because reduced decision time decreases exposure to natural enemies. This hypothesis is difficult to test but is appealing because the costs of sensory processing are predicted to be highest for organisms that are small and short-lived, features shared by many solitary bees. A pollen-specialist bee reduces its set of cues to those that typically indicate that a host is appropriate and restricts its pollen foraging to species "known" to be good, while ignoring all others.

The hypotheses listed above are not necessarily mutually exclusive, and no single hypothesis accounts for all origins of oligolecty in bees. For example, the neural-limitation model provides a cogent explanation for the observation that host breadth changes little among clades of solitary bees and that transitions to polylecty are very uncommon. It is also consistent with the observation that host shifts among closely related species of bees occur rarely to distantly related hosts. However, it does not make strong predictions about what plants in a community might be favored over others. Furthermore, the hypotheses for the origin of oligolecty are not independent, and it is likely they sometimes work in concert to drive changes in host breadth.

Although it is not clear how they originally got together, bees and flowers have been intimate partners for more than 120 million years. For the most part, this has been a highly profitable union. Bees obviously benefit since their economics are based almost entirely on the floral resources provided by angiosperms. Bees are far more diverse and abundant than their closest living wasp relatives (the pemphredonine subtribe Ammoplanina), so the switch to pollen feeding has obviously worked out quite well for them. As for the angiosperms, bees have contributed greatly to their diversification, since the pollination systems of many modern angiosperms are entirely dependent on bees. Nonetheless, bees have not been an unalloyed blessing for flowers. Their high demand for floral resources is the prime trait that often makes them excellent pollinators, but that same high demand has its downside. Consume too much pollen and plant reproductive success is likely to suffer (Box 12-5). In a sense, bees are like an overly demanding lover: they are great to have around at times, but if left without boundaries, they can take over your life and ruin it. In that sense, many floral adaptations can be seen as evolutionary attempts to erect such boundaries.

CHAPTER 13

Solitary Bees and Agricultural Pollination

As we have seen in earlier chapters of this book, solitary bees are fascinating creatures. In this chapter, we take up the question of whether solitary bees are contributing to the pollination of our crops. From a purely anthropocentric view, are solitary bees useful? For many years, the credit for much of agricultural crop pollination has been given to a single, managed pollinator—*Apis mellifera* (the European honey bee). Honey bees can be easily transported into agricultural habitats, and colonies consist of thousands of workers, so an agricultural field or orchard can be flooded with foraging female bees. But are honey bees really doing all the hard work? Since the 1970s, a small group of bee biologists have been arguing that other bees, including wild solitary and social bees, are also important, but underappreciated, crop pollinators. Suzanne Batra, Peter Kevan, Jim Cane, Phil Torchio, Vince Tepedino, Stephen Buchmann, and others have long argued that we should more fully understand the contribution of wild bees to commercial agricultural pollination before giving all the credit to the honey bee. More recently, thanks to the work of a whole new generation of bee biologists, we are now able to more accurately quantify the contribution of wild bees to agricultural pollination, and we are finding that, in many crops, wild bees are contributing significantly to pollination. Wild solitary and social bees are abundant and diverse in many agroecosystems, and these wild pollinators are sometimes more effective pollinators (on a per-visit basis) than honey bees. In this chapter, we delve deeply into the challenging question of how one quantifies pollinator importance and the contribution of different pollinator species to crop pollination. As we will see below, an increasing body of evidence demonstrates that wild bees are making significant economic contributions to the production of our crops. Solitary bees are not only fascinating; they are a fundamental part of sustainable agriculture.

The human diet is highly dependent on pollination. Klein and colleagues (2007) reviewed the literature on global crop pollination and determined that, of the 124 major agricultural crops produced around the world, 87 (70%) benefit from animal (mostly bee) pollination. In terms of volume of production, the proportion of our diet that depends on pollination drops to 35% because the bulk of human caloric intake comes from wind-pollinated crops, such as wheat, corn, and rice. Crops that Klein and colleagues (2007) identified as globally important and highly dependent on insect pollination included fruits, such as apple (*Malus*), apricot (*Prunus*), avocado (*Persea*), blueberry (both highbush and lowbush, *Vaccinium*), caneberries (*Rubus*), cantaloupe (*Cucumis*), cherry (*Prunus*), cranberry (*Vaccinium*), durian (*Durio*), feijoa (*Feijoa*), kiwifruit (*Actinidia*), loquat (*Eriobotrya*), mango (*Mangifera*), naranjillo (*Solanum*), passion fruit (*Passiflora*), pawpaw (*Asimina*), peach (*Prunus*), pear (*Pyrus*), plum (*Prunus*), starfruit (*Averrhoa*), watermelon (*Citrullus*); nuts, including almond (*Prunus*), cashew (*Anacardium*), macadamia nuts (*Macadamia*); vegetables, such as pumpkins, squashes, and gourds (*Cucurbita*); spices and condiments, most importantly vanilla (*Vanilla*); stimulants, such as coffee (*Coffea*); and edible oils, such as turnip and rape seed (*Brassica*) (Klein et al. 2007, supplementary table 2).

The crops listed above are *directly dependent* (DD crops) on insect pollinators. However, there are other crops that are *indirectly dependent* (ID crops) on insect pollinators. These crops include a variety of field crops and vegetables in which seed production is dependent on pollinators. The most economically important of these crops is alfalfa, which is grown as animal forage from seeds produced through bee pollination (Box 13-1), but also vegetables, such as broccoli, carrots, celery, cauliflower, and onions. Even some biofuels have been shown to be dependent on insect pollinators (Gardiner et al. 2010). The fruits, vegetables, nuts, spices, and stimulants that are either directly or indirectly dependent on insect pollination provide the most tasty, interesting, and vitamin- and antioxidant-rich component of our diet.

BOX 13-1: *NOMIA MELANDERI* (HALICTIDAE)

Alfalfa (*Medicago sativa*, Fabaceae) was originally domesticated in Asia Minor over 6,000 years ago as food for horses. Since then, alfalfa cultivation has spread throughout Europe, the Americas, and Asia as food for domesticated animals, including dairy and beef cattle, horses, sheep, and goats. Today, alfalfa is the third-largest crop in the United States, with nearly 76 million tons produced per year (Pitts-Singer and Cane 2011). To grow alfalfa for animal consumption requires seed production, which is entirely dependent on insect pollination. Annual alfalfa seed production is done on a monumental scale. Approximately 80,000 tons of alfalfa seed are produced each year, and the United States alone accounts for 35,000 tons (43%) of alfalfa seed production. In the United States, most alfalfa production occurs in the western states, including California, Idaho, Nevada, Oregon, Utah, Washington, and Wyoming. Not all bees are effective alfalfa pollinators. Alfalfa, like many legumes, has a complex floral structure that requires "tripping" for successful pollination. When the flower is "tripped," the fused column of anthers and stigma snaps abruptly out from between the paired keel flowers and strikes the head or body of the bee, thus transferring pollen. Honey bees, and many other bees, do not like to be pummeled with the reproductive parts of flowers and they learn to access nectar without tripping the flower. Thus, they are ineffective alfalfa pollinators.

Other bees have no problem with the legume flower, and these bees make highly effective alfalfa pollinators.

Nomia melanderi (Halictidae), the alkali bee, is a solitary, ground-nesting, univoltine bee that has been used extensively for alfalfa pollination in the western United States. *N. melanderi* is the only ground-nesting, solitary bee that has ever been managed for crop pollination. And it has been managed on an impressive scale. In the Touchet Valley, near Walla Walla, Washington, an estimated 17 million female alkali bees emerge each spring in man-made nesting sites (bee beds) that are scattered across the valley (Cane 2008). Across the entire Touchet Valley, an estimated 20 hectares are dedicated to alkali bee nesting beds. Nesting beds range from 400 square meters to 61,000 square meters, and nest sites can have as many as 700 nests per square meter, although a typical density would be more like 300 nests per square meter. Nesting beds are constructed to take advantage of the natural soil preferences of the alkali bee. They are composed of silty, well-drained, compact soil, devoid of vegetation and covered with a crusty layer of salt (usually NaCl). Subsoil layers are moistened through a series of underground drainage pipes or trenches constructed around the nesting beds. The bees like the soil moist so that it can be easily excavated. Sites that dry out typically lose their nesting populations of *N. melanderi*. Newly constructed nest beds can be populated by transferring 1-cubic-foot soil cores (loaded with *N. melanderi* prepupae) from existing nesting beds.

Alkali bees are highly effective as alfalfa pollinators. They have a natural preference for legumes (like several other *Nomia* species) and are efficient at tripping the flowers and gathering the pollen that is deposited loosely on their bodies. They are estimated to pollinate 2,240 kilograms of clean seed per hectare (as opposed to 168 kg of clean seed when managed pollinators are not present). Nesting beds of alkali bees have been in existence for over 50 years at some sites in the Touchet Valley (Cane 2008).

While the alkali bee continues to be used in the Touchet Valley of Washington, its use has decreased across the western United States. It has been largely replaced by the alfalfa leaf-cutter bee (*Megachile rotundata*). Advantages of *M. rotundata* include the fact that they can be more easily transported from site to site (they nest in tubes), and they can be reared, screened for parasites, and sold commercially to alfalfa growers. *M. rotundata* is now cultivated on a huge scale in Canada and the United States for alfalfa pollination.

ESTIMATING THE ECONOMIC VALUE OF POLLINATION

But are these crops, and the contribution of bee pollination to their production, worth anything in cold, hard cash? Gallai and colleagues (2009) combined economic data from 2005 Food and Agriculture Organization (FAO) records for the top 100 agricultural crops grown around the world with an estimate of the dependence of each crop on animal-mediated pollination. The dependence ratio is the estimated proportion of crop production that would be lost in the absence of animal pollination. For a wind-pollinated crop (e.g., maize), the dependence ratio would equal zero, meaning that there would be no loss of production if all the world's pollinators were to disappear (what a horrible thought). Other crops are much more highly dependent on pollinators. Melon (*Cucumis*), for example, is nearly entirely dependent on pollination for fruit production. Gallai and colleagues (2009) gave this crop a dependence ratio of 0.95 (95% of melon production would disappear in the absence of pollinators). Fruits vary widely in their level of pollinator dependence. Citrus fruits, for example, are at the low end of the scale (with a dependence ratio of 0.05), whereas more pollinator-dependent fruits, such as apples, plums, peaches, and pears are at the high end (with a dependence ratio of 0.65). In some crops, varieties vary in their dependence on pollinators for fruit production (e.g., persimmons). And for some crops (e.g., cinnamon), we simply don't know how much fruit production is dependent on pollination. The dependence ratios used by different studies can be widely variable (as pointed out by Gallai et al. 2009), so these numbers should be interpreted cautiously.

Combining the total economic value of each crop with the pollinator dependence of that crop, Gallai and colleagues (2009) calculated that the total annual value of pollinators to the global agricultural economy was €153 billion ($170 billion), or 9.5% of the value of food produced for human consumption in 2005. Regions of the world varied widely in their dependence on insect pollinators. East Asia, for example, produces a large proportion of crops that are insect-pollinated, whereas South and Central America produce crops that are, typically, much less dependent on insect pollination. North America and Europe are similar in their reliance on insect pollination. Gallai and colleagues (2009) also estimated the vulnerability of different crop classes across different regions of the world to reduced pollination. Vulnerability was calculated as simply the economic value of insect-pollinated crops divided by the total economic value of all crops. Vulnerability of 10% would mean that, for a crop class or region of the world, 10% of agricultural production is vulnerable to the loss of pollinators. Stimulant crops are the most threatened by pollinator decline (vulnerability of 39%), followed by nuts (31%), fruits (23%), edible oils (16%), and vegetables (12%). African stimulant crops (primarily coffee) are highly vulnerable to pollinator loss, as is nut production in North Africa, Europe, and the United States. Fruit production is particularly vulnerable in Asia and Eastern Europe. The vulnerability of nuts to pollinator loss is particularly clear in crops like almonds, where the entire industry is dependent on the supply of migratory honey bee colonies in the United States. Intuitively, the Gallai and colleagues (2009) results seem entirely reasonable.

ESTIMATING THE ECONOMIC VALUE OF WILD BEES

Much of the credit for this $170-billion value of insect pollination is given to a single, managed pollinator species: *Apis mellifera*, the European honey bee. A widely quoted, but incorrect, statistic is that "one-third of the human diet is dependent on the honey bee." This statement derives from the (more or less) correct estimation that one-third of a healthy human diet is due to insect-pollinated crops (Klein et al. 2007) combined with the incorrect estimation that all agricultural pollination is carried out by the honey bee. As we will see below, the honey bee is not the only pollinator, and certainly not the only bee, contributing to pollination in agricultural habitats.

There are a surprising number of studies that have attempted to quantify the economic contribution of honey bees and/or wild bees to crop pollination; these studies have used varying methods and derived widely varying results. A total of eight studies have focused just on estimating the economic value of honey bees to the US economy (Calderone 2012). An additional study (Losey and Vaughan 2006) utilized similar data to estimate the economic contribution of wild bees. Economic estimates from these studies vary widely, with Burgett and colleagues (2004) estimating that honey bees contribute just $170.36 million to the US economy, while Levin (1983) estimated that the combined impact of pollination to the US economy was $49.2 billion. The most divergent estimates were determined using widely differing methods (reviewed in Calderone 2012). Burgett and colleagues (2004), for example, simply calculated the economic contribution of *Apis* to crop pollination as the sum of the fees paid to beekeepers. This method is guaranteed to ignore any contribution from wild pollinators (since they are typically not rented) as well as to underestimate the impact of the honey bee (since not all colonies that contribute to crop pollination are rented). Levin (1983) combined the gross value of directly dependent crops (DD crops, mentioned above), the gross value of indirectly dependent crops (ID crops, mentioned above), and 10% of the value of beef and dairy production resulting from the consumption of legume hay by cattle. Levin (1983) made no effort to distinguish between the contribution of honey bees and wild bees; all the economic benefits of pollination were attributed to the honey bee.

A number of studies, including Robinson and colleagues (1989), Morse and Calderone (2000), and, most recently, Calderone (2012), utilized a standardized approach in which the total value of directly dependent plus indirectly dependent crops was combined with the dependence ratio (the degree to which the crop is dependent on animal pollination) to calculate the income that would be lost in the absence of pollination. In addition, these studies attempted to account for the fact that not all of the pollination in every crop system results exclusively from the activity of honey bees. Thus, they added an additional variable in the equation that quantifies the proportion of crop pollination due to honey bees. The proportion of wild bee pollination can range from low (0.1 in apples) to nearly 100% (0.9 in strawberries), according to the tables in these studies.

The math underlying these calculations is fairly simple. Take the total annual economic value of each crop as determined by the US Department of Agriculture (USDA) or the Food and Agricultural Organization (FAO). Call this value V. Combine this value with the dependence of each crop on insect pollinators (the dependence ratio described above). Call this value D. Then add the estimated proportion of the insect pollinators that are honey bees (P) or wild bees (1-P).

For the economic value of honey bees, we have the following equation:

$$V_{hb} = \Sigma \, (V \times D \times P)$$

For the economic value of wild bees, we have the following equation:

$$V_{np} = \Sigma \, [V \times D \times (1\text{-}P)]$$

Table 13-1 summarizes the data for crops directly dependent on insect pollination, and Table 13-2 summarizes the data for crops that are indirectly dependent on insect pollination from the most recent study (Calderone 2012).

The estimates of who is actually doing the pollination (P and 1-P) are particularly problematic for many reasons. First, they are based primarily on expert opinion, with very little in the way of actual quantitative data. Most of these values trace back to anecdotal comments made in McGregor (1976), so these are essentially 40-year-old, back-of-the-envelope estimates. Second, they seem to be recycled over and over again. Robinson and colleagues (1989), Morse and Calderone (2000), Calderone (2012), and Losey and Vaughan (2006) all use (more or less) the same values. For the vast majority of crops, the assumption is that 90% of the pollination is done by honey bees and 10% by wild bees. Oddly enough, and for no clear reason, for strawberries, the ratio is reversed, with wild bees accounting for 90% of the pollination and honey bees accounting for the other 10% (Table 13-1). Squash and pumpkin fall into the same category (90% for the wild bees and 10% for the honey bees) because there is a single, ground-nesting, native pollinator (*Peponapis pruinosa*) that is likely responsible for much of squash and pumpkin pollination (Box 13-2). It is difficult to know where these numbers come from. Wild bees can often outnumber honey bees in crop systems (Winfree et al. 2008, Adamson et al. 2012), especially in early-spring fruit trees, such as apple, cherry, peach, and pear; and, as we will see below, wild bees have been shown to be more effective pollinators than honey bees on a per-visit basis in many crops. Finally, these values obviously do not allow for variation across farms in the abundance and importance of wild pollinators (Isaacs and Kirk 2010). For example, small, organic apple orchards in upstate New York are likely to have substantially more native pollinator diversity and abundance than large, industrial apple orchards in Washington State. Wild bees are likely contributing significantly more to the small organic orchard in New York than they are in the large, industrial orchard in Washington. In fact, the New York apple orchard may not even have any honey bee colonies; many New York growers have stopped using the honey bee for apple pollination, so the contribution of wild bees may range from 1.0 (in New York) to zero (in Washington). It is, admittedly, difficult to account for this geographic variation in the importance of wild bees given that we have so little information on wild bee abundance across agricultural habitats. The point is that we should be very cautious about the economic estimates derived from such a simple view of the world. As we will see below, there are ways of more accurately capturing the contribution of wild pollinators; they just require substantially more information on both the abundance and per-visit effectiveness of the wild and managed bees.

In the end, the economic calculations appear to be converging, more or less. Based on the most recent study (Calderone 2012), the value of insect pollination to the US economy is estimated to be $15.12 billion. The portion attributed to honey bees is $11.68 billion (77% of the total), and the portion attributed to wild bees is $3.44 billion (23% of the total). These numbers are close to those calculated by Losey and Vaughan (2006; $14.6 billion) and even come close

Table 13-1. Economic data for directly dependent (DD) crops for 2010.

Values of total economic value (V), dependence on insect pollinators (D), dependence on honey bee pollination (P) are explained in text (modified from Calderone 2012). OR = Oregon, CA = California.

Crop	Total economic value per annum (V; 1000s $)	Dependence on insect pollinators (D)	Economic value of insect pollination per annum (1000s $)	Dependence on honey bee pollination (P)	Economic value of honey bee pollination per annum (Vhb; 1000s $)	Dependence on wild bee pollination (1-P)	Economic value of wild bee pollination per annum (Vnp; 1000s $)
Berries							
blackberry	33,291	0.8	26,632.8	0.9	23,969.52	0.1	2,663.28
blueberry [cultivated]	593,407	1	593,407	0.9	534,066.3	0.1	59,340.7
blueberry [wild]	50,600	1	50,600	0.9	45,540	0.1	5,060
raspberry [black (OR)]	2,185	0.8	1,748	0.9	1,573.2	0.1	174.8
raspberry [red]	56,426	0.8	45,140.8	0.9	40,626.72	0.1	4,514.08
raspberry [all (CA)]	200,288	0.8	160,230.4	0.9	144,207.36	0.1	16,023.04
cranberry	316,486	1	316,486	0.9	284,837.4	0.1	31,648.6
strawberry	2,245,319	0.2	449,063.8	0.1	44,906.38	0.9	404,157.42
boysenberry	1,834	0.8	1,467.2	0.9	1,320.48	0.1	146.72
grapefruit	285,993	0.8	228,794.4	0.9	205,914.96	0.1	22,879.44
lemon	380,634	0.2	76,126.8	0.1	7,612.68	0.9	68,514.12
orange	1,934,982	0.3	580,494.6	0.9	522,445.14	0.1	58,049.46
tangelo	6,780	0.4	2,712	0.9	2,440.8	0.1	271.2
tangerine [& mandarin]	276,135	0.5	138,067.5	0.9	124,260.75	0.1	13,806.75
Cucurbits							
muskmelon [cantaloupe]	314,379	0.8	251,503.2	0.9	226,352.88	0.1	25,150.32
cucumber [fresh]	193,643	0.9	174,278.7	0.9	156,850.83	0.1	17,427.87
cucumber [pickled]	184,525	0.9	166,072.5	0.9	149,465.25	0.1	16,607.25
muskmelon [honeydew]	49,608	0.8	39,686.4	0.9	35,717.76	0.1	3,968.64
pumpkin	116,539	0.9	104,885.1	0.1	10,488.51	0.9	94,396.59
squash	203,592	0.9	183,232.8	0.1	18,323.28	0.9	164,909.52
watermelon	492,035	0.7	344,424.5	0.9	309,982.05	0.1	34,442.45

	Total value		Total value of pollination		Value of honey bees		Value of wild bees
Grapes							
grape	3,626,760	0.1	362,676	0.1	36,267.6	0.9	326,408.4
Legumes							
peanut	901,347	0.1	90,134.7	0.2	18,026.94	0.8	72,107.76
soybean	38,915,328	0.1	3,891,532.8	0.5	1,945,766.4	0.5	1,945,766.4
Nuts and seeds							
almond	2,838,500	1	2,838,500	1	2,838,500	0	0
Macadamia nut	30,000	0.9	27,000	0.9	24,300	0.1	2,700
canola	486,865	0.5	243,432.5	0.9	219,089.25	0.1	24,343.25
cotton [seed]	1,003,861	0.2	200,772.2	0.8	160,617.76	0.2	40,154.44
rapeseed	975	1	975	0.9	877.5	0.1	97.5
sunflower	582,448	1	582,448	0.9	524,203.2	0.1	58,244.8
Tree fruits							
apple	2,220,817	1	2,220,817	0.9	1,998,735.3	0.1	222,081.7
apricot	47,486	0.7	33,2402	0.8	26,592.16	0.2	6,648.04
avocado	322,108	1	322,108	0.9	289,897.2	0.1	32,210.8
cherry [sweet]	721,154	0.9	649,038.6	0.9	584,134.74	0.1	64,903.86
cherry [tart]	40,516	0.9	36,464.4	0.9	32,817.96	0.1	3,646.44
kiwifruit	24,961	0.9	22,464.9	0.9	20,213.41	0.1	2,246.49
nectarine	129,075	0.6	77,445	0.3	61,956	0.2	15,489
olive	113,360	0.1	11,336	0.1	1,133.6	0.9	10,202.4
peach	614,908	0.6	368,944.8	0.8	295,155.84	0.2	73,788.96
pear	381,695	0.7	267,186.5	0.9	240,467.85	0.1	26,718.65
plum	78,422	0.7	54,895.4	0.9	49,405.86	0.1	5,489.54
prune	149,860	0.7	104,902	0.9	94,411.8	0.1	10,490.2
prune and plum	4,915	0.7	3,440.5	0.9	3,096.45	0.1	344.05
TOTALS	$61,174,042.0		$16,344,809.0		$12,356,574.07		$3,988,234.93

Table 13-2. Economic data for indirectly dependent (ID) crops for 2010.

Values of total economic value (V), dependence on insect pollinators (D), dependence on honey bee pollination (P) are explained in text. Modified from Calderone (2012).

Crop	Total economic value per annum (V; 1000s $)	Dependence on insect pollinators (D)	Economic value of insect pollination per annum (1000s $)	Dependence on honey bee pollination (P)	Economic value of honey bee pollination per annum (Vhb; 1000s $)	Dependence on wild bee pollination (1-P)	Economic value of wild bee pollination per annum (Vnp; 1000s $)
Field crops							
alfalfa	7,519,469	1	7,519,469	0.33	2,507,286.13	0.67	5,012,182.87
cotton	7,317,704	0.2	1,463,540.8	0.8	1,170,832.64	0.2	292,708.16
Vegetables							
asparagus	90,777	1	90,777	0.9	81,699.3	0.1	9,077.7
broccoli	648,886	1	648,886	0.9	583,997.4	0.1	64,888.6
carrot	597,362	1	597,362	0.9	537,625.8	0.1	59,736.2
carrot	29,608	1	29,608	0.9	26,647.2	0.1	2,960.8
cauliflower	247,456	1	247,456	0.9	222,710.4	0.1	24,745.6
celery	398,854	1	398,854	0.8	319,083.2	0.2	79,770.8
onion	1,455,103	1	1,455,103	0.9	1,309,592.7	0.1	145,510.3
sugarbeet	1,968,389	0.1	196,838.9	0.2	39,367.78	0.8	157,471.12
	Total value		*Total value of pollination*		*Value of honey bees*		*Value of wild bees*
TOTALS	$20,273,608		$12,647,894.70		$6,798,842.55		$5,849,052.15

BOX 13-2: *PEPONAPIS PRUINOSA* (APIDAE)

Plants in the genus *Cucurbita* (pumpkins, squashes, and gourds) are high-value, obligately insect-pollinated crops that have been cultivated by humans for over 10,000 years. The economic value of pumpkins, squashes, and gourds was estimated by Gallai and colleagues (2009) to be €3.7 billion globally in 2005. Major production occurs in Asia, Europe, North Africa, and North America. Pumpkin and squash production in the United States alone was estimated by Calderone (2012) to be worth $320 million in 2010.

The history of squash cultivation extends back ~10,000 years in the Americas and involves at least six (and possibly seven) unique domestication events, most of which occurred before the arrival of Europeans. Squash domestication by Meso-American societies occurred in the same time period (~10,000 BP) and geographic range (northern and central Mexico) as the domestication of other important crops, including maize (*Zea mays*), peppers (*Capsicum annuum*), common beans (*Phaseolus vulgaris*), and cotton (*Gossypium hirsutum*) (Lentz et al. 2008). Wild species of *Cucurbita* would have been an attractive food source for early New World hunters and gatherers because of the relatively large, conspicuous fruits that can be easily gathered during the dry and/or winter months. The seeds are nutritious and palatable when extracted from the bitter and sometimes highly unpalatable fruit. When dried, gourds served as important storage vessels for water and food.

Two solitary bee species (*Peponapis pruinosa* and *Xenoglossa strenua*) have greatly expanded their geographic ranges in association with human cultivation of *Cucurbita* (especially domesticated varieties, such as *C. pepo*). *P. pruinosa* and *X. strenua* now occur across much of continental North America from Central Mexico to the Canadian province of Ontario westward to Oregon. These two species of squash bees likely spread along with the human cultivation of *Cucurbita*, providing an amazing example of a pollinator mutualism allowing for both domestication and geographic range expansion (López-Uribe et al. 2016).

P. pruinosa are highly effective pollinators of squash, pumpkins, and gourds (Artz et al. 2011, Artz and Nault 2011, Julier and Roulston 2009, Michelbacher et al. 1968, Shuler et al. 2005). Indeed, they are such good pollinators that Michelbacher and colleagues (1968) recommended introducing them into Europe for cucurbit pollination. In surveys of pumpkin pollinators in New York (Artz et al. 2011, Artz and Nault 2011) and Virginia (including neighboring regions of Maryland and West Virginia; Julier and Roulston 2009, Shuler et al. 2005), *Apis mellifera*, *Bombus impatiens*, and *P. pruinosa* comprised the vast majority (>90%) of floral visitors. In New York, *Peponapis* comprised roughly 45% of the visits to pumpkins, and in Virginia, *Peponapis* was present on 23 of 25 farms and the most abundant visitor to pumpkin flowers in 15 of the 25 farms. Over the 25 farms studied in Virginia, *P. pruinosa* visited flowers 3 times more frequently than honey bees. Based on studies by Tepedino (1981) and Artz and Nault (2011), female *P. pruinosa* are equal to honey bees in per-visit pollen deposition, but because they fly so much earlier in the day, they are likely responsible for much of the pollination at sites where they are abundant. When *P. pruinosa* is present, honey bees may be essentially irrelevant for pollination (Petersen et al. 2013).

Canada, and Bermuda combined (€14.4 billion or $15.89 billion; Table 3 in Gallai et al. [2009]). The Gallai and colleagues (2009) numbers seem a bit low compared to the Calderone (2012) values, given that Gallai and colleagues combine US and Canadian agricultural production into a single estimate for North America. In any case, the good news is that the wild bees, many of which are solitary, are at least being included in the equation. We suspect that the models used in these studies may be substantially underestimating the importance of wild bees to agricultural pollination. As pointed out above, the values of P (and therefore 1-P) are not based on careful studies of the actual contribution of wild bees to crop pollination. Only very recently have the economic contributions of wild bees been carefully quantified, as we will see below.

A FRAMEWORK FOR MEASURING THE ACTUAL IMPORTANCE OF WILD BEES IN AGRICULTURAL POLLINATION

Prior to the colony collapse disorder (CCD) crisis of 2007, when extremely high rates of overwintering honey bee mortality were reported across the country, there was little impetus to quantify the impact of wild pollinators vis-à-vis the honey bee. Honey bees were cheap and available crop pollinators that could be moved conveniently into any agricultural setting. However, with a steady decline in the number of managed honey bee colonies in the United States over the past fifty years and annual losses of honey bee colonies approaching 40% (vanEngelsdorp and Meixner 2010), there is now a substantial economic incentive to understand just how important the wild bees are for crop pollination. These wild pollinators provide us with "insurance" against continuing declines in honey bees (Winfree et al. 2007b), so this is really an issue of national and global food security.

A number of studies (Aizen and Harder 2009, Breeze et al. 2011, Calderone 2012) have pointed out that the steady decline in managed honey bee colonies in the United States (and Europe) has coincided with a steady *increase* in the production of pollinator-dependent crops in these same regions (Fig. 13-1). The most likely explanation for why we have not seen a precipitous decline in crop pollination is that wild bees are already filling the gap; we just did not have the quantitative data to show it.

But how can we actually estimate the relative contribution of wild and domesticated bees to agricultural pollination? An obvious experimental approach would be to remove one of those pollinators (say the honey bee) and see how

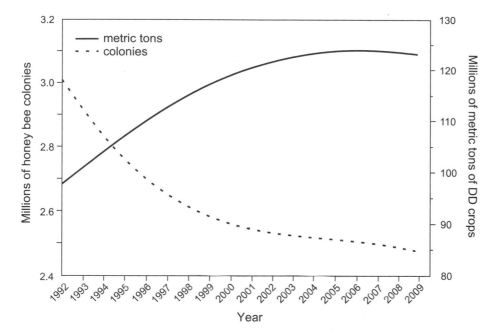

Figure 13-1. Relationship between the number of managed honey bee colonies in the United States and the total production (in metric tons) of directly dependent (DD) crops over the period 1992–2009. Note that the gap between number of honey bee colonies and the total production of directly dependent crops is likely filled by the contribution of wild bees to crop production (redrawn from Calderone 2012, Fig. S4).

crop pollination is impacted. Unfortunately, this simple and informative experiment can't actually be conducted for most crops because honey bees are pretty much everywhere. So one needs to be more creative. Fortunately, pollination biologists think a lot about this kind of thing, and they have come up with a wonderfully simple way to quantify *pollinator importance* (see Ne'eman et al. 2009 for a review).

In its simplest form, one can think of pollinator importance as having two components. The first component is how frequently any particular pollinator lands on a flower and contacts the reproductive parts of the flower. This is ideally measured as visitation rate but is often approximated by abundance. The logic is that an abundant pollinator will likely be more important, overall, than a rare pollinator. A second component of the pollinator importance equation is per-visit effectiveness. How effective is each species at depositing pollen grains, setting seeds, or effecting pollination? This is a much more difficult thing to quantify, because it involves carefully designed field experiments in which one presents a previously unvisited ("virgin") flower (usually placed on the end of a stick) to a foraging bee. The bee is allowed to visit the flower and then the flower is taken back to the lab and examined for pollen deposition, pollen tube growth, and seed or fruit set. Such experiments are extremely time-consuming, and when there are a lot of species visiting the crop, it can be impossible to obtain a sufficient amount of data on them all. One shortcut has been to lump pollinator species into "functional groups" (i.e., groups of species that are likely to have the same level of per-visit effectiveness based on their body size, mode of visitation, social behavior, phylogenetic affinities, or overall biology). Data can then be collected on one species (presumably the most common) and applied to all the species in that functional group. The exact species composition of each "functional group" needs to be carefully considered when adopting this approach.

The "pollinator importance" equation, in its simplest form, looks like this:

$$pollinator\ importance = visitation\ rate \times per\text{-}visit\ effectiveness$$

Note that a rare visitor may have equal importance to a common visitor if the rare visitor is, per visit, much more effective as a pollinator. Note also that some authors (Vázquez et al. 2005) have argued that visitation rate is all you really need to know because variance in per-visit effectiveness is so much lower than variance in visitation rate. In some cases, this is indeed the approach taken because gathering per-visit effectiveness data can be extremely time-consuming. The beauty of the pollinator-importance equation is that it allows one to quantify the relative importance of two (or more) species visiting a single crop. In addition, one can quantify the importance of an entire pollinator community vis-à-vis another pollinator community. The pollinator-importance equation provides the best method for empirically determining the importance of honey bees versus wild bees, as the following examples illustrate.

WILD BEES AS CROP POLLINATORS—CASE STUDIES FROM ECONOMICALLY IMPORTANT CROPS

Apple (*Malus domestica*, Rosaceae)

Apples are an enormously valuable, heavily pollinator-dependent crop. Globally, apple production is valued at over €24 billion, with much of the production occurring in China (€8 billion), the European Union (€6.3 billion), and North America (€2 billion; Gallai et al. 2009). US apple production is valued at $3.1 billion annually based on USDA-NASS (http://www.nass.usda.gov/) data from 2012, and New York apple production alone was worth an estimated $250 million in 2012 (http://www.agriculture.ny.gov/agfacts.html). Studies in North America have documented a diverse fauna of wild bees (Blitzer et al. 2016, Gardner and Ascher 2006, Mallinger and Gratton 2014, Martins et al. 2015, Russo et al. 2015, Watson et al. 2011), with over 120 wild bee species collected in or around flowering apple trees in central New York alone (Russo et al. 2015). The wild bee fauna of apples is dominated by solitary, ground-nesting bees in the genera *Andrena* (Fig. 13-2) and *Colletes*, solitary stem- and cavity-nesting bees in the genus *Osmia*, a diversity of ground-nesting solitary and weakly social halictid bees (*Lasioglossum, Halictus, Agapostemon, Augochlora*), and social bumble bees (*Bombus*).

A number of studies indicate that native bees may be better apple pollinators on a per-visit basis than honey bees. Native bee species have been shown to carry more pollen (Kendall and Solomon 1973), to carry more compatible fruit

Figure 13-2. Habitus drawing of female *Andrena crataegi* (Andrenidae: Andreninae), an important wild pollinator of apple in the eastern United States. Original artwork by Frances Fawcett.

pollen (Kendall 1973), to transfer fruit pollen at a higher rate (Park et al. 2016, Thomson and Goodell 2001), to yield higher fruit set per visit (Vicens and Bosch 2000a), to forage in more inclement weather (Vicens and Bosch 2000b), and to show a stronger preference for *Malus* flowers than honey bees (Johnson 1984, Kendall and Solomon 1973, Vicens and Bosch 2000a).

Wild bee abundance and species richness (diversity) have been shown to positively impact seed set in apples (Blitzer et al. 2016, Mallinger and Gratton 2014, Martins et al. 2015). Based on work in central New York apple orchards, honey bee abundance did not strongly correlate with seed set (Blitzer et al. 2016).

Species richness and the abundance of wild bees is clearly impacted by both the landscape surrounding apple orchards and the level of pesticide use. Increased natural habitat and decreased pesticide (especially fungicide) use clearly benefit the richness and diversity of native bee species (Park et al. 2015). Apple growers in the northeastern United States and eastern Canada are relying more and more heavily on the diversity of wild, mostly solitary, bees in and around their orchards. Most smaller growers and many larger apple producers are no longer renting honey bee colonies for apple pollination. The same cannot be said for large apple orchards in China, where migrant farm workers are now required to hand-pollinate apple trees because of the loss of natural pollinators, presumably due to heavy pesticide use (Partap and Ya 2012).

Apple production is almost certainly supported to a large extent by wild bee pollination. Russo and colleagues (2015) combined abundance data with a proxy for per-visit effectiveness (the proportion of apple pollen carried by various species and species groups) to calculate the combined importance of wild bees vis-à-vis honey bees in central New York apple orchards. Her estimates indicate that wild bees combined are at least as important as honey bees in apple pollination. At smaller orchards and orchards with more natural habitat, wild bees contribute substantially more to apple pollination than honey bees. Based on recent work on the role of wild bees in apple pollination, the estimate used by Calderone (2012) and others (see above) that wild bees contribute just 10% to apple pollination is likely a significant underestimate. Studies in Ontario, New York, Wisconsin, Pennsylvania, and North Carolina have all shown that wild bees are likely equal to or more important than honey bees.

Watermelon (*Citrullus lanatus*, Cucurbitaceae)

Watermelon is an important, pollinator-dependent crop grown around the world. Global production totaled €15 billion in 2005 (Gallai et al. 2009), with major production in East Asia (China, Japan, North and South Korea, and Mongolia). North American watermelon production totaled €270 million in 2005 (Gallai et al. 2009). Watermelons are a wonderful crop from the perspective of pollination research. We know, for example, how many pollen grains are needed to produce a marketable watermelon (500–1,000), and one can quantify the per-visit deposition rate for a variety of bees. Kremen and colleagues (2002) and Winfree and colleagues (2007b, 2008) used this approach to quantify the contribution of wild bees to watermelon pollination in California and the eastern United States.

The wild bee fauna visiting watermelons is not as diverse as the fauna visiting apples. A total of 39 wild bees were reported visiting watermelon in Yolo County, California (Kremen et al. 2002). Kremen and colleagues (2002) calculated visitation rate (based on sampling along a 50-m transect) and per-visit pollen deposition for wild bees in the genera *Melissodes*, *Peponapis*, and *Bombus* (Apidae), *Lasioglossum*, *Halictus* (Halictidae), and *Hylaeus* (Colletidae) as well as managed honey bees. Because species could not be identified reliably on the wing, per-visit pollen deposition was estimated at the generic level. These studies were conducted at organic and conventional farms and at varying distances from natural habitat. At organic farms located near natural habitats, native bees alone provided enough pollen deposition to exceed the 1,000 grains per fruit limit for successful pollination. However, at organic sites located far from natural habitats and at conventional sites, the native bee fauna was insufficient for complete watermelon pollination. This demonstrates clearly how the impact of wild bees can vary according to the habitat context and farm management. Large, heavily managed watermelon fields located far from natural habitat will need honey bees for successful fruit set, whereas smaller, organic orchards located close to natural habitat will not.

A similar study conducted in New Jersey and Pennsylvania by Winfree and colleagues (2007a) indicates that the wild bee fauna is even more important for watermelon pollination in the eastern United States. Surveys of watermelon fields at 23 farms revealed a total of 46 native, wild bees visiting watermelon plants. The fauna is dominated by bumble bees (*Bombus*), large (*Xylocopa*) and small (*Ceratina*) carpenter bees, other large apid bees (*Peponapis, Melissodes, Ptilothrix*), halictid bees (*Augochlora, Augochlorella, Augochloropsis, Agapostemon, Lasioglossum,* and *Halictus*), andrenid bees (*Andrena, Calliopsis*), leaf-cutter bees (*Megachile*), and one colletid (*Hylaeus*). Using visual surveys of bee visitation along 50-meter transects, Winfree and colleagues (2007b) quantified visitation rate and measured per-visit effect as pollen deposited per a single visit. Because of the difficulty of identifying bees on the wing, bee species were lumped into "functional groups," including bumble bees, other large bees, green bees, and small bees, for transect sampling and for quantifying per-visit effectiveness. Pollen deposition was high at all farms, with pollen deposition exceeding the threshold of 1,000 pollen grains per stigma over the course of a single day. Interestingly, while there was a positive correlation between wild bee visitation (visits per flower per day) and pollen deposition, there was no significant correlation between honey bee visitation and pollen deposition. Similar results have been obtained across a wide range of crops (Garibaldi et al. 2011). On a per-visit basis, honey bees were less effective than both bumble bees and other large bees, but more effective than green bees and small bees. Combining visitation rate data and per-visit pollen deposition data, Winfree and colleagues (2007b) showed that native bees as a group were responsible for 62% of pollen deposition in watermelon fields. Native bees alone provided full pollination in 91% of the fields analyzed, meaning honey bees were redundant in a large number of fields. In fact, most farms could have been sufficiently pollinated by just one of the native bee functional groups (e.g., bumble bees alone or all small bees combined). Overall, wild bees provide the bulk of the pollination services in these small, eastern, mixed vegetable farms. Vegetable growers in the eastern United States are therefore likely to be buffered against the ups and downs of honey bee availability. The same cannot be said about vegetable growers in the Central Valley of California or other areas of the country where agricultural intensification precludes reliance on wild pollinators.

Blueberry and cranberry (*Vaccinium* species, Ericaceae)

Various species in the genus *Vaccinium* are economically important, obligately insect-pollinated plants. The two major commercially produced fruits are blueberry (including highbush blueberry [*Vaccinium corymbosum*], lowbush blueberry [*V. angustifolium*], rabbiteye blueberry [*V. ashei*], and common blueberry [*V. myrtillus*]) and cranberry (including American cranberry [*V. macrocarpon*] and common cranberry [*V. oxycoccus*]). The annual economic value of commercially produced cranberries and blueberries was €556 million in 2005 (Gallai et al. 2009). The majority of production (€485 million) is in eastern North America (Canada and the US). *Vaccinium* species require acidic soils, and they are generally found in heath, bog, and woodland habitats dominated by oaks and pines.

Vaccinium, like most ericaceous flowers, has a unique floral morphology. The flowers dangle downward from the plant, and the petals form a bulbous corolla around the reproductive parts of the flower. The corolla has a narrow opening through which the stigma extends. The anthers are poricidal and enclosed within the corolla. To remove pollen, a visitor must either "buzz" the flower by rapidly contracting its thoracic flight musculature or stroke the anthers to extract the pollen. Pollen is ejected downward and out the narrow opening in the corolla onto the body of the visitor. Some visitors (notably honey bees and carpenter bees) rob the flowers by cutting slits in the tubular corolla and extracting nectar without actually contacting either the anthers or the stigma. This type of "dishonest" visitation does not result in the transfer of pollen to the receptive stigma.

Blueberries have a particularly diverse wild bee fauna. Tuell and colleagues (2009) studied the bee fauna in and around fields of highbush blueberries in Michigan using a combination of pan trapping and net collecting. In surveys of 15 farms over three years, they found a total of 166 wild bee species representing 30 genera and 5 families. Of these bee species, 112 were active during the period of blueberry flowering. The majority of the fauna includes ground-nesting, solitary bees in the families Halictidae and Andrenidae. *Andrena carolina*, a specialist on Ericaceae, was the most common wild bee collected during blueberry bloom. The southern blueberry bee, *Habropoda laboriosa*, is an important pollinator of rabbiteye blueberry in the southern United States (Box 13-3).

BOX 13-3: *HABROPODA LABORIOSA* (APIDAE)

Habropoda laboriosa (Apidae: Anthophorini), also known as the southeastern blueberry bee, is an economically important wild pollinator of rabbiteye (*Vaccinium ashei*) and highbush blueberry (*V. corymbosum*) in the southeastern United States (Cane and Payne 1988). This solitary, univoltine, ground-nesting bee ranges from New Jersey and Illinois southward to Mississippi and Florida. Adults are active from early spring (January–April) in the southeastern United States; some populations in Florida are active as early as November. They prefer sandy, well-drained soils (like other species of *Habropoda*). Cane (1994a) found females nesting in soils composed of between 84% and 97% sand in sites in Alabama. Females construct rather deep nests, with cells between 30 and 70 centimeters from the surface.

Based on analysis of floral-visitation data, analysis of pollen in scopal loads, and analysis of pollen-provision masses, *H. laboriosa* is a narrow host-plant specialist on *Vaccinium* (Ericaceae). Ninety-four percent of pollen in scopal loads taken from *H. laboriosa* visiting *Vaccinium* flowers consisted of *Vaccinium* pollen tetrads, and 93% of pollen in pollen-provision masses excavated from recently provisioned cells consisted of *Vaccinium* pollen (Cane 1994a). Other species of *Habropoda* (e.g., *H. pallida*) are also known to have narrow host-plant preferences.

When foraging for pollen, female *H. laboriosa* grasp the inverted flowers with their fore- and midlegs and buzz the poricidal anthers by vibrating their thoracic flight musculature. Pollen is ejected rapidly onto the face (including the labrum and clypeus) and mouthparts (including the mandibles and proboscis) (Fig. 13-3). This region of the head also comes in contact with the protruding stigma, thus effecting pollination when multiple flowers are visited. Pollen is eventually groomed from the head to the scopa on the hindleg. *H. laboriosa* is an extremely efficient harvester of *Vaccinium* pollen—females can remove up to 70% of the pollen from virgin flowers on a single visit (Cane and Payne 1988). *Bombus* species extract *Vaccinium* pollen by buzzing, but they are less efficient than *H. laboriosa*. Cane and Payne (1988) compared the "handling time" (the time spent per flower per visit) of *H. laboriosa* and several species of bumble bees that were present at the same site. Handling time differed dramatically between *H. laboriosa* and the bumble bees. *H. laboriosa* spent just 2.6 seconds (on average) per floral visit, whereas bumble bees spent an average of 7.4 seconds per visit. Handling times were also much more uniform for the oligolectic *H. laboriosa* than for the polylectic bumble bees, presumably because *H. laboriosa* are hardwired for *Vaccinium* visitation, whereas naïve bumble bees need to learn how to manipulate the *Vaccinium* flowers. Because travel times between flowers were the same for *H. laboriosa* and bumble bees, *H. laboriosa* is both faster at harvesting pollen and also likely more effective as a *Vaccinium* pollinator. Foraging bees (including *H. laboriosa*, *Bombus*, and *Xylocopa*) spent more time on previously unvisited (virgin) flowers than flowers that had been visited earlier, indicating that bees "know" when the floral resources are abundant and when they are in limited supply. *H. laboriosa* are such effective pollinators of rabbiteye blueberry in the southeastern United States that Cane (1994b) estimated that each female is worth between $18 and $20 to a commercial blueberry grower.

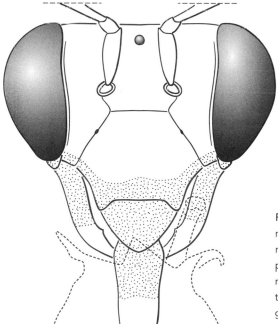

Figure 13-3. Head of female *Habropoda laboriosa* with mouthparts extended. The shaded area indicates the region of the head that accumulates significant *Vaccinium* pollen during a floral visit. Dashed lines indicate the corolla and protruding stigma of the *Vaccinium* flower. Note that the stigmatic surface is in contact with shaded regions of the bee's head, indicating likely stigmatic pollen deposition (redrawn from Cane and Payne 1988, Fig. 3).

Wild blueberry visitors are highly effective pollinators relative to honey bees. Javorek and colleagues (2002) quantified the pollinator effectiveness of wild bees (*Bombus*, *Megachile*, *Andrena*, and *Halictus*) versus honey bees in lowbush blueberry fields in Nova Scotia. They measured foraging rate (flowers visited per minute), pollination percentage (percentage of visits to unvisited [virgin] flowers that yielded fruits), and pollen deposition (pollen deposited per single visit to previously unvisited [virgin] flowers). They then combined the foraging rate and the pollination percentage values to calculate the relative effectiveness of different pollinator species groups. Bumble bees were the fastest visitors, visiting on average 12 flowers per minute while pollen foraging. Honey bees, *Andrena* (presumably mostly *A. carolina*), *Megachile rotundata*, and *Halictus* species were more in the range of 6 to 8 visits per minute. In terms of the efficiency of fruit set, *Andrena*, *Bombus*, and *Halictus* were all substantially better than honey bees. Wild, pollen-foraging bees pollinated between 85% and 95% of the flowers they visited. Honey bees and *Megachile*, both of which were primarily nectar visitors, pollinated just 15–25% of the flowers they visited. Combining the visitation rates with the pollination percentage, Javorek and colleagues (2002) concluded that *Bombus*, *Andrena*, and *Halictus* were 3 to 6 times more effective as pollinators than honey bees. This difference in pollinator effectiveness is largely due to the fact that honey bees (and *Megachile*) are visiting *Vaccinium* flowers for nectar, rather than pollen. Per visit, *Bombus*, *Andrena*, and *Halictus* deposited 25–50 pollen tetrads, whereas honey bees and *Megachile* deposited just 11–12 pollen tetrads. This detailed study clearly demonstrates how effective wild bees can be as agricultural pollinators.

While many wild bees are highly effective pollinators, they may not be sufficiently abundant in large, conventionally managed blueberry fields to fully replace honey bees. Isaacs and Kirk (2010) combined bee flower visitation data for Michigan highbush blueberry fields of varying size with data from Javorek and colleagues (2002) on per-visit effectiveness. While the wild bees were abundant in small fields (comprising 58% of flower visitors), in larger fields they were exceptionally rare (3% of flower visitors). Factoring in per-visit effectiveness, Isaacs and Kirk (2010) estimated that wild bees provide 82% of the pollination services in small fields, but only 12% in the larger fields. Given that most of blueberry production in Michigan occurs in large, conventionally managed fields, many Michigan blueberry growers are essentially dependent on honey bees for their crop pollination. This study illustrates beautifully how the importance of wild bees to crop pollination can change with the scale and management of agricultural habitats. Again, this casts doubt on the economic valuations of wild and managed pollinators based on a single, across-the-board measure of pollinator importance.

Coffee (*Coffea arabica*, *C. canephora*, *C. liberica*, Rubiaceae)

The crops described above are all grown in the temperate regions of the world. Many tropical crops, like coffee, mango, and papaya, also require insect pollination. Of these tropical crops, coffee is by far the most economically important (Gallai et al. 2009). Coffee is native to the highlands of modern-day Ethiopia, southeastern Sudan, and northern Kenya but is grown today in over 80 countries.

Coffee farming directly employs more than 25 million people, many of whom live on subsistence incomes (Ngo et al. 2011). Until the 1970s, almost all cultivated coffee was from sun-intolerant varieties that were grown on small farms embedded within intact tropical forest (Perfecto et al. 1996). The development of sun-tolerant coffee enabled growers to plant much larger fields that could be harvested by tractor. The switch from shade to sun-tolerant varieties has led to significant deforestation in tropical regions of the world. The total economic value of annual coffee production is €11.7 billion in 2005 (Gallai et al. 2009). Major coffee-producing regions of the world include South America (€5.1 billion), southeast Asia (€2.8 billion), and Central America and the Caribbean (€1.8 billion).

Unlike the crops described above, the dependence of coffee on pollinators is relatively low. Gallai and colleagues (2009) ascribe a dependence ratio of 0.25 to coffee—meaning that only 25% of coffee production is dependent on pollinators. Bee pollination results in a larger bean that can be sold at a higher price, but when bees are absent, flowers still set fruit, either through selfing or abiotic (wind) pollination. Coffee is an unusual crop in that flowering occurs in a short window of time (1–4 days), and neighboring fields often undergo synchronous, mass-flowering events (Boreux et al. 2013a,b).

An impressive amount of work has focused on bees and coffee pollination in countries throughout the world (reviewed by Ngo et al. 2011). Because coffee is produced in the tropics, the bee fauna associated with coffee is typically dominated by social bees (honey bees and stingless bees), but there are also solitary bees that contribute to coffee pollination, such as *Amegilla*, *Ceratina*, *Xylocopa*, and a variety of orchid bees (Apidae) and megachilid bees, such as *Megachile* and *Heriades*. Boreux and colleagues (2013b) reviewed studies of bee diversity in coffee plantations in the New World and Old World tropics and found that overall bee species richness varied widely (in Boreux et al. [2013b], see Supplementary Table S1). Some studies report just 3 species visiting coffee (one study in Mexico), while others reported over 40 species (Costa Rica). The proportion of solitary species ranged from zero (all coffee visitors were social bees, according to several studies in Mexico) to over 70% solitary species (studies in Costa Rica and Indonesia). These numbers do not capture visitation rate, but they do demonstrate that solitary species can comprise a large proportion of the bee fauna visiting coffee flowers.

Klein and colleagues (2003) documented a diverse fauna of solitary and social bees in studies of coffee plantations in Indonesia. Based on assessments of per-visit pollinator effectiveness, solitary bees were, on average, more effective pollinators than social bees. They found that species richness, rather than the abundance of any one species, had the most significant impact on fruit set. At plantations with low wild bee species richness (3 species), fruit set was below 60%, whereas at sites with high species richness (20 species), fruit set exceeded 90%.

THE IMPORTANCE OF POLLINATOR DIVERSITY

The studies we described above have managed to combine both pollinator abundance (or visitation rate) with per-visit effectiveness (pollen grains deposited, seeds or fruits set) to quantify the impact of wild pollinators vis-à-vis honey bees. A simpler approach, and one that has been taken in many crop systems, is to simply assume that abundance is what really matters. There is some reason to think that abundance alone will suffice when per-visit effectiveness data are either unavailable or prohibitively difficult to obtain (Vázquez et al. 2005, Winfree et al. 2015). Since most studies have shown that wild bees are generally more effective pollinators than honey bees, you could call this the "let's give the honey bees a break and call them equal" approach.

Studies—especially in the eastern United States, where farms and orchards are managed on a smaller scale than, say, the Central Valley of California—have shown that wild bee visitation can be higher than honey bee visitation, even when honey bee colonies are brought into fields (Adamson et al. 2012, Winfree et al. 2008). Winfree and colleagues (2008) examined the relative abundance of wild bees and honey bees in four crops grown commercially in New Jersey and neighboring eastern Pennsylvania. Their study focused on muskmelon, pepper, watermelon, and tomato. They documented a total of 54 wild bee species in surveys of 29 farms over 2 years and found that wild bees provide the majority of floral visitation in tomato and pepper and more or less equal visitation to honey bees in watermelon and muskmelon. Over all crops combined, wild bees provided 62% of floral visitation in spite of the fact that honey bees were rented at the majority of farms surveyed. There was no detectable impact of farm management or landscape complexity on wild bee species richness and abundance. Adamson and colleagues (2012) conducted a similar survey of wild bee and honey bee visitation in farms in the vicinity of Blacksburg, Virginia. Their studies focused on four pollinator-dependent crops: apple, blueberry, caneberry (raspberry, blackberry, and black raspberry), and cucurbits (squash, cucumber, cantaloupe, and watermelon). They conducted visual surveys of pollinator visitation and collected bees using aerial netting and pan trapping. Over the entire growing season, they documented 105 species of wild bees visiting these crop plants, with 37–59 species visiting each crop. Wild bee visitation was greater than honey bee visitation in all crops surveyed. Wild bees accounted for 77% of visits to apple flowers, 75% of visits to blueberry flowers, 68% of visits to caneberry flowers, and 83% of visits to cucurbit flowers, in spite of the fact that two-thirds of farms had honey bee colonies present. The most common bees were *Andrena*, *Bombus*, *Osmia*, *Lasioglossum*, *Peponapis*, *Agapostemon*, and *Melissodes*. Important solitary bee pollinators included *Andrena barbara*, *A. carlini*, *A. vicina* (Andrenidae), *L. leucozonium* (Halictidae), and *P. pruinosa* (Apidae). Other crops that have been shown to benefit from wild bee pollination include tomato (Greenleaf and Kremen 2006a, Morandin et al. 2001a,b), sunflower (Greenleaf and Kremen 2006b), and canola (Morandin and Winston 2005).

There is also increasing evidence that pollinator diversity per se is important for fruit set. Studies from diverse crop systems have supported the idea that a more diverse pollinator fauna provides better pollination services than a simple fauna consisting of just one or a few dominant species. Results from cucurbits (Hoehn et al. 2008), radishes (Albrecht et al. 2012), highbush blueberry (Rogers et al. 2014), and apples (Blitzer et al. 2016, Mallinger and Gratton 2014, Martins et al. 2015) have all documented the benefits of bee diversity for crop pollination. A diverse fauna is thought to provide more effective pollination for a number of reasons. First, a diverse fauna is more likely than a simple fauna to include at least a few of the more highly effective species. In the case of pollinators, this would mean that increased diversity translates into improved community effectiveness simply because there is more chance of having one or more really effective pollinators in a diverse community than in a simple community. Second, interactions among community members may improve pollinator effectiveness of one or more of the community members. This effect has been documented in sunflower pollination in which the movement of honey bees across rows and among flowers was influenced by the presence of wild bees (Greenleaf and Kremen 2006b). This would seem to be a relatively minor factor in most pollinator communities, where direct interactions among flower visitors are rare. Finally, diverse pollinator communities might provide better pollination because of what has been termed *niche complementarity*. Pollinator niche complementarity can arise for a number of reasons. Pollinators may differ in the way they manipulate the flower, with some pollinators mostly nectaring and others mostly gathering pollen. Pollinators may differ in the time of day they visit flowers or the temperatures at which they forage. Pollinators may even differ in the region of the host plant they are most likely to visit. Diversity in these traits may be what underlies the increased effectiveness of a diverse pollinator community. This is the most widely cited mechanism for the empirical observation that pollinator species richness and functional group diversity impact fruit and/or seed set (Albrecht et al. 2012, Blitzer et al. 2016, Brittain et al. 2013, Brittain and Potts 2011, Hoehn et al. 2008, Rogers et al. 2014).

Diverse pollinator communities are also likely to be buffered more than simple pollinator communities against the loss of species or against shifts in flowering phenology. This buffering effect is clearly crucial to human crop pollination, given that there is increasing evidence of declines in and extinction of some wild bee species (Bartomeus et al. 2013, Biesmeijer et al. 2006, Cameron et al. 2011, Ollerton et al. 2014). A diverse pollinator community will also be more robust in the face of shifting flowering phenology associated with a warming planet (Bartomeus et al. 2013).

One of the most compelling demonstrations of the importance of wild pollinators for agricultural pollination comes from a meta-analysis of pollinator communities across 600 sites, 41 crops, and all continents where crops are grown (Garibaldi et al. 2011). Crops analyzed included fruits, seeds, nuts, and stimulants. For the 41 crops analyzed, the authors quantified bee visitation rates and wild bee species richness across sites and related these values to per-visit pollen deposition and/or fruit set. Fruit set increased with wild bee visitation in all crops, but honey bee visitation impacted fruit set in only 14% of the crops. Fruit set increased twice as strongly with wild bee visitation than with honey bee visitation. The impact of both wild bee species richness and abundance on fruit set was independent of honey bee visitation, indicating that honey bees cannot replace the impact of either wild bee diversity or abundance. The authors concluded that "although honey bees are generally viewed as a substitute for wild pollinators, our results demonstrate that they neither maximize pollination, nor fully replace the contributions of diverse, wild-insect assemblages to fruit set for a broad range of crops and agricultural practices on all continents with farmland."

Increasing evidence indicates that wild pollinators (many of which are solitary, ground-, stem-, and wood-nesting bees) are providing a valuable, but underappreciated, service as crop pollinators. In some cases, these solitary bees can be managed, but in others, we are simply relying on the diversity of wild pollinators living in and around our agricultural habitats. These wild bees are providing economically important levels of pollination, and they are doing it for free. However, the effectiveness of these wild pollinators can be negatively impacted by a number of factors, including loss of habitat and floral resources, pathogens (sometimes introduced from managed pollinators), pesticides, invasive species, and climate change. Some of these factors, such as pesticides and habitat loss, are likely to have similarly negative impacts on both honey bees and wild, solitary bees. Others are likely to have specific impacts. For example, the ectoparasites that impact honey bees are largely restricted to honey bees and their relatives (e.g., *Varroa*). In the next chapter, we discuss some of the threats to wild, mostly solitary, pollinators.

CHAPTER 14

Threats to Solitary Bees and Their Biological Conservation

Colony collapse disorder (CCD), a syndrome originally reported in managed honey bee operations in the spring of 2007, spurred a massive increase in public and scientific interest in bee conservation—and not just honey bee conservation. One could argue that the bee conservation movement began with the 2007 CCD crisis. The high overwintering mortality rate observed among managed honey bee colonies was both dramatic and frightening; large, industrial beekeeping operations across the United States were impacted, and colony losses approached 100% in some apiaries. The CCD story garnered considerable media attention, with headlines like "Bees vanish, and scientists race for reasons" (Alexei Barrionuevo, *New York Times*, April 24, 2007). Explanations for the massive die-offs of managed bees ranged from the plausible (a new pathogen or parasite) to the absurd (cell phones and bible prophecies). A number of very high-profile scientists and policy makers raised the possibility that the decline in honey bee colonies could threaten global food production. The general public embraced the idea that a pollinator crisis might be brewing. We now know that the decline of honey bees, particularly in North America and Europe, is not due to any single factor. Many factors, including pathogens and parasites, pesticide exposure, long-distance transportation, limited floral resources, and even the fluctuating market for honey seem to underlie the declines we are seeing in managed honey bee colonies (Goulson et al. 2015, vanEngelsdorp and Meixner 2010).

It is becoming increasingly clear that honey bees are not the only bees that are having problems. In Europe, 45% of the bumble bee species are considered to be in decline (Nieto et al. 2014). The range of one critically endangered species (*Bombus cullumanus*) has shrunk dramatically because of habitat fragmentation and reduction in the amount of flowering clover, its main forage plant (Rasmont et al. 2005). An additional seven bumble bee species are listed as endangered, and eight species are considered vulnerable (Nieto et al. 2014). In North America, four bumble bee species are considered imperiled (*B. affinis*, *B. occidentalis*, *B. pensylvanicus*, and *B. terricola*), and one species (*B. franklini*) has potentially already gone extinct. As in honey bees, bumble bee decline seems to be caused by multiple interacting factors, including habitat loss, climate change (Kerr et al. 2015), pesticide exposure, and spread of novel pathogens (Cameron et al. 2011). Evidence of above-average mortality in managed honey bee colonies and documented cases of bumble bee decline drove home the point that pollinator loss could have dramatic impacts on humans, primarily in the area of crop pollination.

The intense media, policy, and research focus on honey bees and bumble bees overshadowed any real consideration of a much broader question: What is the conservation status of bees in general? Solitary bees far outnumber social bees, yet the heightened attention to pollinators highlighted that we know nothing about solitary bee populations and their extinction risk, nor exactly how important they are to crop pollination (Chapter 13). There are surprisingly few studies that have evaluated historical trends in solitary bee abundance. These studies have primarily been conducted in Europe and North America. Trends in other parts of the world are almost entirely unknown (Teichroew et al. 2017).

Europe may have the best-known bee fauna in the world. The European bee fauna consists of nearly 2,000 described species occurring in a diversity of habitats. There are well-curated collections of bees in natural history museums that can be used to assess long-term historical trends in social and solitary bees. In the United Kingdom, the Bees, Wasps, and Ants Recording Society (http://www.bwars.com/) provides a comprehensive data set for analyzing historical trends in bees, and these data sets have been employed to assess changes in bee distributions over time (Biesmeijer et al. 2006) as well as to document regional extirpations (Ollerton et al. 2014).

The conservation status of the European bee fauna was most recently assessed in an International Union for the Conservation of Nature (IUCN) report published under the title *European Red List of Bees* (Nieto et al. 2014). This study focused on a geographical region from Iceland in the west to the Urals in the east, and the land bounded by the Arctic Ocean to the north and the Mediterranean to the south. The Canary Islands, Madeira, and the Azores were also included in the survey. The geographic range, the habitat and ecological requirements, historical trends in abundance, and obvious threats to long-term persistence were assessed for each described species. Data are surprisingly poor for most species. Nearly 60% of species were deemed "data deficient," meaning that there was insufficient information to make a conclusive statement about the conservation status of these species. The most data-deficient groups include Andrenidae (307 species), Megachilinae (242 species), and non-social, especially cleptoparasitic, Apidae (192 species). In spite of these limitations, the authors were able to identify a list of 77 bee species (out of the 1,942 total species known) that were under threat. This list included 7 critically endangered, 46 endangered, and 24 vulnerable species (Table 14-1). Of the 7 species that are critically endangered, 2 are cleptoparasites (*Ammobates dusmeti* and *Nomada siciliensis*), 4 are solitary, ground- or stem-nesting bees (*Andrena labiatula, A. ornata, A. tridentata, Megachile cypricola*), and only 1 is eusocial (*Bombus cullumanus*). The majority of the threatened species in Europe are solitary (n=45). Eusocial (n=17), and cleptoparasitic species (n=9) comprised a smaller percentage of all threatened species. However, while eusocial species

Table 14-1. Threatened and endangered bees in Europe (Nieto et al. 2014).

CR = critically endangered; EN = endangered; VU = vulnerable; LC = least concern.

EU27 = 27 states of the European Union (a smaller geographic region than all of Europe).

Family	Genus	Species	Sociality	Europe	EU27	Endemic to Europe	Endemic to EU27
Apidae	Ammobates	dusmeti	cleptoparasitic	CR	CR	Yes	Yes
Andrenidae	Andrena	labiatula	solitary	CR	CR	Yes	No
Andrenidae	Andrena	ornata	solitary	CR	EN	No	No
Andrenidae	Andrena	tridentata	solitary	CR	CR	No	No
Apidae	Bombus	cullumanus	eusocial	CR	CR	No	No
Megachilidae	Megachile	cypricola	solitary	CR	CR	No	No
Apidae	Nomada	siciliensis	cleptoparasitic	CR	CR	Yes	Yes
Apidae	Ammobates	melectoides	cleptoparasitic	EN	EN	Yes	No
Apidae	Ammobatoides	abdominalis	cleptoparasitic	EN	EN	No	No
Andrenidae	Andrena	cornta	solitary	EN	EN	No	No
Colletidae	Colletes	wolfi	solitary	EN	EN	Yes	Yes
Melittidae	Dasypoda	braccata	solitary	EN	EN	No	No
Andrenidae	Andrena	magna	solitary	EN	EN	No	No
Andrenidae	Andrena	stepposa	solitary	EN	EN	Yes	No
Andrenidae	Andrena	stigmatica	solitary	EN	EN	No	No
Apidae	Bombus	armeniacus	eusocial	EN	EN	No	No
Apidae	Bombus	brodmannicus	eusocial	EN	EN	No	No
Apidae	Bombus	fragrans	eusocial	EN	EN	No	No
Apidae	Bombus	inexspectatus	eusocial	EN	EN	Yes	No
Apidae	Bombus	mocsaryi	eusocial	EN	EN	No	No
Apidae	Bombus	reinigiellus	eusocial	EN	EN	Yes	Yes
Apidae	Bombus	zonatus	eusocial	EN	EN	No	No
Colletidae	Colletes	anchusae	solitary	EN	EN	No	No
Colletidae	Colletes	caspicus	solitary	EN	EN	No	No
Colletidae	Colletes	collaris	solitary	EN	EN	No	No
Colletidae	Colletes	graeffei	solitary	EN	EN	No	No

(continued)

Table 14-1. (Continued)

Family	Genus	Species	Sociality	Europe	EU27	Endemic to Europe	Endemic to EU27
Colletidae	Colletes	merceti	solitary	EN	EN	Yes	Yes
Colletidae	Colletes	meyeri	solitary	EN	EN	No	No
Colletidae	Colletes	punctatus	solitary	EN	EN	No	No
Colletidae	Colletes	sierrensis	solitary	EN	EN	Yes	No
Melittidae	Dasypoda	frieseana	solitary	EN	EN	Yes	No
Melittidae	Dasypoda	spinigera	solitary	EN	EN	No	No
Melittidae	Dasypoda	suripes	solitary	EN	EN	No	No
Andrenidae	Flavipanurgus	granadensis	solitary	EN	EN	Yes	Yes
Halictidae	Halictus	carinthiacus	eusocial	EN	EN	Yes	No
Halictidae	Halictus	microcardia	eusocial	EN	EN	Yes	Yes
Halictidae	Halictus	semitectus	eusocial	EN	EN	No	No
Megachilidae	Icteranthidium	cimbiciforme	solitary	EN	EN	No	No
Halictidae	Lasioglossum	breviventre	solitary	EN	EN	Yes	No
Halictidae	Lasioglossum	laeve	solitary	EN	EN	No	No
Halictidae	Lasioglossum	quadrisignatum	solitary?	EN	EN	No	No
Halictidae	Lasioglossum	sexmaculatum	solitary?	EN	EN	No	No
Halictidae	Lasioglossum	sexnotatulum	solitary?	EN	EN	No	No
Halictidae	Lasioglossum	soror	solitary?	EN	EN	No	No
Halictidae	Lasioglossum	subfasciatum	solitary?	EN	EN	No	No
Halictidae	Lasioglossum	virens	solitary?	EN	EN	No	No
Melittidae	Melitta	melanura	solitary	EN	EN	No	No
Apidae	Nomada	italica	cleptoparasitic	EN	EN	No	No
Apidae	Nomada	pulchra	cleptoparasitic	EN	EN	No	No
Megachilidae	Osmia	maritima	solitary	EN	EN	No	No
Apidae	Parammobatodes	minutus	cleptoparasitic	EN	EN	No	No
Megachilidae	Trachusa	interrupta	solitary	EN	EN	No	No

Andrenidae	Andrena	nanaeformis	solitary	LC	VU	No	No
Andrenidae	Andrena	transitoria	solitary	VU	VU	No	No
Apidae	Biastes	truncatus	solitary	VU	VU	No	No
Apidae	Bombus	alpinus	eusocial	VU	VU	Yes	No
Apidae	Bombus	corfusus	eusocial	VU	VU	No	No
Apidae	Bombus	distinguendus	eusocial	VU	VU	No	No
Apidae	Bombus	gerstaeckeri	eusocial	VU	VU	No	No
Apidae	Bombus	hyperboreus	eusocial	VU	VU	No	No
Apidae	Bombus	muscorum	eusocial	VU	VU	No	No
Apidae	Bombus	polaris	eusocial	VU	VU	No	No
Apidae	Bombus	pomorum	eusocial	VU	VU	No	No
Megachilidae	Coelioxys	elongatula	deptoparasitic	VU	VU	No	No
Colletidae	Colletes	chengtehensis	solitary	VU	EN	No	No
Colletidae	Colletes	dimidiatus	solitary	VU	VU	Yes	Yes
Colletidae	Colletes	floralis	solitary	VU	VU	No	No
Colletidae	Colletes	fodiens	solitary	VU	VU	No	No
Colletidae	Colletes	impunctatus	solitary	VU	VU	No	No
Colletidae	Colletes	mcricei	solitary	VU	VU	Yes	Yes
Colletidae	Colletes	perezi	solitary	VU	VU	No	No
Colletidae	Colletes	puchellus	solitary	VU	VU	Yes	Yes
Halictidae	Halictus	leucaneneus	eusocial	VU	VU	No	No
Melittidae	Melitta	hispanica	solitary	VU	VU	Yes	Yes
Melittidae	Melitta	kastiliensis	solitary	VU	VU	Yes	Yes
Apidae	Nomada	noskiewiczi	cleptoparasitic	VU	VU	Yes	Yes
Halictidae	Systropha	planidens	solitary	VU	VU	No	No

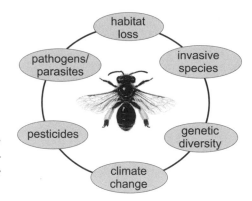

Figure 14-1. Threats to solitary bees. The bee at the center of the figure is a female *Macropis europaea* (Melittidae). Original artwork by Birgitte Rubæk.

(mostly bumble bees) are well known, the conservation status of cleptoparasitic bees is extremely poorly known. The families Melittidae (18.9%) and Colletidae (12.8%) show the highest proportion of threatened species, and Megachilidae (1.1%) shows the lowest.

There are important lessons to be learned from the IUCN's assessment of the European bee fauna. First, bumble bees are not the only threatened bees. The list of threatened bees in Europe includes many solitary, ground-nesting bees as well as cleptoparasitic bees. Second, the threats to wild bees are multifaceted and include changes in land use, urban and agricultural intensification, pesticide use, and even climate change, and these factors may interact synergistically to drive bee populations downward. Finally, the IUCN report clearly documents how little we know about the conservation status of solitary bees. For Europe, which has the most thoroughly documented bee fauna on earth, we can assess the status of only 40% of the bee fauna.

The primary threats to the long-term viability of solitary bee populations include (1) habitat loss and fragmentation, (2) exposure to pesticides, (3) the impact of invasive plants and bees, (4) pathogen spillover from commercially managed bees, (5) low genetic variability and limited inter-population gene flow, and (6) climate change (Fig. 14-1). We discuss each of these threats below.

HABITAT LOSS AND FRAGMENTATION

Habitat change—by loss, conversion, or fragmentation—is the single greatest impact that humans have on bee populations and communities. Agriculture, urbanization, logging, and fire alter the distribution of floral resources, the availability of nesting substrates, the levels of parasite and predator populations, and the genetic structure of bee populations. These anthropogenic changes are not always detrimental to bee populations; some forms (or levels) of disturbance enhance bee diversity by opening up habitats for early-successional flowering plants and enhancing floral resource availability. Furthermore, bee species that differ in behavior or ecology may respond in very different ways to the same disturbance event. One can think of habitat loss as creating ecological "winners" (those species that do well in the modified habitat) and ecological "losers" (those species that drop out with changes in habitat) (McKinney and Lockwood 1999). It is not always easy to predict a priori who the "winners" and "losers" might be (Winfree et al. 2009, Williams et al. 2010, Bartomeus et al. 2017).

Here we cover the major conclusions for each of the primary categories of habitat loss and point out some counterintuitive results. These exceptions often provide important insights into the heterogeneous response of wild bees to habitat alteration.

Agriculture

Agricultural land is an umbrella term that includes arable land that is replanted annually, cropland with perennial crops, and pasture for domesticated animals. The proportion of land converted to agricultural purposes is massive. In 2005, an estimated 40% of the land on earth was being used for agriculture, up from 11% in 1700 (Foley et al. 2005). Although land conversion to agricultural uses will continue to increase, the rate of conversion is slowing, because most of the land suitable for agriculture has already been converted to that purpose. Continued human population growth will increase the incentive to use available agricultural areas more intensively, and this change will further alter bee species richness and abundance. Agricultural ecosystems are generally (but not always) much simplified compared to the ecosystems they replace.

A study by Kleijn and colleagues (2015) indicates the magnitude of the impact that this ecosystem simplification has already had on bee richness and abundance. Across 90 studies on all continents except Antarctica (where there are no bees!), the proportion of bees in fields where there was crop agriculture (annual and perennial crops) was an average of only 12.6% of the bee species known to occur in the region. Almost 90% of the bee species in the region were never collected in crop fields. Furthermore, most bees collected on crop flowers were common species; only 2% of the regional pool of bee species were collected more than 5% of the time. In other words, 95% of the bees in an area are either not found or are very rare in agricultural fields. An important caveat is that the bee fauna of a region is never represented in a limited area. Nevertheless, the study (Kleijn et al. 2015) strongly suggests that bee communities in agricultural habitats are generally less abundant and less species-rich, and they have a very different composition than those in surrounding natural habitats.

Annual crops

Important annual crops that are bee-pollinated include squash and pumpkin, alfalfa, watermelon, cucumber, sunflower, canola, rape seed, and eggplant. Annual crop production involves frequent mechanical tilling, annual crop rotation, and heavy pesticide, herbicide, and fungicide use. For this reason, annual crop fields are a harsh environment for many bees. Plowing can destroy the nests of ground-nesting bees except when nests are exceptionally deep (Kim et al. 2006, Shuler et al. 2005). Crop rotation leads to unpredictable (from the bees' perspective) resource availability. In the eastern United States, squash and pumpkin are normally rotated on an annual basis with corn. Squash bee (*Peponapis pruinosa*) populations do exceptionally well in years when cucurbit crops are planted but are at a complete loss in years when corn is planted.

Annual crops can be beneficial to some bees, especially those that have a natural preference for the crop plant. In Africa, eggplant (*Solanum melongena*) serves as an important host plant for two solitary bee species: *Xylocopa caffra* (Apidae) and *Macronomia rufipes* (Halictidae) (Gemmill-Herren and Ochieng 2008). In Europe, a diverse fauna of legume specialists (*Melitta leporina* [Melittidae], *Melitturga clavicornis* and *Andrena ovatula* [Andrenidae], and *Rhophitoides canus* [Halictidae]) benefits from the annual cultivation of alfalfa (lucerne, *Medicago sativa*) (Pesenko and Radchenko 1993). In North America, a diverse fauna of solitary bees benefits from the annual cultivation of sunflower (*Helianthus annuus*; Hurd et al. 1980).

The scale of agricultural production also matters. Many studies have documented an edge effect on wild bee abundance and diversity. Bee species richness and abundance decline with increasing distance from field margins (Morandin and Winston 2005, Ricketts et al. 2004). Greater habitat heterogeneity at field edges can promote bee biodiversity in various ways, such as extending how long plants are in flower over the season and proximity to the appropriate nesting sites and floral resources that bees require (Diekötter and Crist 2013, Winfree et al. 2007b).

In general, the "winners" in annual crop fields include generalist bees and bees that build deep nests. "Losers" include bees that build very shallow nests and habitat and host-plant specialists (Williams et al. 2010, Bommarco et al. 2010, De Palma et al. 2015, Wood et al. 2016, Wood and Roberts 2017).

BOX 14-1: *ANDRENA VAGA*—A SAND-LOVING WILLOW SPECIALIST

One common aspect of solitary bee nesting aggregations is that they are highly ephemeral. Local populations appear for a few years, grow to high densities, and then abruptly disappear. At a local scale, one might think these bees are highly vulnerable to extinction. However, the rapid colonization of new, ephemeral habitats may allow meta-populations to persist over a much larger geographic scale, in spite of the short-term local population persistence.

An extremely well-studied bee in Central Europe, *Andrena vaga* (Andrenidae), illustrates this aspect of solitary bee biology. In many ways, *A. vaga* is a "typical" solitary, ground-nesting, univoltine bee. It occurs throughout much of western Europe, from France to the Ural Mountains and from the Mediterranean to Sweden. Like many solitary bees, *A. vaga* has particular soil preferences (females prefer sandy soil) and narrow host-plant preferences (females are specialists on willows, *Salix*). This species has been described as a "pioneer species" that inhabits riparian habitats where sandy soil and willows are common. In the early spring, *A. vaga* forms large, conspicuous, dense, nesting aggregations of sometimes thousands of nests, and while these aggregations can remain viable for over 60 years, they are also highly vulnerable to catastrophic flooding in the spring (Fellendorf et al. 2004).

Like most ground-nesting bees, *A. vaga* is attacked by a variety of predators and parasites, including a cleptoparasitic bee (*Nomada lathburiana*, Apidae) and the parasitoid beefly (*Bombylius major*, Bombyliidae). A multiyear study by Bischoff (2003) documented a steady decline in *A. vaga* population size in a large population along the Rhine Valley, apparently due to heavy beefly parasitism.

How does a species that has narrow host-plant and soil requirements and lives in an ephemeral habitat prone to catastrophic flooding and extinction due to parasite attack survive over the long term? In spite of their low reproductive rate and habitat and host-plant specialization, *A. vaga* can apparently persist in the disturbed riparian habitats through frequent dispersal at the start of each nesting season. Based on mark-recapture studies, approximately half of the females emerging at a nest site in the spring disperse to other, nearby suitable habitat patches (Bischoff 2003). This observation is supported by two population genetic studies—one conducted in Lower Saxony, Germany (Exeler et al. 2008) and the other in the Czech Republic and Slovakia (Černá et al. 2013a)—that demonstrate substantial gene flow among isolated, patchily distributed nest sites of *A. vaga*. While there was a noticeable level of inbreeding within discrete populations, presumably due to a fairly high level of philopatry (females remaining in the same site from which they emerged), there was substantial gene flow among populations due to early-spring foundress dispersal.

A. vaga inhabits ephemeral, highly disturbed habitats but manages to maintain stable overall abundance by having a high dispersal rate and a well-developed capacity to colonize new nest sites with suitable soil characteristics and sufficient willow populations. So while small, local populations may go extinct at a relatively high rate, new populations are founded on a regular basis. This pattern of short-term population persistence but frequent colonization of new habitats (and/or recolonization of previously occupied habitats) may be a common feature of many solitary, ground-nesting bees (Franzén and Nilsson 2013). And while local extinction of small populations may give the impression of vulnerability, some ground-nesting solitary bees are exquisitely adapted to exploit a constantly changing, sometimes highly fragmented landscape (Fellendorf et al. 2004, Visscher et al. 1994).

Perennial crops

Perennial crops—for example, early-spring-flowering fruit trees, such as apricots, peaches, pears, plums, prunes, cherries, and apples, as well as flowering bushes, such as highbush blueberry—would seem to provide an abundant and predictable food supply for wild bee populations. Apple orchards in eastern North America host up to 120 wild bee species, and wild bees are abundant in and around apple orchards (Chapter 13). Blueberry fields also host a diverse wild bee fauna (Tuell et al. 2009).

But not all early-spring-flowering trees are the same when it comes to supporting bee populations. Almond (*Prunus dulcis*) is a tree crop that is so profitable that almond orchards are replacing many vineyards in northern California. Almonds are native to the Middle East but are currently grown around the world, including Australia, Spain, and the western United States. Except for a few varieties, almond flowers are self-incompatible and require a pollinator visit to set seed. Bees visit the flowers readily for both nectar and pollen, but flowering takes place so early in the spring that little native vegetation is in flower, and few bee species are active (Mandelik and Roll 2009, Saunders and Luck 2013). In northern California, honey bees accounted for 70% of the visits to almond flowers, and 19 species of wild bees made only 11% of the visits (Klein et al. 2012). Finally, almonds are often planted in large orchards, and growers keep the orchards free of other potential floral hosts with herbicides or mechanical clearing. It is not surprising that almonds are almost entirely dependent on honey bees for pollination (Ward et al. 2010). Almonds are a mass-blooming floral resource that many bees could potentially use. Unfortunately, a combination of early flowering and intensive application of agrochemicals deters solitary bee populations from exploiting this ever-expanding floral resource.

One can view mass-flowering crops as both an incredible resource for solitary bees, because of the massive presentation of abundant floral rewards, but also a hazardous "death zone" if insecticides, herbicides, fungicides, and plant-growth regulators are used. This may well be what makes agricultural settings so inhospitable for many solitary bee species. Natural habitat surrounding croplands may be a "source" of wild bee diversity, whereas the managed agricultural habitat is a "sink" where bees suffer high levels of mortality.

Grazing

Large-mammal grazing comprises the largest proportion of agricultural land use around the world (Fleischner 1994). Domesticated animals raised for milk, meat, or wool include cows, sheep, goats, and llamas. The most immediate impact of grazing mammals on wild bees is altered floral resource availability, but grazing mammals can also alter nesting resources for bees (Sjödin et al. 2008, Vulliamy et al. 2006).

The impact of grazing on wild bee diversity and abundance seems to depend largely on the intensity of grazing. Low-density cattle ranching can be compatible with high levels of bee diversity and abundance. Minckley (2014) studied bee diversity and abundance at Chihuahuan Desert habitats along the US–Mexico border that had been grazed since the 1800s. Bee diversity was exceptionally high; over 295 species were collected in a 10-square-kilometer area. The fauna was composed almost entirely of solitary, ground-nesting bees, and more than 30% of the pollen-collecting species were host-plant specialists. Bee abundance was negatively impacted at sites that had been more recently grazed, but species richness and overall composition were not impacted by grazing history. Host-plant specialists were no more significantly impacted by grazing than host-plant generalists. This study suggests that low-intensity grazing, even in a biodiversity hotspot for bees, is compatible with maintaining healthy wild bee populations.

However, not all forms of grazing are compatible with wild bee conservation. Where cattle density is higher than available forage can support and fodder is artificially augmented, bees are negatively affected (Le Feon et al. 2010). Similarly, continuous grazing has greater negative effects on bees than grazing over part of the growing season (Le Feon et al. 2010, Sjödin 2007). Sheep and goats graze closer to the ground than cattle and appear to affect bees more severely (Yoshihara et al. 2008), although it is not clear if this response is related to reduced floral abundance or changed floral-host composition. Studies of bee communities in regions grazed by cattle have concluded that grazing had no effect on bee population density or species richness (Elwell et al. 2016, Steffan-Dewenter and Leschke 2003), a positive effect (Vulliamy et al. 2006), a negative effect (Yoshihara et al. 2008), and a mixed effect (Lázaro et al. 2016, Minckley 2014). These disparate conclusions are difficult to reconcile because the studies differed substantially in sampling intensity, sampling method (pan traps, net collecting, visual observation, etc.), and location. Some studies compare the entire bee fauna (Elwell et al. 2016, Lázaro et al. 2016, Minckley 2014, Vulliamy et al. 2006, Yoshihara et al. 2008) and others a subset of the fauna (e.g., aboveground nesting species [Steffan-Dewenter and Leschke 2003]).

Making generalizations about which bee species are "winners" and "losers" in grazed environments is difficult. However, one study suggests that ground-nesting bees may benefit in grazed habitats. Vulliamy and colleagues (2006) found that bee abundance (but not species richness) peaked in more heavily grazed sites because heavily trampled sites had more open, exposed soil for ground nesters.

Urbanization

Urban and suburban habitats would seem to be poor sites for maintaining wild bee diversity. Urban sites are heavily disturbed, floral resources are patchily distributed, and much of the ground is paved, making it completely inaccessible to ground-nesting bees and worthless from the perspective of floral diversity. Urban and suburban sites are also dominated by horticultural and non-native plants that may not provide floral resources for native, wild bee species. It is not surprising that non-native bees (e.g. *Anthidium manicatum*, *Osmia cornifrons*, and *Megachile sculpturalis*) do exceptionally well in urban and suburban habitats.

However, studies of wild bees in urban and suburban habitats have revealed a surprising ability of wild bees to persist in these environments. Small parks, vacant lots, green roofs, and private gardens in the most densely populated cities in the world (including Berlin, Chicago, New York, Toronto, Montreal, and Paris) host a surprising diversity of bee species (Colla et al. 2009, Geslin et al. 2016, Matteson et al. 2008, Normandin et al. 2017, Saure 1996, Tonietto et al. 2011). Studies that have compared urban areas to nearby parks, green areas, or reserves have found that, in the city, the species richness of bees decreases (Fetridge et al. 2008, Hennig and Ghazoul 2012, Matteson et al. 2008), is the same (Baldock et al. 2015, Hinners et al. 2012), or increases (Sirohi et al. 2015). Sirohi and colleagues (2015) netted 48 bee species at churchyards, roadsides, small gardens, and roundabouts in the urban core of Southampton, England, including two species that are considered nationally rare. Species richness of bees was greater in the urban core than reserves sampled outside of the city, and altogether their collections yielded about 22% of the total species and 58% of the genera of solitary bees in the United Kingdom.

Buildings, fencing, and other "built environments" (Adler and Tanner 2013) are pocked with holes, cracks, and cavities that can serve as nesting substrates for cavity-nesting bees. In the short-grass prairie (Hinners et al. 2012) and the Sonoran Desert (Cane et al. 2006), cavity-nesting bees were more abundant in urban areas than areas outside the city, suggesting that nest-site availability severely constrains populations of these species outside of cities.

As the intensity of urbanization increases, bee faunas shift toward widespread, pollen-generalist species, many of which are non-native (Banaszak-Cibicka and Żmihorski 2011, Jedrzejewska-Szmek and Zych 2013, Lowenstein et al. 2014, Threlfall et al. 2015). An interesting study done in Chicago, Illinois, compared bees and floral visitation to a "phytometer" along a suburban-to-urban gradient (Lowenstein et al. 2014). Phytometers are potted plants (usually one species) placed at multiple sites to control for flower density and differences in how attractive different floral hosts are to bees. Interestingly, bee diversity and visitation to phytometers increased as the percent of the area that was impervious (asphalt, concrete) increased. However, the bee fauna also shifted at sites that were the most human-modified to species that are geographically widespread, ecologically tolerant, and pollen generalists (the communal bee *Agapostemon virescens* and honey bees; Lowenstein et al. 2014). Pollen analysis from foraging female *A. virescens* indicated that they increased the number of hosts they visited in urban sites—a testament to their ecological flexibility and ability to live in highly modified environments.

Although it is often noted that pollen-specialist bee species are sensitive to urbanization (Fetridge et al. 2008, Hernandez et al. 2009), specialists can persist in urban areas where their preferred host occurs. *Lasioglossum (Sphecodogastra) oenotherae*, a host-plant specialist on sundrops (*Oenothera fruticosa* and *O. pilosella*) that grow in suburban gardens, was found only in urban and suburban habitats throughout its recorded range in eastern North America, in spite of considerable efforts to find it in more natural habitats (Zayed and Packer 2007). Cane and colleagues (2006) sampled the bee fauna of creosote bush (*Larrea tridentata*) growing in urban and suburban sprawl in and around the city of Tucson, Arizona. The creosote bush bee fauna is unusually diverse, with both oligolectic and polylectic species. In small urban fragments, specialist bees were rare, but specialist bees were able to persist in the larger urban fragments. In summary, as with agricultural habitats, urbanization can lead to ecological "winners" and ecological "losers." Cavity nesters, host-plant generalists, and invasive species are all likely "winners" in urban habitats, but this is far from a hard-and-fast rule, as the examples above illustrate.

Fire

Although fire is an important natural source of disturbance in many ecosystems, human activities change the frequency and intensity at which it occurs. Fire can have both positive and negative effects on wild, solitary bees. Fires have a devastating impact on bees that nest in above-ground, woody substrates. Nests and the developing offspring of renters, above-ground builders, and wood nesters (Chapter 6) are all highly vulnerable to fire (Grundel et al. 2010, Potts et al. 2003). Immediately after a burn, there are clear drops in above-ground nesters, but no obvious change in below-ground nesters, which are insulated from the intense heat of a wildfire. Cane and Neff (2011) examined the sensitivity of two stem-nesting bees (*Megachile rotundata* and *Osmia lignaria*) to fire. The egg stage was very sensitive to high temperatures (above 38°C), but the pupal stage could survive up to 46°C.

But fire can also provide benefits for wild bee populations. Although less complex structurally, post-burn areas are more open, and the soil is richer in nutrients. Kerrakin (a compound in smoke) strongly stimulates seed germination across plant groups (Flematti et al. 2004), and some plants germinate only after fire; plant species richness increases and the composition shifts toward annual plants that favor recently disturbed habitats.

Some bee species may even be dependent on fire. *Rediviva peringueyi* (Melittidae), one of the fascinating "oil bees" in the Cape region of southern Africa, is entirely dependent on an orchid (*Pterygodium catholicum*) for floral oils (Pauw 2007). Seeds of *P. catholicum* germinate only after fires. Both the bee and its host plant showed peak abundance at sites that were recently burned but declined in abundance as sites gradually recovered post-burn.

As for other types of disturbances, fire creates "winners" and "losers." Above-ground nesting bees are clearly the "losers" in this case. Winners include species that specialize upon early-successional plant species and species that prefer open habitats to shaded forest understory.

The types of habitat conversion that we have listed above can dramatically alter the abundance and species richness of solitary bees. However, it is clear that these disturbances do not universally lead to decreased solitary bee abundance and species richness. Many bee species do surprisingly well in frequently disturbed habitats, and the plants that first colonize such habitats may provide acceptable floral resources for a long list of wild bee species. In comparisons of bees along a gradient of urbanization (Fortel et al. 2014) and grazing (Lázaro et al. 2016), the diversity and abundance of bees was far higher at intermediate levels of disturbance than when disturbance was on a small scale or infrequent and on a large scale, frequent, or unusually intense. It is clear that the response of different bee species to habitat alteration is highly variable. Habitat alteration can lead to eradication of some species and proliferation of others. Those species that do well in the new habitat can be viewed as the ecological "winners," while those that do poorly in the new habitat are clearly the "losers." Predicting whether a species is a "winner" or "loser" is very difficult, but in an emerging pattern, geographically widespread pollen-generalist species persist in the most human-modified habitats, and host-plant specialists are the most seriously impacted by habitat alteration.

PESTICIDES

Pesticides are one of the most obvious threats to solitary bee populations, especially in agricultural and urban and suburban habitats where they are extensively used. We know surprisingly little about the impact of pesticides on solitary bees, because most of the previous research has focused on managed, eusocial bees (e.g., honey bees and bumble bees). We know from studies of these managed, eusocial bees that foraging workers (and their colonies) are exposed to diverse agrochemicals, including fungicides, insecticides, miticides, herbicides, and plant growth regulators (Chauzat et al. 2006, Mullin et al. 2010, Krupke et al. 2012, Long and Krupke 2016). Mullin and colleagues (2010), in studies of honey bee colonies across California, Florida, and Pennsylvania, detected 121 different pesticides in samples of wax, pollen, and worker bees. Subsequent studies have documented high levels of both fungicides (David et al. 2016) and neonicotinoids (Botías et al. 2015) in pollen from both agricultural plants (oil seed rape, in the case of David et al. 2016) and wildflowers growing near agricultural fields (Botías et al. 2015).

We know much less about exposure of wild, mostly solitary, bees in agroecosystems. In one of the few studies to quantify pesticide exposure to wild bees in agroecosystems, Hladik and colleagues (2016) documented pesticide contamination in foraging adult wild bees in wheat fields and grassland habitats in Colorado. They screened for 136 pesticides and detected 19 pesticides and pesticide metabolites in the 54 samples of pooled adult bees they analyzed. Among the pesticides they detected were neonicotinoid insecticides, fungicides, and herbicides. Not surprisingly, foraging bees collected near crops with high pesticide use (such as corn, sorghum, and millet) had the highest levels of pesticide contamination.

But what effect do these pesticides have on solitary bees? One way to answer this question is to distinguish between the impacts of pesticides on the behavior and reproductive success of individual bees (individual level effects) and the resulting changes in the diversity, abundance, and composition of communities (community level effects; Fig. 14-2). Individual-level effects include both direct mortality of foraging adult bees (what we call lethal effects in Fig. 14-2) and more subtle effects on foraging, memory, learning, and reproductive success (what we call sublethal effects in Fig. 14-2). Sublethal effects are manifest

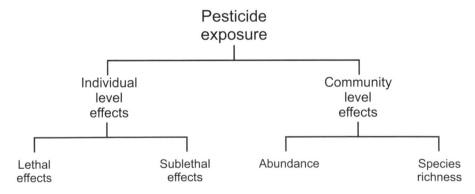

Figure 14-2. Impacts of pesticide exposure on solitary bees. Pesticides can have both individual-level and community-level effects. Individual-level effects include direct effects on adult longevity (lethal effects) as well as indirect effects on foraging, memory, learning, nest finding, and offspring production (sublethal effects). Community-level effects include decreases in overall abundance as well as diminished species richness (i.e., decreased diversity).

primarily in altered adult behavior but not outright death. Think of the sublethal effects of pesticides on bees as equivalent to how we feel after a night of heavy drinking. We are not dead, but we feel pretty awful, and our performance at work is greatly diminished. These impacts at the individual level likely underlie the changes that have been detected at the community level, including changes in abundance and species richness of wild bee communities (Fig. 14-2).

A standard approach to calculating the lethal effect of a pesticide on bees (or any organism) is to calculate the LD50 of the active ingredient. LD50 corresponds to the dose that would kill (under highly controlled laboratory conditions) 50% of the individuals exposed to the chemical within 48 hours. LD50 is typically expressed as microgram per bee (µg/bee). One can calculate the LD50 based on topical application of the pesticide ("contact" LD50) or via oral exposure ("oral" LD50). Typically, but not always, the oral dose is more toxic than the topical application. Honey bees are widely used to measure "bee toxicity" for virtually all pesticides, because they are easy to rear and one can obtain LD50 values quickly in the laboratory. Agrochemical companies are required by law to determine the LD50 of each new pesticide released on the market, and LD50 values for honey bees are widely available via a variety of sources, including the US EPA Pesticide Ecotoxicity Database (website: http://www.ipmcenters .org/Ecotox/), the French Agritox database (website: http://www.dive.afssa.fr /agritox/index.php), and the International Union of Pure and Applied Chemistry Pesticide Properties Database (website: http://sitem.herts.ac.uk/aeru/iupac).

LD50 values for honey bees vary widely among pesticides. Highly toxic pesticides, such as clothianidin (a neonicotinoid), have extremely low LD50 values; an extremely small dose was needed to kill 50% of the individuals tested within 48 hours. In the case of clothianidin, contact LD50 (0.04426 µg/bee) is extremely low, but not as low as the oral LD50 (0.00379 µg/bee). Less-toxic pesticides can have a very high LD50 value (i.e., it takes a lot of the pesticide to have any measurable impact on bee mortality). Fungicides and herbicides generally have relatively high LD50 values (in the range of 50 to over 200 µg/bee). While LD50 values are widely determined based on honey bees, only a few studies have determined LD50 values for solitary bees. Biddinger and colleagues (2013) compared toxicity of five pesticides commonly used in apple orchards on the honey bee and *Osmia cornifrons*, a solitary bee that is used for apple pollination. The toxicity of each pesticide to these two species was highly species-specific, meaning the response of one species did not necessarily mirror the response of the other. For example, *Apis mellifera* was considerably more sensitive to imidacloprid than *O. cornifrons*, but *O. cornifrons* was more susceptible to acetamiprid than honey bees. A meta-analysis conducted by Arena and Sgolastra (2014) came to the same conclusion. They gathered LD50 values for 19 bee species other than the honey bee and compared these values to the LD50 widely reported for honey bees. Their sample of taxa included solitary bees (mason bees, leaf-cutter bees, and the alkali bee, *Nomia melanderi*) and social bees (bumble bees and stingless bees). The LD50 values reported for these other bees varied substantially (by as much as 1,000-fold) from the values reported for honey bees, indicating that the honey bee is not necessarily providing a good measure of pesticide toxicity to bees in general. These studies demonstrate the profound lack of information on the direct, lethal effects of pesticides on wild, solitary bees.

The sublethal effects of pesticides are even harder to quantify than the direct, lethal effects. Nevertheless, a number of studies have clearly documented sublethal effects of a diverse array of pesticides on solitary bees (Fig. 14-2). These studies have documented indirect effects on foraging and reproduction from fungicides (Artz and Pitts-Singer 2015, Ladurner et al. 2008), organophosphates (Alston et al. 2007), and neonicotinoids (Abbott et al. 2008, Rundlöf et al. 2015, Sandrock et al. 2014). Indirect effects that have been detected in both cage and field experiments include impaired nest-finding ability (Artz and Pitts-Singer 2015), reduced offspring production and male-biased offspring production (Sandrock et al. 2014), and suppressed foraging activity (Rundlöf et al. 2015).

But do these studies of individual reproductive success translate to changes in abundance and species richness of wild bees in agroecosystems? A number of studies have shown a clear faunistic response to pesticide intensity in solitary bees. In a now classic and widely cited study, Peter Kevan (1975) documented a dramatic decline in both wild bee abundance and species richness in blueberry fields in New Brunswick, Canada, following widespread aerial application of the organophosphate fenitrothion for control of spruce budworm. More recent studies have been conducted across a variety of crops, including canola (Morandin and Winston 2005), highbush blueberry (Tuell and Isaacs 2010), mixed agricultural habitats in Italy (Brittain et al. 2010), and apples (Mallinger et al. 2015, Park et al. 2015). Park and colleagues (2015) documented a significant interaction between the proportion of natural habitat around each apple orchard and the response of the bee community to varying pesticide levels. Orchards surrounded by a greater proportion of natural habitat appear to be buffered from the negative effects of pesticides. Natural habitat surrounding orchards may be providing important resources for wild bees, such as nesting and foraging resources, but it may also be a refuge from pesticide exposure in the relatively toxic orchard center. When pesticides were parsed into fungicides and insecticides, Park and colleagues (2015) found that fungicides, not insecticides, had the most significant negative impact on the wild bee fauna. This is surprising given the relatively low LD50 values reported for fungicides based on the honey bee studies we described above. Either the LD50 values calculated for honey bees are not representative of LD50 values for the wild bee fauna of apple orchards (as suggested by Biddinger et al. 2013) or the effects of fungicides under field conditions are not represented by the laboratory-based LD50 values that have been calculated.

INVASIVE PLANTS AND BEES

Human activity is making the biological world more homogenous. Rapid transportation of wood packing material, heavy equipment, soil, and plants has allowed many plants and insects to expand their ranges into new, previously unoccupied areas. Some of these non-native organisms become invasive species when they are released into a paradise free of pathogens, parasites, and predators that allows them to establish massive populations and disrupt ecological interactions among native local flora and fauna. Others reproduce and spread but have little impact on the native communities (adventive species) or pose little harm because they are restricted to urban or other human-modified areas (naturalized species). The impacts of non-native organisms, both plants and other insects, on native bees range from insignificant or at least undetectable to dramatic.

Invasive plants

Alien invasive plants are widespread in virtually every terrestrial habitat on earth. The horticultural trade is a major source of invasive plants, and in some regions of the world, the local flora can be dominated by non-natives. In a typical British home garden, for example, non-natives comprise 67% of the plant species (Thompson et al. 2003). Many of these non-native plants produce showy, abundant floral displays that are attractive to native solitary bees. The impacts of native plants on bees can be varied and complex.

Introduced plants sometimes augment floral resources by providing more pollen or nectar rewards than other species or by blooming earlier or later than the native flora. This shift in the quantity and timing of floral resource availability may have a positive or negative impact on native bees. *Impatiens glandulifera* has spread widely and is so attractive to native bees in Central Europe that, at some sites, there was a marked reduction in bee visitation to a native plant, *Stachys palustris* (Chittka and Schürkens 2001, but see references in Stout and Morales 2009). *Tamarix aphylla* and *T. ramosissima* inhabit riparian areas and are considered among the worst invasive species in western North America. The minute flowers are produced in abundance and open in the midsummer gap between the early-spring and late-summer flowering of native plants (Andersen and Nelson 2013). In some areas, *Tamarix* greatly increases floral resources (Tepedino et al. 2009) during the driest part of the year. One negative effect of *Tamarix* invasion for solitary bees is that *Tamarix* allows social bees to persist in areas where they would normally be absent or rare due to the lack of midsummer floral resources. The greater abundance of social bees increases the probability of competition with the native, solitary bee community.

Non-native plants can strongly affect solitary bee abundance if they invade otherwise intact habitats and displace native plants. Starthistle, *Centaurea solstitialis*, arrived in North America early in the 1800s. It is a highly aggressive invader of grasslands in California and Oregon that grows in dense stands, is toxic to livestock, and greatly depletes soil moisture (Gutierrez et al. 2005). Starthistle is avoided by most native solitary bees, but it attracts native social species, the introduced honey bee and two introduced cavity-nesting bees, *Megachile apicalis* and *M. rotundata* (McIver et al. 2009). Starthistle provides a frightening example of how a non-native plant and non-native pollinators (honey bees and non-native *Megachile*) can be mutually beneficial at the expense of both the native flora and fauna. Not only does starthistle displace native host plants by competing for open space, water, and nutrients, it also benefits a non-native bee (*M. apicalis*) that usurps active nests and uses abandoned nests of native bee species (Barthell et al. 1998). The array of interactions would have been difficult to predict when starthistle first arrived in California in the 1800s, yet it provides a cautionary tale for intentional or accidental introductions of bees and plants.

Finally, invasive plants that alter the frequency and intensity of disturbance can profoundly affect solitary bees. Grasslands and deserts worldwide have been dramatically altered because of exotic grasses (Chapin et al. 2000). For example, buffelgrass (*Pennisetum ciliare*), a xeric-adapted bunchgrass, has been spread through warm deserts worldwide to increase forage for cattle. In Australia and North America, buffelgrass increases the frequency of wildfires, which kill native vegetation and open more habitat for buffelgrass invasion. This positive feedback has eliminated large swaths of native vegetation (Arriaga et al. 2004). Not only does buffelgrass offer no floral resources to wild bees, the wildfires it supports also eliminate nesting resources, such as resins, wood, and hollow stems used by solitary bees. There has been no attempt to document the impact of buffelgrass on solitary bee diversity and abundance, but it is likely significant because buffelgrass has invaded areas of high bee species richness.

BOX 14-2: *COLLETES HEDERAE*—AN IVY SPECIALIST WITH A RAPIDLY EXPANDING RANGE IN EUROPE

Colletes hederae is a mysterious bee. A member of the *C. succinctus* species group (Kuhlmann et al. 2007), it was only recently described as distinct from a closely related species, *C. halophilus* (Schmidt and Westrich 1993). Since the original description in 1993, *C. hederae* has undergone a rapid range expansion (Dellicour et al. 2014). When originally described, the range was known to include much of France and continue eastward to northern Italy. However, by 2002, it had expanded its range eastward to the Adriatic Sea and northward into southern England. By 2006, it had spread westward to southern Spain and eastward to Belgium, Germany, and as far as Greece. Since the original description in 1993, the range of this bee has expanded roughly 10-fold in geographic area (Dellicour et al. 2014). The arrival in southern England is especially well documented thanks to the efforts of the Bee Wasp and Ant Recording Society (BWARS). In 1993, the year *C. hederae* was originally described, the BWARS started surveying sites in southern England, and they continued these surveys on an annual basis. In 2001, *C. hederae* was first collected at a site that had been surveyed previously, providing a very precise estimate of when the bee first arrived. *C. hederae* is now common throughout much of southern England, including the coastal counties of Cornwall, Devon, Dorset, Hampshire, Sussex, and Kent, and increasingly northward and inland. Large nesting aggregations consisting of tens of thousands of individuals occur throughout this range.

What factors could explain the rapid range expansion of *C. hederae*? Dellicour and colleagues (2014) provide a number of hypotheses. First, this species is a host-plant specialist on ivy (*Hedera helix*), a widespread and abundant host plant ranging from Spain and North Africa eastward to Greece and western Turkey. Closely related species within the *C. succinctus* species group specialize on much less widely distributed host plants. The rapid expansion of *C. hederae* could represent a post-glacial recolonization of Europe from an Italian refuge facilitated by the broad geographic range of its host plant (Kuhlmann et al. 2007). Second, *C. hederae* has unusually effective dispersal capabilities. Based on an analysis of molecular variation in populations across the geographic range of the species, Dellicour and colleagues (2014) showed that colonization of southern England occurred via repeated dispersal across the English Channel. The dispersal to England was not a singular, rare event—it has occurred repeatedly—suggesting that these bees have a remarkable dispersal capability compared to many solitary, oligolectic, ground-nesting bees. Third, population growth rates in this species are rapid. Females can produce up to 18 brood cells over the roughly one month of adult activity (from late September to early November). This is roughly three times the per-capita offspring production of closely related *C. halophilus* (Sommeijer et al. 2012). Finally, climate change, specifically the increasing frequency of summer heat waves in Europe over the past two decades, may be driving this species northward.

    C. hederae illustrates an important point about solitary bees. Even host-plant specialists that we normally think of as showing low dispersal capabilities and a high level of philopatry can undergo rapid range expansion when the preferred host plant is abundant and widely distributed.

Invasive bees

A surprising number of bees—76 species in total as of 2017—have been introduced accidentally or intentionally to previously unoccupied regions of the world (Table 14-2; modified from Russo 2016). These 76 species represent 26 genera from 5 of the 7 bee families. There are no known non-native species of Melittidae or Stenotritidae and relatively few species of Andrenidae, Halictidae, and Colletidae. The vast majority of invasive species fall into 2 families: Apidae and Megachilidae.

    Most introduced bees were accidentally translocated to their new home, but nearly one-quarter were introduced intentionally by humans (Russo 2016). Honey bees were intentionally introduced practically everywhere in the world where there are pollinator-dependent crops. Several related species of *Apis* have also been intentionally or accidentally introduced to new regions. Other species that have been intentionally introduced for crop pollination include *Megachile rotundata*; *Osmia cornifrons*, *O. cornuta*, and *O. ribifloris*; *Anthophora villosula*; *Nomia melanderi*; and several species of *Bombus*. The vast majority of accidentally introduced species are above-ground cavity nesters (e.g., species in the genera *Hylaeus*, *Osmia*, *Megachile*, and *Ceratina*) or wood nesters (e.g., *Lithurgus*, *Xylocopa*) that have spread along with the international trade of wood and plant materials. There is even one brood parasitic bee (*Coelioxys coturnix*) that has expanded its range into the New World along with its presumed host (*Megachile rotundata*). Non-native bees are often first detected near ports or coastal areas. Ontario, Canada, is a major point of entry for commodities coming from

Table 14-2. Non-native bee species, grouped by family, that are known from around the world.

Modes of introduction include accidental (A) and intentional (I). Additional columns indicate likely year of introduction, likely geographic source, and current non-native range. Question marks indicate uncertainty about the exact date of arrival or geographic source.

Data originally summarized and analyzed by Russo (2016). Reports of non-native species introduced into North America since 2016 include Martins et al. (2017), Normandin et al. (2017), and Gibbs and Dathe (2017).

Family	Species	Mode of introduction	Year	Origin	Invasive range
Colletidae	Chilicola rostrata	A	2008	Argentina	Chile
	Hylaeus communis	A	2017	Europe	eastern North America
	Hylaeus variegatus	A	1990	North Africa	New York City
	Hylaeus albonitens	A	1995	Australia	Hawaii
	Hylaeus hyalinatus	A	1990	Europe	eastern North America
	Hylaeus leptocephalus	A	1990	Europe	eastern North America
	Hylaeus pictipes	A	2017	Europe	eastern North America
	Hylaeus punctatus	A	1980	Europe	North and South America
	Hylaeus strenuus	A	2007	Asia	Hawaii
Andrenidae	Andrena wilkella	A	1900s	Europe and N. Asia	eastern North America
Halictidae	Halictus tectus	A	2000	Eurasia	eastern North America
	Lasioglossum eleutherense	A	1990	Bahamas and Cuba	Florida
	Lasioglossum imbrex	A	2013	US	Hawaii
	Lasioglossum impavidum	A	2003	western US	Hawaii
	Lasioglossum leucozonium	A	1900	Eurasia	North America
	Lasioglossum microlepoides	A	2013	continental US	Hawaii
	Lasioglossum zonulum	A	?	Eurasia	North America
	Nomia melanderi	I	1970	North America	New Zealand
Megachilidae	Afranthidium repetitum	A	2000	Africa	Australia
	Anthidium florentinum	A	2017	Europe	eastern North America
	Anthidium manicatum	A	1960	Eurasia, North Africa	widespread
	Anthidium oblongatum	A	1990	Eurasia	eastern North America
	Anthidium vigintiduopunctatum	A	2006	South America	Galápagos
	Chelostoma campanularum	A	1960	Eurasia	eastern North America
	Chelostoma rapunculi	A	1960	Eurasia	eastern North America

	Species	Status	Year	Origin	Destination
	Coelioxys coturnix	A	2000	Eurasia	eastern North America
	Heriades truncorum	A	2010	Eurasia	eastern North America
	Hoplitis adunca	A	2016	Eurasia, Africa	Britain
	Hoplitis anthocopoides	A	1960	Eurasia, Africa	eastern North America
	Lithurgus bractipes	A	?	?	Fiji
	Lithurgus chrysurus	A	1970	Eurasia, N. Africa	eastern North America
	Lithurgus huberi	A	1907	Asia	South America
	Lithurgus scabrosus	A	1907	Europe	Hawaii, Vanuatu
	Megachile apicalis	A	1930	Eurasia	North America
	Megachile australis	A	?	Southeast Asia	Vanuatu, Samoa
	Megachile chlorura	A	1988	Philippines	Hawaii
	Megachile concinna	A	?	Africa	Caribbean
	Megachile pusilla	A	?	Europe, N. Africa	eastern North America
	Megachile ericetorum	A	2000	Eurasia	eastern North America
	Megachile fullawayi	A	1921	Guam	Hawaii
	Megachile gentilis	A	?	western North America	Hawaii
	Megachile lanata	A	1700–1800	India and China	North and South America
	Megachile rotundata	I, A	1920–1940	Eurasia	North America, South America, Australia
	Megachile rufipennis	A	1511–1867	Europe	Antilles
	Megachile sculpturalis	A	1990	east Asia	North America and Europe
	Megachile timberlakei	A	2010	Hawai ?	Galápagos
	Megachile umbripenne	A	2013	Asia	Fiji, Samoa, Hawaii
	Osmia caerulescens	A	1800s	Eurasia	North America, New Zealand
	Osmia cornifrons	I	1960	eastern China, Japan	North America, Europe, Korea
	Osmia cornuta	I	1980	Eurasia	not documented
	Osmia ribifloris	I	1991	western North America	Maine, eastern North America
	Osmia taurus	A	2000	eastern China, Japan	eastern North America
	Pseudoanthidium nana	A	2000	Eurasia	eastern North America
Apidae	*Amegilla pulchra*	A	?	Australia	Fiji
	Anthophora villosula	I	1980	Japan	eastern North America
	Apis cerana	A	2007	Asia	Indoaustralia and Pacific Islands, Russia, Iran
	Apis dorsata	A	?	Asia	Japan

(continued)

Table 14-2. (Continued)

Family	Species	Mode of introduction	Year	Origin	Invasive range
	Apis florea	I	1985	Oman, Asia, Indonesia	Iraq, Sudan
	Apis mellifera	I	1620	Europe	Americas, Australia, globally
	Bombus hortorum	I	1885	United Kingdom	New Zealand
	Bombus impatiens	I	2003	North America	South and Central America
	Bombus lucorum	?	1981	Europe, China	Iceland
	Bombus ruderatus	I	1885	United Kingdom	New Zealand, South America, Canary Islands
	Bombus subterraneous	I	1885	United Kingdom	New Zealand, Chile, China, Israel, etc.
	Bombus terrestris	I	1885	United Kingdom	Asia, Africa, North America, New Zealand, Tasmania
	Braunsapis puangensis	A	2003	Asia, India	Fiji
	Centris nitida	A	2000	western North America, South and Central America	Florida
	Ceratina arizonensis	A	1950	western US	Hawaii
	Ceratina cobaltina	A	1970	Mexico	Texas
	Ceratina dallatorreana	A	1940	Mediterranean	California
	Ceratina dentipes	A	1909	Turkey, Cyprus, S. Asia, Australia	Pacific Islands, Japan
	Ceratina smaragdula	I	1960	Pakistan, India, southeast Asia	Hawaii, Australia
	Euglossa dilemma	A	2000	Mexico and Central America	Florida
	Plebeia frontalis	I?	2010	Mexico, Central and South America	California
	Xylocopa appendiculata	A	2010	Japan and China	California
	Xylocopa augusti	A	2013	Argentina	Chile
	Xylocopa sonorina	I	?	western US	Pacific Islands, Japan
	Xylocopa tabaniformis	A	1990	South Texas	western US
	Xylocopa tranquebarorum	A	2005	Asia	Japan

Europe and Asia, and 16 of 17 exotic species in Canada occur there (Sheffield et al. 2011). Relatively few ground-nesting bees have been introduced.

Some species listed in Table 14-2 (*Megachile sculpturalis* and *Anthidium manicatum*, discussed below) have greatly expanded their ranges since their arrival in the new area. However, most species have established viable populations but have not spread far from where they were first discovered. For some species, suitable habitat limits their spread. *Euglossa dilemma* was first captured in Florida in 2000 and is now common along Florida's east coast (Hinojosa-Díaz et al. 2009). The distances euglossine bees have been known to fly are among the greatest recorded (Pokorny et al. 2015). However, *E. dilemma* is not expected to expand its range beyond those parts of Florida with tropical vegetation (Hinojosa-Díaz et al. 2009). *Anthophora villosula* was introduced near Washington, DC, in 1980 and is now abundant within this metropolitan area (Sam Droege pers. comm.), but it has not spread appreciably despite the presence of seemingly suitable habitat in the urban areas that sprawl continuously to the north and south.

Islands have been particularly frequently hit by non-native bees. Hawaii, Fiji, New Zealand, and the Galápagos have all been recipients of non-native bee species. Hawaii is one of the most frequent destinations for non-native bees. Thirteen of the 76 bees listed in Table 14-2 are non-native species in Hawaii alone. The native bee fauna of Hawaii consists of 60 species of *Hylaeus* (*Nesoprosopis*) (Daly and Magnacca 2003). A large proportion of the native *Hylaeus* are threatened or endangered: 7 species are restricted to endangered habitats, 10 are very rare and potentially endangered, and 10 additional species could be extinct (Magnacca 2007). As of 2017, a total of 27 species of Hawaiian *Hylaeus* are listed as "critically imperiled" or "possibly extinct" in the United States (http://www.xerces.org /pollinator-redlist/). Given the unique and highly vulnerable native bee fauna of Hawaii, invasive plants and bees could be a serious problem there.

Non-native bees potentially impact native species in a variety of ways. Competition for floral and nesting resources is the most obvious, but non-native bees may also transmit pathogens to native species (treated below), and reproductive disruption may occur via interspecific mating with non-native congeners (Goulson 2003, Stout and Morales 2009).

Floral resource competition

The most thoroughly studied interaction between native and non-native bees is inter-specific competition for floral resources. If exotic and native bees depend on the same plant species, native bees should be negatively impacted when resources are limited.

The European honey bee (*Apis mellifera*) is the most widespread and abundant non-native bee on earth. Colonies consist of tens of thousands of workers, and honey bees are broadly polylectic, visiting virtually any flowering plant within their foraging range. They also consume vast quantities of pollen and nectar. Over a three-month period, a single honey bee colony is estimated to collect over 650,000 pollen loads, or 10 kilograms of pollen, during the peak summer months (Winston 1987, Seeley 1995). Cane and Tepedino (2017) estimated the potential impact of honey bees on an average-sized solitary bee in terms of lost reproductive potential. They estimated that a single honey bee colony consumes pollen equivalent to approximately 110,000 solitary bee progeny over a three-month period. A typical apiary (~40 colonies) would consume enough pollen to support approximately 4 million solitary bee offspring. This back-of-the-envelope calculation makes some simplifying assumptions (that solitary bees and honey bees are visiting exactly the same plants for pollen) but demonstrates that the potential impact of honey bees on native, solitary bees could be dramatically reduced fecundity and population size.

But do honey bees really have such a dramatic impact on solitary bee populations? The literature is surprisingly inconclusive. Paini (2004) reviewed 28 studies focused on the impacts of honey bees on native bees. These studies were conducted in a wide variety of locations, including Europe (where *A. mellifera* is native), but also Central and South America, Australia, New Zealand, India, and Japan (where *A. mellifera* is not native). The majority of studies focused on what Paini called *indirect impacts*, such as resource overlap or changes in floral visitation rates. A smaller number of studies focused on measuring *direct impacts*, such as reduced survival, fecundity, and population density. Indirect impacts (such as demonstrating floral resource overlap) are consistent with competition between honey bees and native bees, but resource overlap does not necessarily lead to reduced fecundity or diminished population size in native bees if they are capable of altering their foraging behavior in the presence of honey bees. In other words, indirect effects do not provide definitive evidence that honey bees are suppressing native bee populations.

The majority of studies concluded that honey bees have a negative impact on native solitary and social bees, but as pointed out by Paini, many of these studies lacked sufficient replication to be convincing. Most studies were done at one or two sites or over a single year, for example. Some studies also showed no impact. Roubik and Wolda (2001) monitored bee abundance at light traps on Barro Colorado Island, Panama, for 7 years before the arrival of honey bees and 10 years following their arrival. They found no decrease in abundance of the 15 most common native bee species after the arrival of honey bees. Paini and colleagues (2005) conducted an elegant study in which they compared the fecundity of a native leaf-cutter bee (*Megachile* sp. 323) before and after the installation of apiaries in a region of western Australia. They found no significant change in reproduction and fecundity with the arrival of the honey bee colonies.

Why has evidence for competition among native bees and honey bees been so elusive, in spite of the apparent potential for floral resource competition? One possibility is that native bees may be able to alter their foraging in the presence of honey bees in order to avoid competition. Many native bee species are broad floral generalists. This foraging flexibility allows individuals to adjust their host-plant use in response to the changes in the floral community over the season and across years. Another possibility is that honey bees, because they communicate within the colony about the location of floral resources, can shift rapidly to the most rewarding floral resources, even when these resources are located far from the nest. This ability may allow honey bees to shift rapidly to the most profitable floral resources (O'Neal and Waller 1984, Visscher and Seeley 1982) and thereby leave behind some fraction of the resources at less-profitable floral patches for the solitary bees.

Nest-site competition

While competition for floral resources has been the focus of many studies, we know much less about competition for nest sites. Non-native cavity nesters can potentially compete with native species when their nest-site preferences are similar. *Osmia cornifrons* and *O. taurus* are two non-native, cavity-nesting bees that utilize exactly the same nest diameters as a native eastern North American species—*O. lignaria*. Nest-site competition is entirely possible, but we are not aware of any studies that have examined the implications of this kind of resource overlap in native and non-native, cavity-nesting bees. However, there is one case in which a non-native appears to be doing significant damage to native species, and the conflict occurs over nest sites.

Megachile sculpturalis, the giant resin bee, is native to Asia. Female *M. sculpturalis* nest in pre-formed cavities and holes and construct brood cells with plant resins. This species is impressively flexible in its choice of nest sites. In Japan, where it is native, nests were found in the hollow center of a rotten pine tree root, old galleries in the nest of the bee *Anthophora villosula* (Apidae), in mud cells of the wasp *Ancistrocerus fukianus* (Vespidae), in cut bamboo used for fences and plant trellises, and in a folded notebook (Iwata 1933). *M. sculpturalis* was first recorded in North Carolina in 1994 (Mangum and Brooks 1997). By 2015, it had spread north into Canada along the Great Lakes, south to the Gulf Coast of the United States, and west to the Great Plains in Kansas (Parys et al. 2015)—a massive range expansion over a very short period of time. More recently (2008), *M. sculpturalis* was reported in France, and by 2015, it had extended its range over the Alps into Germany (Westrich et al. 2015). It is not clear if the rapid range expansion is entirely due to the natural dispersal capacity of *M. sculpturalis* or if the bee has been accidentally transported on occasion by humans. The latter is entirely possible given the breadth of substrates this species uses for nests. The spread of *M. sculpturalis* has been remarkably well-studied because its large size makes it very conspicuous, it often nests in buildings, and it prefers ornamental plants as floral hosts.

M. sculpturalis poses a major threat to the native bee fauna, especially carpenter bees. This large, aggressive resin bee has been reported to forcibly usurp active nests of a native carpenter bee (*Xylocopa virginica*). Roulston and Malfi (2012) observed an attack on an active carpenter bee nest in which the giant resin bee attacked and physically removed adult carpenter bees guarding the nest entrance. During the attack, the resin bee grabbed the legs of the carpenter bee, bit the carpenter bee in the head, and made repeated attempts to sting it. In addition, the resin bee appeared to use sticky resins to gum up the wings of the carpenter bee. Laport and Minckley (2012) reported a similar example of nest usurpation, but found that *X. virginica* regained possession of the usurped nest the following year. The rapid spread of the giant resin bee throughout the eastern United States (and Europe) and the reports of nest usurpation suggest that this bee may significantly impact native carpenter bee populations across an enormous geographic range. The long-term impact of *M. sculpturalis* on native carpenter bees may prove to be significant.

There are many unknowns about how non-native bees affect native bee faunas. The lack of information on non-native species in natural and semi-natural habitats suggests that most non-native bee species remain closely associated with human-modified ecosystems and are naturalized rather than invasive. However, there have been more studies conducted in urban and suburban habitats than in natural habitats, so we may have a very incomplete understanding of the impacts of non-native bees in undisturbed, natural habitats. Our understanding of the impacts of introduced plants and bees on native bees comes from studies on a very small group of species. It has been astonishingly difficult to demonstrate clear impacts of the widespread, resource-hoarding honey bee on native bees. Carefully designed, long-term experiments and community-level studies would greatly improve our understanding of how native bees are impacted by introduced plants and bees.

Pathogen spillover

One of the greatest threats posed by non-native bee species to native bee communities is the introduction of non-native pathogens—a phenomenon called *pathogen spillover* (Stout and Morales 2009). European honey bees have acquired some of their most devastating pathogens and ectoparasites through contact with other, closely related species of *Apis*. Both the external parasitic mite *Varroa destructor*, the single major source of overwintering mortality in managed European honey bees, and the microsporidian parasite *Nosema ceranae* were obtained through contact with *A. cerana* when *A. mellifera* was imported into Asia (vanEngelsdorp and Meixner 2010). *V. destructor* is now endemic in both managed and feral honey bee colonies around the world (except for Australia).

Managed European honey bee colonies are loaded with pathogens, including fungi, bacteria, microsporidia, trypanosomes, and viruses, and there is increasing evidence that honey bees are a source of pathogens for wild bee species. Two studies (Fürst et al. 2014, McMahon et al. 2015) have demonstrated that honey bee colonies share pathogens, including *Nosema* and several viruses, with wild bumble bee colonies living nearby. Pathogen levels in honey bee colonies were positively correlated with pathogen levels in nearby bumble bee colonies, and analysis of pathogen sequence data suggests that honey bees are the source of the pathogen infection in wild bumble bees. Commercially managed bumble bees, which are heavily loaded with pathogens, are also a source of infection for wild bumble bees (Graystock et al. 2013).

It is not too surprising that honey bees and bumble bees exchange pathogens; they are both eusocial, and the two genera are closely related apid bees. But are more distantly related solitary bees also prone to infection from honey bee pathogens? We have a very limited understanding of the pathogens infecting solitary bees, but some of the viruses and fungi reported from honey bees have also been detected in solitary bees (Singh et al. 2010, Evison et al. 2012). Ravoet and colleagues (2014) conducted a thorough survey of pathogens infecting solitary ground-nesting (*Andrena*) and stem-nesting (*Osmia, Heriades*) bees in the vicinity of honey bee colonies in the Netherlands. Solitary bees were infected with the same species of *Apicystis* and *Nosema* as the honey bee colonies, and the solitary bees shared a total of five viruses with the honey bees. This is the first study to document such extensive coinfection in honey bees and solitary bees.

It is possible that these viruses have been circulating in wild and managed bee populations for a long time.

Honey bees and bumble bees may not be the only source of inter-specific pathogen transmission in solitary bees. Solitary bees themselves, when introduced accidentally or intentionally into new regions of the world, can carry pathogens with them. *Osmia cornifrons*, a Japanese bee managed for orchard pollination, was imported into the United States in the 1970s for apple pollination (Batra 1978b). A study by Hedtke and colleagues (2015) found that North American *O. cornifrons* harbor species of the fungal genus *Ascosphaera* originally described from their native range in Japan. This raises the possibility that the Japanese *Ascosphaera* could spread to related species of *Osmia* in North America. It is possible, but has not yet been confirmed, that the decline of *O. lignaria* in eastern North America is related to the introduction of these novel, Japanese *Ascosphaera*.

If pathogen "spillover" is a threat to wild bees, how is it taking place? An increasing body of evidence suggests pathogen transmission among managed and wild bee species occurs via floral visitation (Graystock et al. 2015). Flowers are the hubs where multiple pollinator species interact. Many bee pathogens are transmitted via oral or fecal contact, and both pollen and nectar have been shown to harbor bee pathogens (Graystock et al. 2015, Figueroa et al. in review). Flower visitation seems to be a likely avenue for pathogen exchange among wild and managed bees.

We know very little about the symptoms and consequences of pathogen infection in wild, unmanaged bees, so it is difficult to know if the documented cases of pathogen infection are actually having impacts on bee longevity and reproductive success. Paxton and colleagues (1997) documented extremely high levels of microsporidian infection in *Andrena scotica*, a communal, ground-nesting bee they studied on the Baltic island of Öland, Sweden. Of the 87 male and 56 female bees collected at first emergence from their nests in the spring, 87% had abundant microsporidian spores in the adipose tissue of the metasoma. In spite of the heavy pathogen loads and high frequency of infection, Paxton and colleagues (1997) were unable to detect any negative impacts of pathogen infection, such as reduced adult body size or decreased longevity.

While limited, our understanding of pathogen impacts on wild bees suggests that managed bees, because they are transported widely around the world and because they harbor multiple viral, bacterial, fungal, and microsporidian pathogens, pose a potential threat to the wild bee fauna. Whether these pathogens impact bees in subtle or dramatic ways is still unclear. Studies of the pathogenicity of the diverse microorganisms that are now circulating in managed and wild bee populations are needed to assess the risk pathogens pose to the conservation and long-term viability of wild bees.

LOW GENETIC VARIABILITY AND LIMITED INTER-POPULATION GENE FLOW

Population genetic studies can provide key information for assessing the conservation status of solitary bees, because they document the levels of genetic variability within and among populations and provide insight into the potential to respond evolutionarily to environmental change. An understanding of how much gene flow occurs among populations (i.e., the genetic "connectivity" among populations), provides information on the conservation status of individual populations. These studies provide information on whether populations are small and isolated, large with considerable immigration and emigration, or are rapidly expanding after a genetic "bottleneck," which can severely diminish overall genetic variability. Population genetic studies also provide an empirical measure of the diploid male frequencies, which are important for assessing the extent of inbreeding in haplodiploid organisms.

All bee species (in fact, all Hymenoptera) are haplodiploid, meaning males are haploid and females are diploid. Sex determination in Hymenoptera is based on allelic variation at a single autosomal locus, the complementary sex determination (csd) locus. In large, outbred populations, this locus typically has high levels of genetic variation (typically 9–20 alleles, but there can be as many as 40; Zayed and colleagues 2007, Zayed and Packer 2005), such that diploid individuals are normally heterozygous (and, hence, female), whereas haploid (hemizygous) individuals are male. This system works beautifully when there are high levels of allelic variation at the csd locus. However, when allelic variation is reduced—such as when populations become fragmented, small, and highly inbred—some diploid individuals can be homozygous at the csd locus. Diploid, homozygous individuals are morphologically male but are sterile and incapable of reproduction. Using standard methods of genotyping (allozymes, microsatellites, single-nucleotide polymorphisms [SNPs]), one can measure the frequency of diploid males. The frequency of diploid males in wild populations is a useful measure of inbreeding (Souza et al. 2010, Zayed et al. 2004).

While haplodiploid species are generally considered to be less susceptible to the negative effects of inbreeding depression, recent theoretical models suggest that haplodiploidy can actually increase the vulnerability to extinction (Zayed and Packer 2005). The logic is as follows: As populations become isolated and genetically depauperate, reduced genetic variability at the csd locus can lead to an increasing frequency of diploid males. Increased diploid males have two negative impacts on the population. First, diploid males are a drain on female reproductive success, because they represent individuals that would have become reproductive females but instead have zero reproductive potential. In addition, females that mate with diploid males produce sterile offspring, reducing female reproductive success even more. Zayed and Packer (2005) modeled inbreeding in haplodiploid organisms and showed theoretically that this can lead to a rapid, downward spiral in population-level reproductive success and local extinction. They termed this the *diploid male extinction vortex* to convey the idea that, once a small population with low levels of genetic variability starts on a downward trajectory, it can lead very rapidly to local population extinction.

Nearly 40 studies have examined the population genetics of solitary bees (Table 14-3). The studies have included a variety of genetic markers, including allozymes (protein polymorphisms), microsatellite markers (highly variable, nucleotide polymorphisms), and direct sequencing of both mitochondrial and nuclear genes. Studies have focused on every family of bee, excluding the tiny family Stenotritidae. These studies have identified species with high levels of genetic variability and substantial gene flow among populations (e.g., *Andrena vaga*; Exeler et al. 2008, Černá et al. 2013a), as well as rare species with low levels of genetic variability and highly fragmented, isolated populations (e.g., *Colletes floralis*; Davis et al. 2010). Most studies document genetic variability across the native range of the species, but other studies have documented genetic variability in rapidly expanding populations (e.g., *C. hederae*; Dellicour et al. 2014) and a non-native, recently introduced species (*Lasioglossum leucozonium* in North America; Zayed et al. 2007). Studies have also compared genetic variability among closely related host-plant specialists and host-plant generalists in order to evaluate if host-plant specialization leads to higher levels of inter-population genetic differentiation.

One can draw some broad conclusions from these studies. First, habitat and host-plant specialists tend to show lower levels of genetic variability and lower levels of inter-population gene flow than habitat or host-plant generalists. Not surprisingly, habitat and host-plant specialist bees appear to exist in much more discrete, isolated populations than habitat and host-plant generalists. Many solitary bee species are highly philopatric, meaning that females are likely to remain at the site from which they emerged. In some species, females even reuse their mother's nest. The small foraging ranges of many solitary bees, coupled with the patchy distribution of host plants and female philopatry, can impose substantial barriers to gene flow among isolated populations. Packer and Zayed conducted a series of studies of closely related pairs of oligolectic and polylectic species in Chile to assess the impact of host-plant specialization on gene flow in solitary bees (Packer et al. 2005, Zayed et al. 2005). In every case, the host-plant specialist showed lower levels of genetic variability (as measured by expected heterozygosity) and more highly fragmented populations (as measured by Fst[1]) than the host-plant generalist. This has important conservation genetic implications for solitary bees and suggests that oligolectic species may be more prone to the negative effects of habitat fragmentation and inbreeding depression than broadly polylectic species. Indeed, one of the few studies of long-term persistence of bee populations found that oligolectic bees were more likely to experience local population extinctions than polylectic species (Biesmeijer et al. 2006).

Second, some bee populations have tolerated extremely low levels of genetic variability. *Lasioglossum* (*Lasioglossum*) *leucozonium* (Zayed et al. 2007) and *Peponapis pruinosa* (López-Uribe et al. 2016) have widespread geographic distributions over much of North America yet show strikingly low levels of genetic variability. *L. leucozonium* is a particularly interesting example of a rapidly expanding, genetically depauperate species in North America. *L. leucozonium* is a solitary, polylectic, ground-nesting bee that occurs in both Europe and North America. McGinley (1986) reviewed the taxonomy of North American species of the subgenus *Lasioglossum* and suspected that this species was likely introduced because the closest relatives of *L. leucozonium* (including *L. zonulum*) are all Old World species, and it seemed to be an outlier relative to the majority of North American *Lasioglossum* sensu stricto. Zayed and colleagues (2007) compared genetic variation based on microsatellite loci and mitochondrial sequence

---

[1] Fst is a widely used measure of genetic differentiation among populations. Higher values of Fst indicate that populations are more highly differentiated genetically, whereas lower values indicate more homogeneous levels of genetic variation.

Table 14-3. Population genetic studies of solitary bees and the types of genetic markers used.

Note that we have focused on studies that quantify genetic variability across multiple populations. We did not include the enormous number of euglossine species analyzed strictly for evaluating the frequency of diploid males (Roubik et al. 1996, Souza et al. 2010, Takahashi et al. 2001, Zayed et al. 2004).

mt = mitochondrial

nuc = nuclear

Family	Species	Methods	Biology	Reference
Andrenidae	*Acamptopoeum submetallicum*	allozymes	host-plant generalist	Packer et al. 2005
Andrenidae	*Andrena fuscipes*	microsatellites	heathland specialist	Exeler et al. 2010
Andrenidae	*Andrena jacobi*	microsatellites	communal bee with inbreeding	Paxton et al. 1996
Andrenidae	*Andrena lapponica*	allozymes	host-plant generalist	Pamilo et al. 1978
Andrenidae	*Andrena scotica*	microsatellites	communal bee with inbreeding	Paxton et al. 2000
Andrenidae	*Andrena vaga*	microsatellites	habitat and host-plant specialist	Exeler et al. 2008
Andrenidae	*Andrena vaga*	microsatellites	habitat and host-plant specialist	Černá et al. 2013a
Andrenidae	*Andrena vaga*	allozymes	habitat and host-plant specialist	Pamilo et al. 1978
Andrenidae	*Andrena wilkella*	allozymes	host-plant generalist	Pamilo et al. 1978
Andrenidae	*Macrotera portalis*	microsatellites	host-plant specialist	Danforth et al. 2003b
Andrenidae	*Nolanomelissa toroi*	allozymes	host-plant specialist	Packer et al. 2005
Apidae	*Amegilla dawsoni*	microsatellites	homogeneous landscape	Beveridge & Simmons 2006
Apidae	*Anthophora plumipes*	microsatellites	widespread, host-plant generalist	Černá et al. 2017
Apidae	*Anthophora villosula*	microsatellites	widespread, host-plant generalist	Černá et al. 2017
Apidae	*Centris chilensis*	allozymes	host-plant generalist	Packer et al. 2005
Apidae	*Centris mixta*	allozymes	host-plant specialist	Packer et al. 2005
Apidae	*Eulaema bombiformis*	mt and nuc genes	neotropical generalist	López-Uribe et al. 2014
Apidae	*Eulaema bombiformis*	microsatellites	neotropical generalist	Suni & Brosi 2012
Apidae	*Eulaema championi*	microsatellites	neotropical generalist	Suni & Brosi 2012
Apidae	*Eulaema cingulata*	mt and nuc genes	neotropical generalist	López-Uribe et al. 2014
Apidae	*Eulaema meriana*	mt and nuc genes	neotropical generalist	López-Uribe et al. 2014
Apidae	*Peponapis pruinosa*	microsatellites	expanding population	López-Uribe et al. 2016

Family	Species	Marker	Category	Reference
Apidae	Svastra sp.	allozymes	host-plant generalist	Lester & Selander 1979
Apidae	Xylocopa virginica	microsatellites	expanding population	Richards et al. 2016
Colletidae	Cadeguala occidentalis	allozymes	host-plant generalist	Packer et al. 2005
Colletidae	Caupolicana quadrifasciata	allozymes	host-part specialist	Packer et al. 2005
Colletidae	Colletes floralis	microsatellites	rare and endangered	Davis et al. 2010
Colletidae	Colletes hederae	nuc genes	expanding population	Dellicour et al. 2014
Colletidae	Colletes inaequalis	microsatellites	male gene flow	López-Uribe et al. 2015
Colletidae	Colletes seminitidus	allozymes	host-plant generalist	Zayed et al. 2005
Colletidae	Colletes seminitidus	allozymes	host-plant generalist	Packer et al. 2005
Colletidae	Colletes stepheni	mt and nuc genes	sand-dune specialist	Wilson et al. 2009, Nelson & Griswold 2015
Colletidae	Colletes succinctus	allozymes	host-plant generalist	Pamilo et al. 1978
Colletidae	Leioproctus rufiventris	allozymes	host-plant specialist	Zayed et al. 2005
Colletidae	Leioproctus rufiventris	allozymes	host-plant specialist	Packer et al. 2005
Halictidae	Lasioglossum leucozonium	microsatellites	invasive species	Zayed et al. 2007
Halictidae	Lasioglossum oenotherae	microsatellites	host-plant specialist	Zayed & Packer 2007
Halictidae	Patellapis doleritica	microsatellites	habitat specialist	Kahnt et al. 2014
Megachilidae	Megachile inermis	allozymes	host-plant generalist	Packer et al. 1995
Megachilidae	Megachile pacifica	allozymes	host-plant generalist	Lester & Selander 1979
Megachilidae	Megachile relativa	allozymes	host-plant generalist	Packer et al. 1995
Megachilidae	Megachile rotundata	allozymes	host-plant generalist	McCorquodale & Owen 1997
Megachilidae	Neofidelia longirostris	allozymes	host-plant generalist	Packer et al. 2005
Megachilidae	Trichothurgus aterrimus	allozymes	host-plant specialist	Packer et al. 2005
Melittidae	Macropis labiata	allozymes	oil-collecting, host-plant specialist	Pamilo et al. 1978
Melittidae	Macropis fulvipes	AFLP	oil-collecting, host-plant specialist	Triponez et al. 2015
Melittidae	Macropis europaea	AFLP	oil-collecting, host-plant specialist	Triponez et al. 2015

data between European and North American populations of *L. leucozonium* and found a striking reduction in genetic variation in North American populations as compared to a French population. Expected heterozygosity was reduced by nearly half in the North American population, and allelic richness was reduced by 76%. There were, on average, between 2 and 3 alleles per locus in North America and an average of 10 alleles per locus in the single European population studied. There was also no detectable population genetic structure among North American populations, which is in striking contrast to the pattern in a closely related native species, *L. (Sphecodogastra) oenotherae* (Zayed and Packer 2007). Consistent with the low level of genetic variability in the North American population, Zayed and colleagues (2007) found a high proportion of diploid males; 30% of males that were genotyped were diploids. This is one of the highest reported levels of diploid males for any solitary bee based on microsatellite data. The genetic results strongly support McGinley's suspicion that *L. leucozonium* is not native to North America. By comparing levels of genetic variability in French and North American populations, Zayed and colleagues (2007) estimated that the original introduction of *L. leucozonium* into North America consisted of a single, mated female bee. This study demonstrates that some bee species can persist in spite of extremely low genetic diversity and a high proportion of diploid males.

One caveat to keep in mind when examining Table 14-3 is that the potentially most vulnerable species from a conservation standpoint are also the rarest and hardest to collect. To conduct a thorough assessment of genetic variability, one needs to collect a fairly large number of individuals from a number of discrete populations. For the rarest bees, this is often impossible, and hence we have no way of assessing their levels of genetic variability. This means that we cannot accurately assess the conservation genetic status of species that are likely the most vulnerable to extinction.

Table 14-3 does not include a single cleptoparasitic bee. Cleptoparasites are rare relative to their hosts, and cleptoparasites on habitat- and host-plant-specialist hosts are likely to be the rarest of the rare. Indeed, we predict that cleptoparasitic bees, when examined from a conservation genetic perspective, may be among the most vulnerable of bees to local extinction. Comparisons of host-cleptoparasite pairs to determine how population genetic parameters of host and parasite compare over the same geographic area are urgently needed.

One group of solitary, neotropical bees that have been the focus of considerable study are the orchid bees (Euglossini). These large, fast-flying, iridescent green, blue, purple, and red bees have been the subject of numerous population genetic studies focused on estimating the frequency of diploid males. An early report, based on an analysis of allozyme data from seven Panamanian orchid bee species, reported extraordinarily high levels of diploid males. From 12% to 100% of males collected using standard odor baits were diploid (Roubik et al. 1996). This, and subsequent reports based on allozyme data (Zayed et al. 2004), raised the possibility that neotropical orchid bee populations were experiencing significant genetic load due to high levels of inbreeding. A follow-up study by Takahashi and colleagues (2001) of 16 orchid bee species, based on 20 allozyme loci, detected a single diploid male out of the 542 males genotyped (i.e., a frequency of 0.2%), a value considerably lower than that obtained by Roubik and colleagues (1996).

How can one reconcile these widely varying estimates of diploid male frequency in orchid bees? Allozymes are difficult genetic markers to work with, and there is some subjectivity involved in scoring allozyme data. In order to resolve this issue, Souza and colleagues (2010) analyzed microsatellite (DNA) data for over 1,000 males from 27 species across an enormous geographic range (southern Mexico to southern Brazil). Microsatellite data are less prone to scoring errors and therefore provide a more conclusive estimate of diploid male frequencies than allozymes. Souza and colleagues (2010) found only 5 diploid males out of the 1,010 that were genotyped (an overall frequency of 0.5%). They found substantial gene flow among populations, high levels of heterogeneity within populations, and low levels of inbreeding. Orchid bees are host-plant generalists and powerful flyers, so unlike ground-nesting, habitat-specialist bees, they show little inter-population genetic differentiation. In retrospect, it seems that the allozyme studies were likely flawed methodologically and that orchid bees are far from falling into the "extinction vortex" predicted by Zayed and Packer (Souza et al. 2010).

CLIMATE CHANGE, DROUGHTS, AND FLOODING

Climate change is a clear threat to biodiversity on earth. Warming temperatures are already causing shifts in the geographic distribution of species as well as the timing of life history events. Plants are blooming earlier, birds are laying eggs earlier in the season, and many species are showing latitudinal and altitudinal shifts in their distributions.

We know relatively little about the impacts of climate change on solitary bees and other pollinators. The impact of climate change on biotic interactions such as plant-pollinator mutualisms are potentially devastating if temporal synchrony between flowering times and pollinator activity is disrupted (Memmott et al. 2007). Climate change is likely to impact solitary bees in at least three ways. First, warming temperatures may dramatically alter the geographic distribution of bee species, potentially leading to substantial range reductions or fragmentation of populations, especially for species that are specialists on geographically constrained habitats, such as mountaintops or dunes. Second, climate change can have significant physiological effects on both plants and bees. Warming temperatures can alter the quantity and quality of floral resources produced by plants, as well as the flight, foraging, and overwintering capabilities of the bees. Third, warming temperature may disrupt the temporal synchrony between flowering and bee activity if the response of bees and plants to warming differ. Spatial or temporal asynchrony between plants and their pollinators could have huge impacts on bee and plant reproductive success and long-term viability.

Niche-modeling methods provide us with a window into how climate change will impact bee distributions in the future. The idea is simple. Take the current distribution of the species and current climatological conditions to understand the ideal climatological "window" for a species at the present time. Then use the future predictions of climate change to estimate the future distribution of the species given its current climatological preferences. One of the few studies that has used this approach for solitary bees modeled the impacts of two climate-change scenarios on the geographic distribution of 12 solitary bee species in the winter and summer rainfall regions of South Africa (Kuhlmann et al. 2012). The winter rainfall region of South Africa (also known as the Cape Floristic Region) is one of the world's biodiversity hotspots for both plants and animals. This is a relatively small area of just over 150,000 square kilometers that hosts an enormous diversity of flowering plants and some ancient bee lineages that are narrowly host-plant specific (e.g., *Fidelia* [Megachilidae], *Scrapter* [Colletidae], and *Rediviva* [Melittidae]). The winter rainfall region is likely to be significantly impacted by climate change, because it is bounded to the east by mountains and to the south by the South Atlantic and Indian oceans; there are not many places for these winter rainfall bees and their host plants to go. Kuhlmann and colleagues (2012) included narrowly endemic habitat and host-plant specialists as well as species with broader sets of ecological requirements. The response of species to the two climate models were variable, with some species showing range expansion (e.g., *Nomia candida* and *Melitta arrogans*) and most species showing a decrease in geographic range. The most significant negative impacts were on winter-rainfall, host-plant specialists (e.g., *Rediviva aurata* and *Fidelia braunsiana*), which showed decreases of 80–90% of their current distribution. In some species (e.g., *Rediviva intermixta*), the current distribution was predicted to shrink by 99.4%; in other words, they are likely to go extinct. This study clearly highlights the vulnerability of bees that occur in areas where emigration to climatically suitable regions is impossible due to pronounced geographic barriers (i.e., oceans and mountain ranges). A similar study conducted on European species of *Colletes* found less dramatic changes in distribution among a sample of six polylectic and six oligolectic species (Roberts et al. 2011). Host-plant specialists were more severely impacted than host-plant generalists. In this case, the opportunity to migrate to new areas was possible, in part, because of the size of the European landmass and lack of geographic barriers to range expansion. All species shifted from where they occur at present under climate change, but the geographic extent of their distributions decreased only slightly.

A second, more subtle, impact of climate change is the impact on the physiology of flowering plants and their bee visitors. Scaven and Rafferty (2013) provide an excellent review of the likely physiological effects of warming temperatures on plants and bees. Increased temperature is known to have profound effects on floral traits relevant to bee visitation, including when flowering occurs, the number and size of flowers, nectar and pollen quality and quantity, and floral scent production. In a striking example of how climate change can impact floral reward quality, Ziska and colleagues (2016) measured the pollen protein content in goldenrod pollen (*Solidago canadensis*, Asteraceae) from museum collections sampled in the period 1842–2014 (spanning a range of CO_2 levels from 280 to 398 ppm). They found a significant decline in protein content over the 172-year period. In addition, they raised goldenrod plants under a range of atmospheric CO_2 levels (280–500 ppm) and found a significant decrease in pollen protein content with increasing CO_2 levels. These results are significant because pollen protein content is a key component of pollen nutritional value to solitary bees (Roulston and Cane 2000, 2002), and goldenrod is the dominant late-summer melittophilous plant in eastern North America, with a host of both generalist and specialist visitors (Fowler 2016). Bee foraging activity, body size, and life span are all sensitive to temperature, and a warming climate could significantly disrupt the normal foraging behavior of bees as well (Scaven and Rafferty 2013).

Finally, climate change may impact the phenological synchrony between plant flowering and bee activity. Only a handful of studies have examined the impact of phenological shifts in flowering plants and their associated visitors (Forrest and Thomson 2011, Memmott et al. 2007), and few studies have included the solitary bees in the analyses (Bartomeus et al. 2011, 2013; Forrest and Thomson 2011). Bartomeus and colleagues (2011) examined the phenology of 10 early-spring eastern North American bee species over a 130-year period, based on specimen records from North American bee collections. Like the early-spring flowering plants that they visit, the bee species showed a similar advance in activity periods. Over the 130-year period, bees and flowering plants have both advanced by an average of 10 days, in association with an increase in mean annual spring temperature. When one looks just at the window of time between 1970 and 2010, both flowering plants and bees are advancing at nearly two days per decade, suggesting that this trend is accelerating in the more recent time period. At least for eastern North America, bees and flowering plants seem to be tracking the same advances in flowering and emergence, so that temporal synchrony between pollinators and flowering plants is maintained. Forrest and Thomson (2011) carefully monitored trap-nesting bees (and wasps) and their floral hosts in the high elevations of the Rocky Mountains and also found that bees and flowering plants maintained phenological overlap under a variety of conditions. Phenological variation may be a constant feature of the relationship of bees and flowering plants, which has favored host generalization among bees and, at the same time, decreased the cost of phenological mismatch among plant flowering and bee activity. How true this is for other regions of the world deserves further study. Arid, Mediterranean climate regions host the greatest diversity of bees, and we know almost nothing about how bees in these regions will respond to phenological shifts in host-plant flowering.

Predictions are that the frequency and intensity of extreme climatic events will increase with climate change (Easterling et al. 2000, Intergovernmental Panel on Climate Change 2015, Westerling et al. 2006). Extreme climatic events often produce substantial changes in community structure and assembly, and provide a bridge to understand how more protracted disturbances predicted to occur from climate change might impact biodiversity. However, the infrequent and stochastic occurrence of such extreme events limits our ability to study them (Smith 2011 and references therein). Those few community studies that have capitalized on major disturbances (e.g., droughts, fires, floods, plagues) have shown that vertebrate and plant communities have undergone dramatic changes (Miriti et al. 2007, Stapp et al. 2004, Thibault and Brown 2008, Turner 1990), yet little is known about how insect communities respond. A number of studies have documented the impacts of catastrophic weather events on bees. These events cannot be directly linked to climate change, but they do illustrate the point that catastrophic events can and do impact bee survival. Flooding (Fellendorf et al. 2004, Visscher et al. 1994), drought (Mayer and Kuhlmann 2004, Minckley et al. 2013), and even hurricanes (Cane 1997) have been shown to impact solitary bee populations, with varying levels of severity. These studies illustrate the fact that both for climate change, which occurs gradually, and for catastrophic weather events, which occur abruptly, the response of solitary bees can be highly variable and difficult to predict.

Much work needs to be done on how climate change will impact bee species in terms of their geographic distributions, their foraging, reproduction, and life history parameters, and their emergence phenology. Bees are a species-rich group, particularly in some deserts, where many species are narrow host-plant specialists and can remain in diapause for multiple years as last-instar larvae. In their capacity as pollinators, their impact on ecosystem function is greater than the number of species seemingly suggests. Thus, there are basic and applied reasons to examine how bees will respond to climate change and the extreme climatic events expected in the near future.

BOX 14-3: SAND DUNE HABITATS AS CRITICALLY THREATENED REFUGES FOR RARE SOLITARY BEES

In the Northern Hemisphere's temperate zone, sand dunes, whether along the coast or in inland areas, are among the most important and vulnerable habitats for solitary bees. A long list of highly specialized psammophiles (sand lovers) have been described. In Europe, species of *Andrena*, *Colletes*, *Melitta*, *Dasypoda*, *Osmia*, and *Megachile*, and even the brood parasite *Coelioxys*, are all restricted to sand dune habitats. In Wales (United Kingdom) alone, there are 17 solitary bee species that occur exclusively in coastal sand dune habitats and another 44 species with a strong preference for dune habitats (Howe et al. 2010). In the Czech Republic, a country with a well-documented solitary bee fauna, a total of 585 bee species have been recorded, and 24 of those are sand dune specialists (Tropek et al. 2013, J. Straka pers. comm.).

In North America, there is also a diverse fauna of solitary bees restricted to dune habitats. Genera such as *Habropoda*, *Martinapis*, *Melissodes*, *Megachile*, *Anthidium*, *Colletes*, *Andrena*, *Calliopsis*, *Perdita*, and *Hesperapis* all include multiple psammophiles. Many of these have narrow host-plant preferences. *Hesperapis oraria*, known only from the sandy barrier islands of the Gulf Coast of the southeastern United States, forages exclusively on *Balduina angustifolia* (Asteraceae), itself a sand specialist (Cane et al. 1996). Some dune-specialist bees appear to be restricted to a single dune system. The Antioch Dunes are located at the junction of the Sacramento and San Joaquin rivers in northern California. Once an impressive dune system with dunes 35 meters high, it has been relegated by sand mining to a series of low, scruffy sandy habitats now protected in a National Wildlife Refuge. It is still believed to host four endemic bees: *Lasioglossum* (*Sphecodogastra*) *antiochensis*, *Perdita hirticeps luteocincta*, *P. scitula antiochensis*, and *Triepeolus antiochensis*, although the latter, and perhaps others, may now be extinct (Rightmyer 2004). The only known host plant of *Lasioglossum antiochensis*, *Oenothera deltoides howelli*, is also endemic to the Antioch Dunes (McGinley 2003).

Given the specialized and geographically restricted nature of psammophile bees, it is no surprise that many are considered at risk. Of the 24 species of dune specialist bees in the Czech Republic, 11 (46%) are considered critically endangered or already extinct (Tropek et al. 2013, J. Straka pers. comm.). There are many threats to the long-term viability of dune pollinator communities. Dunes have historically been viewed as wastelands in need of development for agricultural and urban uses. They are also popular among enthusiasts of off-road vehicles, which can destroy rare and endangered plants as well as nest sites for bees. An emerging threat to sand dunes is mining. Sand is a key ingredient in concrete, asphalt, glass, and electronics, and sand is being mined at an astonishing rate. In 2010, some 28.6 gigatons of sand were extracted for industrial uses around the world, and much of the extraction is unregulated (Torres et al. 2017). Hydraulic fracking, an increasingly popular mode of fossil-fuel extraction, is also contributing to accelerated rates of sand extraction worldwide. In Texas, fracking has led to a huge increase in sand mining in the Monahans sands region (home to the endemic sand dune specialist, *Lasioglossum* [*Sphecodogastra*] *danforthi*, and many other interesting psammophiles). When conflicts arise between oil exploitation and bee conservation in Texas, bees are highly unlikely to come out on top. Unfortunately, this is a common theme the world over. The conservation of sand dune habitats will be essential if we are to preserve this important part of bee diversity.

INTERACTIONS AMONG THESE EFFECTS

The factors we describe above—habitat loss, invasive species, pathogens and parasites, pesticides, climate change, and loss of genetic diversity—do not impact solitary bees in isolation. There are likely to be interactions among these stressors. Pesticides, for example, are not applied individually in crop fields but in combination. LD50 values, which are calculated on an individual basis under highly controlled laboratory settings, may underestimate the toxicity to bees when diverse insecticides, herbicides, and fungicides are applied in combination. Studies, primarily conducted in honey bees, have shown that certain fungicides can elevate the toxicity of neonicotinoids and organophosphates by as much as 1,000-fold (Goulson et al. 2015 and references therein). In addition, many pesticides are applied with other products, called adjuvants (Ciarlo et al. 2012). Adjuvants, such as oils and various surfactants, are added to the pesticide mix to boost efficacy or provide better binding to plant substrates. Chemical companies are not required to list the adjuvents they add to their pesticide mixes, and these adjuvents are not evaluated for direct toxicity to bees.

Pesticides and pathogens have also been shown to interact. Pesticides can impair the immune response of bees (Di Prisco et al. 2013), and a number of studies demonstrate that simultaneous exposure to sublethal doses of pesticides and pathogens can lead to elevated levels of mortality in honey bees (Doublet et al. 2015) and bumble bees (Fauser-Misslin et al. 2014, Baron et al. 2014). As discussed above, pesticides can have significant non-lethal effects on solitary bees, including impaired foraging and nest relocation, altered sex ratios, and reduced overall reproductive success. These impacts could become more severe when reduced habitat quality leads to reduce floral resource availability.

Habitat loss is likely to interact strongly with pesticide use, if natural habitat is replaced by agricultural habitats (Meehan et al. 2011, Meehan and Gratton 2015), as well as pathogen exposure, if loss of natural habitat increases contact with non-native bees. Habitat loss is also likely to lead to increased foraging distance, which can lead to reduced offspring body size (Chapter 8). Reduced offspring body size may render bees more vulnerable to pesticides and/or pathogens. Increased agricultural intensification has been shown to lead to decreased adult female body size in some solitary bees (*Andrena nasonii*; Renauld et al. 2016), most likely because of limited alternative floral resources or increased exposure to pesticides. Decreased female body size may impact long-term viability of solitary bee populations in agroecosystems. In fact, one can envision a different kind of "extinction vortex" than the one described by Zayed and Packer (2005). As females in unfavorable environments become smaller, they produce smaller female (and possibly male) offspring, as well as a more male-biased brood (see Chapter 8). Populations could reach a point at which body size of reproductively active females is so low that they can no longer construct nests, forage for food, or reproduce. At this point, local population extinction could easily follow.

Invasive pests are likely to lead to increased pesticide use in agricultural habitats and potentially increased bee mortality. An example of an invasive pest that is likely impacting bees is the spotted-wing drosophila (*Drosophila suzukii*). *D. suzukii* is an invasive pest from Asia that arrived in California in 2008 (Asplen et al. 2015). It spread across the United States, reaching the Northeast in 2011. Since 2012, *D. suzukii* has caused significant damage in small berry crops, such as blueberries, raspberries, and blackberries, which are bee-pollinated. Growers apply pesticides post-bloom to control this emerging pest, and elevated pesticide use necessitated by this non-native invader will likely impact wild bees living in agricultural areas.

Climate change will likely have wide-ranging impacts. A warming climate may increase exposure to invasive species, lead to the spread of novel pathogens, and decrease habitat availability. Climate change may also lead to increased pesticide use if novel crop pests expand their ranges.

CHAPTER 15

Epilogue: The Scala Naturae

The Scala Naturae (or Great Chain of Being; Lovejoy 1936) is a 2,000-year-old concept that permeates our thinking about evolution even today. The Scala Naturae is a pre-evolutionary view of how organisms on earth are related to each other. It dates to at least Aristotle's History of Animals but was a common element of 17th- and 18th-century classifications of life on earth. According to the Scala Naturae, all organisms could be arranged in a linear series, from primitive (lower) to advanced (higher). Bonnet (1745) provides a typical Scala Naturae in his Echelle des Etre (Fig. 15-1). Inanimate objects (metals, stones, crystals, and salt) were placed below living organisms, plants were placed below animals, invertebrate animals were placed below vertebrate animals, and mammals, particularly humans, are at the top of the ladder. Bonnet's figure does not include two additional levels that were often included in the typical Scala Naturae of the era—angels and God, which were, of course, above humans.

We no longer use the Scala Naturae to depict relationships among organisms. Branching trees (phylogenies) showing the genealogical interrelationships among organisms have replaced the linear concept implicit in the Scala Naturae. Modern biologists now view organisms as a mixture of primitive and advanced traits, and we try not to use the terms *primitive* or *advanced* when referring to species. But old ideas die hard, and one still finds hints of the Scala Naturae throughout modern classifications and even discussions of broader evolutionary patterns. Lower mammals are those that lay eggs, whereas higher mammals are those that nourish their offspring with a placenta. Lower insects include those that lack wings or undergo gradual development, whereas the higher insects include the Holometabola—insects that undergo metamorphosis. The concept of "higher" and "lower" organisms is still very much embedded in our thinking about evolution even today.

What does the Scala Naturae have to do with bees? There has been a tendency to view bee evolution though the lens of the Scala Naturae. In the classic bee Scala Naturae, Colletidae are at the base of the tree, followed shortly thereafter by the Andrenidae and Stenotritidae, all solitary, ground-nesting bees. Evolution proceeds upward from the short-tongued bees through Melittidae to the long-tongued bees (Apidae and Megachilidae) and, finally, to the highly social bees, bumble bees and honey bees. If there is a "pinnacle of bee evolution" in the minds of most people, it is almost certainly the honey bee. The large, elaborate nests, sophisticated modes of communication, including the dance language, and

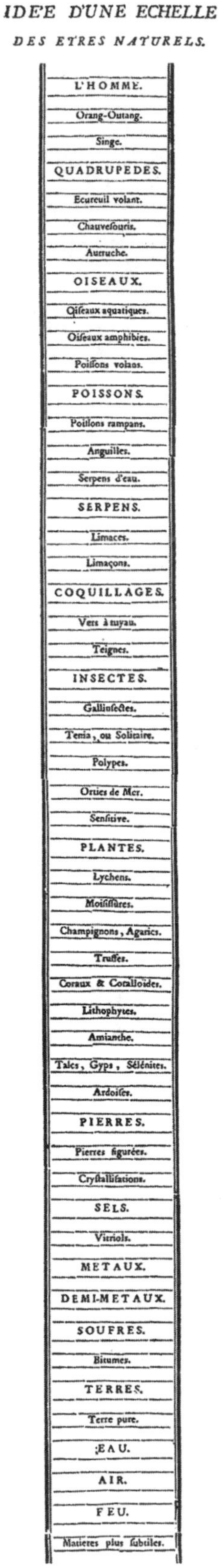

Figure 15-1. "Échelle des être" of Charles Bonnet (1745).

elaborate caste polymorphism must indicate a "higher" status relative to the lowly solitary bees. In the traditional bee Scala Naturae, highly social behavior is a kind of nirvana toward which solitary bees are all striving.

The idea that solitary bees are lower on the ladder of bee evolution is beautifully illustrated in Maurice Maeterlinck's classic, *The Life of the Bee*. Maeterlinck describes in poetic, 19th-century prose the biology of the honey bee, including the life cycle, the hive, foraging behavior, caste polymorphism, queen production, and mating. He reserves the last chapter of the book (entitled "The Progress of the Race") to delve into bee evolution and to trace the origins of the honey bee. Using a classic Scala Naturae for bees, he describes the biology of what was considered at the time to be a "primitive" bee, the genus *Hylaeus* (Colletidae):

> The unfortunate *Prosopis* [=*Hylaeus*)] stands more or less in the same relation to the inhabitants of our hives as the cave-dwellers to the fortunate who live in our great cities. You will probably more than once have seen her fluttering about the bushes, in a deserted corner of your garden, without realizing that you were carelessly watching the venerable ancestor to whom we probably owe most of our flowers and fruits . . . and possibly even our civilization, for in these mysteries all things intertwine. She is nimble and attractive, the variety most common in France being elegantly marked with white on a black background. But this elegance hides an inconceivable poverty. She leads a life of starvation. She is almost naked, whereas her sisters [the honey bees] are clad in a warm and sumptuous fleece. She has not, like the Apidae, baskets to gather the pollen nor, in their default, the tuft of the Andrenae [=Andrenidae], nor the ventral brush of the Gastrilegidae [=Megachilidae]. Her tiny claws must laboriously gather the powder from the calices, which powder she needs must swallow in order to take it back to her lair. She has no implements other than her tongue, her mouth, and her claws; but her tongue is too short, her legs to feeble, and her mandibles without strength. Unable to produce wax, bore holes through wood, or dig in the earth, she contrives clumsy galleries in the tender pith of dry berries; erects a few awkward cells, stores these with a little food for the offspring she never will see; and then, having accomplished this poor task of hers, that tends she knows not whither and of whose aim we are no less ignorant, she goes off and dies in a corner, as solitary as she had lived.

It is perhaps unfair to pick a book published in 1901 to argue that the Scala Naturae is still with us. But Maeterlinck's writing is so beautiful and his description of the "poverty," "starvation," and "nakedness" of the solitary bee[1] so compelling that we could not resist using this quote.

The idea that Colletidae are primitive bees has been with us a long time. Up until the first decade of the 21st-century, the commonly held view in bee phylogeny was that Colletidae were "ancestral" bees. Our new view of bee phylogeny places Colletidae well within the bees, far from the earliest branches of the tree (Chapter 2). Colletids have a remarkable suite of highly derived traits that are

---

[1] His choice of language makes obvious reference to the biblical story of Adam and Eve and their banishment from the Garden of Eden.

interlinked. They provision their brood cells with a liquid mix of mostly nectar and a little pollen that is likely infused with an abundant, coevolved microbial fauna. Larval colletid bees swim across and sometimes into the soupy provisions, dipping their heads "underwater" to stir up the pollen from below. Female colletids line their brood cells with a novel, waterproof, cellophane-like material that keeps this soupy concoction from leaking into the surrounding soil or wood, and they have a specialized structure, the bifid glossa (Chapter 7), for applying the cellophane-like material. In our modern view of bee biology, this novel suite of traits could be viewed as "advanced" relative to many other kinds of bees, including social bees. We are no longer allowed to declare Colletidae the "pinnacle" of bee evolution, but it is certainly tempting. They are definitely not "primitive" bees as Maeterlinck's description would suggest.

The tendency to view social bees as somehow more advanced than solitary bees might come from our own inherent biases about ourselves. We are desperately social animals. Whether the group consists of a few family members or a much larger group (Facebook, Instagram, and Snapchat users), we clearly prefer to develop highly social networks of connections. Perhaps for this reason, we are comfortable with the idea that sociality represents a higher state than solitary living. We can relate to a honey bee colony because it reminds us of ourselves. The white Langstroth hives stacked one upon the other even look a little like a minuscule apartment complex in Brooklyn. A stroll across Times Square on New Year's Eve or a ride on the Paris Metro at rush hour should convince anyone that humans are a lot more like ants (or social bees) than we are like the solitary bees described in this book.

We would like to leave the reader with the idea that social bees are no more advanced or complex than solitary bees. Sociality, like many traits, is an adaptation that works well in certain environments but that can be detrimental in others. Social bees are common in the tropics, where an enormous diversity of flowering plants bloom throughout the year, but are rare in deserts and other seasonal habitats. Sociality can lead to high reproductive success and ecological dominance, but it is a risky path when environments are unpredictable. Solitary bees are exquisitely adapted to certain environments, especially deserts and arid habitats where flowering is temporally constrained and unpredictable. They show remarkable behavioral and anatomical adaptations for excavating nests in soil and wood and for transporting nesting materials to rented cavities. Many solitary bees—the oligolectic, host-plant specialists—are also exquisitely adapted to accessing, gathering, and transporting a narrow range of highly specialized floral rewards. And they have physiological adaptations for dealing with long periods of drought—prolonged diapause and rainfall-induced emergence. So don't think of solitary bees as the lowly, second-class, impoverished relatives of social bees. Think of them as streamlined, rugged individualists that thrive in the face of ecological uncertainty. Solitary bees, more than social bees, follow the "hedgehog concept"—do one thing extremely well or you will end up doing many things badly (Collins 2001). They are the bet-hedgers, the cautious, money-under-the-mattress, spread-the-risk, grit-your teeth survivors. Rugged, focused individualism has been a successful strategy; solitary bees account for more than three-quarters of the bees on earth (see Chapter 2), and the solitary lifestyle has allowed them to persist and thrive for over 120 million years.

We find solitary bees fascinating precisely because of their apparent simplicity. Nests are rapidly constructed, foraging occurs in narrowly defined temporal windows, host plants are narrowly constrained, and the period of adult activity is condensed down to just a few, frantic weeks of activity. We are fascinated by their morphological and behavioral adaptations to nesting, to accessing, gathering, and transporting floral rewards, and to making a living in harsh environments. We hope that we have left the reader with a sense of this fascination and a desire to learn more about these amazing creatures.

Literature Cited

Abbott VA, Nadeau JL, Higo HA, Winston ML. 2008. Lethal and sublethal effects of imidacloprid on *Osmia lignaria* and clothianidin on *Megachile rotundata* (Hymenoptera: Megachilidae). *J. Econ. Entomol.* 101(3):784–796.

Abdalla FC, Cruz-Landim C da. 2001. Dufour glands in the hymenopterans (Apidae, Formicidae, Vespidae): a review. *Rev. Bras. Biol.* 61(1):95–106.

Abrol DP. 1988. Foraging range of subtropical bees, *Megachile flavipes*, *Megachile nana* (Hymenoptera: Megachilidae) and *Apis florea* (Hymenoptera: Apidae). *J. Indian Inst. Sci.* 68:43–48.

Adamson NL, Roulston TH, Fell RD, Mullins DE. 2012. From April to August—Wild bees pollinating crops through the growing season in Virginia, USA. *Environ. Entomol.* 41(4):813–821.

Adler FR, Tanner CJ. 2013. *Urban Ecosystems: Ecological Principles for the Built Environment.* Cambridge: Cambridge University Press.

Adler LS. 2000. The ecological significance of toxic nectar. *Oikos* 91(3):409–420.

Aguiar AJ, Melo GAR. 2007. Taxonomic revision, phylogenetic analysis, and biogeography of the bee genus *Tropidopedia* (Hymenoptera, Apidae, Tapinotaspidini). *Zool. J. Linn. Soc.* 151(3):511–554.

Aguiar AJ, Melo GAR. 2011. Revision and phylogeny of the bee genus *Paratetrapedia* Moure, with description of a new genus from the Andean Cordillera (Hymenoptera, Apidae, Tapinotaspidini). *Zool. J. Linn. Soc.* 162(2):351–442.

Aguiar CM, Garófalo CA. 2004. Nesting biology of *Centris* (*Hemisiella*) *tarsata* Smith (Hymenoptera, Apidae, Centridini). *Rev. Bras. Zool.* 21(3):477–486.

Aizen MA, Harder LD. 2009. The global stock of domesticated honey bees is growing slower than agricultural demand for pollination. *Curr. Biol.* 19(11):915–918.

Albans KR, Aplin RT, Brehcist J, Moore JF, O'Toole C. 1980. Dufour's gland and its role in secretion of nest cell lining in bees of the genus *Colletes* (Hymenoptera: Colletidae). *J. Chem. Ecol.* 6(3):549–564.

Albrecht M, Schmid B, Hautier Y, Muller CB. 2012. Diverse pollinator communities enhance plant reproductive success. *Proc. R. Soc. Lond. B, Biol. Sci.* 279(1748):4845–4852.

Alcock J. 1990. Body size and territorial behavior in the bee *Protoxaea gloriosa* (Fox) (Hymenoptera: Oxaeidae). *Pan-Pac. Entomol.* 66(2):157–161.

Alcock J. 1995. Persistent size variation in the anthophorine bee *Centris pallida* (Apidae) despite a large male mating advantage. *Ecol. Entomol.* 20 (1):1–4.

Alcock J. 1996a. Male size and survival: the effects of male combat and bird predation in Dawson's burrowing bees, *Amegilla dawsoni*. *Ecol. Entomol.* 21(4):309–316.

Alcock J. 1996b. The relation between male body size, fighting, and mating success in Dawson's burrowing bee, *Amegilla dawsoni* (Apidae, Apinae, Anthophorini). *J. Zool.* 239(4):663–674.

Alcock J. 1996c. Provisional rejection of three alternative hypotheses on the maintenance of a size dichotomy in males of Dawson's burrowing bee (Apidae, Apinae, Anthophorini). *Behav. Ecol. Sociobiol.* 39(3):181–188.

Alcock J. 1997. Competition from large males and the alternative mating tactics of small males of Dawson's burrowing bee (*Amegilla dawsoni*) (Apidae, Apinae, Anthophorini). *J. Insect Behav.* 10(1):99–113.

Alcock J. 1998. Sleeping aggregations of the bee *Idiomelissodes duplocincta* (Cockerell) (Hymenoptera: Anthophorini) and their possible function. *J. Kans. Entomol. Soc.* 71(1):74–84.

Alcock J. 2000. The natural history of a miltogrammine fly, *Miltogramma rectangularis* (Diptera: Sarcophagidae). *J. Kans. Entomol. Soc.* 73(4):208–219.

Alcock J. 2013a. Sexual selection and the mating behavior of solitary bees. *Adv. Study Behav.* 45:1–48.

Alcock J. 2013b. Role of body size in the competition for mates by males of *Centris pallida* (Anthophorinae: Hymenoptera). *Southwest. Nat.* 58(4):427–430.

Alcock J, Barrows EM, Gordh G, Hubbard LJ, Kirkendall L et al. 1978. The ecology and evolution of male reproductive behaviour in the bees and wasps. *Zool. J. Linn. Soc.* 64(4):293–326.

Alcock J, Buchmann SL. 1985. The significance of post–insemination display by male *Centris pallida* (Hymenoptera: Anthophoridae). *Z. Für Tierpsychol.* 68(3):231–243.

Alcock J, Eickwort GC, Eickwort KR. 1977a. The reproductive behavior of *Anthidium maculosum* (Hymenoptera: Megachilidae) and the evolutionary significance of multiple copulations. *Behav. Ecol. Sociobiol.* 2(4):385–396.

Alcock J, Houston TF. 1987. Resource defense and alternative mating tactics in the Banksia bee, *Hylaeus alcyoneus* (Erichson). *Ethology* 76(3):177–188.

Alcock J, Houston TF. 1996. Mating systems and male size in Australian hylaeine bees (Hymenoptera: Colletidae). *Ethology* 102(4):591–610.

Alcock J, Jones CE, Buchmann SL. 1976. Location before emergence of the female bee, *Centris pallida*, by its male (Hymenoptera: Anthophoridae). *J. Zool.* 179(2):189–199.

Alcock J, Jones CE, Buchmann SL. 1977b. Male mating strategies in the bee *Centris pallida* Fox (Anthophoridae: Hymenoptera). *Am. Nat.* 111(4):145–155.

Alcock J, Simmons LW, Beveridge M. 2005. Seasonal change in offspring sex and size in Dawson's burrowing bees (*Amegilla dawsoni*) (Hymenoptera: Anthophorini). *Ecol. Entomol.* 30(3):247–254.

Alcock J, Smith AP. 1987. Hilltopping, leks and female choice in the carpenter bee *Xylocopa (Neoxylocopa) varipuncta*. *J. Zool.* 211(1):1–10.

Alexander BA. 1992. An exploratory analysis of cladistic relationships within the superfamily Apoidea, with special reference to sphecid wasps (Hymenoptera). *J. Hymenopt. Res.* 1(1): 25–62.

Alexander BA, Rozen JG Jr. 1987. Ovaries, ovarioles, and oocytes in parasitic bees (Hymenoptera: Apoidea). *Pan-Pac. Entomol.* 63(2):155–164.

Almeida DAO, Martins RP, Buschini MLT. 1997. Behavior and nesting dynamics of the neotropical cavity-nesting specialist bee *Megachile assumptionis* Schrottky, with comparisons to the nearctic *Megachile brevis* Say (Hymenoptera: Megachilidae). *J. Hymenopt. Res.* 6(2):344–352.

Almeida EAB. 2008. Colletidae nesting biology (Hymenoptera: Apoidea). *Apidologie* 39(1): 16–29.

Almeida EAB, Danforth BN. 2009. Phylogeny of colletid bees (Hymenoptera: Colletidae) inferred from four nuclear genes. *Mol. Phylogenet. Evol.* 50(2):290–309.

Almeida EAB, Gibran NS. 2017. Taxonomy of neopasiphaeine bees: review of *Tetraglossula* Ogloblin, 1948 (Hymenoptera: Colletidae). *Zootaxa* 4304:521–544.

Almeida EAB, Pie MR, Brady SG, Danforth BN. 2011. Biogeography and diversification of colletid bees (Hymenoptera: Colletidae): emerging patterns from the southern end of the world. *J. Biogeogr.* 39(3):526–544.

Alqarni AS, Hannan MA, Gonzalez VH, Engel MS. 2012. A new species of *Chalicodoma* from Saudi Arabia with modified facial setae (Hymenoptera, Megachilidae). *ZooKeys* 204:71–83.

Alston DG, Tepedino VJ, Bradley BA, Toler TR, Griswold TL, Messinger SM. 2007. Effects of the insecticide phosmet on solitary bee foraging and nesting in orchards of Capitol Reef National Park, Utah. *Environ. Entomol.* 36(4):811–816.

Alves-dos-Santos I. 2003. Adaptations of bee proboscides for collecting pollen from Pontederiaceae flowers. In *Apoidea Neotropica: Homenagem aos 90 anos de Jesus Santiago Moure*, Melo GAR, Alves-dos-Santos I, eds., pp. 257–263. Criciúma, Editora UNESC.

Alves-dos-Santos I, Gaglianone MC, Naxara SRC. 2009. Male sleeping aggregations of solitary oil-collecting bees in Brazil (Centridini, Tapinotaspidini, and Tetrapediini; Hymenoptera: Apidae). *Genet. Mol. Res.* 8(2):515–524.

Alves-dos-Santos I, Melo GAR, Rozen JG Jr. 2002. Biology and immature stages of the bee tribe Tetrapediini (Hymenoptera: Apidae). *Am. Mus. Novit.* 3377:1–45.

Alves-dos-Santos I, Wittmann D. 1999. The proboscis of the long-tongued *Ancyloscelis* bees (Anthophoridae/Apoidea), with remarks on flower visits and pollen collecting with the mouthparts. *J. Kans. Entomol. Soc.* 72(3):277–288.

Amador GJ, Matherne M, Waller D, Mathews M, Gorb SN, Hu DL. 2017. Honey bee hairs and pollenkitt are essential for pollen capture and removal. *Bioinspir. Biomim.* 12(2):026015.

Andersen DC, Nelson SM. 2013. Floral ecology and insect visitation in riparian *Tamarix* sp. (saltcedar). *J. Arid Environ.* 94:105–112.

Anderson KE, Sheehan TH, Eckholm BJ, Mott BM, DeGrandi-Hoffman G. 2011. An emerging paradigm of colony health: microbial balance of the honey bee and hive (*Apis mellifera*). *Insectes Sociaux* 58(4):431–444.

Anderson WR. 1979. Floral conservatism in neotropical Malpighiaceae. *Biotropica* 11(3):219–223.

Andrietti F, Martinoli A, Rigato F. 1997. Quantitative data concerning the oviposition of *Bombylius fimbriatus* Meigen (Dip.: Bombyliidae), a parasite of *Andrena agilissima* (Scopoli) (Hym.: Andrenidae). *Entomol. Rec. J. Var.* 109(1–2):59–62.

Antropov AV. 2000. Digger wasps (Hymenoptera, Sphecidae) in Burmese amber. *Bull. Nat. Hist. Mus. Lond.* 56(1):59–77.

Anzenberger G. 1977. Ethological study of African carpenter bees of the genus *Xylocopa* (Hymenoptera: Anthophoridae). *Z. Für Tierpsychol.* 44(4):337–374.

Arduser MS, Michener CD. 1987. An African genus of cleptoparasitic halictid bees (Hymenoptera, Halictidae). *J. Kans. Entomol. Soc.* 60(2):324–329.

Arena M, Sgolastra F. 2014. A meta-analysis comparing the sensitivity of bees to pesticides. *Ecotoxicology* 23(3):324–334.

Armbruster WS. 1984. The role of resin in angiosperm pollination: ecological and chemical considerations. *Am. J. Bot.* 71(8):1149–1160.

Armbruster WS, Lee J, Baldwin BG. 2009. Macroevolutionary patterns of defense and pollination in *Dalechampia* vines: adaptation, exaptation, and evolutionary novelty. *Proc. Natl. Acad. Sci. USA* 106(43):18085–18090.

Arriaga L, Castellanos V. AE, Moreno E, Alarcon J. 2004. Potential ecological distribution of alien invasive species and risk assessment: a case study of buffel grass in arid regions of Mexico. *Conserv. Biol.* 18(6):1504–1514.

Artz DR, Hsu CL, Nault BA. 2011. Influence of honey bee, *Apis mellifera*, hives and field size on foraging activity of native bee species in pumpkin fields. *Environ. Entomol.* 40(5):1144–1158.

Artz DR, Nault BA. 2011. Performance of *Apis mellifera*, *Bombus impatiens*, and *Peponapis pruinosa* (Hymenoptera: Apidae) as pollinators of pumpkin. *J. Econ. Entomol.* 104(4):1153–1161.

Artz DR, Pitts-Singer TL. 2015. Effects of fungicide and adjuvant sprays on nesting behavior in two managed solitary bees, *Osmia lignaria* and *Megachile rotundata*. *PLoS One* 10(8): e0135688.

Ascher JS. 2004. Systematics of the bee family Andrenidae *(Hymenoptera: Apoidea)*. PhD thesis. Cornell University.

Ascher JS, Pickering J. 2018. Discover Life bee species guide and world checklist (Hymenoptera: Apoidea: Anthophila). http://www.discoverlife.org/mp/20q?guide=Apoidea_species.

Ashburner M. 1981. Entomophagous and other bizarre Drosophilidae. In *Genetics and Biology of* Drosophila, Ashburner M, Carson HL, Thompson JR Jr, eds., pp. 395–429. London: Academic Press.

Askew RR. 1971. *Parasitic Insects*. New York: American Elsevier Pub. Co.

Asplen MK, Anfora G, Biondi A, Choi D-S, Chu D et al. 2015. Invasion biology of spotted wing Drosophila (*Drosophila suzukii*): a global perspective and future priorities. *J. Pest Sci.* 88(3):469–494.

Augusto SC, Garófalo CA. 2004. Nesting biology and social structure of *Euglossa* (*Euglossa*) *townsendi* Cockerell (Hymenoptera, Apidae, Euglossini). *Insectes Sociaux* 51(4):400–409.

Augusto SC, Garófalo CA. 2009. Bionomics and sociological aspects of *Euglossa fimbriata* (Apidae, Euglossini). *Genet. Mol. Res.* 8(2):525–538.

Augusto SC, Garófalo CA. 2011. Task allocation and interactions among females in *Euglossa carolina* nests (Hymenoptera, Apidae, Euglossini). *Apidologie* 42(2):162–173.

Ayala R, Griswold T. 2005. Nueva especie de abeja del género *Osmia* (Hymenoptera: Megachilidae) de México. *Folia Entomol. Mex.* 44(Suppl. 1):139–145.

Ayasse M, Paxton RJ, Tengö J. 2001. Mating behavior and chemical communication in the Order Hymenoptera. *Annu. Rev. Entomol.* 46:31–78.

Azo'o ME, Tchuenguem Fohouo F, Messi J et al. 2011. Influence of the foraging activity of the entomofauna on okra (*Abelmoschus esculentus*) seed yield. *Int. J. Agric. Biol.* 13(5):761–765.

Baker DB, Engel MS. 2006. A new subgenus of *Megachile* from Borneo with arolia (Hymenoptera: Megachilidae). *Am. Mus. Novit.* 3505:1–12.

Baker HG, Baker I. 1973. Amino-acids in nectar and their evolutionary significance. *Nature* 241(5391):543–545.

Baker HG, Baker I. 1983. Floral nectar sugar constituents in relation to pollinator type. In *Handbook of Experimental Pollination Biology*, Jones CE, Little RJ, eds., pp. 117–141. New York: Van Nostrand Reinhold.

Baldock KCR, Goddard MA, Hicks DM, Kunin WE, Mitschunas N et al. 2015. Where is the UK's pollinator biodiversity? The importance of urban areas for flower-visiting insects. *Proc. R. Soc. Lond. B, Biol. Sci.* 282(1803):20142849.

Banaszak-Cibicka W, Żmihorski M. 2011. Wild bees along an urban gradient: winners and losers. *J. Insect Conserv.* 16(3):331–343.

Bänziger H, Boongird S, Sukumalanand P, Bänziger S. 2009. Bees (Hymenoptera: Apidae) that drink human tears. *J. Kans. Entomol. Soc.* 82(2):135–150.

Barnett JA, Payne RW, Yarrow D. 1990. *Yeasts: Characteristics and Identification*. Cambridge: Cambridge University Press. 2nd ed.

Baron GL, Raine NE, Brown MJ. 2014. Impact of chronic exposure to a pyrethroid pesticide on bumblebees and interactions with a trypanosome parasite. *J. Appl. Ecol.* 51(2):460–469.

Barrows EM. 1980. Results of a survey of damage caused by the carpenter bee *Xylocopa virginica* (Hymenoptera: Anthophoridae). *Proc. Entomol. Soc. Wash.* 82(1):44–47.

Barth FG. 1985. *Insects and Flowers. The Biology of a Partnership*. Princeton: Princeton University Press.

Barthell JF, Frankie GW, Thorp RW. 1998. Invader effects in a community of cavity nesting megachilid bees (Hymenoptera: Megachilidae). *Environ. Entomol.* 27(2):240–247.

Bartomeus I, Ascher JS, Gibbs J, Danforth BN, Wagner DL et al. 2013. Historical changes in northeastern US bee pollinators related to shared ecological traits. *Proc. Natl. Acad. Sci. USA* 110(12):4656–4660.

Bartomeus I, Ascher JS, Wagner D, Danforth BN, Colla S et al. 2011. Climate-associated phenological advances in bee pollinators and bee-pollinated plants. *Proc. Natl. Acad. Sci. USA* 108(51):20645–20649.

Bartomeus I, Cariveau DP, Harrison T, Winfree R. 2017. On the inconsistency of pollinator species traits for predicting either response to land-use change or functional contibution. *Oikos* 127(2):306–315.

Batra LR, Batra SWT, Bohart GE. 1973. The mycoflora of domesticated and wild bees (Apoidea). *Mycopathol. Mycol. Appl.* 49(1):13–44.

Batra SWT. 1966. Social behavior and nests of some nomiine bees in India (Hymenoptera, Halictidæ). *Insectes Sociaux* 13(3):145–153.

Batra SWT. 1968. Behavior of some social and solitary halictine bees within their nests: a comparative study (Hymenoptera: Halictidae). *J. Kans. Entomol. Soc.* 41(1):120–133.

Batra SWT. 1970. Behavior of the alkali bee, *Nomia melanderi*, within the nest (Hymenoptera: Halictidae). *Ann. Entomol. Soc. Am.* 63(2):400–406.

Batra SWT. 1978a. Aggression, territoriality, mating and nest aggregation of some solitary bees (Hymenoptera: Halictidae, Megachilidae, Colletidae, Anthophoridae). *J. Kans. Entomol. Soc.* 51(4):547–559.

Batra SWT. 1978b. *Osmia cornifrons* and *Pithitis smaragdula*, two Asian bees introduced into the United States for crop pollination. *Md. Agric. Exp. Stn. Misc. Publ.* 1:307–312.

Batra SWT. 1980. Ecology, behavior, pheromones, parasites and management of the sympatric vernal bees *Colletes inaequalis, C. thoracicus* and *C. validus. J. Kans. Entomol. Soc.* 53(3): 509–538.

Batra SWT. 1990. Bionomics of *Evylaeus comagenensis* (Knerer and Atwood) (Halictidae), a facultatively polygynous, univoltine, boreal halictine bee. *Proc. Entomol. Soc. Wash.* 92(4): 725–731.

Batra SWT, Norden BB. 1996. Fatty food for their brood: how *Anthophora* bees make and provision their cells (Hymenoptera: Apoidea). *Mem. Entomol. Soc. Wash.* 17:36–44.

Beattie AJ. 1971. Pollination mechanisms in *Viola*. *New Phytol.* 70(2):343–360.

Beattie AJ. 1972. The pollination ecology of *Viola*. 2, Pollen loads of insect-visitors. *Watsonia* 9(1):13–25.

Beekman M, Ratnieks FLW. 2000. Long-range foraging by the honey-bee, *Apis mellifera* L. *Funct. Ecol.* 14(4):490–496.

Bembé B. 2004. Functional morphology in male euglossine bees and their ability to spray fragrances (Hymenoptera, Apidae, Euglossini). *Apidologie* 35(3):283–291.

Bennett B, Breed MD. 1985. The nesting biology, mating behavior, and foraging ecology of *Perdita opuntiae* (Hymenoptera: Andrenidae). *J. Kans. Entomol. Soc.* 58(2):185–194.

Bennett FD. 1966. Notes on the biology of *Stelis (Odontostelis) bilineolata* (Spinola), a parasite of *Euglossa cordata* (Linnaeus) (Hymenoptera: Apoidea: Megachilidae). *J. N. Y. Entomol. Soc.* 74(2):72–79.

Bennett FD. 1972. Observations on *Exaerete* spp. and their hosts *Eulaema terminata* and *Euplusia surinamensis* (Hymen., Apidae, Euglossinae) in Trinidad. *J. N. Y. Entomol. Soc.* 80(3): 118–124.

Bergström G. 2008. Chemical communication by behaviour-guiding olfactory signals. *Chem. Commun.* 34:3959–3979.

Bergström G, Tengö J. 1974. Studies on natural odoriferous compounds: farnesyl-and geranyl esters as main volatile constituents of the secretion from Dufour's gland in 6 species of *Andrena* (Hymenoptera, Apidae). *Chem. Scr.* 5(1):28–38.

Bernardino AS, Gaglianone MC. 2008. Nest distribution and nesting habits of *Xylocopa ordinaria* Smith (Hymenoptera, Apidae) in a resting area in the northern Rio de Janeiro State, Brazil. *Rev. Bras. Entomol.* 52(3):434–440.

Bernays EA. 1989. Host range in phytophagous insects: the potential role of generalist predators. *Evol. Ecol.* 3(4):299–311.

Bernays EA. 2001. Neural limitations in phytophagous insects: implications for diet breadth and evolution of host affiliation. *Annu. Rev. Entomol.* 46:703–727.

Bernhardt P, Walker K. 1996. Observations on the foraging preferences of *Leioproctus (Filiglossa)* Rayment (Hymenoptera: Colletidae) in eastern Australia. *Pan-Pac. Entomol.* 72(3):130–137.

Beveridge M, Simmons LW. 2006. Panmixia: an example from Dawson's burrowing bee (*Amegilla dawsoni*) (Hymenoptera: Anthophorini). *Mol. Ecol.* 15(4):951–957.

Bianchi FA. 1962. Notes on the biology of *Cissites auriculata* (Champion) (Coleoptera: Meloidae). *Proc. Hawaii. Entomol. Soc.* 18(1):111–119.

Biani NB, Mueller UG, Wcislo WT. 2009. Cleaner mites: sanitary mutualism in the miniature ecosystem of neotropical bee nests. *Am. Nat.* 173(6):841–847.

Biani NB, Wcislo WT. 2007. Notes on the reproductive morphology of the parasitic bee *Megalopta byroni* (Hymenoptera: Halictidae), and a tentative new host record. *J. Kans. Entomol. Soc.* 80(4):392–394.

Biddinger DJ, Robertson JL, Mullin C, Frazier J, Ashcraft SA et al. 2013. Comparative toxicities and synergism of apple orchard pesticides to *Apis mellifera* (L.) and *Osmia cornifrons* (Radoszkowski). *PLoS One.* 8(9):e72587.

Biesmeijer JC, Roberts SPM, Reemer M, Ohlemüller R, Edwards M et al. 2006. Parallel declines in pollinators and insect-pollinated plants in Britain and the Netherlands. *Science* 313(5785):351–354.

Bischoff I. 2003. Population dynamics of the solitary digger bee *Andrena vaga* Panzer (Hymenoptera, Andrenidae) studied using mark-recapture and nest counts. *Popul. Ecol.* 45(3):197–204.

Bischoff I, Feltgen K, Breckner D. 2003. Foraging strategy and pollen preferences of *Andrena vaga* and *Colletes cunicularius* (Hymenoptera, Apidae). *J. Hymenopt. Res.* 12(2):220–237.

Bissett J. 1988. Contribution toward a monograph of the genus *Ascosphaera*. *Can. J. Bot.* 66(12):2541–2560.

Blagoveschenskaya NN. 1963. Giant colony of the solitary bee *Dasypoda plumipes* (PZ) (Hymenoptera, Melittidae). *Entomol Obozr.* 42:115–117.

Blaimer, BB, Mawdsley JR, Brady SG. 2018. Multiple origins of sexual dichromatism and aposematism within large carpenter bees. *Evolution.* https://doi.org/10.1111/evo.13558.

Blitzer EJ, Gibbs J, Park MG, Danforth BN. 2016. Pollination services for apple are dependent on diverse wild bee communities. *Agric. Ecosyst. Environ.* 221:1–7.

Blom J, Velthuis HH. 1988. Social behaviour of the carpenter bee *Xylocopa pubescens* (Spinola). *Ethology* 79(4):281–294.

Boesi R, Polidori C, Andrietti F. 2009. Biology of *Lasioglossum (L.) majus* (Hymenoptera: Halictidae), a largely solitary sweat bee with behavioural adaptations to communality. *J. Ethol.* 27(3):361–367.

Bogusch P. 2005. Biology of the cleptoparasitic bee *Epeoloides coecutiens* (Hymenoptera: Apidae: Osirini). *J. Kans. Entomol. Soc.* 78(1):1–12.

Bogusch P, Astapenková A, Heneberg P. 2015. Larvae and nests of six aculeate Hymenoptera (Hymenoptera: Aculeata) nesting in reed galls induced by *Lipara* spp. (Diptera: Chloropidae) with a review of species recorded. *PLoS One* 10(6):e0130802.

Bogusch P, Kratochvíl L, Straka J. 2006. Generalist cuckoo bees (Hymenoptera: Apoidea: *Sphecodes*) are species-specialist at the individual level. *Behav. Ecol. Sociobiol.* 60(3):422–429.

Bohart GE. 1972. Management of wild bees for the pollination of crops. *Annu. Rev. Entomol.* 17(1):287–312.

Bohart GE, Stephen WP, Eppley RK. 1960. The biology of *Heterostylum robustum* (Diptera: Bombyliidae), a parasite of the alkali bee. *Ann. Entomol. Soc. Am.* 53(3):425–435.

Bohart GE, Torchio PF, Maeta Y, Rust RW. 1972. Notes on the biology of *Emphoropsis pallida* Timberlake. *J. Kans. Entomol. Soc.* 45(3):381–392.

Bohart GE, Youssef NN. 1972. Notes on the biology of *Megachile (Megachiloides) umatillensis* Mitchell (Hymenoptera: Megachilidae) and its parasites. *Trans. R. Entomol. Soc. Lond.* 124(1):1–19.

Bohart GE, Youssef NN. 1976. The biology and behavior of *Evylaeus galpinsiae* Cockerell (Hymenoptera: Halictidae). *Wasmann J. Biol.* 34(2):185–234.

Bohart RM, Menke AS. 1976. *Sphecid Wasps of the World: A Generic Revision*. Berkeley: University of California Press.

Bommarco R, Biesmeijer JC, Meyer B, Potts SG, Pöyry J et al. 2010. Dispersal capacity and diet breadth modify the response of wild bees to habitat loss. *Proc. R. Soc. Lond. B, Biol. Sci.* 277(1690):2075–2082.

Bonnet C. 1745. *Traité d'insectologie Ou Observations Sur Quelques Espèces de Vers d'eau Douce, Qui Coupés Par Morceaux, Deviennent Autant d'animaux Complets*, Vol. 2. Paris: Durand.

Boomsma JJ, Isaaks JA. 1985. Energy investment and respiration in queens and males of *Lasius niger* (Hymenoptera: Formicidae). *Behav. Ecol. Sociobiol.* 18(1):19–27.

Boppré M, Colegate SM, Edgar JA. 2005. Pyrrolizidine alkaloids of *Echium vulgare* honey found in pure pollen. *J. Agric. Food Chem.* 53(3):594–600.

Boreux V, Krishnan S, Cheppudira KG, Ghazoul J. 2013a. Impact of forest fragments on bee visits and fruit set in rain-fed and irrigated coffee agro-forests. *Agric. Ecosyst. Environ.* 172:42–48.

Boreux V, Kushalappa CG, Vaast P, Ghazoul J. 2013b. Interactive effects among ecosystem services and management practices on crop production: pollination in coffee agroforestry systems. *Proc. Natl. Acad. Sci. USA* 110(21):8387–8392.

Borg-Karlson AK. 1979. Odour released behaviour of *Eucera longicornis* males (Hymenoptera: Anthophoridae). *Entomol Tidskr.* 100(3–4):125–128.

Borrell BJ. 2004. Suction feeding in orchid bees (Apidae: Euglossini). *Proc. R. Soc. Lond. B, Biol. Sci.* 271(Suppl 4):S164–S166.

Bosch J. 2008. Production of undersized offspring in a solitary bee. *Anim. Behav.* 75(3):809–816.

Bosch J, Kemp WP. 2002. Developing and establishing bee species as crop pollinators: the example of *Osmia* spp. (Hymenoptera: Megachilidae) and fruit trees. *Bull. Entomol. Res.* 92(1):3–16.

Bosch J, Maeta Y, Rust R. 2001. A phylogenetic analysis of nesting behavior in the genus *Osmia* (Hymenoptera: Megachilidae). *Ann. Entomol. Soc. Am.* 94(4):617–627.

Bosch J, Sgolastra F, Kemp WP. 2008. Life cycle ecophysiology of *Osmia* mason bees used as crop pollinators. In *Bee Pollination in Agricultural Ecosystems*, James RR, Pitts-Singer TL, eds., pp. 83–104. Oxford: Oxford University Press.

Bosch J, Vicens N. 2002. Body size as an estimator of production costs in a solitary bee. *Ecol. Entomol.* 27(2):129–137.

Bossert S, Murray EA, Almeida EAB, Brady SG, Blaimer BB et al. 2018. Combining transcriptomes and ultraconserved elements to illuminate the phylogeny of Apidae. *Mol. Phylogenet. Evol.* 130:121–131.

Bossert S, Murray EA, Blaimer BB, Danforth BN. 2017. The impact of GC bias on phylogenetic accuracy using targeted enrichment phylogenomic data. *Mol. Phylogenet. Evol.* 111:149–157.

Botías C, David A, Horwood J, Abdul-Sada A, Nicholls E et al. 2015. Neonicotinoid residues in wildflowers, a potential route of chronic exposure for bees. *Environ. Sci. Technol.* 49(21): 12731–12740.

Boussau B, Walton Z, Delgado JA, Collantes F, Beani L et al. 2014. *Strepsiptera*, phylogenomics and the long branch attraction problem. *PLoS One* 9(10):e107709.

Bowers JE, Dimmitt MA. 1994. Flowering phenology of six woody plants in the northern Sonoran Desert. *Bull. Torrey Bot. Club* 121(3):215–229.

Brach V. 1978. Notes on the biology of *Lithurgus gibbosus* Smith in Florida (Hymenoptera: Megachilidae). *Bull. South. Calif. Acad. Sci.* 77(3):144–147.

Branstetter MG, Danforth BN, Pitts JP, Faircloth BC, Ward PS et al. 2017. Phylogenomic insights into the evolution of stinging wasps and the origins of ants and bees. *Curr. Biol.* 27(7): 1019–1025.

Brauns H. 1926. V. Nachtrag zu "Friese, Bienen Afrikas." *Zool. Jahrbücher Abt. Syst. Geogr. Biol. Tiere.* 52:187–230.

Breeze TD, Bailey AP, Balcombe KG, Potts SG. 2011. Pollination services in the UK: how important are honeybees? *Agric. Ecosyst. Environ.* 142(3):137–143.

Brittain C, Kremen C, Klein A-M. 2013. Biodiversity buffers pollination from changes in environmental conditions. *Glob. Change Biol.* 19(2):540–547.

Brittain C, Potts SG. 2011. The potential impacts of insecticides on the life-history traits of bees and the consequences for pollination. *Basic Appl. Ecol.* 12(4):321–331.

Brittain CA, Vighi M, Bommarco R, Settele J, Potts SG. 2010. Impacts of a pesticide on pollinator species richness at different spatial scales. *Basic Appl. Ecol.* 11(2):106–115.

Brochu K. 2018. Differential impacts of pollen quality and microbial communities on generalist and specialist bees visiting a shared food resource. PhD thesis. Cornell University.

Brooks RW. 1983. Systematics and bionomics of *Anthophora*—the *Bomboides* group and species groups of the New World (Hymenoptera:Apoidea: Anthophoridae). *Univ. Calif. Publ. Entomol.* 98:1–86.

Brooks RW. 1988. Systematics and phylogeny of anthophorine bees (Hymenoptera; Anthophoridae: Anthophorini). *Univ. Kans. Sci. Bull.* 53(9):437–575.

Brothers DJ. 1972. Biology and immature stages of *Pseudomethoca f. frigida*, with notes on other species (Hymenoptera: Mutillidae). *Univ. Kans. Sci. Bull.* 50(1):1–38.

Brothers DJ. 1975. Phylogeny and classification of the aculeate Hymenoptera, with special reference to Mutillidae. *Univ. Kans. Sci. Bull.* 50(11):483–648.

Brothers DJ. 1978. Biology and immature stages of *Myrmosula parvula* (Hymenoptera: Mutillidae). *J. Kans. Entomol. Soc.* 51(4):698–710.

Brothers DJ, Tschuch G, Burger F. 2000. Associations of mutillid wasps (Hymenoptera, Mutillidae) with eusocial insects. *Insectes Sociaux* 47(3):201–211.

Brower LP, Brower JVZ, Westcott PW. 1960. Experimental studies of mimicry. 5. The reactions of toads (*Bufo terrestris*) to bumblebees (*Bombus americanorum*) and their robberfly mimics (*Mallophora bomboides*), with a discussion of aggressive mimicry. *Am. Nat.* 94(878):343–355.

Brünnert U, Kelber A, Zeil J. 1994. Ground-nesting bees determine the location of their nest relative to a landmark by other than angular size cues. *J. Comp. Physiol. A.* 175(3):363–369.

Buchmann SL. 1983. Buzz pollination in angiosperms. In *Handbook of Experimental Pollination Biology*, Jones CE, Little RJ, eds., pp. 73–113. New York: Van Nostrand Reinhold.

Buchmann SL. 1987. The ecology of oil flowers and their bees. *Annu. Rev. Ecol. Syst.* 18: 343–369.

Buchmann SL, Cane JH. 1989. Bees assess pollen returns while sonicating *Solanum* flowers. *Oecologia* 81(3):289–294.

Burdick DJ, Torchio PF. 1959. Notes on the biology of *Hesperapis regularis* (Cresson) (Hymenoptera: Melittidae). *J. Kans. Entomol. Soc.* 32(2):83–87.

Burgett DM, Sukumalanand P. 2000. Flight activity of *Xylocopa* (*Nyctomelitta*) *tranquebarica*: a night flying carpenter bee (Hymenoptera: Apidae). *J. Apic. Res.* 39(1–2):75–83.

Burgett M, Rucker RR, Thurman WN. 2004. Economics and honey bee pollination markets. *Am. Bee J.* 144:269–271.

Burkart A, Schlindwein C, Lunau K. 2013. Assessment of pollen reward and pollen availability in *Solanum stramoniifolium* and *Solanum paniculatum* for buzz-pollinating carpenter bees. *Plant Biol.* 16(2):503–507.

Butler GD. 1967. Biological observations on *Ptilothrix sumichrasti* (Cresson) in southern Arizona. *Pan-Pac. Entomol.* 43(1):8–14.

Calderone NW. 2012. Insect pollinated crops, insect pollinators and US agriculture: trend analysis of aggregate data for the period 1992–2009. *PLoS One* 7(5):e37235.

Cameron SA, Lozier JD, Strange JP, Koch JB, Cordes N et al. 2011. Patterns of widespread decline in North American bumble bees. *Proc. Natl. Acad. Sci. USA* 108(2):662–667.

Cameron SA, Ramírez S. 2001. Nest architecture and nesting ecology of the orchid bee *Eulaema meriana* (Hymenoptera: Apinae: Euglossini). *J. Kans. Entomol. Soc.* 74(3):142–165.

Cameron SA, Whitfield JB, Hulslander CL, Cresko WA, Isenberg SB, King RW. 1996. Nesting biology and foraging patterns of the solitary bee *Melissodes rustica* (Hymenoptera: Apidae) in northwest Arkansas. *J. Kans. Entomol. Soc.* 69(4):260–273.

Camillo E. 2005. Nesting biology of four *Tetrapedia* species in trap-nests (Hymenoptera: Apidae: Tetrapediini). *Rev. Biol. Trop.* 53(1–2):175–186.

Cane JH. 1981. Dufour's gland secretion in the cell linings of bees (Hymenoptera: Apoidea). *J. Chem. Ecol.* 7(2):403–410.

Cane JH. 1983. Chemical evolution and chemosystematics of the Dufour's gland secretions of the lactone-producing bees (Hymenoptera: Colletidae, Halictidae, and Oxaeidae). *Evolution* 37(4):657–674.

Cane JH. 1991. Soils of ground-nesting bees (Hymenoptera: Apoidea): texture, moisture, cell depth and climate. *J. Kans. Entomol. Soc.* 64(4):406–413.

Cane JH. 1994a. Nesting biology and mating behavior of the southeastern blueberry bee, *Habropoda laboriosa* (Hymenoptera: Apoidea). *J. Kans. Entomol. Soc.* 67(3):236–241.

Cane JH. 1994b. Lifetime monetary value of individual pollinators: the bee *Habropoda laboriosa* at rabbiteye blueberry (*Vaccinium ashei* Reade). *Acta Hortic.* 446:67–70.

Cane JH. 1996. Ground-nesting bees: the neglected pollinator resource for agriculture. *Acta Hortic.* 437:309–324.

Cane JH. 1997. Violent weather and bees: populations of the barrier island endemic, *Hesperapis oraria* (Hymenoptera: Melittidae) survive a category 3 hurricane. *J. Kans. Entomol. Soc.* 70(1):73–75.

Cane JH. 2002. Pollinating bees (Hymenoptera: Apiformes) of U.S. alfalfa compared for rates of pod and seed set. *J. Econ. Entomol.* 95(1):22–27.

Cane JH. 2003. Annual displacement of soil in nest tumuli of alkali bees (*Nomia melanderi*) (Hymenoptera: Apiformes: Halictidae) across an agricultural landscape. *J. Kans. Entomol. Soc.* 76(2):172–176.

Cane JH. 2008. A native ground-nesting bee (*Nomia melanderi*) sustainably managed to pollinate alfalfa across an intensively agricultural landscape. *Apidologie* 39(3):315–323.

Cane JH. 2012. Dung pat nesting by the solitary bee, *Osmia* (*Acanthosmioides*) *integra* (Megachilidae: Apiformes). *J. Kans. Entomol. Soc.* 85(3):262–264.

Cane JH. 2017. Specialist bees collect Asteraceae pollen by distinctive abdominal drumming (*Osmia*) or tapping (*Melissodes, Svastra*). *Arthropod-Plant Interact.* 11(3):257–261.

Cane JH, Buchmann SL. 1989. Novel pollen-harvesting behavior by the bee *Protandrena mexicanorum* (Hymenoptera: Andrenidae). *J. Insect Behav.* 2(3):431–436.

Cane JH, Carlson RG. 1984. Dufour's gland triglycerides from *Anthophora, Emphoropsis* (Anthophoridae) and *Megachile* (Megachilidae) bees (Hymenoptera: Apoidea). *Comp. Biochem. Physiol. Part B Comp. Biochem.* 78(3):769–772.

Cane JH, Dobson HE, Boyer B. 2017. Timing and size of daily pollen meals eaten by adult females of a solitary bee (*Nomia melanderi*) (Apiformes: Halictidae). *Apidologie* 48(1):17–30.

Cane JH, Eickwort GC, Wesley FR, Spielholz J. 1983a. Foraging, grooming and mate-seeking behaviors of *Macropis nuda* (Hymenoptera, Melittidae) and use of *Lysimachia ciliata* (Primulaceae) oils in larval provisions and cell linings. *Am. Midl. Nat.* 110(2):257–264.

Cane JH, Gerdin S, Wife G. 1983b. Mandibular gland secretions of solitary bees (Hymenoptera: Apoidea): potential for nest cell disinfection. *J. Kans. Entomol. Soc.* 56(2):199–204.

Cane JH, Griswold T, Parker FD. 2007. Substrates and materials used for nesting by North American *Osmia* bees (Hymenoptera: Apiformes: Megachilidae). *Ann. Entomol. Soc. Am.* 100(3):350–358.

Cane JH, Minckley RL, Kervin L, Roulston TH. 2005. Temporally persistent patterns of incidence and abundance in a pollinator guild at annual and decadal scales: the bees of *Larrea tridentata*. *Biol. J. Linn. Soc.* 85(3):319–329.

Cane JH, Minckley RL, Kervin LJ, Roulston TH, Williams NM. 2006. Complex responses within a desert bee guild (Hymenoptera: Apiformes) to urban habitat fragmentation. *Ecol. Appl.* 16(2):632–644.

Cane JH, Neff JL. 2011. Predicted fates of ground-nesting bees in soil heated by wildfire: thermal tolerances of life stages and a survey of nesting depths. *Biol. Conserv.* 144(11): 2631–2636.

Cane JH, Payne JA. 1988. Foraging ecology of the bee *Habropoda laboriosa* (Hymenoptera: Anthophoridae), an oligolege of blueberries (Ericaceae: *Vaccinium*) in the southeastern United States. *Ann. Entomol. Soc. Am.* 81(3):419–427.

Cane JH, Sampson BJ, Miller SA. 2011. Pollination value of male bees: the specialist bee *Peponapis pruinosa* (Apidae) at summer squash (*Cucurbita pepo*). *Environ. Entomol.* 40(3):614–620.

Cane JH, Sipes S. 2006. Characterizing floral specialization by bees: analytical methods and a revised lexicon for oligolecty. In *Plant-Pollinator Interactions: From Specialization to Generalization*, Waser NM, Ollerton J, eds., pp. 99–122. Chicago and London: University of Chicago Press.

Cane JH, Snelling RR, Kervin LJ. 1996. A new monolectic coastal bee, *Hesperapis oraria* Snelling and Stage (Hymenoptera: Melittidae), with a review of desert and neotropical disjunctives in the southeastern U.S. *J. Kans. Entomol. Soc.* 69(4):238–247.

Cane JH, Tengö JO. 1981. Pheromonal cues direct mate-seeking behavior of male *Colletes cunicularius* (Hymenoptera: Colletidae). *J. Chem. Ecol.* 7(2):427–436.

Cane JH, Tepedino VJ. 2017. Gauging the effect of honey bee pollen collection on native bee communities. *Conserv. Lett.* 10(2):205–210.

Cappellari SC, Harter-Marques B, Aumeier P, Engels W. 2009. *Mecardonia tenella* (Plantaginaceae) attracts oil-, perfume- and pollen-gathering bees in southern Brazil. *Biotropica* 41(6): 721–729.

Cappellari SC, Melo GA, Aguiar AJ, Neff JL. 2012. Floral oil collection by male *Tetrapedia* bees (Hymenoptera: Apidae: Tetrapediini). *Apidologie* 43(1):39–50.

Cardinal S. 2018. Bee (Hymenoptera: Apoidea: Anthophila) diversity through time. In *Insect Biodiversity: Science and Society II*, Adler PH, Foottit RG, eds., pp. 851–867. Hoboken, NJ: John Wiley & Sons.

Cardinal S, Buchmann SL, Russell AL. 2018. The evolution of floral sonication, a pollen foraging behavior used by bees (Anthophila). *Evolution* 72(3):590–600.

Cardinal S, Danforth BN. 2011. The antiquity and evolutionary history of social behavior in bees. *PLoS One* 6:e21086.

Cardinal S, Danforth BN. 2013. Bees diversified in the age of eudicots. *Proc. R. Soc. Lond. B, Biol. Sci.* 280(1755):20122686.

Cardinal S, Packer L. 2007. Phylogenetic analysis of the corbiculate Apinae based on morphology of the sting apparatus (Hymenoptera: Apidae). *Cladistics* 23(2):99–118.

Cardinal S, Straka J, Danforth BN. 2010. Comprehensive phylogeny of apid bees reveals the evolutionary origins and antiquity of cleptoparasitism. *Proc. Natl. Acad. Sci. USA* 107(37):16207–16211.

Caron DM. 1978. Other insects. In *Honey Bee Pests, Predators, and Diseases*, pp. 156–176. Ithaca: Comstock Publishing Associates.

Caron DM, Connor JL. 2010. *Honey Bee Biology and Beekeeping*. Cheshire: Wic-Was Press.

Carrijo TF, Gonçalves RB, Santos RG. 2012. Review of bees as guests in termite nests, with a new record of the communal bee, *Gaesochira obscura* (Smith, 1879) (Hymenoptera, Apidae), in nests of *Anoplotermes banksi* Emerson, 1925 (Isoptera, Termitidae, Apicotermitinae). *Insectes Sociaux* 59(2):141–149.

Carvalho AT, Dötterl S, Schlindwein C. 2014. An aromatic volatile attracts oligolectic bee pollinators in an interdependent bee-plant relationship. *J. Chem. Ecol.* 40(10):1126–1134.

Carvalho AT, Schlindwein C. 2011. Obligate association of an oligolectic bee and a seasonal aquatic herb in semi-arid north-eastern Brazil. *Biol. J. Linn. Soc.* 102(2):355–368.

Cazier MA, Linsley EG. 1963. Territorial behavior among males of *Protoxaea gloriosa* (Fox) (Hymenoptera: Andrenidae). *Can. Entomol.* 95(5):547–556.

Celary W. 2002. The ground-nesting solitary bee, *Dasypoda thoracica* Baer, 1853 (Hymenoptera: Apoidea: Melittidae) and its life history. *Folia Biol. (Praha)* 50(3–4):191–198.

Celary W. 2006. Biology of the solitary ground-nesting bee *Melitta leporina* (Panzer, 1799) (Hymenoptera: Apoidea: Melittidae). *J. Kans. Entomol. Soc.* 79(2):136–145.

Černá K, Munclinger P, Vereecken NJ, Straka J. 2017. Mediterranean lineage endemism, cold-adapted palaeodemographic dynamics and recent changes in population size in two solitary bees of the genus *Anthophora*. *Conserv. Genet.* 18(3):521–538.

Černá K, Straka J, Munclinger P. 2013a. Population structure of pioneer specialist solitary bee *Andrena vaga* (Hymenoptera: Andrenidae) in central Europe: the effect of habitat fragmentation or evolutionary history? *Conserv. Genet.* 14(4):875–883.

Černá K, Zemenova M, Macháčková L, Kolínová Z, Straka J. 2013b. Neighbourhood society: nesting dynamics, usurpations and social behaviour in solitary bees. *PLoS One* 8(9):e73806.

Chapin FS, Zavaleta ES, Eviner VT, Naylor RL, Vitousek PM et al. 2000. Consequences of changing biodiversity. *Nature* 405(6783):234–242.

Chappell MA. 1984. Temperature regulation and energetics of the solitary bee *Centris pallida* during foraging and intermale mate competition. *Physiol. Zool.* 57(2):215–225.

Chauzat M-P, Faucon J-P, Martel A-C, Lachaize J, Cougoule N, Aubert M. 2006. A survey of pesticide residues in pollen loads collected by honey bees in France. *J. Econ. Entomol.* 99(2):253–262.

Chemsak JA. 1985. Observations on adult behavior of *Centris flavofasciata* Friese (Hymenoptera: Anthophoridae). *Pan-Pac. Entomol.* 61(3):265.

Chiappa E, Araya H, Mandujano V, Tosti-Croce E. 2018. Descripción de los estados inmaduros de *Colletes musculus* Friese (Hymenoptera: Colletidae), con notas ecológicas y biológicas. *Rev. Chil. Entomol.* 44(2):123–134.

Chittka L, Schürkens S. 2001. Successful invasion of a floral market. *Nature* 411(6838):653.

Choe D-H, Rust MK. 2007. Use of plant resin by a bee assassin bug, *Apiomerus flaviventris* (Hemiptera: Reduviidae). *Ann. Entomol. Soc. Am.* 100(2):320–326.

Choe JC, Crespi BJ. 1997. *The Evolution of Social Behaviour in Insects and Arachnids.* Cambridge: Cambridge University Press.

Christenhusz MJ, Byng JW. 2016. The number of known plants species in the world and its annual increase. *Phytotaxa* 261(3):201–217.

Ciarlo TJ, Mullin CA, Frazier JL, Schmehl DR. 2012. Learning impairment in honey bees caused by agricultural spray adjuvants. *PLoS One* 7(7):e40848.

Claude-Joseph F. 1926. Recherches biologiques sur les Hymenopteres du Chili (Mellifères). *Ann. Sci. Nat. Zool.* 9(10):113–268.

Clausen CP. 1940. *Entomophagous Insects.* New York and London: McGraw-Hill Book Company.

Clausen CP. 1976. Phoresy among entomophagous insects. *Annu. Rev. Entomol.* 21:343–368.

Clement SL. 1976. The biology of *Dianthidium heterulkei heterulkei* Schwarz, with a description of the larva (Hymenoptera: Megachilidae). *Wasmann J. Biol.* 34(1):9–22.

Cockerell TDA. 1933. The excessive abundance of certain bees. *Am. Nat.* 67(710):286–288.

Cocom Pech ME, May-Itzá W de J, Medina Medina L, Quezada-Euán JJG. 2008. Sociality in *Euglossa* (*Euglossa*) *viridissima* Friese (Hymenoptera, Apidae, Euglossini). *Insectes Soc.* 55(4):428–433.

Cocucci AA, Sérsic A, Roig-Alsina A. 2000. Oil-collecting structures in Tapinotaspidini: their diversity, function and probable origin. *Mitteilungen Münch. Entomol. Ges.* 90:51–74.

Cohen AC. 1995. Extra-oral digestion in predaceous terrestrial Arthropoda. *Annu. Rev. Entomol.* 40:85–103.

Cohen AC. 1998. Solid-to-liquid feeding: the inside(s) story of extra-oral digestion in predaceous Arthropoda. *Am. Entomol.* 44(2):103–117.

Colla SR, Willis E, Packer L. 2009. Can green roofs provide habitat for urban bees (Hymenoptera: Apidae)? *Cities Environ.* 2(1):1–12.

Collins J. 2001. *Good to Great: Why Some Companies Make the Leap and Others Don't.* New York: HarperBusiness.

Compagnucci L. 2006. Dos especies nuevas de *Zikanapis* Moure de la Argentina, con setas modificadas en el clípeo (Hymenoptera: Apoidea: Colletidae). *Rev. Mus. Argent. Cienc. Nat. Nueva Ser.* 8(1):87–94.

Comstock JH. 1909. A note on the habits of the wall-bee, *Chalicodoma muraria*. *Ann. Entomol. Soc. Am.* 2(1):9–10.

Conrad T, Ayasse M. 2015. The role of vibrations in population divergence in the red mason bee, *Osmia bicornis*. *Curr. Biol.* 25(21):2819–2822.

Cook JL. 2014. Review of the biology of parasitic insects in the Order Strepsiptera. *Comp. Parasitol.* 81(2):134–151.

Cooper KW. 1993. The first *Holcopasites* from western California, *H. ruthae* n. sp., and *H. linsleyi*, a new species from southwestern Arizona (Hymenoptera, Nomadinae). *Proc. Entomol. Soc. Wash.* 95(1):113–125.

Corbet SA, Huang S-Q. 2014. Buzz pollination in eight bumblebee-pollinated *Pedicularis* species: does it involve vibration-induced triboelectric charging of pollen grains? *Ann. Bot.* 114(8):1665–1674.

Corbet SA, Willmer PG, Beament JWL, Unwin DM, Prŷs-Jones OE. 1979. Post-secretory determinants of sugar concentration in nectar. *Plant Cell Environ.* 2(4):293–308.

Cordeiro GD, Taniguchi M, Flechtmann CHW, Alves-dos-Santos I. 2011. Phoretic mites (Acari: Chaetodactylidae) associated with the solitary bee *Tetrapedia diversipes* (Apidae: Tetrapediini). *Apidologie* 42(2):128–139.

Coville RE, Frankie GW, Buchmann SL, Vinson SB, Williams HJ. 1986. Nesting and male behavior of *Centris heithausi* (Hymenoptera: Anthophoridae) in Costa Rica with chemical analysis of the hindleg glands of males. *J. Kans. Entomol. Soc.* 59(2):325–336.

Coville RE, Frankie GW, Vinson SB. 1983. Nests of *Centris segregata* (Hymenoptera: Anthophoridae) with a review of the nesting habits of the genus. *J. Kans. Entomol. Soc.* 56(2):109–122.

Crone EE. 2013. Responses of social and solitary bees to pulsed floral resources. *Am. Nat.* 182(4):465–473.

Cross EA, Bohart GE. 1992. The biology of *Imparipes apicola* (Acari: Scutacaridae) and its relationships to the alkali bee, *Nomia melanderi* (Hymenoptera: Halictidae), and to certain fungi in the bee cell ecosystem. *J. Kans. Entomol. Soc.* 65(2):157–173.

Cruden RW. 1972. Pollination biology of *Nemophila menziesii* (Hydrophyllaceae) with comments on the evolution of oligolectic bees. *Evolution* 26(3):373–389.

Crüger H. 1864. A few notes on the fecundation of orchids and their morphology. *J. Proc. Linn. Soc. Lond. Bot.* 8(31):127–135.

Custer CP. 1928. The bee that works in stone: *Perdita opuntiae* Cockerell. *Psyche* 35(2):67–84.

Custer CP. 1929. Notes on cocoons and parasites of *Melissodes obliqua* and nests of *Perdita opuntiae* (Hymenoptera-Apoidea). *Psyche* 36(4):293–295.

Dafni A. 1984. Mimicry and deception in pollination. *Annu. Rev. Ecol. Syst.* 15:259–278.

Dafni A, Ivri Y, Brantjes NBM. 1981. Pollination of *Serapias vomeracea* Briq (Orchidaceae) by imitation of holes for sleeping solitary male bees (Hymenoptera). *Acta Bot. Neerl.* 30(1–2):69–73.

Daly HV. 1966. Biological studies on *Ceratina dallatorreana*, an alien bee in California which reproduces by parthenogenesis (Hymenoptera: Apoidea). *Ann. Entomol. Soc. Am.* 59(6):1138–1154.

Daly HV, Magnacca KN. 2003. *Hawaiian* Hylaeus *(Nesoprosopis) Bees (Hymenoptera: Apoidea)*, Vol. 17. Honolulu: University of Hawaii Press.

Daly HV, Michener CD, Moure JS, Sakagami SF. 1987. The relictual bee genus *Manuelia* and its relation to other Xylocopinae (Hymenoptera: Apoidea). *Pan-Pac. Entomol.* 63(2):102–124.

Danforth BN. 1989a. Nesting behavior of four species of *Perdita* (Hymenoptera: Andrenidae). *J. Kans. Entomol. Soc.* 62(1):59–79.

Danforth BN. 1989b. The evolution of hymenopteran wings: the importance of size. *J. Zool.* 218(2):247–276.

Danforth BN. 1990. Provisioning behavior and the estimation of investment ratios in a solitary bee, *Calliopsis (Hypomacrotera) persimilis* (Cockerell) (Hymenoptera: Andrenidae). *Behav. Ecol. Sociobiol.* 27(3):159–168.

Danforth BN. 1991a. Female foraging and intranest behavior of a communal bee, *Perdita portalis* (Hymenoptera: Andrenidae). *Ann. Entomol. Soc. Am.* 84(5):537–548.

Danforth BN. 1991b. The morphology and behavior of dimorphic males in *Perdita portalis* (Hymenoptera: Andrenidae). *Behav. Ecol. Sociobiol.* 29(4):235–247.

Danforth BN. 1994. Taxonomic review of *Calliopsis* subgenus *Hypomacrotera* (Hymenoptera: Andrenidae), with special emphasis on the distributions and host plant associations. *Pan-Pac. Entomol.* 70:283–300.

Danforth BN. 1996. Phylogenetic analysis and taxonomic revision of the *Perdita* subgenera *Macrotera*, *Macroteropsis*, *Macroterella*, and *Cockerellula* (Hymenoptera: Andrenidae). *Univ. Kans. Sci. Bull.* 55(16):635–692.

Danforth BN. 1999. Emergence dynamics and bet hedging in a desert bee, *Perdita portalis*. *Proc. R. Soc. Lond. B, Biol. Sci.* 266(1432):1985–1994.

Danforth BN. 2002. Evolution of sociality in a primitively eusocial lineage of bees. *Proc. Natl. Acad. Sci. USA* 99(1):286–290.

Danforth BN, Brady S, Sipes S, Pearson A. 2004. Single-copy nuclear genes recover Cretaceous-age divergences in bees. *Syst. Biol.* 53(2):309–326.

Danforth BN, Cardinal S, Praz C, Almeida EA, Michez D. 2013. The impact of molecular data on our understanding of bee phylogeny and evolution. *Annu. Rev. Entomol.* 58:57–78.

Danforth BN, Conway L, Ji S. 2003a. Phylogeny of eusocial *Lasioglossum* reveals multiple losses of eusociality within a primitively eusocial clade of bees (Hymenoptera: Halictidae). *Syst. Biol.* 52(1):23–36.

Danforth BN, Desjardins CA. 1999. Male dimorphism in *Perdita portalis* (Hymenoptera, Andrenidae) has arisen from preexisting allometric patterns. *Insectes Sociaux* 46(1):18–28.

Danforth BN, Eardley C, Packer L, Walker K, Pauly A, Randrianambinintsoa FJ. 2008. Phylogeny of Halictidae with an emphasis on endemic African Halictinae. *Apidologie* 39(1):86–101.

Danforth BN, Eickwort GC. 1997. The evolution of social behavior in the augochlorine sweat bees (Hymenoptera: Halictidae) based on a phylogenetic analysis of the genera. In *The Evolution of Social Behavior in Insects and Arachnids*, Choe JC, Crespi BJ, eds., pp. 270–292. Cambridge: Cambridge University Press.

Danforth BN, Fang J, Sipes S. 2006a. Analysis of family-level relationships in bees (Hymenoptera: Apiformes) using 28S and two previously unexplored nuclear genes: CAD and RNA polymerase II. *Mol. Phylogenet. Evol.* 39(2):358–372.

Danforth BN, Ji S, Ballard LJ. 2003b. Gene flow and population structure in an oligolectic desert bee, *Macrotera* (*Macroteropsis*) *portalis* (Hymenoptera: Andrenidae). *J. Kans. Entomol. Soc.* 76(2):221–235.

Danforth BN, Neff JL. 1992. Male polymorphism and polyethism in *Perdita texana* (Hymenoptera: Andrenidae). *Ann. Entomol. Soc. Am.* 85(5):616–626.

Danforth BN, Poinar GO. 2011. Morphology, classification, and antiquity of *Melittosphex burmensis* (Apoidea: Melittosphecidae) and implications for early bee evolution. *J. Paleontol.* 85(5):882–891.

Danforth BN, Sipes S, Fang J, Brady SG. 2006b. The history of early bee diversification based on five genes plus morphology. *Proc. Natl. Acad. Sci. USA* 103(41):15118–15123.

Danforth BN, Visscher PK. 1993. Dynamics of a host-cleptoparasite relationship: *Holcopasites ruthae* as a parasite of *Calliopsis pugionis* (Hymenoptera: Anthophoridae, Andrenidae). *Ann. Entomol. Soc. Am.* 86(6):833–840.

Danks HV. 1971. Nest mortality factors in stem-nesting aculeate Hymenoptera. *J. Anim. Ecol.* 40(1):79–82.

David A, Botías C, Abdul-Sada A, Nicholls E, Rotheray EL et al. 2016. Widespread contamination of wildflower and bee-collected pollen with complex mixtures of neonicotinoids and fungicides commonly applied to crops. *Environ. Int.* 88:169–178.

Davidowitz G. 2002. Does precipitation variability increase from mesic to xeric biomes? *Glob. Ecol. Biogeogr.* 11(2):143–154.

Davidson A. 1896. *Alcidamea producta* Cress. and its parasites. *Entomol. News* 7:216–218.

Davies GB, Brothers DJ. 2006. Morphology of *Scrapter* (Hymenoptera: Anthophila: Colletidae), with description of three new species and taxonomic status of five Cockerell taxa. *Afr. Invertebr.* 47:135–138.

Davis CC, Schaefer H, Xi Z, Baum DA, Donoghue MJ, Harmon LJ. 2014. Long-term morphological stasis maintained by a plant–pollinator mutualism. *Proc. Natl. Acad. Sci. USA* 111(16):5914–5919.

Davis ES, Murray TE, Fitzpatrick U, Brown MJ, Paxton RJ. 2010. Landscape effects on extremely fragmented populations of a rare solitary bee, *Colletes floralis*. *Mol. Ecol.* 19(22): 4922–4935.

De Luca PA, Vallejo-Marin M. 2013. What's the "buzz" about? The ecology and evolutionary significance of buzz-pollination. *Curr. Opin. Plant Biol.* 16(4):429–435.

De Palma A, Kuhlmann M, Roberts SP, Potts SG, Börger L et al. 2015. Ecological traits affect the sensitivity of bees to land-use pressures in European agricultural landscapes. *J. Appl. Ecol.* 52(6):1567–1577.

Debevec AH, Cardinal S, Danforth BN. 2012. Identifying the sister group to the bees: a molecular phylogeny of Aculeata with an emphasis on the superfamily Apoidea. *Zool. Scr.* 41(5):527–535.

Dehon M, Perrard A, Engel MS, Nel A, Michez D. 2017. Antiquity of cleptoparasitism among bees revealed by morphometric and phylogenetic analysis of a Paleocene fossil nomadine (Hymenoptera: Apidae). *Syst. Entom.* 42(3):543–554.

Dellicour S, Mardulyn P, Hardy OJ, Hardy C, Roberts SPM, Vereecken NJ. 2014. Inferring the mode of colonization of the rapid range expansion of a solitary bee from multilocus DNA sequence variation. *J. Evol. Biol.* 27(1):116–132.

Dennis DS, Lavigne RJ. 2007. Hymenoptera as prey of robber flies (Diptera: Asilidae) with new prey records. *J. Entomol. Res. Soc.* 9(3):23–42.

Dewulf A, De Meulemeester T, Dehon M, Engel M, Michez D. 2014. A new interpretation of the bee fossil *Melitta willardi* Cockerell (Hymenoptera, Melittidae) based on geometric morphometrics of the wing. *ZooKeys* 389:35–48.

Di Prisco G, Cavaliere V, Annoscia D, Varricchio P, Caprio E et al. 2013. Neonicotinoid clothianidin adversely affects insect immunity and promotes replication of a viral pathogen in honey bees. *Proc. Natl. Acad. Sci. USA* 110(46):18466–18471.

Diekötter T, Crist TO. 2013. Quantifying habitat-specific contributions to insect diversity in agricultural mosaic landscapes. *Insect Conserv. Divers.* 6(5):607–618.

Dobson HE. 1988. Survey of pollen and pollenkitt lipids–chemical cues to flower visitors? *Am. J. Bot.* 75(2):170–182.

Dobson HE, Peng Y-S. 1997. Digestion of pollen components by larvae of the flower-specialist bee *Chelostoma florisomne* (Hymenoptera: Megachilidae). *J. Insect Physiol.* 43(1):89–100.

Dod DD. 1965. *Tolumnia henekenii*—bee orchid pollinated by bee. *Am. Orchid Soc. Bull.* 4:792–794.

Dodson CH. 1965. Studies in orchid pollination: the genus *Coryanthes*. *Am. Orchid Soc. Bull.* 34:680–687.

Dodson CH, Dressler RL, Hills HG, Adams RM, Williams NH. 1969. Biologically active compounds in orchid fragrances. *Science* 164(3885):1243–1249.

Dodson CH, Frymire GP. 1961. Natural pollination of orchids. *Mo. Bot. Gard. Bull.* 49(2): 133–139.

Doublet V, Labarussias M, Miranda JR, Moritz RF, Paxton RJ. 2015. Bees under stress: sublethal doses of a neonicotinoid pesticide and pathogens interact to elevate honey bee mortality across the life cycle. *Environ. Microbiol.* 17(4):969–983.

Dressler RL. 1968. Pollination by euglossine bees. *Evolution* 22(1):202–210.

Dressler RL. 1982. Biology of the orchid bees (Euglossini). *Annu. Rev. Ecol. Syst.* 13:373–394.

Duffield RM, Wheeler JW, Eickwort GC. 1984. Sociochemicals of bees. In *Chemical Ecology of Insects*, Bell WJ, Carde RT, eds., pp. 387–428. Dordrecht: Springer.

Dukas R. 2001. Effects of perceived danger on flower choice by bees. *Ecol. Lett.* 4(4):327–333.

Dukas R, Morse DH. 2003. Crab spiders affect flower visitation by bees. *Oikos* 101(1):157–163.

Easterling DR, Meehl GA, Parmesan C, Changnon SA, Karl TR, Mearns LO. 2000. Climate extremes: observations, modeling, and impacts. *Science* 289(5487):2068–2074.

Eberhard WG. 1991. Copulatory courtship and cryptic female choice in insects. *Biol. Rev.* 66(1): 1–31.

Ebermann E, Hall M. 2003. First record of sporothecae within the mite Family Scutacaridae (Acari, Tarsonemina). *Zool. Anz.* 242(4):367–375.

Ebermann E, Hall M, Hausl-Hofstätter U, Jagersbacher-Baumann JM, Kirschner R et al. 2013. A new phoretic mite species with remarks to the phenomenon "Sporothecae" (Acari, Scutacaridae; Hymenoptera, Aculeata). *Zool. Anz.* 252(2):234–242.

Eckhardt M, Haider M, Dorn S, Müller A. 2014. Pollen mixing in pollen generalist solitary bees: a possible strategy to complement or mitigate unfavourable pollen properties? *J. Anim. Ecol.* 83(3):588–597.

Ehrlich PR, Raven PH. 1964. Butterflies and plants: a study in coevolution. *Evolution* 18(4): 586–608.

Eickwort GC. 1967. Aspects of the biology of *Chilicola ashmeadi* in Costa Rica (Hymenoptera: Colletidae). *J. Kans. Entomol. Soc.* 40(1):42–73.

Eickwort GC. 1969. A comparative morphological study and generic revision of the augochlorine bees (Hymenoptera: Halictidae). *Univ. Kans. Sci. Bull.* 48(13):325–524.

Eickwort GC. 1975. Gregarious nesting of the mason bee *Hoplitis anthocopoides* and the evolution of parasitism and sociality among megachilid bees. *Evolution* 29(1):142–150.

Eickwort GC. 1977. Aspects of the nesting biology and descriptions of immature stages of *Perdita octomaculata* and *P. halictoides* (Hymenoptera: Andrenidae). *J. Kans. Entomol. Soc.* 50(4):577–599.

Eickwort GC. 1994. Evolution and life-history patterns of mites associated with bees. In *Mites*, Houck MA, ed., pp. 218–251. Boston: Springer.

Eickwort GC, Eickwort KR. 1972. Aspects of the biology of Costa Rican halictine bees, III. *Sphecodes kathleenae*, a social cleptoparasite of *Dialictus umbripennis* (Hymenoptera: Halictidae). *J. Kans. Entomol. Soc.* 45(4):529–541.

Eickwort GC, Eickwort JM, Gordon J, Eickwort MA, Wcislo WT. 1996. Solitary behavior in a high-altitude population of the social sweat bee *Halictus rubicundus* (Hymenoptera: Halictidae). *Behav. Ecol. Sociobiol.* 38(4):227–233.

Eickwort GC, Ginsberg HS. 1980. Foraging and mating behavior in Apoidea. *Annu. Rev. Entomol.* 25:421–446.

Eickwort GC, Kukuk PF, Wesley FR. 1986. The nesting biology of *Dufourea novaeangliae* (Hymenoptera: Halictidae) and the systematic position of the Dufoureinae based on behavior and development. *J. Kans. Entomol. Soc.* 59(1):103–120.

Eickwort GC, Matthews RW, Carpenter J. 1981. Observations on the nesting behavior of *Megachile rubi* and *M. texana* with a discussion of the significance of soil nesting in the evolution of megachilid bees (Hymenoptera: Megachilidae). *J. Kans. Entomol. Soc.* 54(3):557–570.

Eickwort GC, Rozen JG Jr. 1997. The entomological evidence. *J. Forensic Sci.* 42(3):394–397.

Ellis AG, Johnson SD. 2010. Floral mimicry enhances pollen export: the evolution of pollination by sexual deceit outside of the Orchidaceae. *Am. Nat.* 176(5):E143–151.

Ellis JD, Munn PA. 2005. The worldwide health status of honey bees. *Bee World.* 86(4):88–101.

Eltz T, Bause C, Hund K, Quezada-Euan JJG, Pokorny T. 2015a. Correlates of perfume load in male orchid bees. *Chemoecology* 25(4):193–199.

Eltz T, Küttner J, Lunau K, Tollrian R. 2015b. Plant secretions prevent wasp parasitism in nests of wool-carder bees, with implications for the diversification of nesting materials in Megachilidae. *Front. Ecol. Evol.* 2(86):1–7.

Eltz T, Sager A, Lunau K. 2005. Juggling with volatiles: exposure of perfumes by displaying male orchid bees. *J. Comp. Physiol. A.* 191:575–581.

Eltz T, Zimmerman Y, Haftmann J, Twele R, Francke W et al. 2007. Enfleurage, lipid recycling and the origin of perfume collection in orchid bees. *Proc. R. Soc. Lond. B, Biol. Sci.* 274: 2843–2848.

Elwell SL, Griswold T, Elle E. 2016. Habitat type plays a greater role than livestock grazing in structuring shrubsteppe plant–pollinator communities. *J. Insect Conserv.* 20(3):515–525.

Emlen ST, Oring LW. 1977. Ecology, sexual selection, and the evolution of mating systems. *Science* 197(4300):215–223.

Engel MS. 2000. Classification of the bee tribe Augochlorini (Hymenoptera: Halictidae). *Bull. Am. Mus. Nat. Hist.* 250:1–89.

Engel MS. 2001. A monograph of the Baltic amber bees and evolution of the Apoidea (Hymenoptera). *Bull. Am. Mus. Nat. Hist.* 259:1–192.

Engel MS. 2006. A new genus of cleptoparasitic bees from the West Indies (Hymenoptera: Halictidae). *Acta Zool. Cracoviensia-Ser. B Invertebr.* 49(1–2):1–8.

Engel MS, Brooks RW. 1998. The nocturnal bee genus *Megaloptidia* (Hymenoptera: Halictidae). *J. Hymenopt. Res.* 7(1):1–14.

Engel MS, Perkovsky EE. 2006. An Eocene bee in Rovno amber, Ukraine (Hymenoptera: Megachilidae). *Am. Mus. Novit.* 3506:1–11.

Erbar C. 2014. Nectar secretion and nectaries in basal angiosperms, magnoliids, and non-core eudicots and a comparison with core eudicots. *Plant Divers. Evol.* 131:63–143.

Erenler HE, Orr MC, Gillman MP, Parkes BR, Rymer H, Maes J-M. 2016. Persistent nesting by *Anthophora* Latreille, 1803 (Hymenoptera: Apidae) bees in ash adjacent to an active volcano. *Pan-Pac. Entomol.* 92(2):67–78.

Erickson EH, Buchmann SL. 1983. Electrostatics of pollination. In *Handbook of Experimental Pollination Biology*, Jones CE, Little RJ, eds., pp. 173–184. New York: Van Nostrand Reinhold.

Erickson EH, Werner FG. 1974. Bionomics of Nearctic bee-associated Meloidae (Coleoptera)—life histories and nutrition of certain Meloinae. *Ann. Entomol. Soc. Am.* 67(3):394–400.

Etl F, Franschitz A, Aguiar AJ, Schönenberger J, Dötterl S. 2017. A perfume-collecting male oil bee? Evidences of a novel pollination system involving *Anthurium acutifolium* (Araceae) and *Paratetrapedia chocoensis* (Apidae, Tapinotaspidini). *Flora* 232:7–15.

Evans HE. 1966. The behavior patterns of solitary wasps. *Annu. Rev. Entomol.* 11:123–154.

Evans HE, O'Neill KM. 1988. *The Natural History and Behavior of North American Bee-wolves*. Ithaca: Comstock Publishing Associates.

Evans HE, West-Eberhard MJ. 1970. *The Wasps*. Ann Arbor: University of Michigan Press.

Evans JD, Schwarz RS. 2011. Bees brought to their knees: microbes affecting honey bee health. *Trends Microbiol.* 19(12):614–620.

Eves JD. 1970. Biology of *Monodontomerus obscurus* Westwood, a parasite of the alfalfa leaf-cutting bee, *Megachile rotundata* (Fabricius) (Hymenoptera: Torymidae; Megachilidae). *Melanderia* 4:1–18.

Evison SE, Roberts KE, Laurenson L, Pietravalle S, Hui J et al. 2012. Pervasiveness of parasites in pollinators. *PLoS One* 7(1):e30641.

Exeler N, Kratochwil A, Hochkirch A. 2008. Strong genetic exchange among populations of a specialist bee, *Andrena vaga* (Hymenoptera: Andrenidae). *Conserv. Genet.* 9(5):1233–1241.

Exeler N, Kratochwil A, Hochkirch A. 2010. Does recent habitat fragmentation affect the population genetics of a heathland specialist, *Andrena fuscipes* (Hymenoptera: Andrenidae)? *Conserv. Genet.* 11(5):1679–1687.

Fabre JH. 1914. *The Mason-Bees* (Translated by AT de Mattos). London: Hodder and Stoughton.

Fabre JH. 1921. The bee hunting *Philanthus*. In *More Hunting Wasps* de Mattos AT, trans. pp. 243–284. New York: Dodd, Mead, and Company.

Faegri K. 1986. The solanoid flower. *Trans. Bot. Soc. Edinb.* 45(suppl. 1):51–59.

Fauser-Misslin A, Sadd BM, Neumann P, Sandrock C. 2014. Influence of combined pesticide and parasite exposure on bumblebee colony traits in the laboratory. *J. Appl. Ecol.* 51(2): 450–459.

Fellendorf M, Mohra C, Paxton RJ. 2004. Devasting effects of river flooding to the ground-nesting bee, *Andrena vaga* (Hymenoptera: Andrenidae), and its associated fauna. *J. Insect Conserv.* 8(4):311–312

Fetridge ED, Ascher JS, Langellotto GA. 2008. The bee fauna of residential gardens in a suburb of New York City (Hymenoptera: Apoidea). *Ann. Entomol. Soc. Am.* 101(6):1067–1077.

Field J. 1992. Intraspecific parasitism as an alternative reproductive tactic in nest-building wasps and bees. *Biol. Rev.* 67(1):79–126.

Field J. 1996. Patterns of provisioning and iteroparity in a solitary halictine bee, *Lasioglossum* (*Evylaeus*) *fratellum* (Perez), with notes on *L*. (*E*.) *calceatum* (Scop.) and *L*. (*E*.) *villosulum* (K.). *Insectes Sociaux* 43(2):167–182.

Figueroa LL, Grab H, Graystock P, McFrederick QS, McArt S (in review). Landscape simplification and interaction network structure drive pathogen prevalence in bee communities. *Science*.

Fleischner TL. 1994. Costs of livestock grazing in western North America. *Conserv. Biol.* 8(3): 629–644.

Flematti GR, Ghisalberti EL, Dixon KW, Trengove RD. 2004. A compound from smoke that promotes seed germination. *Science* 305(5686):977–977.

Flores-Prado L, Chiappa E, Niemeyer HM. 2008. Nesting biology, life cycle, and interactions between females of *Manuelia postica*, a solitary species of the Xylocopinae (Hymenoptera: Apidae). *N. Z. J. Zool.* 35(1):93–102.

Foley JA, Defries R, Asner GP, Barford C, Bonan G et al. 2005. Global consequences of land use. *Science* 309(5734):570–574.

Forero D, Choe D-H, Weirauch C. 2011. Resin gathering in Neotropical resin bugs (Insecta: Hemiptera: Reduviidae): functional and comparative morphology. *J. Morphol.* 272(2): 204–229.

Forrest JR, Thomson JD. 2011. An examination of synchrony between insect emergence and flowering in Rocky Mountain meadows. *Ecol. Monogr.* 81(3):469–491.

Forsgren E, Olofsson TC, Vásquez A, Fries I. 2010. Novel lactic acid bacteria inhibiting *Paenibacillus larvae* in honey bee larvae. *Apidologie* 41(1):99–108.

Fortel L, Henry M, Guilbaud L, Guirao AL, Kuhlmann M et al. 2014. Decreasing abundance, increasing diversity and changing structure of the wild bee community (Hymenoptera: Anthophila) along an urbanization gradient. *PLoS One* 9(8):e104679.

Fowler J. 2016. Specialist bees of the northeast: host plants and habitat conservation. *Northeast. Nat.* 23(2):305–320.

Fraberger RJ, Ayasse M. 2007. Mating behavior, male territoriality and chemical communication in the european spiral-horned bees, *Systropha planidens* and *S. curvicornis* (Hymenoptera: Halictidae). *J. Kans. Entomol. Soc.* 80(4):348–360.

Frankie GW, Thorp RW, Pawelelk JC, Hernandez J, Coville RE. 2009. Urban bee diversity in a small residential garden in Northern California. *J. Hymenopt. Res.* 18(2):368–379.

Franzén M, Larsson M. 2007. Pollen harvesting and reproductive rates in specialized solitary bees. *Ann. Zool. Fenn.* 44 (6):405–414.

Franzén M, Larsson M, Nilsson SG. 2009. Small local population sizes and high habitat patch fidelity in a specialised solitary bee. *J. Insect Conserv.* 13(1):89–95.

Franzén M, Nilsson SG. 2013. High population variability and source–sink dynamics in a solitary bee species. *Ecology* 94(6):1400–1408.

Free JB. 1955. The behaviour of robber honeybees. *Behaviour* 7(1):233–239.

Fregonezi JN, Turchetto C, Bonatto SL, Freitas LB. 2013. Biogeographical history and diversification of *Petunia* and *Calibrachoa* (Solanaceae) in the Neotropical Pampas grassland. *Bot. J. Linn. Soc.* 171(1):140–153.

Friedman WE. 2009. The meaning of Darwin's "abominable mystery." *Am. J. Bot.* 96:1–18.

Friese H. 1912. Die Seidenbienen (*Colletes*) von Zentral-Europa. *Arch. Für Naturgeschichte* 78(7):149–161.

Friese H. 1923. *Die Europäischen Bienen Apidae, Das Leben Und Wirken Unserer Blumenwespen.* Berlin und Leipzig: Walter de Gruyter.

Frohlich DR. 1983. On the nesting biology of *Osmia (Chenosmia) bruneri* (Hymenoptera: Megachilidae). *J. Kans. Entomol. Soc.* 56(2):123–130.

Frohlich DR, Parker FD. 1983. Nest building behavior and development of the sunflower leafcutter bee: *Eumegachile (Sayapis) pugnata* (Say) (Hymenoptera: Megachilidae). *Psyche* 90(3):193–209.

Frohlich DR, Parker FD. 1985. Observations on the nest-building and reproductive behavior of a resin-gathering bee: *Dianthidium ulkei* (Hymenoptera: Megachilidae). *Ann. Entomol. Soc. Am.* 78(6):804–810.

Frohlich DR, Tepedino VJ. 1986. Sex ratio, parental investment, and interparent variability in nesting success in a solitary bee. *Evolution* 40(1):142–151.

Fu F, Kocher SD, Nowak MA. 2015. The risk-return trade-off between solitary and eusocial reproduction. *Ecol. Lett.* 18(1):74–84.

Fürst MA, McMahon DP, Osborne JL, Paxton RJ, Brown MJF. 2014. Disease associations between honeybees and bumblebees as a threat to wild pollinators. *Nature* 506(7488): 364–366.

Gaglianone MC. 2000. Behavior on flowers, structures associated to pollen transport and nesting biology of *Perditomorpha brunerii* and *Cephalurgus anomalus* (Hymenoptera: Colletidae, Andrenidae). *Rev. Biol. Trop.* 48(1):89–99.

Gaglianone MC. 2005. Nesting biology, seasonality, and flower hosts of *Epicharis nigrita* (Friese, 1900) (Hymenoptera: Apidae: Centridini), with a comparative analysis for the genus. *Stud. Neotropical Fauna Environ.* 40(3):191–200.

Gallai N, Salles J-M, Settele J, Vaissière BE. 2009. Economic valuation of the vulnerability of world agriculture confronted with pollinator decline. *Ecol. Econ.* 68(3):810–821.

Gardiner MA, Tuell JK, Isaacs R, Gibbs J, Ascher JS, Landis DA. 2010. Implications of three biofuel crops for beneficial arthropods in agricultural landscapes. *BioEnergy Res.* 3(1):6–19.

Gardner KE, Ascher JS. 2006. Notes on the native bee pollinators in New York apple orchards. *J. N. Y. Entomol. Soc.* 114(1):86–91.

Garibaldi LA, Aizen MA, Klein AM, Cunningham SA, Harder LD. 2011. Global growth and stability of agricultural yield decrease with pollinator dependence. *Proc. Natl. Acad. Sci. USA* 108(14):5909–5914.

Garófalo CA. 1980. Reproductive aspects and evolution of social behavior in bees (Hymenoptera, Apoidea). *Rev. Bras. Genética.* 3:139–152.

Garófalo CA. 1985. Social structure of *Euglossa cordata* nests (Hymenoptera: Apidae: Euglossini). *Entomol. Gen.* 11(1–2):77–83.

Garófalo CA, Camillo E, Campos MJO, Serrano JC. 1992. Nest re-use and communal nesting in *Microthurge corumbae* (Hymenoptera, Megachilidae), with special reference to nest defense. *Insectes Sociaux* 39(3):301–311.

Garófalo CA, Camillo E, Campos MJO, Zucchi R, Serrano JC. 1981. Bionomical aspects of *Lithurgus corumbae* (Hymenoptera, Megachilidae), including evolutionary considerations on the nesting behavior of the genus. *Rev. Bras. Genet.* 4(2):165–182.

Garófalo CA, Rozen JG Jr. 2001. Parasitic behavior of *Exaerete smaragdina* with descriptions of its mature oocyte and larval instars (Hymenoptera: Apidae: Euglossini). *Am. Mus. Novit.* 3349:1–26.

Gaskett AC. 2011. Orchid pollination by sexual deception. *Biol. Rev.* 86(1):33–75.

Gathmann A, Tscharntke T. 2002. Foraging ranges of solitary bees. *J. Anim. Ecol.* 71(5):757–764.

Gebhardt M, Rohr G. 1987. Zur bionomie der sandbienen *Andrena clarkella* (Kirby), *A. cineraria* (L.), *A. fuscipes* (Kirby) und ihren Kuckucksbienen (Hymenoptera: Apoidea). *Drosera.* 87:89–114.

Gemmill-Herren B, Ochieng AO. 2008. Role of native bees and natural habitats in eggplant (*Solanum melongena*) pollination in Kenya. *Agric. Ecosyst. Environ.* 127(1):31–36.

Gerber HS, Klostermeyer EC. 1970. Sex control by bees: a voluntary act of egg fertilization during oviposition. *Science* 167(3914):82–84.

Gerlach G. 2011. The genus *Coryanthes*: a paradigm in ecology. *Lankesteriana* 11:253–264.

Gerling D, Hurd PD Jr, Hefetz A. 1981. In-nest behavior of the carpenter bee, *Xylocopa pubescens* (Hymenoptera: Anthophoridae). *J. Kans. Entomol. Soc.* 54(2):209–218.

Gerling D, Hurd PD Jr, Hefetz A. 1983. Comparative behavioral biology of two Middle East species of carpenter bees (*Xylocopa* Latreille) (Hymenoptera: Apoidea). *Smithson. Contrib. Zool.* (369):1–33.

Gerling D, Velthuis HHW, Hefetz A. 1989. Bionomics of the large carpenter bees of the genus *Xylocopa. Annu. Rev. Entomol.* 34(1):163–190.

Geslin B, Le Féon V, Kuhlmann M, Vaissière BE, Dajoz I. 2016. The bee fauna of large parks in downtown Paris, France. *Ann. Société Entomol. Fr.* 51(5–6):487–493.

Gess SK. 1996. *The Pollen Wasps: Ecology and Natural History of the Masarinae.* Cambridge: Harvard University Press.

Gess SK, Gess FW. 2004. Distributions of flower associations of pollen wasps (Vespidae: Masarinae) in southern Africa. *J. Arid Environ.* 57(1):17–44.

Gess SK, Gess FW. 2014. *Wasps and Bees in Southern Africa.* Pretoria: South African National Biodiversity Institute.

Gess SK, Roosenschoon PA. 2017. Notes on the nesting of three species of Megachilinae in the Dubai Desert Conservation Reserve, UAE. *J. Hymenopt. Res.* 54(1):43–56.

Giacomini JJ, Leslie J, Tarpy DR, Palmer-Young EC, Irwin RE, Adler LS. 2018. Medicinal value of sunflower pollen against bee pathogens. *Scientific Reports* 8:14394.

Gibbs J. 2009. A new cleptoparasitic *Lasioglossum* (Hymenoptera, Halictidae) from Africa. *J. Hymenopt. Res.* 18(1):74–79.

Gibbs J, Albert J, Packer L. 2012a. Dual origins of social parasitism in North American *Dialictus* (Hymenoptera: Halictidae) confirmed using a phylogenetic approach. *Cladistics* 28(2): 195–207.

Gibbs J, Brady SG, Kanda K, Danforth BN. 2012b. Phylogeny of halictine bees supports a shared origin of eusociality for *Halictus* and *Lasioglossum* (Apoidea: Anthophila: Halictidae). *Mol. Phylogenet. Evol.* 65(3):926–939.

Gibbs J, Dathe HH. 2017. First records of *Hylaeus* (*Paraprosopis*) *pictipes* Nylander, 1852 (Hymenoptera: Colletidae) in North America. *Check List* 13(3):2116.

Gibbs J, Packer L, Dumesh S, Danforth BN. 2013. Revision and reclassification of *Lasioglossum* (*Evylaeus*), *L.* (*Hemihalictus*) and *L.* (*Sphecodogastra*) in eastern North America (Hymenoptera: Apoidea: Halictidae). *Zootaxa* 3672(1):1–117.

Giblin-Davis RM, Norden BB, Batra SWT, Eickwort GC. 1990. Commensal nematodes in the glands, genitalia, and brood cells of bees (Apoidea). *J. Nematol.* 22(2):150–161.

Giles V, Ascher JS. 2006. A survey of the bees of the Black Rock Forest Preserve, New York. *J. Hymenopt. Res.* 15(2):208–231.

Gilliam M, Buchmann SL, Lorenz BJ. 1984. Microbial flora of the larval provisions of the solitary bees, *Centris pallida*. *Apidologie* 15(1):1–10.

Gilliam M, Buchmann SL, Lorenz BJ, Schmazel RJ. 1990. Bacteria belonging to the genus *Bacillus* associated with three species of solitary bees. *Apidologie* 21(2):99–105.

Giovanetti M, Lasso E. 2005. Body size, loading capacity and rate of reproduction in the communal bee *Andrena agilissima* (Hymenoptera; Andrenidae). *Apidologie* 36(3):439–447.

Golonka AM, Vilgalys R. 2013. Nectar inhabiting yeasts in Virginian populations of *Silene latifolia* (Caryophyllaceae) and coflowering species. *Am. Midl. Nat.* 169(2):235–258.

Gonçalves RB. 2016. A molecular and morphological phylogeny of the extant Augochlorini (Hymenoptera, Apoidea) with comments on implications for biogeography. *Syst. Entomol.* 41(2):430–440.

Gonzalez VH, Griswold TL. 2013. Wool carder bees of the genus *Anthidium* in the Western Hemisphere (Hymenoptera: Megachilidae): diversity, host plant associations, phylogeny, and biogeography. *Zool. J. Linn. Soc.* 168(2):221–425.

Gonzalez VH, Griswold T, Praz CJ, Danforth BN. 2012. Phylogeny of the bee family Megachilidae (Hymenoptera: Apoidea) based on adult morphology. *Syst. Entomol.* 37(2):261–286.

Gonzalez VH, Ospina M, Palacios E, Trujillo E. 2011. Nesting habitats and rates of cell parasitism in some bee species of the genera *Ancyloscelis*, *Centris* and *Euglossa* (Hymenoptera: Apidae) from Colombia. *Bol. Mus. Entomol. Univ. Val.* 8(2):23–29.

Gonzálvez FG, Rodríguez-Gironés MA. 2013. Seeing is believing: information content and behavioural response to visual and chemical cues. *Proc. R. Soc. Lond. B, Biol. Sci.* 280(1763): 20130886.

Gonzálvez FG, Santamaría L, Corlett RT, Rodríguez-Gironés MA. 2013. Flowers attract weaver ants that deter less effective pollinators. *J. Ecol.* 101(1):78–85.

Goodell K. 2003. Food availability affects *Osmia pumila* (Hymenoptera: Megachilidae) foraging, reproduction, and brood parasitism. *Oecologia* 134(4):518–27.

Gotlieb A, Pisanty G, Rozen JG Jr, Müller A, Röder G et al. 2014. Nests, floral preferences, and immatures of the bee *Haetosmia vechti* (Hymenoptera: Megachilidae: Osmiini). *Am. Mus. Novit.* 3808:1–20.

Gottsberger G. 2016. Generalist and specialist pollination in basal angiosperms (ANITA grade, basal monocots, magnoliids, Chloranthaceae and Ceratophyllaceae): what we know now. *Plant Div. Evol.* 131(4):263–362.

Goukon K, Maeta Y. 2016. Overwintering stages of an autumnal bee, *Colletes collaris* Dours (Hymenoptera, Colletidae). *New Entomol.* 65(3,4):105–106.

Goulson D. 2003. Effects of introduced bees on native ecosystems. *Annu. Rev. Ecol. Evol. Syst.* 34(1):1–26.

Goulson D, Nicholls E, Botias C, Rotheray EL. 2015. Bee declines driven by combined stress from parasites, pesticides, and lack of flowers. *Science* 347(6229):1255957.

Graenicher S. 1906. A contribution to our knowledge of the visual memory of bees. *Bull. Wis. Nat. Hist. Soc.* 4:135–142.

Graenicher S. 1927. On the biology of the parasitic bees of the genus *Coelioxys* (Hymenoptera: Megachilidae). *Entomol. News.* 38(8):231–235.

Grandi G. 1936. Morfologia ed etologia comparate di insetti a regime specializzato. XII *Macrosiagon ferrugineum flabellatum* F. *Boll. Ist. Ent. R. Univ. Bologna.* 9:33–64

Grant V. 1950. The flower constancy of bees. *Bot. Rev.* 16(7):379–398.

Graystock P, Blane EJ, McFrederick QS, Goulson D, Hughes WOH. 2016. Do managed bees drive parasite spread and emergence in wild bees? *Int. J. Parasitol.: Parasites Wildl.* 5(1):64–75.

Graystock P, Goulson D, Hughes WOH. 2015. Parasites in bloom: flowers aid dispersal and transmission of pollinator parasites within and between bee species. *Proc. R. Soc. Lond. B, Biol. Sci.* 282(1813):20151371.

Graystock P, Yates K, Evison SEF, Darvill B, Goulson D, Hughes WOH. 2013. The Trojan hives: pollinator pathogens, imported and distributed in bumblebee colonies. *J. Appl. Ecol.* 50(5):1207–1215.

Green TW, Bohart GE. 1975. The pollination ecology of *Astragalus cibarius* and *Astragalus utahensis* (Leguminosae). *Am. J. Bot.* 62(4):379–386.

Greenleaf SS, Kremen C. 2006a. Wild bee species increase tomato production and respond differently to surrounding land use in Northern California. *Biol. Conserv.* 133(1):81–87.

Greenleaf SS, Kremen C. 2006b. Wild bees enhance honey bees' pollination of hybrid sunflower. *Proc. Natl. Acad. Sci. USA* 103(37):13890–13895.

Greenleaf SS, Williams NM, Winfree R, Kremen C. 2007. Bee foraging ranges and their relationship to body size. *Oecologia* 153(3):589–596.

Greiner B, Ribi WA, Warrant EJ. 2004. Retinal and optical adaptations for nocturnal vision in the halictid bee *Megalopta genalis*. *Cell Tissue Res.* 316(3):377–390.

Grigarick AA, Stange LA. 1968. The pollen-collecting bees of the Anthidiini of California (Hymenoptera: Megachilidae). *Bull. Calif. Insect Surv.* 9:1–113.

Grissell EE. 2007. Torymidae (Hymenoptera: Chalcidoidea) associated with bees (Apoidea), with a list of chalcidoid bee parasitoids. *J. Hymenopt. Res.* 16(2):234–265.

Griswold T. 1986. Notes on the nesting biology of *Protosmia* (*Chelostomopsis*) *rubifloris* (Cockerell) (Hymenoptera: Megachilidae). *Pan-Pac. Entomol.* 62(1):84–87.

Griswold TL, Miller W. 2010. A revision of *Perdita* (*Xerophasma*) Timberlake (Hymenoptera: Andrenidae). *Zootaxa.* 2517:1–14.

Griswold TL, Parker FD. 1988. New *Perdita* (*Perdita*) oligoleges of *Mentzelia*, with notes on related species of the Ventralis group (Hymenoptera: Andrenidae). *Pan-Pac. Entomol.* 64(1):43–52.

Groom SV, Rehan SM. 2018. Climate-mediated behavioural variability in facultatively social bees. *Biol. J. Linn. Soc.* 125(1):165–170.

Grundel R, Jean RP, Frohnapple KJ, Glowacki GA, Scott PE, Pavlovic NB. 2010. Floral and nesting resources, habitat structure, and fire influence bee distribution across an open forest gradient. *Ecol. Appl.* 20(6):1678–1692.

Guédot C, Bosch J, Kemp WP. 2009. Relationship between body size and homing ability in the genus *Osmia* (Hymenoptera; Megachilidae). *Ecol. Entomol.* 34(1):158–161.

Guédot C, Pitts-Singer TL, Buckner JS, Bosch J, Kemp WP. 2006. Olfactory cues and nest recognition in the solitary bee *Osmia lignaria*. *Physiol. Entomol.* 31(2):110–119.

Guirguis GN, Brindley WA. 1974. Insecticide susceptibility and response to selected pollens of larval alfalfa leafcutting bees, *Megachile pacifica* (Panzer) (Hymenoptera: Megachilidae). *Environ. Entomol.* 3(4):691–694.

Gutbier A. 1914. Über einige Hymenopterennester aus Turkestan. *Z. Für Wiss. Insektenbiologie.* 10:339–345.

Gutierrez AP, Pitcairn MJ, Ellis CK, Carruthers N, Ghezelbash R. 2005. Evaluating biological control of yellow starthistle (*Centaurea solstitialis*) in California: a GIS based supply-demand demographic model. *Biol. Control.* 34(2):115–131.

Gwynne DT. 1981. Nesting biology of the bumblebee wolf *Philanthus bicinctus* Mickel (Hymenoptera: Sphecidae). *Am. Midl. Nat.* 105(1):130–138.

Habermannová J, Bogusch P, Straka J. 2013. Flexible host choice and common host switches in the evolution of generalist and specialist cuckoo bees (Anthophila: *Sphecodes*). *PLoS One* 8(5):e64537.

Hafernik J, Saul-Gershenz L. 2000. Beetle larvae cooperate to mimic bees. *Nature* 405:35–36.

Haider M, Dorn S, Müller A. 2013. Intra-and interpopulational variation in the ability of a solitary bee species to develop on non-host pollen: implications for host range expansion. *Functional Ecology* 27(1):255–263.

Haider M, Dorn S, Müller A. 2014a. Better safe than sorry? A Fabaceae species exhibits unfavourable pollen properties for developing bee larvae despite its hidden anthers. *Arthropod-Plant Interact.* 8(3):221–231.

Haider M, Dorn S, Sedivy C, Müller A. 2014b. Phylogeny and floral hosts of a predominantly pollen generalist group of mason bees (Megachilidae: Osmiini). *Biol. J. Linn. Soc.* 111(1):78–91.

Hannan MA, Maeta Y, Miyanaga R. 2013. Nesting biology and life cycle of *Nomia* (*Acunomia*) *chalybeata* Smith on Iriomote Island, southernmost Archipelago of Japan, with notes on the simultaneous occurrence of diapausing and non-diapausing prepupae within the same nests (Hymenoptera: Halictidae). *J. Saudi Soc. Agric. Sci.* 12(2):91–99.

Hanson T, Ascher JS. 2018. An unusually large nesting aggregation of the digger bee *Anthophora bomboides* Kirby, 1838 (Hymenoptera: Apidae) in the San Juan Islands, Washington State. *Pan-Pac. Entomol.* 94(1):4–16.

Harder LD. 1982. Measurement and estimation of functional proboscis length in bumblebees (Hymenoptera: Apidae). *Can. J. Zool.* 60(5):1073–1079.

Harder LD. 1983. Functional differences of the proboscides of short-and long-tongued bees (Hymenoptera, Apoidea). *Can. J. Zool.* 61(7):1580–1586.

Harder LD, Barclay RMR. 1994. The functional significance of poricidal anthers and buzz pollination: controlled pollen removal from *Dodecatheon. Funct. Ecol.* 8(4):509–517.

Harder LD, Johnson SD. 2007. Function and evolution of aggregated pollen in angiosperms. *Int. J. Plant Sci.* 169:59–78.

Harder LD, Thomson JD. 1989. Evolutionary options for maximizing pollen dispersal of animal-pollinated plants. *Am. Nat.* 133(3):323–344.

Harder LD, Wilson WG. 1994. Floral evolution and male reproductive success: optimal dispensing schedules for pollen dispersal by animal-pollinated plants. *Evol. Ecol.* 8(6):542–559.

Hargreaves AL, Harder LD, Johnson SD. 2009. Consumptive emasculation: the ecological and evolutionary consequences of pollen theft. *Biol. Rev.* 84(2):259–276.

Hazir C, Thomas WK, Scheuhl E, Keskin N, Giblin-Davis R et al. 2010. Diversity and distribution of nematodes associated with wild bees in Turkey. *Nematology* 12(1):65–80.

He C-L, Zhu C-D. 2018. Nesting biology of *Xylocopa xinjiangensis* (Hymenoptera: Apidae: Xylocopinae). *J. Insect Sci.* 18(4):1–8.

Hedges SB, Kumar S. 2009. *The Timetree of Life*. Oxford: Oxford University Press.

Hedtke SM, Blitzer EJ, Montgomery GA, Danforth BN. 2015. Introduction of non-native pollinators can lead to trans-continental movement of bee-associated fungi. *PLoS One* 10(6): e0130560.

Hedtke SM, Patiny S, Danforth BN. 2013. The bee tree of life: a supermatrix approach to apoid phylogeny and biogeography. *BMC Evol. Biol.* 13:138.

Hefetz A. 1987. The role of Dufour's gland secretions in bees. *Physiol. Entomol.* 12(3):243–253.

Hefetz A. 1992. Individual scent marking of the nest entrance as a mechanism for nest recognition in *Xylocopa pubescens* (Hymenoptera: Anthophoridae). *J. Insect Behav.* 5(6):763–772.

Hefetz A, Fales HM, Batra SW. 1979. Natural polyesters: Dufour's gland macrocyclic lactones form brood cell laminesters in *Colletes* bees. *Science* 204(4391):415–417.

Heinrich B. 1993. *The Hot-Blooded Insects: Mechanisms and Evolution of Thermoregulation*. Cambridge: Harvard University Press.

Hennig EI, Ghazoul J. 2012. Pollinating animals in the urban environment. *Urban Ecosyst.* 15(1):149–166.

Herbst P. 1922. Zur Biologie der Gattung *Chilicola* Spin. (Apidae, Hymen.). *Entomologische Mitteilungen* 12(2):63–68.

Hernandez JL, Frankie GW, Thorp RW. 2009. Ecology of urban bees: a review of current knowledge and directions for future study. *Cities Environ.* 2(1):3–15.

Herrera CM. 1987. Components of pollinator "quality": comparative analysis of a diverse insect assemblage. *Oikos* 50(1):79–90.

Herrera CM. 1995. Floral biology, microclimate, and pollination by ectothermic bees in an early-blooming herb. *Ecology* 76(1):218–228.

Herrera-Flórez AF. 2014. A new species of *Grotea* Cresson (Hymenoptera, Ichneumonidae, Labeninae) from Colombia. *ZooKeys* 389:27–33.

Heyneman AJ. 1983. Optimal sugar concentrations of floral nectars—dependence on sugar intake efficiency and foraging costs. *Oecologia* 60(2):198–213.

Hicks CH. 1933. Observations on *Dianthidium ulkei* (Cresson) (Hymenoptera: Megachilidae). *Entomol. News.* 44:75–79.

Hinners SJ, Kearns CA, Wessman CA. 2012. Roles of scale, matrix, and native habitat in supporting a diverse suburban pollinator assemblage. *Ecol. Appl.* 22(7):1923–1935.

Hinojosa-Díaz IA, Feria-Arroyo TP, Engel MS. 2009. Potential distribution of orchid bees outside their native range: the cases of *Eulaema polychroma* (Mocsary) and *Euglossa viridissima* Friese in the USA (Hymenoptera: Apidae). *Divers. Distrib.* 15(3):421–428.

Hladik ML, Vandever M, Smalling KL. 2016. Exposure of native bees foraging in an agricultural landscape to current-use pesticides. *Sci. Total Environ.* 542:469–477.

Hocking B. 1949. *Hornia minutipennis* Riley: a new record and some notes on behavior (Coleoptera, Meloidae). *Can. Entomol.* 81(3):61–66.

Hoehn P, Tscharntke T, Tylianakis JM, Steffan-Dewenter I. 2008. Functional group diversity of bee pollinators increases crop yield. *Proc. R. Soc. Lond. B, Biol. Sci.* 275(1648):2283–2291.

Hogendoorn K. 1991. Intraspecific competition in the carpenter bee *Xylocopa pubescens* and its implications for the evolution of sociality. *Proc. Exp. Appl. Entomol.* 2:123–128.

Hogendoorn K, Velthuis HHW. 1993. The sociality of *Xylocopa pubescens*: does a helper really help? *Behav. Ecol. Sociobiol.* 32(4):247–257.

Hölldobler B, Wilson EO. 1990. *The Ants*. Cambridge: Harvard University Press.

Hook AW, Oswald JD, Neff JL. 2010. *Plega hagenella* (Neuroptera: Mantispidae) parasitism of *Hylaeus* (*Hylaeopsis*) sp. (Hymenoptera: Colletidae) reusing nests of *Trypoxylon manni* (Hymenoptera: Crabronidae) in Trinidad. *J. Hymenopt. Res.* 19(1):77–83.

Houck MA, O'Connor BM. 1991. Ecological and evolutionary significance of phoresy in the Astigmata. *Annu. Rev. Entomol.* 36:611–636.

Houston TF. 1975. Nests, behaviour and larvae of the bee *Stenotritus pubescens* (Smith) and behaviour of some related species (Hymenoptera: Apoidea: Stenotritinae). *Aust. J. Entomol.* 14(2):145–154.

Houston TF. 1981. Alimentary transport of pollen in a paracolletine bee (Hymenoptera: Colletidae). *Aust. Entomol.* 8(4):57–59.

Houston TF. 1983. An extraordinary new bee and adaptation of palpi for nectar-feeding in some Australian Colletidae and Pergidae (Hymenoptera). *Aust. J. Entomol.* 22(3):263–270.

Houston TF. 1984. Biological observations of bees in the genus *Ctenocolletes* (Hymenoptera: Stenotritidae). *Rec. West. Aust. Mus.* 11(2):153–172.

Houston TF. 1987. A second contribution to the biology of *Ctenocolletes* bees (Hymenoptera: Apoidea: Stenotritidae). *Rec. West. Aust. Mus.* 13(2):189–201.

Houston TF. 1991a. Ecology and behaviour of the bee *Amegilla* (*Asaropoda*) *dawsoni* (Rayment) with notes on a related species (Hymenoptera: Anthophoridae). *Rec. West. Aust. Mus.* 15(3):591–609.

Houston TF. 1991b. Two new and unusual species of the bee genus *Leioproctus* Smith (Hymenoptera: Colletidae), with notes on their behaviour. *Rec. West. Aust. Mus.* 15(1):83–96.

Houston TF. 2018. *A Guide to Native Bees of Australia*. Clayton South, VIC, Australia: CSIRO.

Houston TF, Ladd PG. 2002. Buzz pollination in the Epacridaceae. *Aust. J. Bot.* 50(1):83–91.

Houston TF, Lamont BB, Radford S, Errington SG. 1993. Apparent mutualism between *Verticordia nitens* and *V. aurea* (Myrtaceae) and their oil-ingesting bee pollinators (Hymenoptera, Colletidae). *Aust. J. Bot.* 41(3):369–380.

Houston TF, Maynard GV. 2012. An unusual new paracolletine bee, *Leioproctus* (*Ottocolletes*) *muelleri* subgen. & sp. nov. (Hymenoptera: Colletidae): with notes on nesting biology and in-burrow nest guarding by macrocephalic males. *Aust. J. Entomol.* 51(4):248–257.

Houston TF, Thorp RW. 1984. Bionomics of the bee *Stenotritus greavesi* and ethological characteristics of Stenotritidae (Hymenoptera). *Rec. West. Aust. Mus.* 11(4):375–385.

Howe MA, Knight GT, Clee C. 2010. The importance of coastal sand dunes for terrestrial invertebrates in Wales and the UK, with particular reference to aculeate Hymenoptera (bees, wasps and ants). *J. Coast. Conserv.* 14(2):91–102.

Howell DJ. 1974. Bats and pollen: physiological aspects of the syndrome of chiropterophily. *Comp. Biochem. Physiol. A Physiol.* 48(2):263–276.

Howell JF. 1967. Biology of *Zodion obliquefasciatum* (Macq.) (Diptera: Conopidae). *Tech. Bull. Wash. Agric. Exp. Stn.* 51:1–33.

Hu S, Dilcher DL, Jarzen DM, Taylor DW. 2008. Early steps of angiosperm–pollinator coevolution. *Proc. Natl. Acad. Sci. USA* 105(1):240–245.

Hurd PD Jr. 1957. Notes on the autumnal emergence of the vernal desert bee, *Hesperapis fulvipes* Crawford (Hymenoptera, Apoidea). *J. Kans. Entomol. Soc.* 30(1):10.

Hurd PD Jr. 1958. Observations on the nesting habits of some of the New World carpenter bees with remarks on their importance in the problem of species formation (Hymenoptera: Apoidea). *Ann. Entomol. Soc. Am.* 51:365–375.

Hurd PD Jr, LaBerge WE, Linsley EG. 1980. Principal sunflower bees of North America with emphasis on the southwestern United States (Hymenoptera: Apoidea). *Smithson. Contrib. Zool.* 310:1–158.

Hurd PD Jr, Linsley EG. 1950. Some insects associated with nests of *Dianthidium dubium dilectum* Timberlake, with a list of recorded parasites and inquilines of *Dianthidium* in North America. *J. N. Y. Entomol. Soc.* 58(4):247–250.

Hurd PD Jr, Linsley EG. 1963. Pollination of the unicorn plant (Martyniaceae) by an oligolectic, corolla-cutting bee (Hymenoptera: Apoidea). *J. Kans. Entomol. Soc.* 36(4):248–252.

Hurd PD Jr, Linsley EG. 1975. The principal *Larrea* bees of the southwestern United States (Hymenoptera, Apoidea). *Smithson. Contrib. Zool.* 193:1–74.

Hurd PD Jr, Michener CD. 1955. The megachiline bees of California (Hymenoptera: Megachilidae). *Bull. Calif. Insect Surv.* 3:1–248.

Hurd PD Jr, Moure JS. 1963. *A Classification of the Large Carpenter Bees (Xylocopini) (Hymenoptera: Apoidea).* Berkeley: University of California Press.

Hurd PD Jr, Powell JA. 1958. Observations on the nesting habits of *Colletes stepheni* Timberlake. *Pan-Pac. Entomol.* 34(3):147–153.

Immelman K, Eardley C. 2000. Gathering of grass pollen by solitary bees (Halictidae, *Lipotriches*) in South Africa. *Zoosystematics Evol.* 76(2):263–68.

Inglis GD, Sigler L, Goette MS. 1993. Aerobic microorganisms associated with alfalfa leafcutter bees (*Megachile rotundata*). *Microb. Ecol.* 26(2):125–143.

Inouye DW. 1980. The terminology of floral larceny. *Ecology* 61(5):1251–1253.

Intergovernmental Panel on Climate Change. 2015. *Climate Change 2014: Mitigation of Climate Change.* Vol. 3. Cambridge: Cambridge University Press.

Isaacs R, Kirk AK. 2010. Pollination services provided to small and large highbush blueberry fields by wild and managed bees: field size and blueberry pollination. *J. Appl. Ecol.* 47(4):841–849.

Iwata K. 1933. Studies on the nesting habits and parasites of *Megachile sculpturalis* Smith (Hymenoptera, Megachilidae). *Mushi* 6:4–24.

Iwata K. 1976. *Evolution of Instinct: Comparative Ethology of Hymenoptera.* New Delhi: Amerind Publiishing Company.

Iwata K, Sakagami SF. 1966. Gigantism and dwarfism in bee eggs in relation to the mode of life, with notes on the number of ovarioles. *Jpn. J. Ecol.* 16(1):4–16.

Jander R. 1976. Grooming and pollen manipulation in bees (Apoidea): the nature and evolution of movements involving the foreleg. *Physiol. Entomol.* 1(3):179–194.

Janzen DH. 1966. Notes on the behavior of the carpenter bee *Xylocopa fimbriata* in Mexico (Hymenoptera: Apoidea). *J. Kans. Entomol. Soc.* 39(4):633–641.

Javorek SK, Mackenzie KE, Vander Kloet SP. 2002. Comparative pollination effectiveness among bees (Hymenoptera: Apoidea) on lowbush blueberry (Ericaceae: *Vaccinium angustifolium*). *Ann. Entomol. Soc. Am.* 95(3):345–351.

Jayasingh DB, Freeman BE. 1980. The comparative population dynamics of eight solitary bees and wasps (Aculeata; Apocrita; Hymenoptera). *Biotropica* 12(3):214–219.

Jedrzejewska-Szmek K, Zych M. 2013. Flower-visitor and pollen transport networks in a large city: structure and properties. *Arthropod-Plant Interact.* 7(5):503–516.

Jennings JT, Austin AH. 2004. Biology and host relationships of aulacid and gasteruptiid wasps (Hymenoptera: Evanoioidea): a review. In *Perspectives on Biosystematics and Biodiversity*, Rajmohana K, Sudheer K, Girish Kumar P, Santhosh S, eds., pp. 187–215. University of Calicut: Systematic Entomology Research Scholars Association.

Jesus BMV, Garófalo CA. 2000. Nesting behaviour of *Centris* (*Heterocentris*) *analis* (Fabricius) in southeastern Brazil (Hymenoptera, Apidae, Centridini). *Apidologie* 31(4):503–515.

Johnson LK. 1983. *Trigona fulviventris.* In *Costa Rican Natural History*, Janzen DH, ed., pp. 770–772. Chicago: University of Chicago Press.

Johnson MD. 1981. Observations on the biology of *Andrena* (*Melandrena*) *dunningi* Cockerell (Hymenoptera: Andrenidae). *J. Kans. Entomol. Soc.* 54(1):32–40.

Johnson MD. 1984. The pollen preferences of *Andrena* (*Melandrena*) *dunningi* Cockerell (Hymenoptera: Andrenidae). *J. Kans. Entomol. Soc.* 57:34–43.

Johnson MD. 1990. Female size and fecundity in the small carpenter bee, *Ceratina calcarata* (Robertson) (Hymenoptera: Anthophoridae). *J. Kans. Entomol. Soc.* 63(3):414–419.

Julier HE, Roulston TH. 2009. Wild bee abundance and pollination service in cultivated pumpkins: farm management, nesting behavior and landscape effects. *J. Econ. Entomol.* 102(2):563–573.

Kahnt B, Montgomery GA, Murray EA, Kuhlmann M, Pauw A et al. 2017. Playing with extremes: origins and evolution of exaggerated female forelegs in South African *Rediviva* bees. *Mol. Phylogenet. Evol.* 115:95–105.

Kahnt B, Soro A, Kuhlmann M, Gerth M, Paxton RJ. 2014. Insights into the biodiversity of the Succulent Karoo hotspot of South Africa: the population genetics of a rare and endemic halictid bee, *Patellapis doleritica*. *Conserv. Genet.* 15(6):1491–1502.

Kaiser W. 1995. Rest at night in some solitary bees—a comparison with the sleep-like state of honey bees. *Apidologie* 26(3):213–230.

Kaltenpoth M. 2009. Actinobacteria as mutualists: general healthcare for insects? *Trends Microbiol.* 17(12):529–535.

Kaltenpoth M, Engl T. 2014. Defensive microbial symbionts in Hymenoptera. *Funct. Ecol.* 28(2):315–327.

Katayama E. 2001. Nesting biology of Japanese leaf-cutting bee *Megachile willughbiella munakatai*. *New Entomol.* 50:21–27.

Kathirithamby J. 2009. Host-parasitoid associations in Strepsiptera. *Annu. Rev. Entomol.* 54:227–249.

Kearns CA, Thomson JD. 2001. *The Natural History of Bumblebees: A Sourcebook for Investigations*. Boulder: University Press of Colorado.

Kelber A, Warrant EJ, Pfaff M, Wallén R, Theobald JC et al. 2006. Light intensity limits foraging activity in nocturnal and crepuscular bees. *Behav. Ecol.* 17(1):63–72.

Kemp WP, Bosch J, Dennis B. 2004. Oxygen consumption during the life cycles of the prepupa-wintering bee *Megachile rotundata* and the adult-wintering bee *Osmia lignaria* (Hymenoptera: Megachilidae). *Ann. Entomol. Soc. Am.* 97(1):161–170.

Kendall DA. 1973. The viability and compatibility of pollen on insects visiting apple blossom. *J. Appl. Ecol.* 10(3):847–853.

Kendall DA, Solomon ME. 1973. Quantities of pollen on the bodies of insects visiting apple blossom. *J. Appl. Ecol.* 10(2):627–634.

Kerfoot WB. 1967a. Correlation between ocellar size and the foraging activities of bees (Hymenoptera; Apoidea). *Am. Nat.* 101(917):65–70.

Kerfoot WB. 1967b. The lunar periodicity of *Sphecodogastra texana*, a nocturnal bee (Hymenoptera: Halictidae). *Anim. Behav.* 15(4):479–486.

Kerfoot WB. 1967c. Nest architecture and associated behavior of the nocturnal bee, *Sphecodogastra texana* (Hymenoptera: Halictidae). *J. Kans. Entomol. Soc.* 40(1):84–93.

Kerr JT, Pindar A, Galpern P, Packer L, Potts SG et al. 2015. Climate change impacts on bumblebees converge across continents. *Science* 349(6244):177–180.

Kevan PG. 1975. Forest application of the insecticide fenitrothion and its effect on wild bee pollinators (Hymenoptera: Apoidea) of lowbush blueberries (*Vaccinium* spp.) in Southern New Brunswick, Canada. *Biol. Conserv.* 7(4):301–309.

Kim J, Williams N, Kremen C. 2006. Effects of cultivation and proximity to natural habitat on ground-nesting native bees in California sunflower fields. *J. Kans. Entomol. Soc.* 79(4):309–320.

Kim J-Y. 1997. Female size and fitness in the leaf-cutter bee *Megachile apicalis*. *Ecol. Entomol.* 22(3):275–282.

Kim J-Y. 1999. Influence of resource level on maternal investment in a leaf-cutter bee (Hymenoptera: Megachilidae). *Behav. Ecol.* 10(5):552–56.

Kimsey LS. 1980. The behavior of male orchid bees (Apidae, Hymenoptera, Insecta) and the question of leks. *Anim. Behav.* 28:996–1004.

Kimsey LS. 1984. The behavioural and structural aspects of grooming and related activities in euglossine bees (Hymenoptera: Apidae). *J. Zool.* 204(4):541–550.

Kimsey LS. 2006. California cuckoo wasps in the family Chrysididae (Hymenoptera). *Univ. Calif. Publ. Entomol.* 125:1–319.

Kirk WDJ. 1996. *Thrips*, Vol. 25. Slough, England: Richmond Publishing Co.

Kleijn D, Winfree R, Bartomeus I, Carvalheiro LG, Henry M et al. 2015. Delivery of crop pollination services is an insufficient argument for wild pollinator conservation. *Nat. Commun.* 6:7414.

Klein A-M, Brittain C, Hendrix SD, Thorp R, Williams N, Kremen C. 2012. Wild pollination services to California almond rely on semi-natural habitat. *J. Appl. Ecol.* 49(3):723–732.

Klein A-M, Steffan–Dewenter I, Tscharntke T. 2003. Fruit set of highland coffee increases with the diversity of pollinating bees. *Proc. R. Soc. Lond. B, Biol. Sci.* 270(1518):955–961.

Klein A-M, Steffan-Dewenter I, Tscharntke T. 2004. Foraging trip duration and density of mega-chilid bees, eumenid wasps and pompilid wasps in tropical agroforestry systems. *J. Anim. Ecol.* 73(3):517–525.

Klein A-M, Vaissiere BE, Cane JH, Steffan-Dewenter I, Cunningham SA et al. 2007. Importance of pollinators in changing landscapes for world crops. *Proc. R. Soc. Lond. B, Biol. Sci.* 274(1608):303–313.

Klimov PB, Bochkov AV, O'Connor BM. 2006. Host specificity and multivariate diagnostics of cryptic species in predacious cheyletid mites of the genus *Cheletophyes* (Acari: Cheyletidae) associated with large carpenter bees. *Biol. J. Linn. Soc.* 87(1):45–58.

Klimov PB, O'Connor BM. 2007. Ancestral area analysis of chaetodactylid mites (Acari: Chaetodactylidae), with description of a new early derivative genus and six new species from the Neotropics. *Ann. Entomol. Soc. Am.* 100(6):810–829.

Klimov PB, O'Connor BM, Knowles LL. 2007a. Museum specimens and phylogenies elucidate ecology's role in coevolutionary associations between mites and their bee hosts. *Evolution* 61(6):1368–1379.

Klimov PB, Vinson SB, O'Connor BM. 2007b. Acarinaria in associations of apid bees (Hymenoptera) and chaetodactylid mites (Acari). *Invertebr. Syst.* 21(2):109–136.

Klostermeyer EC, Gerber HS. 1969. Nesting behavior of *Megachile rotundata* (Hymenoptera: Megachilidae) monitored with an event recorder. *Ann. Entomol. Soc. Am.* 62(6):1321–1325.

Knerer G. 1993. Periodizitat und strategie der schmarotzer einer socialen schmalbiene, *Evylaeus malachurus* (K.) (Apoidea: Halictidae). *Zool. Anz.* 190:41–63.

Knerer G, Atwood CE. 1966. Parasitization of social halictine bees in southern Ontario. *Proc. Entomol. Soc. Ont.* 97:103–110.

Knerer G, Schwarz M. 1976. Halictine social evolution: the Australian enigma. *Science* 194 (4263):445–448.

Koch H, Schmid-Hempel P. 2011. Socially transmitted gut microbiota protect bumble bees against an intestinal parasite. *Proceedings of the National Academy of Sciences* 108(48): 19288–19292.

Kölreuter JG. 1761–1766. *Vorläufige Nachricht von Einigen Das Geschlect Der Pflanzen Betreffenden Versuchen Und Beobachtungen: Nebst Forsetzungen 1,2 and 3.* Leipzig: Johann Friedrich Gledischens Buchhandlung.

Koptur S. 1992. Extrafloral nectary-mediated interactions between insects and plants. In *Insect-Plant Interactions*, Vol. IV, Bernays EA, ed., pp. 81–129. Boca Raton: CRC Press.

Kremen C, Williams NM, Thorp RW. 2002. Crop pollination from native bees at risk from agricultural intensification. *Proc. Natl. Acad. Sci. USA* 99(26):16812–16816.

Krenn HW, Mauss V, Plant J. 2002. Evolution of the suctorial proboscis in pollen wasps (Masarinae, Vespidae). *Arthropod Struct. Dev.* 31(2):103–120.

Krenn HW, Plant JD, Szucsich NU. 2005. Mouthparts of flower-visiting insects. *Arthropod Struct. Dev.* 34(1):1–40.

Krombein KV. 1967. *Trap Nesting Bees and Wasps.* Washington, DC: Smithsonian Press.

Krombein KV. 1969. *Life History Notes on Some Egyptian Solitary Wasps and Bees and Their Associates (Hymenoptera: Aculeata).* Washington, DC: Smithsonian Institution Press.

Krombein KV, Hurd PD Jr, Smith DR, Burks BD. 1979. *Catalog of Hymenoptera in America North of Mexico.* Washington, DC: Smithsonian Institution Press.

Krombein KV, Norden BB. 1997a. Bizarre nesting behavior of *Krombeinictus nordenae* Leclercq (Hymenoptera: Sphecidae, Crabroninae). *J. South Asian Nat. Hist.* 2(2):145–154.

Krombein KV, Norden BB. 1997b. Nesting behavior of *Krombeinictus nordenae* Leclercq, a sphecid wasp with vegetarian larvae (Hymenoptera: Sphecidae: Crabroninae). *Entomol. Soc. Wash.* 99(1):42–49.

Kronenberg S, Hefetz A. 1984a. Role of labial glands in nesting behaviour of *Chalicodoma sicula* (Hymenoptera; Megachilidae). *Physiol. Entomol.* 9(2):175–179.

Kronenberg S, Hefetz A. 1984b. Comparative analysis of Dufour's gland secretions of two carpenter bees (Xylocopinae: Anthophoridae) with different nesting habits. *Comp. Biochem. Physiol. Part B Comp. Biochem.* 79(3):421–425.

Krunić M, Stanisavljević L, Pinzauti M, Felicioli A. 2005. The accompanying fauna of *Osmia cornuta* and *Osmia rufa* and effective measures of protection. *Bull. Insectology* 58(2):141–152.

Krupke CH, Hunt GJ, Eitzer BD, Andino G, Given K. 2012. Multiple routes of pesticide exposure for honey bees living near agricultural fields. *PLoS One* 7(1):e29268.

Kuhlmann M. 2006. Scopa reduction and pollen collecting of bees of the *Colletes fasciatus*-group in the winter rainfall area of South Africa (Hymenoptera: Colletidae). *J. Kans. Entomol. Soc.* 79(2):165–175.

Kuhlmann M. 2012. Revision of the South African endemic bee genus *Redivivoides* Michener, 1981 (Hymenoptera: Apoidea: Melittidae). *Eur. J. Taxon.* 34:1–34.

Kuhlmann M. 2014. Nest architecture and use of floral oil in the oil-collecting South African solitary bee *Rediviva intermixta* (Cockerell) (Hymenoptera: Apoidea: Melittidae). *J. Nat. Hist.* 48(43–44):2633–2644.

Kuhlmann M, Else GR, Dawson A, Quicke DL. 2007. Molecular, biogeographical and phenological evidence for the existence of three western European sibling species in the *Colletes succinctus* group (Hymenoptera: Apidae). *Org. Divers. Evol.* 7(2):155–165.

Kuhlmann M, Gess FW, Koch F, Gess SK. 2011. Southern African osmiine bees: taxonomic notes, two new species, a key to *Wainia*, and biological observations (Hymenoptera: Anthophila: Megachilidae). *Zootaxa* 3108:1–24.

Kuhlmann M, Guo D, Veldtman R, Donaldson J. 2012. Consequences of warming up a hotspot: species range shifts within a centre of bee diversity. *Divers. Distrib.* 18(9):885–897.

Kuhlmann M, Hollens H. 2015. Morphology of oil-collecting pilosity of female *Rediviva* bees (Hymenoptera: Apoidea: Melittidae) reflects host plant use. *J. Nat. Hist.* 49(9–10):561–573.

Kuhlmann M, Timmermann K. 2009. Nest architecture and floral hosts of the South African monolectic solitary bee *Othinosmia* (*Megaloheriades*) *schultzei* (Hymenoptera: Megachilidae). *Entomol. Gen.* 32(1):1–9.

Kuhlmann M, Timmermann K. 2011. Nest architecture of the monolectic South African solitary bee, *Samba (Prosamba) spinosa* Eardley (Hymenoptera: Apoidea: Melittidae). *Afr. Entomol.* 19(1):141–145.

Kukuk PF, Schwarz M. 1987. Intranest behavior of the communal sweat bee *Lasioglossum* (*Chilalictus*) *erythrurum* (Hymenoptera: Halictidae). *J. Kans. Entomol. Soc.* 60(1):58–64.

Kukuk PF, Schwarz M. 1988. Macrocephalic male bees as functional reproductives and probable guards. *Pan-Pac. Entomol.* 64:131–137.

Kullenberg B. 1956. On the scents and colours of *Ophyrys* flowers and their specific pollinators among aculeate Hymenoptera. *Sven. Bot. Tidskr.* 44:446–464.

Kullenberg B, Bergström G. 1976. The pollination of *Ophrys* orchids. *Bot. Not.* 129:11–29.

Kurczewski FE, Miller RE. 1983. Nesting behavior of *Philanthus sanbornii* in Florida (Hymenoptera: Sphecidae). *Fla. Entomol.* 66(1):199–206.

Labandeira CC, Johnson KR, Wilf P. 2002. Impact of the terminal Cretaceous event on plant–insect associations. *Proc. Natl. Acad. Sci. USA* 99(4):2061–2066.

Labandeira CC, Kvacek J, Mostovski MB. 2007. Pollination drops, pollen, and insect pollination of Mesozoic gymnosperms. *Taxon* 56(3):663–695.

LaBerge WE. 1956. A revision of the bees of the genus *Melissodes* in North and Central America. Parts I, II. (Hymenoptera, Apidae). *Univ. Kans. Sci. Bull.* 37(2):911–1194.

LaBerge WE. 1961. A revision of the bees of the genus *Melissodes* in North and Central America: Part III (Hymenoptera, Apidae). *Univ. Kans. Sci. Bull.* 42(5):283–661.

LaBerge WE. 1970. A new genus with three new species of eucerine bees from Mexico (Hymenoptera: Anthophoridae). *J. Kans. Entomol. Soc.* 43(3):321–328.

LaBerge WE. 1978. *Andrena (Callandrena) micheneriana*, a remarkable new bee from Arizona and Mexico (Apoidea: Andrenidae). *J. Kans. Entomol. Soc.* 51(4):592–596.

LaBerge WE. 1986. A revision of the bees of the genus *Andrena* of the Western Hemisphere. Part XI. Minor subgenera and subgeneric key. *Trans. Am. Entomol. Soc.* 111:441–567.

LaBerge WE. 1987. A revision of the bees of the genus *Andrena* of the Western Hemisphere. Part XII. Subgenera *Leucandrena*, *Ptilandrena*, *Scoliandrena*, and *Melandrena*. *Trans. Am. Entomol. Soc.* 112(3):191–248.

LaBerge WE. 1989. A review of the bees of the genus *Pectinapis* (Hymenoptera: Anthophoridae). *J. Kans. Entomol. Soc.* 62(4):524–527.

Lach L. 2008. Argentine ants displace floral arthropods in a biodiversity hotspot. *Divers. Distrib.* 14(2):281–290.

Ladurner E, Bosch J, Kemp WP, Maini S. 2008. Foraging and nesting behavior of *Osmia lignaria* (Hymenoptera: Megachilidae) in the presence of fungicides: cage studies. *J. Econ. Entomol.* 101(3):647–653.

Lampert KP, Pasternak V, Brand P, Tollrian R, Leese F, Eltz T. 2014. "Late" male sperm precedence in polyandrous wool-carder bees and the evolution of male resource defence in Hymenoptera. *Anim. Behav.* 90:211–217.

Laport RG, Minckley RL. 2012. Occupation of active *Xylocopa virginica* nests by the recently invasive *Megachile sculpturalis* in upstate New York. *J. Kans. Entomol. Soc.* 85(4):384–386.

Laport RG, Minckley RL, Ramsey J. 2012. Phylogeny and cytogeography of the North American creosote bush (*Larrea tridentata*, Zygophyllaceae). *Syst. Bot.* 37(1):153–164.

Larkin LL, Neff JL, Simpson BB. 2006. Phylogeny of the *Callandrena* subgenus of *Andrena* (Hymenoptera: Andrenidae) based on mitochondrial and nuclear DNA data: polyphyly and convergent evolution. *Mol. Phylogenet. Evol.* 38(2):330–343.

Larkin LL, Neff JL, Simpson BB. 2008. The evolution of a pollen diet: host choice and diet breadth of *Andrena* bees (Hymenoptera: Andrenidae). *Apidologie* 39(1):133–145.

Laroca S, Michener CD, Hofmeister RM. 1989. Long mouthparts among "short-tongued" bees and the fine structure of the labium in *Niltonia* (Hymenoptera, Colletidae). *J. Kans. Entomol. Soc.* 62(3):400–410.

Larsen ON, Gleffe G, Tengö J. 1986. Vibration and sound communication in solitary bees and wasps. *Physiol. Entomol.* 11(3):287–296.

Larsson M. 2005. Higher pollinator effectiveness by specialist than generalist flower-visitors of unspecialized *Knautia arvensis* (Dipsacaceae). *Oecologia* 146(3):394–403.

Larsson M, Franzén M. 2007. Critical resource levels of pollen for the declining bee *Andrena hattorfiana* (Hymenoptera, Andrenidae). *Biol. Conserv.* 134(3):405–414.

Lau JA, Galloway LF. 2004. Effects of low-efficiency pollinators on plant fitness and floral trait evolution in *Campanula americana* (Campanulaceae). *Oecologia* 141:577–583.

Lázaro A, Tscheulin T, Devalez J, Nakas G, Petanidou T. 2016. Effects of grazing intensity on pollinator abundance and diversity, and on pollination services. *Ecol. Entomol.* 41(4): 400–412.

Le Feon V, Schermann-Legionnet A, Delettre Y, Aviron S, Billeter R et al. 2010. Intensification of agriculture, landscape composition and wild bee communities: a large scale study in four European countries. *Agric. Ecosyst. Environ.* 137(1–2):143–150.

Lello E de. 1971a. Adnexal glands of the sting apparatus of bees: anatomy and histology, I (Hymenoptera: Colletidae and Andrenidae). *J. Kans. Entomol. Soc.* 44(1):5–13.

Lello E de. 1971b. Adnexal glands of the sting apparatus of bees: anatomy and histology, II (Hymenoptera: Halictidae). *J. Kans. Entomol. Soc.* 44(1):14–20.

Lello E de. 1971c. Anatomia e histologia das glândulas do ferrão das abelhas. III. (Hymenoptera: Megachilidae, Melittidae). *Cien. Cult.* 23(3):253–258.

Lello E de. 1971d. Glândulas anexas ao aparelho de ferrão das abelhas, anatomia e histologia IV (Hymenoptera: Anthophoridae). *Cien. Cult.* 23(6):765–772.

Lello E de. 1976. Adnexal glands of the sting apparatus in bees: anatomy and histology, V (Hymenoptera: Apidae). *J. Kans. Entomol. Soc.* 49(1):85–99.

Lentz DL, Pohl MD, Alvarado JL, Tarighat S, Bye R. 2008. Sunflower (*Helianthus annuus* L.) as a pre-Columbian domesticate in Mexico. *Proc. Natl. Acad. Sci. USA* 105(17):6232–6237.

Lester LJ, Selander RK. 1979. Population genetics of haplodiploid insects. *Genetics* 92(4): 1329–1345.

Levin MD. 1983. Value of bee pollination to US agriculture. *Am. Entomol.* 29(4):50–51.

Levin MD, Haydak MH. 1957. Comparative value of different pollens in the nutrition of *Osmia lignaria*. *Bee World* 38(9):221–226.

Leys R, Cooper SJ, Schwarz MP. 2000. Molecular phylogeny of the large carpenter bees, genus *Xylocopa* (Hymenoptera: Apidae), based on mitochondrial DNA sequences. *Mol. Phylogenet. Evol.* 17(3):407–418.

Leys R, Cooper SJ, Schwarz MP. 2002. Molecular phylogeny and historical biogeography of the large carpenter bees, genus *Xylocopa* (Hymenoptera: Apidae). *Biol. J. Linn. Soc.* 77(2):249–266.

Leys R, Hogendoorn K. 2008. Correlated evolution of mating behaviour and morphology in large carpenter bees (*Xylocopa*). *Apidologie* 39(1):119–132.

Lind H. 1968. Nest provisioning cycle and daily routine of behavior in *Dasypoda plumipes*. *Entomol. Medd.* 36(4):343–372.

Lindquist EE. 1985. Discovery of sporothecae in adult female *Trochometridium* Cross, with notes on analogous structures in *Siteroptes* Amerling (Acari: Heterostigmata). *Exp. Appl. Acarol.* 1(1):73–85.

Linsley EG. 1958. The ecology of solitary bees. *Hilgardia* 27:543–599.

Linsley EG. 1960. Ethology of some bee- and wasp-killing robber flies of southeastern Arizona and western New Mexico (Diptera: Asilidae). *Univ. Calif. Publ. Entomol.* 16(7):357–392.

Linsley EG. 1962. Sleeping aggregations of aculeate Hymenoptera—II. *Ann. Entomol. Soc. Am.* 55(2):148–164.

Linsley EG, Cazier MA. 1970. Some competitive relationships among matinal and late afternoon foraging activities of caupolicanine bees in southeastern Arizona (Hymenoptera, Colletidae). *J. Kans. Entomol. Soc.* 43(3):251–261.

Linsley EG, MacSwain JW. 1942a. Bionomics of the meloid genus *Hornia* (Coleoptera). *Univ. Calif. Publ. Entomol.* 7:189–206.

Linsley EG, MacSwain JW. 1942b. The parasites, predators, and inquiline associates of *Anthophora linsleyi*. *Am. Midl. Nat.* 27(2):402–417.

Linsley EG, MacSwain JW. 1943. Observations in the life history of *Trichodes ornatus* (Coleoptera, Cleridae), a larval predator in the nests of bees and wasps. *Ann. Entomol. Soc. Am.* 36(4):589–601.

Linsley EG, MacSwain JW. 1952. Notes on some effects of parasitism upon a small population of *Diadasia bituberculata* (Cresson) (Hymenoptera: Anthophoridae). *Pan-Pac. Entomol.* 28(3):131–135.

Linsley EG, MacSwain JW. 1955. The habits of *Nomada opacella* Timberlake with notes on other species (Hymenoptera: Anthophoridae). *Wasmann J. Biol.* 13:253–276.

Linsley EG, MacSwain JW. 1956. Observations on the habits of *Stylops pacifica* Bohart. *Univ. Calif. Publ. Entomol.* 11(7):395–430.

Linsley EG, MacSwain JW. 1957. The nesting habits, flower relationships, and parasites of some North American species of *Diadasia* (Hymenoptera: Anthophoridae). *Wasmann J. Biol.* 15:199–235.

Linsley EG, MacSwain JW. 1958. The significance of floral constancy among bees of the genus *Diadasia* (Hymenoptera, Anthophoridae). *Evolution* 12(2):219–223.

Linsley EG, MacSwain JW, Raven PH. 1963. Comparative behavior of bees and Onagraceae II. *Oenothera* bees of the Great Basin. *Univ. Calif. Publ. Entomol.* 33(2):25–50.

Linsley EG, MacSwain JW, Smith RF. 1952a. The life history and development of *Rhipiphorus smithi* with notes on their phylogenetic significance (Coleoptera: Rhipiphoridae). *Univ. Calif. Publ. Entomol.* 9(4):291–306.

Linsley EG, MacSwain JW, Smith RF. 1952b. The bionomics of *Diadasia consociata* Timberlake and some biological relationships of emphorine and anthophorine bees. *Univ. Calif. Publ. Entomol.* 9(3):267–290.

Litman JR, Danforth BN, Eardley CD, Praz CJ. 2011. Why do leafcutter bees cut leaves? New insights into the early evolution of bees. *Proc. R. Soc. Lond. B, Biol. Sci.* 278(1724):3593–3600.

Litman JR, Griswold T, Danforth BN. 2016. Phylogenetic systematics and a revised generic classification of anthidiine bees (Hymenoptera: Megachilidae). *Mol. Phylogenet. Evol.* 100: 183–198.

Litman JR, Praz CJ, Danforth BN, Griswold TL, Cardinal S. 2013. Origins, evolution, and diversification of cleptoparasitic lineages of long-tongued bees. *Evolution* 67(10):2982–2998.

Loew E. 1884. Beobachtungen über den Blümenbesuch von Insekten an Freilandplanzen des Botanischen Gartens zu Berlin. *Jahrb. K. Bot. Gart. Bot. Mus. Zu Berln.* 3:69–118, 253–259.

Lokvam J, Braddock JF. 1999. Anti-bacterial function in the sexually dimorphic pollinator rewards of *Clusia grandiflora* (Clusiaceae). *Oecologia* 119(4):534–540.

Lomholdt O. 1982. On the origin of the bees (Hymenoptera: Apidae, Sphecidae). *Insect Syst. Evol.* 13(2):185–190.

Long EY, Krupke CH. 2016. Non-cultivated plants present a season-long route of pesticide exposure for honey bees. *Nat. Commun.* 7:11629.

Loonstra AJ. 2012. Het ondergrondse leven van de gewone sachembij, *Anthophora plumipes* (Hymenoptera, Apidae). *Entomol. Ber.* 72(1–2):41–51.

López-Uribe MM, Cane JH, Minckley RL, Danforth BN. 2016. Crop domestication facilitated rapid geographical expansion of a specialist pollinator, the squash bee *Peponapis pruinosa*. *Proc. R. Soc. Lond. B, Biol. Sci.* 283(1833):20160443.

López-Uribe MM, Morreale SJ, Santiago CK, Danforth BN. 2015. Nest suitability, fine-scale population structure and male-mediated dispersal of a solitary ground nesting bee in an urban landscape. *PLoS One* 10(5):e0125719.

López-Uribe MM, Zamudio KR, Cardoso CF, Danforth BN. 2014. Climate, physiological tolerance and sex-biased dispersal shape genetic structure of Neotropical orchid bees. *Mol. Ecol.* 23(7):1874–1890.

Losey JE, Vaughan M. 2006. The economic value of ecological services provided by insects. *Bioscience* 56(4):311–323.

Lovejoy AO. 1936. *The Great Chain of Being: A Study of the History of an Idea*. Cambridge, Massachusetts: Harvard University Press.

Lovell JH. 1913. The origin of the oligotropic habit among bees (Hymen.). *Entomol. News.* 24:104–112.

Lovell JH. 1914. The origin of oligotropism (Hymen.). *Entomol. News* 25:314–321.

Lovell JH. 1918. *The Flower and the Bee*. New York: C. Scribner's and Sons.

Lowenstein DM, Matteson KC, Xiao I, Silva AM, Minor ES. 2014. Humans, bees, and pollination services in the city: the case of Chicago, IL (USA). *Biodivers. Conserv.* 23(11):2857–2874.

Lucia M, Telleria MC, Ramello PJ, Abrahamovich AH. 2017. Nesting ecology and floral resource of *Xylocopa augusti* Lepeletier de Saint Fargeau (Hymenoptera, Apidae) in Argentina. *Agric. For. Entomol.* 19(3):281–293.

Luckow M, Hopkins HC. 1995. A cladistic analysis of *Parkia* (Leguminosae: Mimosoideae). *Am. J. Bot.* 82(10):1300–1320.

Lunau K, Piorek V, Krohn O, Pacini E. 2015. Just spines—mechanical defense of malvaceous pollen against collection by corbiculate bees. *Apidologie* 46(2):144–149.

Machado IC, Vogel S, Lopes AV. 2002. Pollination of *Angelonia cornigera* Hook (Scrophulariaceae) by long-legged, oil-collecting bees in NE Brazil. *Plant Biol.* 4(3):352–359.

MacIvor JS, Moore AE. 2013. Bees collect polyurethane and polyethylene plastics as novel nest materials. *Ecosphere* 4(12):art155.

MacIvor JS, Salehi B. 2014. Bee species-specific nesting material attracts a generalist parasitoid: implications for co-occurring bees in nest box enhancements. *Environ. Entomol.* 43(4):1027–1033.

MacSwain JW. 1958. Longevity of some anthophorid bee larvae (Hymenoptera: Apoidea). *Pan-Pac. Entomol.* 34:40.

MacSwain JW, Raven PH, Thorp RW. 1973. Comparative behavior of bees and Onagraceae: IV. *Clarkia* bees of the western United States. *Univ. Calif. Publ. Entomol.* 70:1–80.

Mader D. 1980. Zur Substrat-Gebundenheit von Nestbauten der solitaren Urbiene *Colletes daviesanus* (Hymenoptera: Colletidae). *Entomol. Gen.* 6:57–63.

Mader D. 1999. *Geologische und biologische Entomoökologie der rezenten Seidenbiene Colletes*. Köln: Logabook.

Madras-Majewska B, Zajdel B, Grygo M. 2011. Section analysis of after born mason bee (*Osmia rufa* L.) material. *Ann. Wars. Univ. Life Sci.* 108(49):103–108.

Maeta Y, Sakagami SF, Michener CD. 1985. Laboratory studies on the life cycle and nesting biology of *Braunsapis sauteriella*, a social xylocopine bee (Hymenoptera: Apidae). *Sociobiology* 10(1):27–41.

Maeta Y, Sasaki Y, Fujimoto G. 1988. *Andrena postomias* of the Gakuonji temple in Hyogo prefecture. *Insectarium* 25(2):50–57.

Magallón S, Gomez-Acevedo S, Sanchez-Reyes LL, Hernandez-Hernandez T. 2015. A meta-calibrated time-tree documents the early rise of flowering plant phylogenetic diversity. *New Phytol.* 207(2):437–453.

Magnacca KN. 2007. Conservation status of the endemic bees of Hawai'i, *Hylaeus* (*Nesoprosopis*) (Hymenoptera: Colletidae). *Pac. Sci.* 61(2):173–190.

Magnacca KN, Danforth BN. 2006. Evolution and biogeography of native Hawaiian *Hylaeus* bees (Hymenoptera: Colletidae). *Cladistics* 22(5):393–411.

Magnacca KN, Danforth BN. 2007. Low nuclear DNA variation supports a recent origin of Hawaiian *Hylaeus* bees (Hymenoptera: Colletidae). *Mol. Phylogenet. Evol.* 43(3):908–915.

Magwere T, Pamplona R, Miwa S, Martinez-Diaz P, Portero-Otin M et al. 2006. Flight activity, mortality rates, and lipoxidative damage in *Drosophila*. *J. Gerontol. Ser. A.* 61(2):136–145.

Mallinger RE, Gratton C. 2014. Species richness of wild bees, but not the use of managed honeybees, increases fruit set of a pollinator-dependent crop. *J. Appl. Ecol.* 52(2):323–330.

Mallinger RE, Werts P, Gratton C. 2015. Pesticide use within a pollinator-dependent crop has negative effects on the abundance and species richness of sweat bees, *Lasioglossum* spp., and on bumble bee colony growth. *J. Insect Conserv.* 19(5):999–1010.

Malyshev S. 1925. The nesting habits of *Anthophora* Latr. *Tr. Leningr Obshch. Estest.* 55:137–183.

Malyshev S. 1930. Nistgewohnheiten der steinbienen, *Lithurgus* Latr. (Apoidea). *Z. Für Morphol. Ökol. Tiere.* 19(1):116–134.

Malyshev SI. 1923. La nidification des *Colletes* Latr. *Rev. Russe Entom.* 18:103–124.

Malyshev SI. 1935. Nesting habits of solitary bees. A comparative study. *Eos* 11:201–309.

Malyshev SI. 1968. *Genesis of the Hymenoptera and the Phases of Their Evolution*. London: Methuen.

Mandelik Y, Roll U. 2009. Diversity patterns of wild bees in almond orchards and their surrounding landscape. *Isr. J. Plant Sci.* 57(3):185–191.

Mangum WA, Brooks RW. 1997. First records of *Megachile (Callomegachile) sculpturalis* Smith (Hymenoptera: Megachilidae) in the continental United States. *J. Kans. Entomol. Soc.* 70(2):140–142.

Martins AC, Melo GAR, Renner SS. 2014. The corbiculate bees arose from New World oil-collecting bees: implications for the origin of pollen baskets. *Mol. Phylogenet. Evol.* 80: 88–94.

Martins KT, Gonzalez A, Lechowicz MJ. 2015. Pollination services are mediated by bee functional diversity and landscape context. *Agric. Ecosyst. Environ.* 200:12–20.

Martins KT, Normandin É, Ascher JS. 2017. *Hylaeus communis* (Hymenoptera: Colletidae), a new exotic bee for North America with generalist foraging and habitat preferences. *Can. Entomol.* 149(3):377–390.

Martins RP, Almeida DA de. 1994. Is the bee, *Megachile assumptionis* (Hymenoptera: Megachilidae), a cavity-nesting specialist? *J. Insect Behav.* 7(5):759–765.

Martins RP, Antonini Y. 1994. The biology of *Diadasina distincta* (Holmberg, 1903) (Hymenoptera: Anthophoridae). *Proc. Entomol. Soc. Wash.* 96(3):553–560.

Martins RP, Antonini Y, da Silveira FA, West SA. 1999. Seasonal variation in the sex allocation of a Neotropical solitary bee. *Behav. Ecol.* 10(4):401–408.

Martins RP, Guerra STM, Barbeitos MS. 2001. Variability in egg-to-adult development time in the bee *Ptilothrix plumata* and its parasitoids. *Ecol. Entomol.* 26(6):609–616.

Martins RP, Guimarães FG, Dias CM. 1996. Nesting biology of *Ptilothrix plumata* Smith, with a comparison to other species in the genus (Hymenoptera: Anthophoridae). *J. Kans. Entomol. Soc.* 69(1):9–16.

Martinson VG, Danforth BN, Minckley RL, Rueppell O, Tingek S, Moran NA. 2011. A simple and distinctive microbiota associated with honey bees and bumble bees. *Mol. Ecol.* 20(3):619–628.

Mason LG. 1986. Free-loaders, free-lancers and bushwackers: sexual dimorphism and seasonal changes in pre-capture behavior of ambush bugs. *Am. Midl. Nat.* 116(2):323–327.

Masonick P, Amy M, Franckenberg S, Tabitsch W, Weirauch C. 2017. Molecular phylogenetics and biogeography of the ambush bugs (Hemiptera: Reduviidae: Phymatinae). *Mol. Phylogenet. Evol.* 114:225–233.

Mateus S, Noll FB. 2004. Predatory behavior in a necrophagous bee *Trigona hypogea* (Hymenoptera; Apidae, Meliponini). *Naturwissenschaften* 91(2):94–96.

Matteson KC, Ascher JS, Langellotto GA. 2008. Bee richness and abundance in New York City urban gardens. *Ann. Entomol. Soc. Am.* 101(1):140–150.

Matthews RW. 1965. Biology of *Heriades carinata* Cresson (Hymenoptera, Megachilidae). *Contrib. Am. Entomol. Inst. USA.* 1:1–33.

Matthews RW. 1968. *Microstigmus comes*: sociality in a sphecid wasp. *Science* 160(3829): 787–788.

Matthews RW, Fischer RL. 1964. A modified trap-nest for twig-nesting Aculeata. *Proc. North Cent. Branch Entomol. Soc. Am.* 19:79–81.

Matthews RW, González JM, Matthews JR, Deyrup LD. 2009. Biology of the parasitoid *Melittobia* (Hymenoptera: Eulophidae). *Annu. Rev. Entomol.* 54:251–266.

Mawdsley JR. 2001. Comparative ecology of the genus *Lecontella* Wolcott and Chapin (Coleoptera: Cleridae: Tillinae), with notes on chemically defended species of the beetle family Cleridae. *Proc. Entomol. Soc. Wash.* 104:164–167.

Mawdsley JR. 2004. Pollen transport by North American *Trichodes* Herbst (Coleoptera: Cleridae). *Proc. Entomol. Soc. Wash.* 106(1):199–201.

Mayer C, Kuhlmann M. 2004. Synchrony of pollinators and plants in the winter rainfall area of South Africa—observations from a drought year. *Trans. R. Soc. South Afr.* 59(2):55–57.

Maynard GV, Rao S. 2010. The solitary bee, *Leioproctus* (*Leioproctus*) *nigrofulvus* (Cockerell 1914) (Hymenoptera: Colletidae), in SE Australia: unique termite-mound-nesting behavior and impacts of bushfires on local populations. *Pan-Pac. Entomol.* 86(1):14–19.

McAuliffe JR. 1988. Markovian dynamics of simple and complex desert plant communities. *Am. Nat.* 131(4):459–490.

McAuslane HJ, Vinson SB, Williams HJ. 1990. Change in mandibular and mesosomal gland contents of male *Xylocopa micans* (Hymenoptera: Anthophoridae) associated with mating system. *J. Chem. Ecol.* 16(6):1877–1885.

McCorquodale DB, Owen RE. 1997. Allozyme variation, relatedness among progeny in a nest, and sex ratio in the leafcutter bee, *Megachile rotundata* (Fabricius) (Hymenoptera: Megachilidae). *Can. Entomol.* 129(2):211–219.

McFrederick QS, Cannone JJ, Gutell RR, Kellner K, Plowes RM, Mueller UG. 2013a. Specificity between lactobacilli and hymenopteran hosts is the exception rather than the rule. *Appl. Environ. Microbiol.* 79(6):1803–1812.

McFrederick QS, Mueller UG, James RR. 2014. Interactions between fungi and bacteria influence microbial community structure in the *Megachile rotundata* larval gut. *Proc. R. Soc. Lond. B, Biol. Sci.* 281(1779):20132653.

McFrederick QS, Roulston TH, Taylor DR. 2013b. Evolution of conflict and cooperation of nematodes associated with solitary and social sweat bees. *Insectes Sociaux* 60(3):309–317.

McFrederick QS, Taylor DR. 2013. Evolutionary history of nematodes associated with sweat bees. *Mol. Phylogenet. Evol.* 66(3):847–856.

McFrederick QS, Wcislo WT, Taylor DR, Ishak HD, Dowd SE, Mueller UG. 2012. Environment or kin: whence do bees obtain acidophilic bacteria? *Mol. Ecol.* 21(7):1754–1768.

McGinley RJ. 1980. Glossal morphology of the Colletidae and recognition of the Stenotritidae at the family level (Hymenoptera: Apoidea). *J. Kans. Entomol. Soc.* 53(3):539–552.

McGinley RJ. 1986. Studies of Halictinae (Apoidea: Halictidae), I: Revision of New World *Lasioglossum* Curtis. *Smithson. Contrib. Zool.* 429:1–294.

McGinley RJ. 2003. Studies of Halictinae (Apoidea: Halictidae), II: Revision of *Sphecodogastra* Ashmead, floral specialists of Onagraceae. *Smithson. Contrib. Zool.* 610:1–55.

McGinley RJ, Rozen JG Jr. 1987. Nesting biology, immature stages, and phylogenetic placement of the Palaearctic bee *Pararhophites* (Hymenoptera, Apoidea). *Am. Mus. Novit.* 2903:1–21.

McGregor SE. 1976. *Insect Pollination of Cultivated Crop Plants: Agriculture Handbook 496.* Washington, DC: Agricultural Research Service, US Department of Agriculture.

McIver J, Thorp R, Erickson K. 2009. Pollinators of the invasive plant, yellow starthistle (*Centaurea solstitialis*), in north-eastern Oregon, USA. *Weed Biol. Manag.* 9(2):137–45.

McKinney ML, Lockwood JL. 1999. Biotic homogenization: a few winners replacing many losers in the next mass extinction. *Trends Ecol. Evol.* 14(11):450–453.

McMahon DP, Fürst MA, Caspar J, Theodorou P, Brown MJ, Paxton RJ. 2015. A sting in the spit: widespread cross-infection of multiple RNA viruses across wild and managed bees. *J. Anim. Ecol.* 84(3):615–624.

Medler JT. 1964. *Anthophora* (*Clisodon*) *terminalis* Cresson in trap-nests in Wisconsin (Hymenoptera: Anthophoridae). *Can. Entomol.* 96(10):1332–1336.

Meehan TD, Gratton C. 2015. A consistent positive association between landscape simplification and insecticide use across the Midwestern US from 1997 through 2012. *Environ. Res. Lett.* 10(11):114001.

Meehan TD, Werling BP, Landis DA, Gratton C. 2011. Agricultural landscape simplification and insecticide use in the midwestern United States. *Proc. Natl. Acad. Sci. USA* 108(28): 11500–11505.

Meiners JM, Griswold TL, Harris DJ, Ernest SM. 2017. Bees without flowers: before peak bloom, diverse native bees find insect-produced honeydew sugars. *Am. Nat.* 190(2):281–291.

Melander AL, Brues CT. 1903. Guests and parasites of the burrowing bee *Halictus*. *Biol. Bull.* 5(1):1–27.

Mello MLS, Garófalo CA. 1986. Structural dimorphism in the cocoons of a solitary bee, *Lithurgus corumbae* (Hymenoptera, Megachilidae) and its adaptive significance. *Zool. Anz.* 217(3–4):195–206.

Melo GAR, Gaglianone MC. 2005. Females of *Tapinotaspoides*, a genus in the oil-collecting bee tribe Tapinotaspidini, collect secretions from non-floral trichomes (Hymenoptera, Apidae). *Rev. Bras. Entomol.* 49(1):167–168.

Melo GAR. 1999. Phylogenetic relationships and classification of the major lineages of Apoidea (Hymenoptera): with emphasis on the crabronid wasps. *Sci. Pap. Nat. Hist. Mus. Univ. Kans.* 14:1–55.

Memmott J, Craze PG, Waser NM, Price MV. 2007. Global warming and the disruption of plant-pollinator interactions. *Ecol. Lett.* 10(8):710–717.

Mendelson TC, Shaw KL. 2005. Sexual behaviour: rapid speciation in an arthropod. *Nature* 433(7024):375–376.

Menezes C, Vollet-Neto A, Contrera FAFL, Venturieri GC, Imperatriz-Fonseca VL. 2013. The role of useful microorganisms to stingless bees and stingless beekeeping. In *Pot-Honey*, pp. 153–171. New York: Springer.

Messer AC. 1984. *Chalicodoma pluto*: the world's largest bee rediscovered living communally in termite nests (Hymenoptera: Megachilidae). *J. Kans. Entomol. Soc.* 57(1):165–168.

Messer AC. 1985. Fresh dipterocarp resins gathered by megachilid bees inhibit growth of pollen-associated fungi. *Biotropica* 17(2):175–176.

Mevi-Schütz J, Erhardt A. 2005. Amino acids in nectar enhance butterfly fecundity: a long-awaited link. *Am. Nat.* 165(4):411–419.

Michelbacher AE, Hurd PD Jr, Linsley EG. 1968. The feasibility of introducing squash bees (*Peponapis* and *Xenoglossa*) into the Old World. *Bee World* 49(4):159–167.

Michelette ERF, Camargo JMF, Rozen JG Jr. 2000. Biology of the bee *Canephorula apiformis* and its cleptoparasite *Melectoides bellus*: nesting habits, floral preferences, and mature larvae (Hymenoptera, Apidae). *Am. Mus. Novit.* 3308:1–23.

Michener CD. 1944. Comparative external morphology, phylogeny, and a classification of the bees (Hymenoptera). *Bull. Am. Mus. Nat. Hist.* 82(6):151–326.

Michener CD. 1947. Some observations on *Lasioglossum (Hemihalictus) lustrans* (Hymenoptera: Halictidae). *J. N. Y. Entomol. Soc.* 55(1):49–50.

Michener CD. 1948. The generic classification of the anthidiine bees (Hymenoptera, Megachilidae). *Am. Mus. Novit.* 138:1–29.

Michener CD. 1953a. Comparative morphological and systematic studies of bee larvae: with a key to the families of hymenopterous larvae. *Univ. Kans. Sci. Bull.* 35(8):987–1102.

Michener CD. 1953b. The biology of a leafcutter bee (*Megachile brevis*) and its associates. *Univ. Kans. Sci. Bull.* 35(3):1659–1748.

Michener CD. 1954. Bees of Panama. *Bull. Am. Mus. Nat. Hist.* 104:1–176.

Michener CD. 1955. Some biological observations on *Hoplitis pilosifrons* and *Stelis lateralis* (Hymenoptera, Megachilidae). *J. Kans. Entomol. Soc.* 28(3):81–87.

Michener CD. 1960. Notes on the behavior of Australian colletid bees. *J. Kans. Entomol. Soc.* 33(1):22–31.

Michener CD. 1962. The genus *Ceratina* in Australia, with notes on its nests (Hymenoptera: Apoidea). *J. Kans. Entomol. Soc.* 35(4):414–421.

Michener CD. 1964. Evolution of the nests of bees. *Integr. Comp. Biol.* 4(2):227–239.

Michener CD. 1965. A classification of the bees of the Australian and South Pacific regions. *Bull. Am. Mus. Nat. Hist.* 130:1–362.

Michener CD. 1968. Nests of some African megachilid bees, with description of a new *Hoplitis* (Hymenoptera, Apoidea). *J. Entomol. Soc. South. Afr.* 31(2):337–359.

Michener CD. 1974. *The Social Behavior of the Bees: A Comparative Study.* Cambridge: Harvard University Press.

Michener CD. 1978. The parasitic groups of Halictidae (Hymenoptera: Apoidea). *Univ. Kans. Sci. Bull.* 51(10):291–339.

Michener CD. 1979. Biogeography of the bees. *Ann. Mo. Bot. Gard.* 66(3):277–347.

Michener CD. 1981. Classification of the bee family Melittidae, with a review of species of Meganomiinae. *Contrib. Am. Entomol. Inst.* 18(3):1–135.

Michener CD. 2007. *The Bees of the World*. Baltimore: Johns Hopkins University Press. 2nd ed.

Michener CD, Brothers DJ. 1971. A simplified observation nest for burrowing bees. *J. Kans. Entomol. Soc.* 44(2):236–239.

Michener CD, Fraser A. 1978. A comparative anatomical study of mandibular structure in bees (Hymenoptera: Apoidea). *Univ. Kans. Sci. Bull.* 51(14):463–482.

Michener CD, Greenberg L. 1980. Ctenoplectridae and the origin of long-tongued bees. *Zool. J. Linn. Soc.* 69(3):183–203.

Michener CD, Grimaldi DA. 1988. A *Trigona* from late Cretaceous amber of New Jersey (Hymenoptera, Apidae, Meliponinae). *Am. Mus. Novit.* 2917:1–10.

Michener CD, Lange RB. 1958. Observations on the ethology of Neotropical anthophorine bees (Hymenoptera: Apoidea). *Univ. Kans. Sci. Bull.* 39:69–96.

Michener CD, Ordway E. 1963. The life history of *Perdita maculigera maculipennis* (Hymenoptera: Andrenidae). *J. Kans. Entomol. Soc.* 36(1):34–45.

Michener CD, Rettenmeyer CW. 1956. The ethology of *Andrena erythronii* with comparative data on other species (Hymenoptera, Andrenidae). *Univ. Kans. Sci. Bull.* 16:645–684.

Michener CD, Szent Ivany JH. 1960. Observations on the biology of a leaf-cutter bee "*Megachile frontalis*" in New Guinea. *Papua N. Guin. Agric. J.* 13(1):22–35.

Michener GR, Michener CD. 1999. Mating behavior of *Dianthidium curvatum* (Hymenoptera, Megachilidae) at a nest aggregation. *Univ. Kans. Nat. Hist. Mus. Spec. Publ.* 24:37–43.

Michez D, De Meulemeester T, Rasmont P, Nel A, Patiny S. 2009a. New fossil evidence of the early diversification of bees: *Paleohabropoda oudardi* from the French Paleocene (Hymenoptera, Apidae, Anthophorini). *Zool. Scr.* 38(2):171–181.

Michez D, Eardley C. 2007. Monographic revision of the bee genus *Melitta* Kirby 1802 (Hymenoptera: Apoidea: Melittidae). *Ann. Société Entomol. Fr.* 43(4):379–440.

Michez D, Eardley CD, Timmermann K, Danforth BN. 2010. Unexpected polylecty in the bee genus *Meganomia* (Hymenoptera: Apoidea: Melittidae). *J. Kans. Entomol. Soc.* 83(3): 221–230.

Michez D, Nel A, Menier J-J, Rasmont P. 2007. The oldest fossil of a melittid bee (Hymenoptera: Apiformes) from the early Eocene of Oise (France). *Zool. J. Linn. Soc.* 150(4):701–709.

Michez D, Patiny S. 2005. World revision of the oil-collecting bee genus *Macropis* Panzer 1809 (Hymenoptera: Apoidea: Melittidae) with a description of a new species from Laos. *Ann. Société Entomol. Fr.* 41(1):15–28.

Michez D, Patiny S, Danforth BN. 2009b. Phylogeny of the bee family Melittidae (Hymenoptera: Anthophila) based on combined molecular and morphological data. *Syst. Entomol.* 34(3):574–597.

Michez D, Patiny S, Rasmont P, Timmermann K, Vereecken NJ. 2008. Phylogeny and host-plant evolution in Melittidae s.l., (Hymenoptera: Apoidea). *Apidologie* 39(1):146–162.

Michez D, Vanderplanck M, Engel MS. 2012. Fossil bees and their plant associates. In *Evolution of Plant-Pollinator Relationships*, Patiny S, ed., pp. 103–164. Cambridge: Cambridge University Press.

Mihajlovic L, Krunic MD, Richards KW. 1989. Hyperparasitism of *Physocephala vittata* (F.) (Diptera: Conopidae) by *Habrocytus conopidarum* (Boucek) (Hymenoptera: Pteromalidae), a pest of *Megachile rotundata* (F.) (Hymenoptera: Megachilidae) in Yugoslavia. *J. Kans. Entomol. Soc.* 62(3):418–420.

Milet-Pinheiro P, Schlindwein C. 2008. Comunidade de abelhas (Hymenoptera, Apoidea) e plantas em uma área do Agreste pernambucano, Brasil. *Rev. Bras. Entomol.* 52(4):625–636.

Milet-Pinheiro P, Schlindwein C. 2010. Mutual reproductive dependence of distylic *Cordia leucocephala* (Cordiaceae) and oligolectic *Ceblurgus longipalpis* (Halictidae, Rophitinae) in the Caatinga. *Ann. Bot.* 106(1):17–27.

Miliczky ER. 1985. Observations on the nesting biology of *Tetralonia hamata* Bradley with a description of its mature larva (Hymenoptera: Anthophoridae). *J. Kans. Entomol. Soc.* 58(4):686–700.

Minckley RL. 1994. Comparative morphology of the mesosomal "gland" in male large carpenter bees (Apidae: Xylocopini). *Biol. J. Linn. Soc.* 53(3):291–308.

Minckley RL. 1998. Cladistic analysis and classification of the subgenera and genera of the large carpenter bees, tribe Xylocopini (Hymenoptera: Apidae). *Sci. Pap. Nat. Hist. Mus. Univ. Kans.* 9:1–47.

Minckley RL. 2008. Faunal composition and species richness differences of bees (Hymenoptera: Apiformes) from two North American regions. *Apidologie* 39(1):176–188.

Minckley RL. 2014. Maintenance of richness despite reduced abundance of desert bees (Hymenoptera: Apiformes) to persistent grazing. *Insect Conserv. Divers.* 7(3):263–273.

Minckley RL, Buchmann SL. 1990. Territory site selection of male *Xylocopa (Neoxylocopa) varipuncta* Patton (Hymenoptera: Anthophoridae). *J. Kans. Entomol. Soc.* 63(2):329–339.

Minckley RL, Buchmann SL, Wcislo WT. 1991. Bioassay evidence for a sex attractant pheromone in the large carpenter bee, *Xylocopa varipuncta* (Anthophoridae: Hymenoptera). *J. Zool.* 224(2):285–291.

Minckley RL, Cane JH, Kervin L. 2000. Origins and ecological consequences of pollen specialization among desert bees. *Proc. R. Soc. Lond. B, Biol. Sci.* 267(1440):265–271.

Minckley RL, Cane JH, Kervin L, Roulston TH. 1999. Spatial predictability and resource specialization of bees (Hymenoptera: Apoidea) at a super abundant, widespread resource. *Biol. J. Linn. Soc.* 67(1):119–147.

Minckley RL, Cane JH, Kervin L, Yanega D. 2003. Biological impediments to measures of competition among introduced honey bees and desert bees (Hymenoptera: Apiformes). *J. Kans. Entomol. Soc.* 76(2):306–319.

Minckley RL, Danforth BN. 2019. Sources and frequency of brood loss in solitary bees. *Apidologie* (in press).

Minckley RL, Roulston TH. 2006. Incidental mutualisms and pollen specialization among bees. In *Plant-Pollinator Interactions: From Specialization to Generalization*, Waser NM, Ollerton J, eds., pp. 69–98. Chicago: University of Chicago Press.

Minckley RL, Roulston TH, Williams NM. 2013. Resource assurance predicts specialist and generalist bee activity in drought. *Proc. R. Soc. Lond. B, Biol. Sci.* 280(1759):20122703.

Minckley RL, Wcislo WT, Yanega D, Buchmann SL. 1994. Behavior and phenology of a specialist bee (*Dieunomia*) and sunflower (*Helianthus*) pollen availability. *Ecology* 75(5):1406–1419.

Miriti MN, Rodríguez-Buriticá S, Wright SJ, Howe HF. 2007. Episodic death across species of desert shrubs. *Ecology* 88(1):32–36.

Misof B, Liu S, Meusemann K, Peters RS, Donath A et al. 2014. Phylogenomics resolves the timing and pattern of insect evolution. *Science* 346(6210):763–767.

Mitchell TB. 1980. *A Generic Revision of the Megachiline Bees of the Western Hemisphere*. Raleigh: Department of Entomology, North Carolina State University.

Mitra A. 2013. Function of the Dufour's gland in solitary and social Hymenoptera. *J. Hymenopt. Res.* 35(1):33–58.

Moldenke AR. 1976. Evolutionary history and diversity of the bee faunas of Chile and Pacific North America. *Wasmann J. Biol.* 34(2):147–178.

Moldenke AR. 1979. Host-plant coevolution and the diversity of bees in relation to the flora of North America. *Phytologia* 43(4):357–419.

Morandin LA, Laverty TM, Kevan PG. 2001a. Bumble bee (Hymenoptera: Apidae) activity and pollination levels in commercial tomato greenhouses. *J. Econ. Entomol.* 94(2):462–467.

Morandin LA, Laverty TM, Kevan PG. 2001b. Effect of bumble bee (Hymenoptera: Apidae) pollination intensity on the quality of greenhouse tomatoes. *J. Econ. Entomol.* 94(1):172–179.

Morandin LA, Winston ML. 2005. Wild bee abundance and seed production in conventional, organic, and genetically modified canola. *Ecol. Appl.* 15(3):871–881.

Mordechai YRB, Cohen R, Gerling D, Moscovitz E. 1978. The biology of *Xylocopa pubescens* Spinola (Hymenoptera: Anthophoridae) in Israel. *Isr. J. Entomol.* 12:107–121.

Morse DH. 2007. *Predator Upon a Flower: Life History and Fitness in a Crab Spider*. Cambridge: Harvard University Press.

Morse RA, Calderone NW. 2000. The value of honey bees as pollinators of US crops in 2000. *Bee Cult.* 128(3):1–15.

Motten AF, Campbell DT, Alexander DE. 1981. Pollination effectiveness of specialist and generalist visitors to a North Carolina population of *Claytonia virginica*. *Ecology* 62(5):1278–1287.

Moure JS. 1963. Sobre a provavél ocorrencia de metandria em algumas especies de abelhas do genero *Centris* Fabricius, 1804 (Hymenoptera, Apoidea). *Cienc. Cult.* 15:183.

Müller A. 1995. Morphological specializations in central European bees for the uptake of pollen from flowers with anthers hidden in narrow corolla tubes (Hymenoptera: Apoidea). *Entomol. Gen.* 20(1–2):43–57.

Müller A. 1996a. Convergent evolution of morphological specializations in Central European bee and honey wasp species as an adaptation to the uptake of pollen from nototribic flowers (Hymenoptera, Apoidea and Masaridae). *Biol. J. Linn. Soc.* 57(3):235–252.

Müller A. 1996b. Host plant specialization in the western palearctic anthidiine bees (Hymenoptera: Apoidea: Megachilidae). *Ecol. Monogr.* 66(2):235–235.

Müller A. 2006. Unusual host plant of *Hoplitis pici*, a bee with hooked bristles on its mouthparts (Hymenoptera: Megachilidae: Osmiini). *Eur. J. Entomol.* 103(2):497–500.

Müller A. 2012. *Osmia* (*Orientosmia*) *maxschwarzi* sp. n., a new Palaearctic osmiine bee with extraordinarily long mouthparts (Hymenoptera, Apiformes, Megachilidae). *Mitteilungen Schweiz. Entomol. Ges.* 85(1):27–35.

Müller A. 2018. Palaearctic osmiine bees—systematics and biology of a fascinating group of solitary bees. *Palaearctic osmiine bees.* http://blogs.ethz.ch/osmiini/.

Müller A, Diener S, Schnyder S, Stutz K, Sedivy C, Dorn S. 2006. Quantitative pollen requirements of solitary bees: implications for bee conservation and the evolution of bee–flower relationships. *Biol. Conserv.* 130(4):604–615.

Müller A, Krebs A, Amiet F. 1997. *Bienen: Mitteleuropäische Gattungen, Lebensweise, Beobachtung.* Augsburg: Naturbuch-Verlag.

Müller A, Kuhlmann M. 2003. Narrow flower specialization in two European bee species of the genus *Colletes* (Hymenoptera: Apoidea: Colletidae). *Eur. J. Entomol.* 100(4):631–636.

Müller A, Kuhlmann M. 2008. Pollen hosts of western Palaearctic bees of the genus *Colletes* (Hymenoptera: Colletidae): the Asteraceae paradox. *Biol. J. Linn. Soc.* 95(4):719–733.

Müller A, Mauss V. 2016. Palaearctic *Hoplitis* bees of the subgenera *Formicapis* and *Tkalcua* (Megachilidae, Osmiini): biology, taxonomy and key to species. *Zootaxa* 4127(1):105–120.

Müller A, Topfl W, Amiet F. 1996. Collection of extrafloral trichome secretions for nest wool impregnation in the solitary bee *Anthidium manicatum*. *Naturwissenschaften* 83(5): 230–232.

Müller H. 1872. Anwendung der Darwinischen Lehre auf Bienen. *Verh Naturhist Ver Preuss Rheinl. Westphalen* 29:1–96.

Müller H. 1883. *The Fertilisation of Flowers.* London: MacMillan and Company.

Mullin CA, Frazier M, Frazier JL, Ashcraft S, Simonds R, Pettis JS. 2010. High levels of miticides and agrochemicals in North American apiaries: implications for honey bee health. *PLoS One* 5(3):e9754.

Münster-Swendsen M. 1970. The nesting behaviour of the bee *Panurgus banksianus* Kirby (Hymenoptera, Andrenidae, Panurginae). *Insect Syst. Evol.* 1(2):93–101.

Murray EA, Bossert S, Danforth BN. 2018. Pollinivory and the diversification dynamics of bees. *Biology Letters* 14(11):20180530.

Ne'eman G, Jürgens A, Newstrom-Lloyd L, Potts SG, Dafni A. 2009. A framework for comparing pollinator performance: effectiveness and efficiency. *Biol. Rev.* 85(3):435–451.

Ne'eman G, Shavit O, Shaltiel L, Shmida A. 2006. Foraging by male and female solitary bees with implications for pollination. *J. Insect Behav.* 19(3):383–401.

Neff JL. 1984. Observations on the biology of *Eremapis parvula* Ogloblin, an anthophorid bee with a metasomal scopa (Hymenoptera: Anthophoridae). *Pan-Pac. Entomol.* 60(2): 155–162.

Neff JL. 2003. Nest and provisioning biology of the bee *Panurginus polytrichus* Cockerell (Hymenoptera: Andrenidae), with a description of a new *Holcopasites* species (Hymenoptera: Apidae), its probable nest parasite. *J. Kans. Entomol. Soc.* 76(2):203–216.

Neff JL. 2004. Hooked hairs and not so narrow tubes: two new species of *Colletes* Latreille from Texas (Hymenoptera: Apoidea: Colletidae). *J. Hymenopt. Res.* 13(2):250–261.

Neff JL. 2008. Components of nest provisioning behavior in solitary bees (Hymenoptera: Apoidea). *Apidologie* 39(1):30–45.

Neff JL, Danforth BN. 1991. The nesting and foraging behavior of *Perdita texana* (Cresson) (Hymenoptera: Andrenidae). *J. Kans. Entomol. Soc.* 64(4):394–405.

Neff JL, Rozen JG Jr. 1995. Foraging and nesting biology of the bee *Anthemurgus passiflorae* (Hymenoptera: Apoidea), descriptions of its immature stages, and observations on its floral host (Passifloraceae). *Am. Mus. Novit.* 3138:1–19.

Neff JL, Simpson BB. 1981. Oil-collecting structures in the Anthophoridae (Hymenoptera): morphology, function, and use in systematics. *J. Kans. Entomol. Soc.* 54(1):95–123.

Neff JL, Simpson BB. 1990. The roles of phenology and reward structure in the pollination biology of wild sunflower (*Helianthus annuus* L., Asteraceae). *Isr. J. Bot.* 39(1–2):197–216.

Neff JL, Simpson BB. 1991. Nest biology and mating behavior of *Megachile fortis* in central Texas (Hymenoptera: Megachilidae). *J. Kans. Entomol. Soc.* 64:324–336.

Neff JL, Simpson BB. 1992. Partial bivoltinism in a ground-nesting bee: the biology of *Diadasia rinconis* in Texas (Hymenoptera, Anthophoridae). *J. Kans. Entomol. Soc.* 65(4):377–392.

Neff JL, Simpson BB. 1997. Nesting and foraging behavior of *Andrena* (*Callandrena*) *rudbeckiae* Robertson (Hymenoptera: Apoidea: Andrenidae) in Texas. *J. Kans. Entomol. Soc.* 70(2):100–113.

Neff JL, Simpson BB. 2017. Vogel's great legacy: the oil flower and oil-collecting bee syndrome. *Flora* 232:104–116.

Neff JL, Simpson BB, Dorr LJ. 1982. The nesting biology of *Diadasia afflicta* Cress. (Hymenoptera: Anthophoridae). *J. Kans. Entomol. Soc.* 55(3):499–518.

Neff JL, Simpson BB, Evenhuis NL, Dieringer G. 2003. Character analysis of adaptation for tarsal pollen collection in the Bombyliidae (Insecta: Diptera): the benefits of putting your foot in your mouth. *Zootaxa* 157:1–14

Nel A, Petrulevicius JF. 2003. New Palaeogene bees from Europe and Asia. *Alcheringa Australas. J. Palaeontol.* 27(4):277–293.

Nelson RA, Griswold TL. 2015. The floral hosts and distribution of a supposed creosote bush specialist, *Colletes stepheni* Timberlake (Hymenoptera: Colletidae). *J. Melittology* 49:1–12.

Ness JH. 2006. A mutualism's indirect costs: the most aggressive plant bodyguards also deter pollinators. *Oikos* 113(3):506–514.

Ngo HT, Mojica AC, Packer L. 2011. Coffee plant–pollinator interactions: a review. *Can. J. Zool.* 89(8):647–660.

Nicolson SW. 2011. Bee food: the chemistry and nutritional value of nectar, pollen and mixtures of the two. *Afr. Zool.* 46(2):197–204.

Nicolson SW, Human H. 2013. Chemical composition of the "low quality' pollen of sunflower (*Helianthus annuus*, Asteraceae). *Apidologie* 44:144–152.

Nicolson SW, Thornburg RW. 2007. Nectar chemistry. In *Nectaries and Nectar*, Nicolson SW, Nepi M, Pacini E, eds., pp. 215–264. Netherlands: Springer.

Nieto A, Roberts SP, Kemp J, Rasmont P, Kuhlmann M et al. 2014. *European Red List of Bees*. Luxembourg: Publication Office of the European Union.

Nilsson LA, Rabakonandrianina E. 1988. Chemical signalling and monopolization of nectar resources by territorial *Pachymelus limbatus* (Hymenoptera Anthophoridae) male bees in Madagascar. *J. Zool.* 215(3):475–489.

Norden BB. 1984. Nesting biology of *Anthophora abrupta* (Hymenoptera: Anthophoridae). *J. Kans. Entomol. Soc.* 57(2):243–262.

Norden BB, Batra SWT. 1985. Male bees sport black mustaches for picking up parsnip perfume (Hymenoptera: Anthophoridae). *Proc. Entomol. Soc. Wash.* 87(2):317–322.

Norden BB, Batra SWT, Fales HM, Hefetz A, Shaw GJ. 1980. *Anthophora* bees: unusual glycerides from maternal Dufour's glands serve as larval food and cell lining. *Science* 207(4435):1095–1097.

Norden BB, Krombein KV, Deyrup MA, Edirisinghe JP. 2003. Biology and behavior of a seasonally aquatic bee, *Perdita* (*Alloperdita*) *floridensis* Timberlake (Hymenoptera: Andrenidae: Panurginae). *J. Kans. Entomol. Soc.* 76(2):236–249.

Normandin É, Vereecken NJ, Buddle CM, Fournier V. 2017. Taxonomic and functional trait diversity of wild bees in different urban settings. *PeerJ* 5:e3051.

North F, Lillywhite H. 1980. The function of burrow turrets in a gregariously nesting bee. *Southwest. Nat.* 25(3):373–378.

Nylinder S, Swenson U, Persson C, Janssens SB, Oxelman B. 2012. A dated species-tree approach to the trans-Pacific disjunction of the genus *Jovellana* (Calceolariaceae, Lamiales). *Taxon* 61(2):381–391.

Ochoa R, O'Connor BM. 2000. Revision of the genus *Horstiella* (Acari: Acaridae): mites associated with Neotropical *Epicharis* bees (Hymenoptera: Apidae). *Ann. Entomol. Soc. Am.* 93(4):713–737.

O'Connor BM. 1993. The mite community associated with *Xylocopa latipes* (Hymenoptera: Anthophoridae: Xylocopinae) with description of a new type of acarinarium. *Int. J. Acarol.* 19(2):159–166.

Ohl M, Bleidorn C. 2006. The phylogenetic position of the enigmatic wasp family Heterogynaidae based on molecular data, with description of a new, nocturnal species (Hymenoptera: Apoidea). *Syst. Entomol.* 31(2):321–337.

Ohl M, Linde D. 2003. Ovaries, ovarioles, and oocytes in apoid wasps, with special reference to cleptoparasitic species (Hymenoptera: Apoidea: "Sphecidae"). *J. Kans. Entomol. Soc.* 76(2): 147–159.

Okazaki K. 1992. Nesting habits of the small carpenter bee, *Ceratina dentipes*, in Hengchun Peninsula, southern Taiwan (Hymenoptera: Anthophoridae). *J. Kans. Entomol. Soc.* 65(2): 190–195.

Oldroyd BP, Wongsiri S. 2009. *Asian Honey Bees: Biology, Conservation, and Human Interactions.* Cambridge: Harvard University Press.

Oliveira R, Carvalho AT, Schlindwein C. 2012. Territorial or wandering: how males of *Protodiscelis palpalis* (Colletidae, Paracolletinae) behave in searching for mates. *Apidologie* 43(6): 674–684.

Oliveira R, Carvalho AT, Schlindwein C. 2013. Plasticity in male territoriality of a solitary bee under different environmental conditions. *J. Insect Behav.* 26(5):690–694.

Oliveira R, Schlindwein C. 2010. Experimental demonstration of alternative mating tactics of male *Ptilothrix fructifera* (Hymenoptera, Apidae). *Anim. Behav.* 80(2):241–247.

Ollerton J, Erenler H, Edwards M, Crockett R. 2014. Extinctions of aculeate pollinators in Britain and the role of large-scale agricultural changes. *Science* 346(6215):1360–1362.

Ollerton J, Winfree R, Tarrant S. 2011. How many flowering plants are pollinated by animals? *Oikos* 120(3):321–326.

Olofsson TC, Vásquez A. 2008. Detection and identification of a novel lactic acid bacterial flora within the honey stomach of the honeybee *Apis mellifera*. *Curr. Microbiol.* 57(4): 356–363.

Olofsson TC, Vásquez A. 2009. Phylogenetic comparison of bacteria isolated from the honey stomachs of honey bees *Apis mellifera* and bumble bees *Bombus* spp. *J. Apic. Res.* 48(4):233–237.

O'Neal RJ, Waller GD. 1984. On the pollen harvest by the honey bee (*Apis mellifera* L.) near Tucson, Arizona (1976–1981). *Desert Plants* 6(2):81–109.

O'Neill KM. 2001. *Solitary Wasps: Behavior and Natural History.* Ithaca: Comstock Publishing Associates.

O'Neill KM, Bjostad L. 1987. The male mating strategy of the bee *Nomia nevadensis* (Hymenoptera: Halictidae): leg structure and mate guarding. *Pan-Pac. Entomol.* 63(3):207–217.

O'Neill KM, O'Neill RP. 2001. Correlates of feeding duration in the robber fly *Effersia staminea* (Williston) (Diptera: Asilidae). *J. Kans. Entomol. Soc.* 74(2):79–82.

Ordway E. 1964. *Sphecodes pimpinellae* and other enemies of *Augochlorella* (Hymenoptera: Halictidae). *J. Kans. Entomol. Soc.* 37(2):139–152.

Orr MC, Griswold T, Pitts JP, Parker FD. 2016. A new bee species that excavates sandstone nests. *Curr. Biol.* 26(17):R792–R793.

Osten T. 1989. Vergleichend-funktionsmorphologische Untersuchungen des Paarungsverhaltens von *Platynopoda* und *Mesotrichia* (Hymenoptera: Xylocopini). *Stuttg. Beitrage Naturkunde Ser. A.* 433:1–18.

O'Toole C. 2013. *Bees: A Natural History.* Buffalo: Firefly Books.

O'Toole C, Raw A. 1991. *Bees of the World.* New York: Facts on File.

Pacini E, Hesse M. 2005. Pollenkitt—its composition, forms and functions. *Flora-Morphol. Distrib. Funct. Ecol. Plants.* 200(5):399–415.

Packer L. 2004. Taxonomic and behavioural notes on Patagonian Xeromelissinae with the description of a new species (Hymenoptera: Colletidae). *J. Kans. Entomol. Soc.* 77(4):805–820.

Packer L. 2005. A new species of *Geodiscelis* (Hymenoptera: Colletidae: Xeromelissinae) from the Atacama Desert of Chile. *J. Hymenopt. Res.* 14(1):84–91.

Packer L. 2008. Phylogeny and classification of the Xeromelissinae (Hymenoptera: Apoidea, Colletidae) with special emphasis on the genus *Chilicola*. *Syst. Entomol.* 33(1):72–96.

Packer L. 2010. *Keeping the Bees*. Toronto: Harper Collins.

Packer L, Dzinas A, Strickler K, Scott V. 1995. Genetic differentiation between two host "races" and two species of cleptoparasitic bees and between their two hosts. *Biochem. Genet.* 33(3–4):97–109.

Packer L, Zayed A, Grixti JC, Ruz L, Owen RE et al. 2005. Conservation genetics of potentially endangered mutualisms: reduced levels of genetic variation in specialist versus generalist bees. *Conserv. Biol.* 19(1):195–202.

Paini DR. 2004. Impact of the introduced honey bee (*Apis mellifera*) (Hymenoptera: Apidae) on native bees: a review. *Austral Ecol.* 29(4):399–407.

Paini DR, Williams MR, Roberts JD. 2005. No short-term impact of honey bees on the reproductive success of an Australian native bee. *Apidologie* 36(4):613–621.

Pamilo P, Varvio-Aho S-L, Pekkarinen A. 1978. Low enzyme gene variability in Hymenoptera as a consequence of haplodiploidy. *Hereditas* 88(1):93–99.

Park MG, Blitzer EJ, Gibbs J, Losey JE, Danforth BN. 2015. Negative effects of pesticides on wild bee communities can be buffered by landscape context. *Proc. R. Soc. Lond. B, Biol. Sci.* 282(1809):20150299.

Park MG, Raguso RA, Losey JE, Danforth BN. 2016. Per-visit pollinator performance and regional importance of wild *Bombus* and *Andrena* (*Melandrena*) compared to the managed honey bee in New York apple orchards. *Apidologie* 47(2):145–160.

Parker AJ, Williams NM, Thomson JD. 2016. Specialist pollinators deplete pollen in the spring ephemeral wildflower *Claytonia virginica*. *Ecol. Evol.* 6(15):5169–5177.

Parker FD. 1981. How efficient are bees in pollinating sunflowers? *J. Kans. Entomol. Soc.* 54(1):61–67.

Parker FD. 1984. The nesting biology of *Osmia* (*Trichinosmia*) *latisulcata* Michener. *J. Kans. Entomol. Soc.* 57(3):430–436.

Parker FD. 1987. Nests of *Callanthidium* from block traps (Hymenoptera: Megachilidae). *Pan-Pac. Entomol.* 63:125–129.

Parker FD, Cane JH, Frankie GW, Vinson SB. 1987. Host records and nest entry by *Dolichostelis*, a kleptoparasitic anthidiine bee. *Pan-Pac. Entomol.* 63(2):172–177.

Parker FD, Griswold TL. 1982. Biological notes on *Andrena* (*Callandrena*) *haynesi* Viereck and Cockerell (Hymenoptera: Andrenidae). *Pan-Pac. Entomol.* 58(4):284–287.

Parker FD, Potter HW. 1973. Biological notes on *Lithurgus apicalis* Cresson (Hymenoptera: Megachilidae). *Pan-Pac. Entomol.* 49(4):294–299.

Parker FD, Tepedino VJ. 1982a. A nest and pollen-collection records of *Osmia sculleni* Sandhouse, a bee with hooked hairs on the mouthparts (Hymenoptera: Megachilidae). *J. Kans. Entomol. Soc.* 55(2):329–334.

Parker FD, Tepedino VJ. 1982b. The behavior of female *Osmia marginata* Michener in the nest (Hymenoptera: Megachilidae). *Pan-Pac. Entomol.* 58:231–235.

Partap U, Ya T. 2012. The human pollinators of fruit crops in Maoxian County, Sichuan, China: a case study of the failure of pollination services and farmers' adaptation strategies. *Mt. Res. Dev.* 32(2):176–186.

Parys K, Tripodi A, Sampson B. 2015. The giant resin bee, *Megachile sculpturalis* Smith: new distributional records for the mid- and Gulf-south USA. *Biodivers. Data J.* 3:e6733.

Pascarella JB, Waddintgon KD, Neal PR. 1999. The bee fauna (Hymenoptera: Apoidea) of Everglades National Park, Florida and adjacent areas: distribution, phenology and biogeography. *J. Kans. Entomol. Soc.* 72(1):32–45.

Pasteels JJ, Pasteels JM. 1974. Étude au microscope électronique à balayage des scopas abdominales chez de nombreuses especes d'abeilles (Apoidea, Megachilidae). *Tissue Cell* 6(1):65–83.

Pasteels JJ, Pasteels JM. 1976. Étude au microscope électronique à balayage des scopas collectrices de pollen chez les Colletidae et Oxaeidae (Hymenoptera: Apoidea: Andrenidae). *Arch. Biol. (Bruxelles)* 87:79–102.

Pasteels JJ, Pasteels JM. 1979. Étude au microscope électronique à balayage des scopas collectrices de pollen chez les Andrenidae (Hymenoptera: Apoidea: Andrenidae). *Arch. Biol. (Bruxelles)* 90:113–130.

Pasteels JM, Pasteels JJ. 1975. Étude au microscope électronique à balayage des scopas collectrices de pollen chez les Fideliidae (Hymenoptera, Apoidea). *Arch. Biol. (Bruxelles)* 86:453–466.

Patiny S, Michez D, Danforth BN. 2007. Phylogenetic relationships and host-plant evolution within the basal clade of Halictidae (Hymenoptera, Apoidea). *Cladistics* 24(3):255–269.

Pauly A. 1984. Contribution à l'étude des genres afrotropicaux de Nomiinae (Hymenoptera Apoidea Halictidae). *Rev. Zool. Afr.* 98(4):693–702.

Pauly A. 1997. *Paraseladonia*, nouveau genre cleptoparasite Afrotropical (Hymenoptera, Apoidea, Halictidae). *Bull. Ann. Soc. R. Belge Entomol.* 133:91–99.

Pauw A. 2006. Floral syndromes accurately predict pollination by a specialized oil-collecting bee (*Rediviva peringueyi*, Melittidae) in a guild of South African orchids (Coryciinae). *Am. J. Bot.* 93(6):917–926.

Pauw A. 2007. Collapse of a pollination web in small conservation areas. *Ecology* 88(7):1759–1769.

Pauw A, Kahnt B, Kuhlmann M, Michez D, Montgomery GA et al. 2017. Long-legged bees make adaptive leaps: linking adaptation to coevolution in a plant–pollinator network. *Proc. R. Soc. Lond. B, Biol. Sci.* 284(1862):20171707.

Paxton RJ. 2005. Male mating behaviour and mating systems of bees: an overview. *Apidologie* 36(2):145–156.

Paxton RJ, Fries I, Pieniazek NJ, Tengö J. 1997. High incidence of infection of an undescribed microsporidium (*Microspora*) in the communal bee *Andrena scotica* (Hymenoptera, Andrenidae). *Apidologie* 28(3–4):129–141.

Paxton RJ, Giovanetti M, Andrietti F, Scamoni E, Scanni B. 1999a. Mating in a communal bee, *Andrena agilissima* (Hymenoptera Andrenidae). *Ethol. Ecol. Evol.* 11(4):371–382.

Paxton RJ, Kukuk PF, Tengö J. 1999b. Effects of familiarity and nestmate number on social interactions in two communal bees, *Andrena scotica* and *Panurgus calcaratus* (Hymenoptera, Andrenidae). *Insectes Sociaux* 46(2):109–118.

Paxton RJ, Tengö J. 1996. Intranidal mating, emergence, and sex ratio in a communal bee *Andrena jacobi* Perkins 1921 (Hymenoptera: Andrenidae). *J. Insect Behav.* 9(3):421–440.

Paxton RJ, Thoren PA, Gyllenstrand N, Tengö J. 2000. Microsatellite DNA analysis reveals low diploid male production in a communal bee with inbreeding. *Biol. J. Linn. Soc.* 69(4):483–502.

Paxton RJ, Thorén PA, Tengö J, Estoup A, Pamilo P. 1996. Mating structure and nestmate relatedness in a communal bee, *Andrena jacobi* (Hymenoptera, Andrenidae), using microsatellites. *Mol. Ecol.* 5(4):511–519.

Peinert M, Wipfler B, Jetschke G, Kleinteich T, Gorb SN et al. 2016. Traumatic insemination and female counter-adaptation in Strepsiptera (Insecta). *Sci. Rep.* 6:25052.

Pekar S, Coddington JA, Blackledge TA. 2011. Evolution of stenophagy in spiders (Araneae): evidence based on the comparative analysis of spider diets. *Evolution* 66(3):776–806.

Pellegrino G, Luca A, Beelusci F, Musacchio A. 2012. Comparative analysis of floral scents in four sympatric speceis of *Serapias* L. (Orchidaceae): clues on their pollination strategies. *Plant Syst. Evol.* 298:1837–1843.

Pellmyr O, Thompson JN. 1996. Sources of variation in pollinator contribution within a guild: the effects of plant and pollinator factors. *Oecologia* 107(4):595–604.

Pemberton RW, Wheeler GS. 2006. Orchid bees don't need orchids: evidence from the naturalization of an orchid bee in Florida. *Ecology* 87(8):1995–2001.

Peñalver E, Labandeira CC, Barron E, Delclos X, Nel P et al. 2012. Thrips pollination of Mesozoic gymnosperms. *Proc. Natl. Acad. Sci. USA* 109(22):8623–8628.

Perfecto I, Rice RA, Greenberg R, Van der Voort ME. 1996. Shade coffee: a disappearing refuge for biodiversity. *BioScience* 46(8):598–608.

Perkins RCL. 1899. Hymenoptera, Aculeata. In *Fauna Hawaiiensis*, Vol. 1, Sharp D, ed., pp. 1–115. Cambridge: Cambridge University Press.

Perkins RCL. 1912. Notes, with descriptions of new species, on aculeate Hymenoptera of the Australian region. *J. Nat. Hist.* (8)9:96–121.

Pesenko YA. 1975. On the fauna and ecology of Apoidea (Hymenoptera) of the Lower Don. 6. A review of the trophic links. *Entomol. Rev.* 54:53–59.

Pesenko YA, Banaszak J, Radchenko VG, Cierzniak T. 2000. *Bees of the Family Halictidae (Excluding* Sphecodes*) of Poland.* Bydgoszcz: Pedagogical University.

Pesenko YA, Radchenko VD. 1993. The use of bees (Hymenoptera, Apoidea) for alfalfa pollination: the main directions and modes, with methods of evaluation of populations of wild bees and pollinator efficiency. *Entomol. Rev.* 72(2):101–119.

Peters DS. 1974. Uber die Untergattung *Haetosmia* Popov 1952 (Insecta: Hymenoptera: Megachilidae: *Osmia*). *Senckenberg. Biol.* 55:293–309.

Peters RS, Krogmann L, Mayer C, Donath A, Gunkel S et al. 2017. Evolutionary history of the Hymenoptera. *Curr. Biol.* 27(7):1013–1018.

Petersen JD, Reiners S, Nault BA. 2013. Pollination services provided by bees in pumpkin fields supplemented with either *Apis mellifera* or *Bombus impatiens* or not supplemented. *PLoS One* 8(7):e69819.

Peterson JH, Roitberg BD. 2006a. Impact of resource levels on sex ratio and resource allocation in the solitary bee, *Megachile rotundata*. *Environ. Entomol.* 35:1404–1410.

Peterson JH, Roitberg BD. 2006b. Impacts of flight distance on sex ratio and resource allocation to offspring in the leafcutter bee, *Megachile rotundata*. *Behav. Ecol. Sociobiol.* 59:589–596.

Phillips JK, Klostermeyer EC. 1978. Nesting behavior of *Osmia lignaria propinqua* Cresson (Hymenoptera: Megachilidae). *J. Kans. Entomol. Soc.* 51(1):91–108

Pick RA, Schlindwein C. 2011. Pollen partitioning of three species of Convolvulaceae among oligolectic bees in the Caatinga of Brazil. *Plant Syst. Evol.* 293(1–4):147–159.

Pilgrim EM, Von Dohlen CD, Pitts JP. 2008. Molecular phylogenetics of Vespoidea indicate paraphyly of the superfamily and novel relationships of its component families and subfamilies. *Zool. Scr.* 37(5):539–560.

Pimentel M, Antonini Y, Martins R, Lachance M, Rosa C. 2005. *Candida riodocensis* and *Candida cellae*, two new yeast species from the *Starmerella* clade associated with solitary bees in the Atlantic rain forest of Brazil. *FEMS Yeast Res.* 5(9):875–879.

Pitts-Singer TL, Cane JH. 2011. The alfalfa leafcutting bee, *Megachile rotundata*: the world's most intensively managed solitary bee. *Annu. Rev. Entomol.* 56:221–237.

Plant JD, Paulus HF. 2016. Evolution and phylogeny of bees: review and cladistic analysis in light of morphological evidence (Hymenoptera: Apoidea). *Zoologica* 161:1–364.

Poinar GO, Danforth BN. 2006. A fossil bee from early Cretaceous Burmese amber. *Science* 314(5799):614.

Pokorny T, Loose D, Dyker G, Quezada-Euan JJG, Eltz T. 2015. Dispersal ability of male orchid bees and direct evidence for long-range flights. *Apidologie* 46(2):224–237.

Polidori C, Scanni B, Scamoni E, Giovanetti M, Andrietti F, Paxton RJ. 2005. Satellite flies (*Leucophora personata*, Diptera: Anthomyiidae) and other dipteran parasites of the communal bee *Andrena agilissima* (Hymenoptera: Andrenidae) on the island of Elba, Italy. *J. Nat. Hist.* 39(29):2745–2758.

Ponchau O, Iserbyt S, Verhaeghe J-C, Rasmont P. 2006. Is the caste-ratio of the oligolectic bumblebee *Bombus gerstaeckeri* Morawitz (Hymenoptera: Apidae) biased to queens? *Ann. Soc. Entomol. Fr.* 42:207–214.

Porter JC. 1951. Notes on the digger-bee *Anthophora occidentalis*, and its inquilines. *Iowa State J. Sci.* 26(1):23–30.

Portman ZM, Neff JL, Griswold T. 2016. Taxonomic revision of *Perdita* subgenus *Heteroperdita* Timberlake (Hymenoptera: Andrenidae), with descriptions of two ant-like males. *Zootaxa* 4214(1):1–97.

Portugal-Araújo V de. 1958. A contribution to the bionomics of *Lestrimelitta cubiceps* (Hymenoptera, Apidae). *J. Kans. Entomol. Soc.* 31(3):203–211.

Potts SG, Vulliamy B, Dafni A, Ne'eman G, O'Toole C et al. 2003. Response of plant-pollinator communities to fire: changes in diversity, abundance and floral reward structure. *Oikos* 101(1):103–112.

Pouyanne A. 1917. La fécondation des *Ophrys* par les insectes. *Bull Soc Hist Nat Afr Nord.* 8(1):6–7.

Prager SM. 2008. Behaviour and life history of a large carpenter bee (*Xylocopa virginica*) in the northern extent of its range. PhD thesis. Brock University.

Praz CJ, Müller A, Danforth BN, Griswold TL, Widmer A, Dorn S. 2008a. Phylogeny and biogeography of bees of the tribe Osmiini (Hymenoptera: Megachilidae). *Mol. Phylogenet. Evol.* 49(1):185–197.

Praz CJ, Müller A, Dorn S. 2008b. Specialized bees fail to develop on non-host pollen: do plants chemically protect their pollen? *Ecology* 89(3):795–804.

Prentice MA. 1998. The comparative morphology and phylogeny of apoid wasps (Hymenoptera: Apoidea). PhD thesis. University of California.

Prosi R, Wiesbauer H, Müller A. 2016. Distribution, biology and habitat of the rare European osmiine bee species *Osmia (Melanosmia) pilicornis* (Hymenoptera, Megachilidae, Osmiini). *J. Hymenopt. Res.* 52(1):1–36.

Pulawski WJ. 2018. Catalog of Sphecidae *sensu lato*. https://www.calacademy.org/scientists/projects/catalog-of-sphecidae.

Qu D-C, Maeta Y, Goubara M, Nakatsuka KJ, Kitamura K. 2002. Reproductive strategy in the two species of cleptoparasitic astigmatid mites, *Chaetodactylus nipponicus* and *Tortonia sp.* (Acari: Chaetodactylidae and Suidasiidae), infesting *Osmia cornifrons* (Hymenoptera: Megachilidae). I. Invasion/infestation patterns and partitive use of the host food. *Jpn. J. Entomol. New Ser.* 5(4):121–141.

Radchenko VG, Pesenko YA. 1994. *Biology of Bees*. St. Petersburg: Russian Academy of Sciences [In Russian; English summary, pp. 314–331].

Radchenko VG, Pesenko YA. 1996. Protobee and its nests: a new hypothesis concerning the early evolution of Apoidea (Hymenoptera). *Entomol. Rev.* 75(2):140–162.

Raguso RA. 2004. Why are some floral nectars scented? *Ecology* 85(6):1486–1494.

Rajotte EG. 1979. Nesting, foraging and pheromone response of the bee *Colletes validus* Cresson and its association with lowbush blueberries. (Hymenoptera: Colletidae) (Ericaceae: *Vaccinium*). *J. Kans. Entomol. Soc.* 52(2):349–361.

Ramírez SR, Dressler RL, Ospina M. 2002. Abejas euglosinas (Hymenoptera: Apidae) de la región Neotropical: listado de especies con notas sobre su biología. *Biota Colomb.* 3(1):7–118.

Ramírez SR, Eltz T, Fujiwara MK, Gerlach G, Goldman-Huertas B et al. 2011. Asynchronous diversification in a specialized plant-pollinator mutualism. *Science* 333(6050):1742–1746.

Ramírez-Arriaga E, Cuadriello-Aguilar JI, Hernández EM. 1996. Nest structure and parasite of *Euglossa atroveneta* Dressler (Apidae: Bombinae: Euglossini) at Unión Juárez, Chiapas, México. *J. Kans. Entomol. Soc.* 69(2):144–152.

Ramos KS, Rozen JG Jr. 2014. *Psaenythisca*, a new genus of bees from South America (Apoidea: Andrenidae: Protandrenini) with a description of the nesting biology and immature stages of one species. *Am. Mus. Novit.* 3800:1–32.

Rasmont P, Pauly A, Terzo M, Patiny S, Michez D et al. 2005. *The Survey of Wild Bees (Hymenoptera, Apoidea) in Belgium and France*. Rome: Food and Agricultural Organization.

Rau P. 1928. The nesting habits of the little carpenter-bee, *Ceratina calcarata*. *Ann. Entomol. Soc. Am.* 21(3):380–396.

Rau P, Rau N. 1916. The sleep of insects; an ecological study. *Ann. Entomol. Soc. Am.* 9(3): 227–274.

Ravoet J, De Smet L, Meeus I, Smagghe G, Wenseleers T, de Graaf DC. 2014. Widespread occurrence of honey bee pathogens in solitary bees. *J. Invertebr. Pathol.* 122:55–58.

Raw A. 1972. The biology of the solitary bee *Osmia rufa* (L.) (Megachilidae). *Trans. R. Entomol. Soc. Lond.* 124(3):213–229.

Raw A. 1976. The behaviour of males of the solitary bee *Osmia rufa* (Megachilidae) searching for females. *Behaviour* 56(3):279–285.

Raw A. 1992. Solitary bees (Hymenoptera: Megachilidae), restricted to identical resources for nesting, recognized their own nests: an example of genetically determined personal scents? *Entomol. U.K.* 111(2):79–87.

Rayment T. 1935. *A Cluster of Bees*. Sydney: Endeavour Press.

Rayment T. 1959. A new genus of bees in the family Colletidae. *Aust. Zool.* 12(6):324–329.

Rehan SM, Leys R, Schwarz MP. 2012. A mid-Cretaceous origin of sociality in Xylocopine bees with only two origins of true worker castes indicates severe barriers to eusociality. *PLoS One* 7(4):e34690.

Rehan SM, Leys R, Schwarz MP. 2013. First evidence for a massive extinction event affecting bees close to the K-T boundary. *PLoS One* 8(10):e76683.

Rehan SM, Richards MH. 2010. The influence of maternal quality on brood sex allocation in the small carpenter bee, *Ceratina calcarata*: bee maternal quality and sex allocation. *Ethology* 116(9):876–887.

Renauld M, Hutchinson A, Loeb G, Poveda K, Connelly H. 2016. Landscape simplification constrains adult size in a native ground-nesting bee. *PLoS One* 11(3):e0150946.

Renner SS, Schaefer H. 2010. The evolution and loss of oil-offering flowers: new insights from dated phylogenies for angiosperms and bees. *Philos. Trans. R. Soc. B Biol. Sci.* 365(1539): 423–435.

Ribble DW. 1965. A revision of the banded subgenera of *Nomia* in America (Hymenoptera, Halictidae). *Univ. Kans. Sci. Bull.* 45(3):277–357

Richards MH. 1994. Social evolution in the genus *Halictus*: a phylogenetic approach. *Insectes Sociaux* 41(3):315–325.

Richards MH, Vickruck JL, Botezatu A, Pickering G. 2016. Do bees smell? Hydrocarbon profiles and microsatellites reveal social recognition in carpenter bees. *Integr. Comp. Biol.* 56:E356.

Richards MH, von Wettberg EJ, Rutgers AC. 2003. A novel social polymorphism in a primitively eusocial bee. *Proc. Natl. Acad. Sci. USA* 100(12):7175–7180.

Ricketts TH, Daily GC, Ehrlich PR, Michener CD. 2004. Economic value of tropical forest to coffee production. *Proc. Natl. Acad. Sci. USA* 101(34):12579–12582.

Rightmyer M, Deyrup M, Ascher J, Griswold T. 2011. *Osmia* species (Hymenoptera, Megachilidae) from the southeastern United States with modified facial hairs: taxonomy, host plants, and conservation status. *ZooKeys* 148:257–278.

Rightmyer MG. 2004. Phylogeny and classification of the parasitic bee tribe Epeolini (Hymenoptera: Apidae, Nomadinae). *Sci. Pap. Nat. Hist. Mus. Univ. Kans.* 33:1–51.

Riveros AJ, Hernández EJ, Wcislo WT. 2009. Nesting biology of *Euglossa dodsoni* Moure (Hymenoptera: Euglossinae) in Panama. *J. Kans. Entomol. Soc.* 82(2):210–214.

Roberts RB. 1969. Biology of the bee genus *Agapostemon* (Hymenoptera: Halictidae). *Kans. Univ. Sci. Bull.* 48(16):698–719.

Roberts RB. 1971. Biology of the crepuscular bee *Ptiloglossa guinnae* n. sp. with notes on associated bees, mites, and yeasts. *J. Kans. Entomol. Soc.* 44(3):283–294.

Roberts RB. 1973. Nest architecture and immature stages of the bee *Oxaea flavescens* and the status of Oxaeidae (Hymenoptera). *J. Kans. Entomol. Soc.* 46(4):437–446.

Roberts RB, Vallespir SR. 1978. Specialization of hairs bearing pollen and oil on the legs of bees (Apoidea: Hymenoptera). *Ann. Entomol. Soc. Am.* 71(4):619–627.

Roberts S, Potts S, Biesmeijer K, Kuhlmann M, Kunin W, Ohlemüller R. 2011. Assessing continental-scale risks for generalist and specialist pollinating bee species under climate change. *BioRisk* 6:1–18.

Robertson C. 1899. Flowers and Insects. XIX. *Bot. Gaz.* 28(1):27–45.

Robertson C. 1914. Origin of oligotropy of bees (Hym.). *Entomol. News* 25(1):67–73.

Robertson C. 1918. Proterandry and flight of bees (Hym.). *Entomol. News* 29(9):340–342.

Robertson C. 1925. Heterotropic bees. *Ecology* 6(4):412–436.

Robinson WS, Nowogrodzki R, Morse RA. 1989. The value of honey bees as pollinators of US crops. *Am. Bee J.* 129:411–423, 477–487.

Rocha Filho LC da, Serrano JC, Garófalo CA. 2016. First report of the cleptoparasitic wasp *Huarpea wagneriella* (du Buysson) (Hymenoptera: Sapygidae) attacking nests of the orchid bee *Eufriesea violacea* (Blanchard) (Hymenoptera: Apidae). *J. Apic. Res.* 55(3):251–252.

Rocha-Filho LC, Garófalo CA. 2016. Natural history of *Tetrapedia diversipes* (Hymenoptera: Apidae) in an Atlantic semideciduous forest remnant surrounded by coffee crops, *Coffea arabica* (Rubiaceae). *Ann. Entomol. Soc. Am.* 109(2):183–197.

Rodríguez-Gironés MA, Gonzálvez FG, Llandres AL, Corlett RT, Santamaría L. 2013. Possible role of weaver ants, *Oecophylla smaragdina*, in shaping plant-pollinator interactions in south-east Asia. *J. Ecol.* 101(4):1000–1006.

Rogers SR, Tarpy DR, Burrack HJ. 2014. Bee species diversity enhances productivity and stability in a perennial crop. *PLoS One* 9(5):e97307.

Roig-Alsina A. 1993. The evolution of the apoid endophallus, its phylogenetic implications, and functional significance of the genital capsule (Hymenoptera, Apoidea). *Boll. Zool.* 60:169–183.

Roig-Alsina A, Michener CD. 1993. Studies of the phylogeny and classification of long-tongued bees (Hymenoptera: Apoidea). *Univ. Kans. Sci. Bull.* 55(4):123–162.

Rolón G, Cilla G. 2012. Adobe wall biodeterioration by the *Centris muralis* Burmeister bee (Insecta: Hymenoptera: Apidae) in a valuable colonial site, the Capayán ruins (La Rioja, Argentina). *Int. Biodeterior. Biodegrad.* 66(1):33–38.

Romiguier J, Cameron SA, Woodard SH, Fischman BJ, Keller L, Praz CJ. 2016. Phylogenomics controlling for base compositional bias reveals a single origin of eusociality in corbiculate bees. *Mol. Biol. Evol.* 33(3):670–678.

Rooijakkers EF, Sommeijer MJ. 2009. Gender specific brood cells in the solitary bee *Colletes halophilus* (Hymenoptera; Colletidae). *J. Insect Behav.* 22(6):492–500.

Rosa C, Lachance M, Silva J, Teixeira A, Marini M et al. 2003. Yeast communities associated with stingless bees. *FEMS Yeast Res.* 4(3):271–275.

Rosa CA, Viana EM, Martins RP, Antonini Y, Lachance M-A. 1999. *Candida batistae*, a new yeast species associated with solitary digger nesting bees in Brazil. *Mycologia*. 91(3):428–433.

Rosenheim JA. 1987. Host location and exploitation by the cleptoparasitic wasp *Argochrysis armilla*: the role of learning (Hymenoptera: Chrysididae). *Behav. Ecol. Sociobiol.* 21(6): 401–406.

Ross KG, Matthews RW. 1989. New evidence for eusociality in the sphecid wasp *Microstigmus comes*. *Anim. Behav.* 38(4):613–619.

Rossi BH, Nonacs P, Pitts-Singer TL. 2010. Sexual harassment by males reduces female fecundity in the alfalfa leafcutting bee, *Megachile rotundata*. *Anim. Behav.* 79(1):165–171.

Roubik DW. 1982. Obligate necrophagy in a social bee. *Science* 217:1059–1060.

Roubik DW. 1989. *Ecology and Natural History of Tropical Bees*. Cambridge: Cambridge University Press.

Roubik DW. 2006. Stingless bee nesting biology. *Apidologie* 37(2):124–143.

Roubik DW, Hanson PE. 2004. *Orchid Bees of Tropical America: Biology and Field Guide*. Heredia: Editorial Instituto Nacional de Biodiversidad.

Roubik DW, Michener CD. 1980. The seasonal cycle and nests of *Epicharis zonata*, a bee whose cells are below the wet-season water table (Hymenoptera, Anthophoridae). *Biotropica* 12(1):56–60.

Roubik DW, Michener CD. 1984. Nesting biology of *Crawfordapis* in Panama (Hymenoptera, Colletidae). *J. Kans. Entomol. Soc.* 57(4):662–671.

Roubik DW, Weight LA, Bonilla MA. 1996. Population genetics, diploid males, and limits to social evolution of euglossine bees. *Evolution* 50(2):931–935.

Roubik DW, Wolda H. 2001. Do competing honey bees matter? Dynamics and abundance of native bees before and after honey bee invasion. *Popul. Ecol.* 43(1):53–62.

Roulston TH. 1997. Hourly capture of two species of *Megalopta* (Hymenoptera: Apoidea; Halictidae) at black lights in Panama with notes on nocturnal foraging by bees. *J. Kans. Entomol. Soc.* 70(3):189–196.

Roulston TH, Cane JH. 2000. Pollen nutritional content and digestibility for animals. *Plant Syst. Evol.* 222(1–4):187–209.

Roulston TH, Cane JH. 2002. The effect of pollen protein concentration on body size in the sweat bee *Lasioglossum zephyrum* (Hymenoptera: Apiformes). *Evol. Ecol.* 16(1):49–65.

Roulston TH, Cane JH, Buchmann SL. 2000. What governs protein content of pollen: pollinator preferences, pollen–pistil interactions, or phylogeny? *Ecol. Monogr.* 70(4):617–643.

Roulston TH, Malfi R. 2012. Aggressive eviction of the eastern carpenter bee (*Xylocopa virginica* (Linnaeus) from its nest by the giant resin bee (*Megachile sculpturalis* Smith). *J. Kans. Entomol. Soc.* 85(4):387–388.

Rozen JG Jr. 1958. Monographic study of the genus *Nomadopsis* Ashmead (Hymenoptera: Andrenidae). *Univ. Calif. Publ. Entomol.* 15:1–202.

Rozen JG Jr. 1965a. The larvae of the Anthophoridae (Hymenoptera, Apoidea). Part 1, Introduction, Eucerini, and Centridini (Anthophorinae). *Am. Mus. Novit.* 2233:1–28.

Rozen JG Jr. 1965b. The biology and immature stages of *Melitturga clavicornis* (Latreille) and of *Sphecodes albilabris* (Kirby): and the recognition of the Oxaeidae at the family level (Hymenoptera, Apoidea). *Am. Mus. Novit.* 2224:1–18.

Rozen JG Jr. 1967a. Review of the biology of panurgine bees, with observations on North American forms (Hymenoptera, Andrenidae). *Am. Mus. Novit.* 2297:1–44.

Rozen JG Jr. 1967b. The immature instars of the cleptoparasitic genus *Dioxys* (Hymenoptera: Megachilidae). *J. N. Y. Entomol. Soc.* 236–248.

Rozen JG Jr. 1968. Biology and immature stages of the aberrant bee genus *Meliturgula* (Hymenoptera, Andrenidae). *Am. Mus. Novit.* 2331:1–18.

Rozen JG Jr. 1969a. The larvae of the Anthophoridae (Hymenoptera, Apoidea). Part 3, The Melectini, Ericrocini, and Rhathymini. *Am. Mus. Novit.* 2382:1–24.

Rozen JG Jr. 1969b. The biology and description of a new species of African *Thyreus*, with life history notes on two species of *Anthophora* (Hymenoptera: Anthophoridae). *J. N. Y. Entomol. Soc.* 77:51–60.

Rozen JG Jr. 1970. Biology, immature stages, and phylogenetic relationships of fideliine bees, with the description of a new species of *Neofidelia* (Hymenoptera, Apoidea). *Am. Mus. Novit.* 2427:1–26.

Rozen JG Jr. 1973. Life history and immature stages of the bee *Neofidelia* (Hymenoptera, Fidelidae). *Am. Mus. Novit.* 2519:1–14.

Rozen JG Jr. 1977a. The ethology and systematic relationships of fideliine bees, including a description of the mature larva of *Parafidelia* (Hymenoptera, Apoidea). *Am. Mus. Novit.* 2637:1–15.

Rozen JG Jr. 1977b. Immature stages of and ethological observations on the cleptoparasitic tribe Nomadini (Apoidea, Anthophoridae). *Am. Mus. Novit.* 2638:1–16.

Rozen JG Jr. 1977c. Biology and immature stages of the bee genus *Meganomia* (Hymenoptera, Melittidae). *Am. Mus. Novit.* 2630:1–14.

Rozen JG Jr. 1978. The relationships of the bee subfamily Ctenoplectrinae as revealed by its biology and mature larva (Apoidea: Melittidae). *J. Kans. Entomol. Soc.* 51(4):637–652.

Rozen JG Jr. 1984a. Comparative nesting biology of the bee tribe Exomalopsini (Apoidea, Anthophoridae). *Am. Mus. Novit.* 2798:1–37.

Rozen JG Jr. 1984b. Nesting biology of diphaglossine bees (Hymenoptera, Colletidae). *Am. Mus. Novit.* 2786:1–33.

Rozen JG Jr. 1987a. Nesting biology and immature stages of a new species in the bee genus *Hesperapis* (Hymenoptera, Apoidea, Melittidae, Dasypodinae). *Am. Mus. Novit.* 2887:1–13.

Rozen JG Jr. 1987b. Nesting biology of the bee *Ashmeadiella holtii* and its cleptoparasite, a new species of *Stelis* (Apoidea: Megachilidae). *Am. Mus. Novit.* 2900:1–10.

Rozen JG Jr. 1989a. Morphology and systematic significance of first instars of the cleptoparasitic bee tribe Epeolini (Anthophoridae, Nomadinae). *Am. Mus. Novit.* 2957:1–19.

Rozen JG Jr. 1989b. Life history studies of the "primitive" panurgine bees (Hymenoptera, Andrenidae, Panurginae). *Am. Mus. Novit.* 2962:1–27.

Rozen JG Jr. 1990. Pupa of the bee *Pararhophites orobinus* (Hymenoptera: Apoidea: Megachilidae). *J. N. Y. Entomol. Soc.* 98(3):379–382.

Rozen JG Jr. 1991. Evolution of cleptoparasitism in anthophorid bees as revealed by their mode of parasitism and first instars (Hymenoptera, Apoidea). *Am. Mus. Novit.* 3029:1–36.

Rozen JG Jr. 1993. Nesting biologies and immature stages of the rophitine bees (Halictidae) with notes on the cleptoparasite *Biastes* (Anthophoridae) (Hymenoptera; Apoidea). *Am. Mus. Novit.* 3066:1–28.

Rozen JG Jr. 1994. Biologies of the bee genera *Ancylandrena* (Andrenidae: Andreninae) and *Hexepeolus* (Apidae: Nomadinae), and phylogenetic relationships of *Ancylandrena* based on its mature larva (Hymenoptera: Apoidea). *Am. Mus. Novit.* 3108:1–19.

Rozen JG Jr. 1996. A new species of the bee *Heterosarus* from Dominican amber (Hymenoptera: Andrenidae; Panurginae). *J. Kans. Entomol. Soc.* 69(4):346–352.

Rozen JG Jr. 1997. South American rophitine bees (Hymenoptera: Halictidae: Rophitinae). *Am. Mus. Novit.* 3206:1–27.

Rozen JG Jr. 2000. Systematic and geographic distributions of Neotropical cleptoparasitic bees, with notes on their modes of parasitism. In *An. IV Encontro Sobre Abelhas*, Bitondi MMG, Hartfelder K et al., eds., pp. 204–210. Ribeirão Preto, Brazil.

Rozen JG Jr. 2003a. Eggs, ovariole numbers, and modes of parasitism of cleptoparasitic bees, with emphasis on neotropical species (Hymenoptera: Apoidea). *Am. Mus. Novit.* 3413:1–36.

Rozen JG Jr. 2003b. A new tribe, genus, and species of South American panurgine bee (Andrenidae, Panurginae), oligolectic on *Nolana* (Nolanaceae). In *Apoidea Neotropica: Homenagem aos 90 anos de Jesus Santiago Moure*, Melo GAR, Alves-dos-Santos I, eds., pp. 93–106. Criciúma, Editora UNESC.

Rozen, JG Jr. 2007 Investigations on the biologies and immature stages of the cleptoparasitic bee genera *Radoszkowskiana* and *Coelioxys* and their *Megachile* hosts (Hymenoptera: Megachilini). *Am. Mus. Novit.* 3573:1–43.

Rozen JG Jr. 2008a. Biology and immature stages of the bee *Nomioides patruelis* (Halictidae: Halictinae: Nomioidini) and of its cleptoparasite, *Chiasmognathus pashupati* (Apidae: Nomadinae: Ammobatini), with a preliminary phylogeny of the Halictidae based on mature larvae (Apoidea). *Am. Mus. Novit.* 3604:1–23.

Rozen JG Jr. 2008b. The solitary bee *Calliopsis zebrata*: biological and distributional notes and description of its larva (Hymenoptera: Andrenidae: Panurginae). *Am. Mus. Novit.* 3632:1–12.

Rozen JG Jr. 2010. Immatures of the Old World oil-collecting bee *Ctenoplectra cornuta* (Apoidea: Apidae: Apinae: Ctenoplectrini). *Am. Mus. Novit.* 3699:1–14.

Rozen JG Jr. 2011. Immatures of exomalopsine bees with notes on nesting biology and a tribal key to mature larvae of noncorbiculate, nonparasitic Apinae (Hymenoptera: Apidae). *Am. Mus. Novit.* 3726:1–52.

Rozen JG Jr. 2013. Mature larvae of calliopsine bees: *Spinoliella, Callonychium,* and *Arhysosage* including biological notes, and a larval key to calliopsine genera (Hymenoptera: Apoidea: Andrenidae: Panurginae). *Am. Mus. Novit.* 3782:1–27.

Rozen JG Jr. 2016a. *Hesperapis rhodocerata*: behavioral biology, egg, and larval instars, including behavioral and larval comparisons with *H. larreae* (Hymenoptera: Melittidae: Dasypodainae). *Am. Mus. Novit.* 3856:1–19.

Rozen JG Jr. 2016b. Mature larvae of euglossine bees, a comparative study (Apoidea: Apidae: Euglossini). *Am. Mus. Novit.* 3861:1–16.

Rozen JG Jr. 2018. Nesting biologies and mature larvae of oxaeine bees (Apoidea: Andrenidae). *Am. Mus. Novit.* 3893:1–32.

Rozen JG Jr, Buchmann SL. 1990. Nesting biology and immature stages of the bees *Centris caesalpiniae, C. pallida,* and the cleptoparasite *Ericrocis lata* (Hymenoptera: Apoidea: Anthophoridae). *Am. Mus. Novit.* 2985:1–30.

Rozen JG Jr, Eickwort KR, Eickwort GC. 1978. The bionomics and immature stages of the cleptoparasitic bee genus *Protepeolus* (Anthophoridae, Nomadinae). *Am. Mus. Novit.* 2640:1–24.

Rozen JG Jr, Favreau MS. 1967. Biological notes on *Dioxys pomonae pomonae* and on its host, *Osmia nigrobarbata* (Hymenoptera: Megachilidae). *J. N. Y. Entomol. Soc.* 75(4):197–203.

Rozen JG Jr, Favreau MS. 1968. Biological notes on *Colletes compactus compactus* and its cuckoo bee, *Epeolus pusillus* (Hymenoptera: Colletidae and Anthophoridae). *J. N. Y. Entomol. Soc.* 76(2):106–111.

Rozen JG Jr, Hall HG. 2011. Nesting and developmental biology of the cleptoparasitic bee *Stelis ater* (Anthidiini) and its host, *Osmia chalybea* (Osmiini) (Hymenoptera: Megachilidae). *Am. Mus. Novit.* 3707:1–38.

Rozen JG Jr, Hall HG. 2012. Nesting biology and immatures of the oligolectic bee *Trachusa larreae* (Apoidea: Megachilidae: Anthidiini). *Am. Mus. Novit.* 3765:1–24.

Rozen JG Jr, Hall HG. 2014. Nest site selection and nesting behavior of the bee *Lithurgopsis apicalis* (Megachilidae: Lithurginae). *Am. Mus. Novit.* 3796:1–24.

Rozen JG Jr, Jacobson R. 1980. Biology and immature stages of *Macropis nuda*, including comparisons to related bees (Apoidea, Melittidae). *Am. Mus. Novit.* 2702:1–11.

Rozen JG Jr, Kamel SM. 2009. Last larval instar and mature oocytes of the Old World cleptoparasitic bee *Stelis murina*, including a review of *Stelis* biology (Apoidea: Megachilidae: Megachilinae: Anthidiini). *Am. Mus. Novit.* 3666:1–19.

Rozen JG Jr, Macneill CD. 1957. Biological observations on *Exomalopsis (Anthophorula) chionura* Cockerell, including a comparison of the biology of *Exomalopsis* with that of other anthophorid groups (Hymenoptera: Apoidea). *Ann. Entomol. Soc. Am.* 50(5):522–529.

Rozen JG Jr, McGinley RJ. 1976. Biology of the bee genus *Conanthalictus* (Halictidae, Dufoureinae). *Am. Mus. Novit.* 2602:1–6.

Rozen JG Jr, McGinley RJ. 1991. Biology and larvae of the cleptoparasitic bee *Townsendiella pulchra* and nesting biology of its host *Hesperapis larreae* (Hymenoptera, Apoidea). *Am. Mus. Novit.* 3005:1–12.

Rozen JG Jr, Melo GAR, Aguiar AJC, Alves-dos-Santos I. 2006. Nesting biologies and immature stages of the Tapinotaspidine bee genera *Monoeca* and *Lanthanomelissa* and of their osirine cleptoparasites *Protosiris* and *Parepeolus* (Hymenoptera: Apidae: Apinae). *Am. Mus. Novit.* 3501:1–60.

Rozen JG Jr, Michener CD. 1968. The biology of *Scrapter* and its cuckoo bee, *Pseudodichroa* (Hymenoptera, Colletidae and Anthophoridae). *Am. Mus. Novit.* 2335:1–13.

Rozen JG Jr, Özbek H. 2005. Egg deposition of the cleptoparasitic bee *Dioxys cincta* (Hymenoptera: Apoidea: Megachilidae). *J. Kans. Entomol. Soc.* 78(3):221–226.

Rozen JG Jr, Özbek H. 2008. Immatures of rophitine bees, with notes on their nesting biology (Hymenoptera: Apoidea: Halictidae). *Am. Mus. Novit.* 3609:1–36.

Rozen JG Jr, Özbek H, Ascher JS, Sedivy C, Praz C et al. 2010a. Nests, petal usage, floral preferences, and immatures of *Osmia (Ozbekosmia) avosetta* (Megachilidae: Megachilinae: Osmiini), including biological comparisons with other Osmiine bees. *Am. Mus. Novit.* 3680:1–22.

Rozen JG Jr, Roig-Alsina A, Alexander BA. 1997. The cleptoparasitic bee genus *Rhopalolemma*: with reference to other Nomadinae (Apidae), and biology of its host *Protodufourea* (Halictidae, Rophitinae). *Am. Mus. Novit.* 3194:1–28.

Rozen JG Jr, Rozen JR, Hall HG. 2011a. Gas diffusion rates through cocoon walls of two bee species (Hymenoptera: Megachilidae). *Ann. Entomol. Soc. Am.* 104(6):1349–1354.

Rozen JG Jr, Ruz L. 1995. South American panurgine bees (Andrenidae: Panurginae), Part II. Adults, immature stages, and biology of *Neffapis longilingua*, a new genus and species with an elongate glossa. *Am. Mus. Novit.* 3136:1–15.

Rozen JG Jr, Smith CS, Cane JH. 2017. Survey of hatching spines of bee larvae including those of *Apis mellifera* (Hymenoptera: Apoidea). *J. Insect Sci.* 17(4):1–10.

Rozen JG Jr, Snelling RR. 1986. Ethology of the bee *Exomalopsis nitens* and its cleptoparasite (Hymenoptera: Anthophoridae). *J. N. Y. Entomol. Soc.* 94(4):480–488.

Rozen JG Jr, Vinson SB, Coville R, Frankie G. 2010b. Biology and morphology of the immature stages of the cleptoparasitic bee *Coelioxys chichimeca* (Hymenoptera: Apoidea: Megachilidae). *Am. Mus. Novit.* 3679:1–26.

Rozen JG Jr, Vinson SB, Coville R, Frankie G. 2011b. Biology of the cleptoparasitic bee *Mesoplia sapphirina* (Ericrocidini) and its host *Centris flavofasciata* (Centridini) (Apidae: Apinae). *Am. Mus. Novit.* 3723:1–36.

Rozen JG Jr, Wyman ES. 2014. Early nesting biology of the wood-nesting adventive bee, *Lithurgus chrysurus* Fonscolombe (Apoidea: Megachilidae: Lithurginae). *Am. Mus. Novit.* 3804:1–12.

Rozen JG Jr, Yanega D. 1999. Nesting biology and immature stages of the South American bee genus *Acamptopoeum* (Hymenoptera: Andrenidae: Panurginae). *Univ. Kans. Nat. Hist. Mus. Spec. Publ.* 24:59–67.

Rundlöf M, Andersson GK, Bommarco R, Fries I, Hederström V et al. 2015. Seed coating with a neonicotinoid insecticide negatively affects wild bees. *Nature* 521(7550):77–80.

Russell AL, Leonard AS, Gillette HD, Papaj DR. 2016. Concealed floral rewards and the role of experience in floral sonication by bees. *Anim. Behav.* 120:83–91.

Russo L. 2016. Positive and negative impacts of non-native bee species around the world. *Insects* 7(4):69.

Russo L, Park M, Gibbs J, Danforth BN. 2015. The challenge of accurately documenting bee species richness in agroecosystems: bee diversity in eastern apple orchards. *Ecol. Evol.* 5(17):3531–3540.

Rust RW. 1980. The biology of *Ptilothrix bombiformis* (Hymenoptera: Anthophoridae). *J. Kans. Entomol. Soc.* 53(2):427–436.

Rust RW. 1988. Biology of *Nomadopsis larreae* (Hymenoptera: Andrenidae), with an analysis of yearly appearance. *Ann. Entomol. Soc. Am.* 81(1):99–104.

Rust RW. 2003. Horned Lark (*Eremophila alpestris* L.) predation on alkali bees, *Nomia melanderi* Cockerell. *West. North Am. Nat.* 63(2):224–228.

Rust RW, Thorp RW. 1973. The biology of *Stelis chlorocyanea*, a parasite of *Osmia nigrifrons* (Hymenoptera: Megachilidae). *J. Kans. Entomol. Soc.* 46(4):548–548.

Rutowski RL, Alcock J. 1980. Temporal variation in male copulatory behaviour in the solitary bee *Nomadopsis puellae* (Hymenoptera: Andrenidae). *Behaviour* 73(3):175–187.

Ruz L, Melo GAR. 1999. Reassessment of the bee genus *Chaeturginus* (Apoidea: Andrenidae, Panurginae), with the description of a new species from southeastern Brazil. *Univ. Kans. Nat. Hist. Mus. Spec. Publ.* 24:231–236.

Sabino WO, da Silva CI, Alves-dos-Santos I. 2017. Mating system and sleeping behaviour of the male and female *Centris* (*Paracentris*) *burgdorfi* Friese (Apidae, Centridini). *J. Insect Behav.* 30(1):103–118.

Sakagami S, Michener CD. 1962. *Nest Architecture of the Sweat Bees (Halictinae)*. Lawrence: University of Kansas Press.

Sakagami SF, Ebmer AW, Tadauchi O. 1996. The halictine bees of Sri Lanka and the vicinity III. *Sudila* (Hymenoptera, Halictidae) Part 1. *Esakia* 36:143–189.

Sakagami SF, Laroca S. 1971. Observations on the bionomics of some neotropical xylocopine bees, with comparative and biofaunistic notes (Hymenoptera, Anthophoridae). *J. Fac. Sci. Hokkaido Univ. Ser. VI Zool.* 18(1):57–127.

Sakagami SF, Maeta Y. 1987. Multifemale nests and rudimentary castes of an "almost" solitary bee *Ceratina flavipes*, with additional observations on multifemale nests of *Ceratina japonica* (Hymenoptera, Apoidea). *Entomol. Soc. Jpn.* 55(3):391–409.

Sakagami SF, Moure JS. 1965. Cephalic polymorphism in some neotropical halictine bees (Hymenoptera-Apoidea). *An. Acad. Bras. Ciênc.* 37(2):303–313.

Sakagami SF, Roubik DW, Zucchi R. 1993. Ethology of the robber stingless bee, *Lestrimelitta limao* (Hymenoptera: Apidae). *Sociobiology* 21:237–277.

Sakagami SF, Zucchi R. 1978. Nests of *Hylaeus* (*Hylaeopsis*) *tricolor*: the first record of non-solitary life in colletid bees, with notes on communal and quasisocial colonies (Hymenoptera: Colletidae). *J. Kans. Entomol. Soc.* 51(4):597–614.

Salt G. 1927. The effects of stylopization in aculeate Hymenoptera. *J. Exp. Zool.* 48(1):223–331

Sampson BJ, Stringer SJ, Cane JH, Spiers JM. 2004. Screenhouse evaluations of a mason bee *Osmia ribifloris* (Hymenoptera: Megachilidae) as a pollinator for blueberries in the southeastern United States. *Small Fruits Rev.* 3(3–4):381–392.

Sandrock C, Tanadini LG, Pettis JS, Biesmeijer JC, Potts SG, Neumann P. 2014. Sublethal neonicotinoid insecticide exposure reduces solitary bee reproductive success. *Agric. For. Entomol.* 16(2):119–128.

Sann M, Niehuis O, Peters RS, Mayer C, Kozlov A et al. 2018. Phylogenomic analysis of Apoidea sheds new light on the sister group of bees. *BMC Evol. Biol.* 18:71.

Sapir Y, Shmida A, Ne'eman G. 2006. Morning floral heat as a reward to the pollinators of *Oncocyclus* irises. *Oecologia* 147:53–69.

Sarzetti L, Genise J, Sanchez MV. 2012. *Trichothurgus bolithophilus* sp. n. (Hymenoptera, Megachilidae) a bee nesting in horse manure pads in Patagonia, Argentina. *J. Hymenopt. Res.* 29(1):1–14.

Sarzetti L, Genise J, Sanchez MV, Farina J, Molina A. 2013. Nesting behavior and ecological preferences of five Diphaglossinae species (Hymenoptera, Apoidea, Colletidae) from Argentina and Chile. *J. Hymenopt. Res.* 33(1):63–82.

Saul-Gershenz LS, Millar JG. 2006. Phoretic nest parasites use sexual deception to obtain transport to their host's nest. *Proc. Natl. Acad. Sci. USA* 103(38):14039–14044.

Saunders ME, Luck GW. 2013. Pan trap catches of pollinator insects vary with habitat. *Aust. J. Entomol.* 52(2):106–113.

Saure C. 1996. Urban habitats for bees: the example of the city of Berlin. In *The Conservation of Bees*, Matheson A, Buchmann SL, O'Toole C, Westrich P, Williams IH, eds., pp. 47–53. New York: Academic Press.

Scaven VL, Rafferty NE. 2013. Physiological effects of climate warming on flowering plants and insect pollinators and potential consequences for their interactions. *Curr. Zool.* 59(3): 418–426.

Schaefer H, Renner SS. 2008. A phylogeny of the oil bee tribe Ctenoplectrini (Hymenoptera: Anthophila) based on mitochondrial and nuclear data: evidence for early Eocene divergence and repeated out-of-Africa dispersal. *Mol. Phylogenet. Evol.* 47(2):799–811.

Schäffler I, Dötterl S. 2011. A day in the life of an oil bee: phenology, nesting, and foraging behavior. *Apidologie* 42(3):409–424.

Schatz B, Wcislo WT. 1999. Ambush predation by the ponerine ant *Ectatomma ruidum* Roger (Formicidae) on a sweat bee *Lasioglossum umbripenne* (Halictidae), in Panama. *J. Insect Behav.* 12(5):641–663.

Schemske DW, Lande R. 1984. Fragrance collection and territorial display by male orchid bees. *Anim. Behav.* 32(3):935–937.

Scheuchl E, Willner W. 2016. *Taschenlexikon Der Wildbienen Mitteleuropas; Alle Arten Im Porträt*. Wiebelsheim: Verlag Quelle & Meyer.

Schlindwein C, Martins CF. 2000. Competition between the oligolectic bee *Ptilothrix plumata* (Anthophoridae) and the flower closing beetle *Pristimerus calcaratus* (Curculionidae) for floral resources of *Pavonia cancellata* (Malvaceae). *Plant Syst. Evol.* 224(3–4):183–194.

Schlindwein C, Pick RA, Martins CF. 2009. Evaluation of oligolecty in the Brazilian bee *Ptilothrix plumata* (Hymenoptera, Apidae, Emphorini). *Apidologie* 40(2):106–116.

Schlindwein C, Wittmann D. 1997. Stamen movements in flowers of *Opuntia* (Cactaceae) favour oligolectic pollinators. *Plant Syst. Evol.* 204(3–4):179–193.

Schlindwein C, Wittmann D, Martins CF, Hamm A, Siqueira JA et al. 2005. Pollination of *Campanula rapunculus* L. (Campanulaceae): how much pollen flows into pollination and into reproduction of oligolectic pollinators? *Plant Syst. Evol.* 250(3–4):147–156.

Schmidt JO. 2016. *Sting of the Wild.* Baltimore: Johns Hopkins University Press.

Schmidt K, Westrich P. 1993. *Colletes hederae* n. sp., eine bisher unerkannte, auf Efeu (*Hedera*) spezialisierte Bienenart (Hymenoptera: Apoidea). *Entomol. Z.* 103(6):89–112.

Schönitzer K, Schuberth J. 1993. Vorkommen und Morphologie der Fovea facialis und der darunterliegenden Drüsen bei Apoidea (Hymenoptera). *Mitteilungen Dtsch. Ges. Allg. Angew. Entomol.* 8(4/6):911–918.

Schoonhoven LM, Van Loon JJA, Dicke M. 2005. *Insect-Plant Biology.* Oxford: Oxford University Press. 2nd ed.

Schuberth J, Schönitzer K. 1993. Vergleichende Morphologie der Fovea facialis und der Stirnseitendrüse bei Apoidea und Specidae (Hymenoptera, Aculeata). *Linzer biologische Beiträge.* 25(1):205–277.

Schulte P, Alegret L, Arenillas I, Arz JA, Barton PJ et al. 2010. The Chicxulub asteroid impact and mass extinction at the Cretaceous-Paleogene boundary. *Science* 327(5970):1214–1218.

Schwarz HF. 1928. Bees of the subfamily Anthidiinae, including some new species and varieties, and some new locality records. *J. N. Y. Entomol. Soc.* 36(4):369–419.

Schwarz HF. 1948. Stingless bees (Meliponidae) of the Western Hemisphere: *Lestrimelitta* and the following subgenera of *Trigona: Trigona, Paratrigona, Schwarziana, Parapartamona, Cephalotrigona, Oxytrigona, Scaura,* and *Mourella. Bull. Am. Mus. Nat. Hist.* 90:1–536.

Schwarz MP, Richards MH, Danforth BN. 2007. Changing paradigms in insect social evolution: insights from halictine and allodapine bees. *Annu. Rev. Entomol.* 52(1):127–150.

Schwarz MP, Tierney SM, Rehan SM, Chenoweth LB, Cooper SJ. 2011. The evolution of eusociality in allodapine bees: workers began by waiting. *Biol. Lett.* 7(2):277–280.

Sedivy C, Dorn S, Müller A. 2013a. Evolution of nesting behaviour and kleptoparasitism in a selected group of osmiine bees (Hymenoptera: Megachilidae). *Biol. J. Linn. Soc.* 108(2):349–360.

Sedivy C, Dorn S, Müller A. 2013b. Molecular phylogeny of the bee genus *Hoplitis* (Megachilidae: Osmiini)—how does nesting biology affect biogeography? *Zool. J. Linn. Soc.* 167(1): 28–42.

Sedivy C, Dorn S, Widmer A, Müller A. 2013c. Host range evolution in a selected group of osmiine bees (Hymenoptera: Megachilidae): the Boraginaceae-Fabaceae paradox. *Biol. J. Linn. Soc.* 108(1):35–54.

Sedivy C, Müller A, Dorn S. 2011. Closely related pollen generalist bees differ in their ability to develop on the same pollen diet: evidence for physiological adaptations to digest pollen. *Funct. Ecol.* 25(3):718–725.

Sedivy C, Praz CJ, Müller A, Widmer A, Dorn S. 2008. Patterns of host-plant choice in bees of the genus *Chelostoma*: the constraint hypothesis of host-range evolution in bees. *Evolution* 62(10):2487–2507.

Seeley TD. 1995. *The Wisdom of the Hive.* Cambridge: Harvard University Press.

Seeley TD. 2010. *Honeybee Democracy.* Princeton: Princeton University Press.

Seidelmann K. 1999. The race for females: the mating system of the red mason bee, *Osmia rufa* (L.) (Hymenoptera: Megachilidae). *J. Insect Behav.* 12(1):13–25.

Seidelmann K. 2006. Open-cell parasitism shapes maternal investment patterns in the red mason bee *Osmia rufa. Behav. Ecol.* 17(5):839–848.

Seidelmann K. 2014. Optimal progeny body size in a solitary bee, *Osmia bicornis* (Apoidea: Megachilidae). *Ecol. Entomol.* 39(5):656–663.

Seidelmann K, Ulbrich K, Mielenz N. 2010. Conditional sex allocation in the red mason bee, *Osmia rufa. Behav. Ecol. Sociobiol.* 64(3):337–347.

Selander RB. 1960. Bionomics, systematics, and phylogeny of *Lytta,* a genus of blister beetles (Coleoptera, Meloidae). *Univ. Ill. Biol. Monogr.* 28:1–295.

Sérsic AN. 2004. Pollination biology in the genus *Calceolaria* L. (Calceolariaceae). *Stapfia* 82:1–121.

Sérsic AN, Cocucci AA. 1999. An unusual kind of nectary in the oil flowers of *Monttea:* its structure and function. *Flora* 194(4):393–404.

Severinghaus LL, Kurtak BH, Eickwort GC. 1981. The reproductive behavior of *Anthidium manicatum* (Hymenoptera: Megachilidae) and the significance of size for territorial males. *Behav. Ecol. Sociobiol.* 9(1):51–58.

Sheffield CS, Dumesh S, Cheryomina M. 2011. *Hylaeus punctatus* (Hymenoptera: Colletidae), a bee species new to Canada, with notes on other non-native species. *J. Entomol. Soc. Ont.* 142:29–43.

Sheffield CS, Heron J. 2018 A new western Canadian record of *Epeoloides pilosulus* (Cresson), with discussion of ecological associations, distribution and conservation status in Canada. *Biodiversity Data Journal* 6:e22837.

Sheffield CS, Pindar A, Packer L, Kevan PG. 2013. The potential of cleptoparasitic bees as indicator taxa for assessing bee communities. *Apidologie* 44(5):501–510.

Sheffield CS, Rigby SM, Smith RF, Kevan PG. 2004. The rare cleptoparasitic bee *Epeoloides pilosula* (Hymenoptera: Apoidea: Apidae) discovered in Nova Scotia, Canada, with distributional notes. *J. Kans. Entomol. Soc.* 77(3):161–164.

Shelly TE, Villalobos EM, Buchmann SL, Cane JH. 1993. Temporal patterns of floral visitation for two bee species foraging on *Solanum*. *J. Kans. Entomol. Soc.* 66(3):319–327.

Shimamoto K, Kasuya E, Yasumoto AA. 2006. Effects of body size on mating in solitary bee *Colletes perforator* (Hymenoptera: Colletidae). *Ann. Entomol. Soc. Am.* 99(4):714–717.

Shimron O, Hefetz A, Tengö J. 1985. Structural and communicative functions of Dufour's gland secretion in *Eucera palestinae* (Hymenoptera; Anthophoridae). *Insect Biochem.* 15(5): 635–638.

Shinn AF. 1967. A revision of the bee genus *Calliopsis* and the biology and ecology of *C. andreniformis* (Hymenoptera, Andrenidae). *Univ. Kans. Sci. Bull.* 46(21):753–939.

Shuler RE, Roulston TH, Farris GE. 2005. Farming practices influence wild pollinator populations on squash and pumpkin. *J. Econ. Entomol.* 98(3):790–795.

Sick M, Ayasse M, Tengö J, Engels W, Lübke G, Francke W. 1994. Host-parasite relationships in six species of *Sphecodes* bees and their halictid hosts: nest intrusion, intranidal behavior, and Dufour's gland volatiles (Hymenoptera: Halictidae). *J. Insect Behav.* 7(1):101–117.

Sihag RC. 1983. Life cycle pattern, seasonal mortality, problem of parasitization and sex ratio pattern in alfalfa pollinating megachilid bees. *Z. Für Angew. Entomol.* 96(1–5):368–379.

Sihag RC. 1993. Behaviour and ecology of the subtropical carpenter bee, *Xylocopa fenestrata* F. 6. Foraging dynamics, crop hosts and pollination potential. *J. Apic. Res.* 32(2):94–101.

Silva CA, Vieira MF. 2015. Flowering and pollinators of three distylous species of *Psychotria* (Rubiaceae) co-occuring in the Brazilian Atlantic Forest. *Rev. Árvore* 39(5):779–789.

Simmons LW, Tomkins JL, Alcock J. 2000. Can minor males of Dawson's burrowing bee, *Amegilla dawsoni* (Hymenoptera: Anthophorini) compensate for reduced access to virgin females through sperm competition? *Behav. Ecol.* 11(3):319–325.

Simpson BB, Neff JL. 1981. Floral rewards: alternatives to pollen and nectar. *Ann. Mo. Bot. Gard.* 68(2):301–322.

Simpson BB, Neff JL. 1983. Evolution and diversity of floral rewards. In *Handbook of Experimental Pollination Biology*, Jones CE, Little RJ, eds., pp. 142–159. New York: Van Nostrand Reinhold.

Singer RB, Cocucci A. 1999. Pollination mechanisms in four sympatric southern Brazilian Epidendroideae orchids. *Lindleyana* 14(1):47–56.

Singh R, Levitt AL, Rajotte EG, Holmes EC, Ostiguy N et al. 2010. RNA viruses in hymenopteran pollinators: evidence of inter-taxa virus transmission via pollen and potential impact on non-apis hymenopteran species. *PLoS One* 5(12):e14357.

Sipes SD, Tepedino VJ. 2005. Pollen-host specificity and evolutionary patterns of host switching in a clade of specialist bees (Apoidea: *Diadasia*). *Biol. J. Linn. Soc.* 86(4):487–505.

Sirohi MH, Jackson J, Edwards M, Ollerton. 2015. Diversity and abundance of solitary and primitively eusocial bees in an urban centre: a case study of Northampton (England). *J. Insect Conserv.* 19(3):487–500.

Sitdikov AA. 1988. Nesting of the bee *Halictus quadricinctus* (F.) (Hymenoptera, Halictidae) in the Udmurt ASSR. *Entomol. Rev.* 67:66–77 [English translation of *Entomol. Obozr.* 3:529–539].

Sjödin NE. 2007. Pollinator behavioural responses to grazing intensity. *Biodivers. Conserv.* 16(7): 2103–2121.

Sjödin NE, Bengtsson J, Ekbom B. 2008. The influence of grazing intensity and landscape composition on the diversity and abundance of flower-visiting insects. *J. Appl. Ecol.* 45(3):763–772.

Skov C, Wiley J. 2005. Establishment of the neotropical orchid bee *Euglossa viridissima* (Hymenoptera: Apidae) in Florida. *Fla. Entomol.* 88:225–227.

Slobodchikoff CN. 1967. Bionomics of *Grotea californica* Cresson, with a description of the larva and pupa. *Pan-Pac. Entomol.* 43(2):161–168.

Smith BH, Carlson RG, Frazier J. 1985. Identification and bioassay of macrocyclic lactone sex pheromone of the halictine bee *Lasioglossum zephyrum*. *J. Chem. Ecol.* 11(10):1447–1456.

Smith BH, Wenzel JW. 1988. Pheromonal covariation and kinship in social bee *Lasioglossum zephyrum* (Hymenoptera: Halictidae). *J. Chem. Ecol.* 14(1):87–94.

Smith MD. 2011. An ecological perspective on extreme climatic events: a synthetic definition and framework to guide future research. *J. Ecol.* 99(3):656–663.

Snelling RR. 1956. Bees of the genus *Centris* in California (Hymenoptera: Anthophoridae). *Pan-Pac. Entomol.* 32(1):1–8.

Snelling RR. 1966. Studies on North American bees of the genus *Hylaeus*. 3. The Nearctic subgenera (Hymenoptera: Colletidae). *Bull. South. Calif. Acad. Sci.* 65(3):164–175.

Snelling RR. 1982. The taxonomy of some neotropical *Hylaeus* and descriptions of new taxa (Hymenoptera: Colletidae). *Bull. South. Calif. Acad. Sci.* 81(1):1–25.

Snelling RR. 1984. Studies on the taxonomy and distribution of American centridine bees (Hymenoptera: Anthophoridae). *Nat. Hist. Mus. Los Angel. Cty. Contrib. Sci.* 347:1–69.

Snodgrass RE. 1985. *Anatomy of the Honey Bee.* Ithaca: Cornell University Press.

Somanathan H, Borges RM. 2001. Nocturnal pollination by the carpenter bee *Xylocopa tenuiscapa* (Apidae) and the effect of floral display on fruit set of *Heterophragma quadriloculare* (Bignoniaceae) in India. *Biotropica* 33(1):78–89.

Somanathan H, Borges RM, Warrant EJ, Kelber A. 2008. Nocturnal bees learn landmark colours in starlight. *Curr. Biol.* 18(21):R996–R997.

Somanathan H, Kelber A, Borges RM, Wallén R, Warrant EJ. 2009. Visual ecology of Indian carpenter bees II: adaptations of eyes and ocelli to nocturnal and diurnal lifestyles. *J. Comp. Physiol. A.* 195(6):571–583.

Somerville DC, Nicol HI. 2006. Crude protein and amino acid composition of honey bee-collected pollen pellets from south-east Australia and a note on laboratory disparity. *Aust. J. Exp. Agric.* 46(1):141–149.

Sommeijer MJ, Neve J, Jacobusse C. 2012. The typical development cycle of the solitary bee *Colletes halophilus*. *Entomol. Ber.* 72(1–2):52–58.

Souza RO, Del Lama MA, Cervini M, Mortari N, Eltz T et al. 2010. Conservation genetics of neotropical pollinators revisited: microsatellite analysis suggests that diploid males are rare in orchid bees. *Evolution* 64(11):3318–3326.

Spear DM, Silverman S, Forrest JR, McPeek MA. 2016. Asteraceae pollen provisions protect *Osmia* mason bees (Hymenoptera: Megachilidae) from brood parasitism. *Am. Nat.* 187(6): 797–803.

Sprengel CK. 1793. *Das entdeckte Geheimniß der Natur im Bau und in der Befruchtung der Blumen.* Berlin: Vieweg.

Stage GI. 1966. Biology and systematics of the American species of the genus *Hesperapis* Cockerell. PhD thesis. University of California.

Stapp P, Antolin MF, Ball M. 2004. Patterns of extinction in prairie dog metapopulations: plague outbreaks follow El Nino events. *Front. Ecol. Environ.* 2:235–240.

Stark RE, Hefetz A, Gerling D, Velthuis HHW. 1990. Reproductive competition involving oophagy in the socially nesting bee *Xylocopa sulcatipes*. *Naturwissenschaften* 77(1):38–40.

Starks PT, Reeve HK. 1999. Condition-based alternative reproductive tactics in the wool-carder bee, *Anthidium manicatum*. *Ethol. Ecol. Evol.* 11(1):71–75.

Steffan SA, Dharampal PS, Danforth BN, Gaines-Day HR, Takizawa Y, and Chikaraishi Y. 2019. Omnivory in bees: elevated trophic positions among all major bee families. *Am. Nat.* (in press).

Steffan-Dewenter I, Leschke K. 2003. Effects of habitat management on vegetation and above-ground nesting bees and wasps of orchard meadows in Central Europe. *Biodivers. Conserv.* 12:1953–1968.

Stehman JR, Semir J. 2001. Biologia reproductiva de *Calibrachoa elegans* (Miers) (Solanceae). *Rev. Bras. Bot.* 24(1):43–49.

Steiner KE, Whitehead VB. 1990. Pollinator adaptation to oil-secreting flowers—*Rediviva* and *Diascia*. *Evolution* 44(6):1701–1707.

Stephen WP. 1956. Notes on the biologies of *Megachile frigida* Smith and *M. inermis* Provancher (Hymenoptera: Megachilidae). *Pan-Pac. Entomol.* 32(3):95–101.

Stephen WP. 1966. *Andrena* (*Cryptandrena*) *viburnella*. I. Bionomics. *J. Kans. Entomol. Soc.* 39(1):42–51.

Stephen WP, Bohart GE, Torchio PF. 1969. *The Biology and External Morphology of Bees— with a Synopsis of the Genera of Northwestern America*. Corvallis: Agricultural Experiment Station, Oregon State University.

Stevens OA. 1948. Native bees. *N. D. Agric. Exp. Stn. Bimon. Bull.* 10:187–194.

Stockhammer KA. 1966. Nesting habits and life cycle of a sweat bee, *Augochlora pura* (Hymenoptera: Halictidae). *J. Kans. Entomol. Soc.* 39(2):157–192.

Stone GN. 1993. Endothermy in the solitary bee *Anthophora plumipes*: independent measures of thermoregulatory ability, costs of warm-up and the role of body size. *J. Exp. Biol.* 174(1):299–320.

Stone GN. 1994. Activity patterns of females of the solitary bee *Anthophora plumipes* in relation to temperature, nectar supplies and body size. *Ecol. Entomol.* 19(2):177–189.

Stone GN. 1995. Female foraging responses to sexual harassment in the solitary bee *Anthophora plumipes*. *Anim. Behav.* 50(2):405–412.

Stone GN, Gilbert F, Willmer P, Potts S, Semida F, Zalat S. 1999. Windows of opportunity and the temporal structuring of foraging activity in a desert solitary bee. *Ecol. Entomol.* 24(2):208–221.

Stort AC, Cruz-Landim C. 1965. Glandulas dos apendices locomotores do genero *Centris* (Hymenoptera, Anthophoridae). *Bol. Inst. Invest. Cienet. Angola* 21(23):5–14.

Stout JC, Morales CL. 2009. Ecological impacts of invasive alien species on bees. *Apidologie* 40(3):388–409.

Stoutamire WP. 1983. Wasp-pollinated species of *Caladenia* (Orchidaceae) in south-western Australia. *Aust. J. Bot.* 31:383–394.

Straka J, Bogusch P. 2007. Description of immature stages of cleptoparasitic bees *Epeoloides coecutiens* and *Leiopodus trochantericus* (Hymenoptera: Apidae: Osirini, Protepeolini) with remarks to their unusual biology. *Entomol. Fenn.* 18(4):242–254.

Straka J, Černá K, Macháčková L, Zemenová M, Keil P. 2014. Life span in the wild: the role of activity and climate in natural populations of bees. *Funct. Ecol.* 28(5):1235–1244.

Straka J, Rezkova K, Batelka J, Kratochvil L. 2011. Early nest emergence of females parasitised by Strepsiptera in protandrous bees (Hymenoptera Andrenidae). *Ethol. Ecol. Evol.* 23(2): 97–109.

Straka J, Rozen JG Jr. 2012. First observations on nesting and immatures of the bee genus *Ancyla* (Apoidea: Apidae: Apinae: Ancylaini). *Am. Mus. Novit.* 3749:1–24.

Strange JP, Koch JB, Gonzalez VH, Nemelka L, Griswold T. 2011. Global invasion by *Anthidium manicatum* (Linnaeus) (Hymenoptera: Megachilidae): assessing potential distribution in North America and beyond. *Biol. Invasions* 13(9):2115–2133.

Streinzer M, Kelber C, Pfabifan S, Kleineidam CJ, Spaethe J. 2015. Sexual dimorphism in the olfactory system of a solitary and a eusocial bee species. *J. Comp. Neurol.* 521:2742–2755.

Strickler K. 1979. Specialization and foraging efficiency of solitary bees. *Ecology* 60(5):998–1009.

Strickler K. 1982. Parental investment per offspring by a specialist bee: does it change seasonally? *Evolution* 36(5):1098–1100.

Stubblefield JW, Seger J. 1994. Sexual dimorphism in the Hymenoptera. In *The Differences Between the Sexes*, Short RV, Balaban E, eds., pp. 71–103. Cambridge: Cambridge University Press.

Stürzl W, Zeil J, Boeddeker N, Hemmi JM. 2016. How wasps acquire and use views for homing. *Curr. Biol.* 26(4):470–482.

Sugiura N. 1991. Male territoriality and mating tactics in the wool-carder bee, *Anthidium septemspinosum* Lepeletier (Hymenoptera: Megachilidae). *J. Ethol.* 9(2):95–103.

Sugiura N. 1994. Parental investment and offspring sex ratio in a solitary bee, *Anthidium septemspinosum* Lepeletier (Hymenoptera: Megachilidae). *J. Ethol.* 12(2):131–139.

Sugiura N, Maeta Y. 1989. Parental investment and offspring sex ratio in a solitary mason bee, *Osmia cornifrons* (Radoszkowski) (Hymenoptera, Megachilidae). *Jpn. J. Entomol.* 57(4): 861–875.

Sung I-H, Dubitzky A, Eardley C, Yamane S. 2009. Descriptions and biological notes of *Cteno-plectra* bees from southeast Asia and Taiwan (Hymenoptera: Apidae: Ctenoplectrini) with a new species from North Borneo. *Entomol. Sci.* 12(3):324–340.

Suni SS, Brosi BJ. 2012. Population genetics of orchid bees in a fragmented tropical landscape. *Conserv. Genet.* 13(2):323–332.

Switzer CM, Hogendoorn K, Ravi S, Combes SA. 2016. Shakers and head bangers: differences in sonication behavior between Australian *Amegilla murrayensis* (blue-banded bees) and North American *Bombus impatiens* (bumblebees). *Arthropod-Plant Interact.* 10(1):1–8.

Takahashi NC, Peruquetti RC, Del Lama MA, de Oliveira Campos LA. 2001. A reanaly-sis of diploid male frequencies in euglossine bees (Hymenoptera: Apidae). *Evolution* 55: 1897–1899.

Tautz J. 2008. *The Buzz about Bees: Biology of a Superorganism*. Berlin, Heidelberg: Springer Verlag.

Taylor JS. 1962. Notes on *Heriades freygessneri* Schletterer (Hymenoptera: Megachilidae). *J. Entomol. Soc. South. Afr.* 25(1):133–139.

Tchuenguem Fohouo F-N, Messi J, Pauly A. 2002. L'activité de butinage des Apoïdes sauvages (Hymenoptera Apoidea) sur les fleurs de maïs à Yaoundé (Cameroun) et réflexions sur la pol-linisation des graminées tropicales. *Biotechnol. Agron. Société Environ.* 6(2):87–98.

Tchuenguem Fohouo F-N, Pauly A, Messi J, Brückner D, Ngamo Tinkeu L, Basga E. 2004. Une abeille afrotropicale spécialisée dans la récolte du pollen de Graminées (Poaceae): *Lipotri-ches notabilis* (Schletterer 1891)(Hymenoptera Apoidea Halictidae). *Ann. Soc. Entomol. Fr.* 40:131–143.

Teichroew JL, Xu J, Ahrends A, Huang ZY, Tan K, Xie Z. 2017. Is China's unparalleled and understudied bee diversity at risk? *Biol. Conserv.* 210:19–28.

Tengö J, Bergström G. 1976. Comparative analyses of lemon-smelling secretions from heads of *Andrena* F. (Hymenoptera, Apoidea) bees. *Comp. Biochem. Physiol. Part B Comp. Biochem.* 55(2):179–188.

Tengö J, Bergström G. 1977. Cleptoparasitism and odor mimetism in bees: do *Nomada* males imitate the odor of *Andrena* females? *Science* 196(4294):1117–1119.

Tepedino VJ. 1981. The pollination efficiency of the squash bee (*Peponapis pruinosa*) and the honey bee (*Apis mellifera*) on summer squash (*Cucurbita pepo*). *J. Kans. Entomol. Soc.* 54(2):359–377.

Tepedino VJ. 1988. Host discrimination in *Monodontomerus obsoletus* Fabricius (Hymenop-tera: Torymidae), a parasite of the alfalfa leafcutting bee *Megachile rotundata* (Fabricius) (Hymenoptera: Megachilidae) *J. N. Y. Entomol. Soc.* 96(1):113–118.

Tepedino VJ, Bradley BA, Griswold TL. 2009. Might flowers of invasive plants increase native bee carrying capacity? Intimations from Capitol Reef National Park, Utah. *Nat. Areas J.* 28(1):44–50.

Tepedino VJ, Frohlich DR. 1984. Fratricide in a parsivoltine bee (*Osmia texana*). *Anim. Behav.* 32(4):1265–1266.

Tepedino VJ, Parker FD. 1988. Alternation of sex ratio in a partially bivoltine bee, *Megachile rotundata* (Hymenoptera: Megachilidae). *Ann. Entomol. Soc. Am.* 81(3):467–476.

Tepedino VJ, Torchio PF. 1982. Phenotypic variability in nesting success among *Osmia lignaria propinqua* females in a glasshouse environment (Hymenoptera: Megachilidae). *Ecol. Ento-mol.* 7(4):453–462.

Tepedino VJ, Torchio PF. 1989. Influence of nest hole selection on sex ratio and progeny size in *Osmia lignaria propinqua* (Hymenoptera: Megachilidae). *Ann. Entomol. Soc. Am.* 82(3):355–360.

Terry I. 2001. Thrips and weevils as dual, specialist pollinators of the Australian cycad *Macro-zamia communis* (Zamiaceae). *Int. J. Plant Sci.* 162(6):1293–1305.

Theobald JC, Coates MM, Wcislo WT, Warrant EJ. 2007. Flight performance in night-flying sweat bees suffers at low light levels. *J. Exp. Biol.* 210(22):4034–4042.

Thibault KM, Brown JH. 2008. Impact of an extreme climatic event on community assembly. *Proc. Natl. Acad. Sci. USA* 105(9):3410–3415.

Thompson K, Austin KC, Smith RM, Warren PH, Angold PG, Gaston KJ. 2003. Urban domestic gardens (I): putting small-scale plant diversity in context. *J. Veg. Sci.* 14(1):71–78.

Thomson JD, Goodell K. 2001. Pollen removal and deposition by honeybee and bumblebee visi-tors to apple and almond flowers. *J. Appl. Ecol.* 38(5):1032–1044.

Thomson JD, Thomson BA. 1992. Pollen presentation and viability schedules in animal-pollinated plants: consequences for reproductive success. In *Ecology and Evolution of Plant Reproduction: New Approaches*, Wyatt R, ed., pp. 1–24. New York, London: Chapman & Hall.

Thorp R, Brooks RW. 1994. A revision of the New World *Trachusa*, subgenera *Ulanthidium* and *Trachusomimus* (Hymenoptera: Megachilidae). *Univ. Kans. Sci. Bull.* 55(8):271–297.

Thorp RW. 1969a. Ecology and behavior of *Anthophora edwardsii* (Hymenoptera: Anthophoridae). *Am. Midl. Nat.* 82(2):321–337.

Thorp RW. 1969b. Systematics and ecology of bees of the subgenus *Diandrena* (Hymenoptera: Andrenidae). *Univ. Calif. Publ. Entomol.* 52:1–146.

Thorp RW. 1979. Structural, behavioral, and physiological adaptations of bees (Apoidea) for collecting pollen. *Ann. Mo. Bot. Gard.* 66(4):788–812.

Thorp RW. 2000. The collection of pollen by bees. *Plant Syst. Evol.* 222:211–223.

Thorp RW, Briggs DL. 1980. Bees collecting pollen from other bees (Hymenoptera: Apoidea). *J. Kans. Entomol. Soc.* 53(1):166–170.

Threlfall CG, Walker K, Williams NSG, Hahs AK, Mata L et al. 2015. The conservation value of urban green space habitats for Australian native bee communities. *Biol. Conserv.* 187(2):240–248.

Tierney SM, Gonzales-Ojeda T, Wcislo WT. 2008a. Nesting biology and social behavior of *Xenochlora* bees (Hymenoptera: Halictidae: Augochlorini) from Perú. *J. Kans. Entomol. Soc.* 81(1):61–72.

Tierney SM, Smith JA, Chenoweth L, Schwarz MP. 2008b. Phylogenetics of allodapine bees: a review of social evolution, parasitism and biogeography. *Apidologie* 39(1):3–15.

Timberlake PH. 1954. A revisional study of the bees of the genus *Perdita* F. Smith: with special reference to the fauna of the Pacific coast (Hymenoptera, Apoidea) Part I. *Univ. Calif. Publ. Entomol.* 9:345–432.

Tinbergen N. 1932. Über die Orientierung des Bienenwolfes (*Philanthus triangulum* Fabr.). *J. Comp. Physiol. A Neuroethol. Sens. Neural. Behav. Physiol.* 16(2):305–334.

Tinbergen N, Kruyt W. 1938. Über die Orientierung des Bienenwolfes (*Philanthus triangulum* Fabr.). III. Die Bevorzugung bestimmter Wegmarken. *Z. Für Vgl. Physiol.* 25(3):292–334.

Tolasch T, Kehl S, Dötterl S. 2012. First sex pheromone of the order Strepsiptera: (3R,5R,9R)-3,5,9-Trimethyldodecanal in *Stylops melittae* Kirby, 1802. *J. Chem. Ecol.* 38(12):1493–1503.

Tomlin AD, Miller JJ. 1989. Physical and behavioral factors governing the pattern and distribution of Rhipiphoridae (Coleoptera) attached to wings of Halictidae (Hymenoptera). *Ann. Entomol. Soc. Am.* 82(6):785–791.

Tonietto R, Fant J, Ascher J, Ellis K, Larkin D. 2011. A comparison of bee communities of Chicago green roofs, parks and prairies. *Landsc. Urban Plan.* 103(1):102–108.

Torchio PF. 1965. Observations on the biology of *Colletes ciliatoides* (Hymenoptera: Apoidea, Colletidae). *J. Kans. Entomol. Soc.* 38(2):182–187.

Torchio PF. 1972. *Sapyga pumila* Cresson, a parasite of *Megachile rotundata* (F.) (Hymenoptera: Sapygidae; Megachilidae). *Melanderia* 10:1–22.

Torchio PF. 1975. The biology of *Perdita nuda* and descriptions of its immature forms and those of its *Sphecodes* parasite (Hymenoptera: Apoidea). *J. Kans. Entomol. Soc.* 48(3):257–279.

Torchio PF. 1984a. The nesting biology of *Hylaeus bisinuatus* Forster and development of its immature forms (Hymenoptera: Colletidae). *J. Kans. Entomol. Soc.* 57(2):276–297.

Torchio PF. 1984b. Discovery of *Osmia tanneri* Sandhouse (Hymenoptera: Megachilidae) nesting in drilled wood trap nests. *J. Kans. Entomol. Soc.* 57(2):350–352.

Torchio PF. 1989a. In-nest biologies and development of immature stages of three *Osmia* species (Hymenoptera: Megachilidae). *Ann. Entomol. Soc. Am.* 82(5):599–615.

Torchio PF. 1989b. Biology, immature development, and adaptive behavior of *Stelis montana*, a cleptoparasite of *Osmia* (Hymenoptera: Megachilidae). *Ann. Entomol. Soc. Am.* 82(5):616–632.

Torchio P.F. 1990. *Osmia ribifloris*, a native bee species developed as a commercially managed pollinator of highbush blueberry (Hymenoptera: Megachilidae). *J. Kans. Entomol. Soc.* 63(3):427–436.

Torchio PF, Burdick DJ. 1988. Comparative notes on the biology and development of *Epeolus compactus* Cresson, a cleptoparasite of *Colletes kincaidii* Cockerell (Hymenoptera: Anthophoridae, Colletidae). *Ann. Entomol. Soc. Am.* 81(4):626–636.

Torchio PF, Rozen JG Jr, Bohart G, Favreau M. 1967. Biology of *Dufourea* and of its cleptoparasite, *Neopasites* (Hymenoptera: Apoidea). *J. N. Y. Entomol. Soc.* 75(3):142–146.

Torchio PF, Tepedino VJ. 1982. Parsivoltinism in three species of *Osmia* bees. *Psyche* 89(3–4): 221–238.

Torchio PF, Trostle GE. 1986. Biological notes on *Anthophora urbana urbana* and its parasite, *Xeromelecta californica* (Hymenoptera: Anthophoridae), including descriptions of late embryogenesis and hatching. *Ann. Entomol. Soc. Am.* 79(3):434–447.

Torchio PF, Trostle GE, Burdick DJ. 1988. The nesting biology of *Colletes kincaidii* Cockerell (Hymenoptera: Colletidae) and development of its immature forms. *Ann. Entomol. Soc. Am.* 81(4):605–625.

Torchio PF, Youssef NN. 1968. The biology of *Anthophora* (*Micranthophora*) *flexipes* and its cleptoparasite, *Zacosmia maculata*, including a description of the immature stages of the parasite (Hymenoptera: Apoidea, Anthophoridae). *J. Kans. Entomol. Soc.* 41(3):289–302.

Toro H. 1985. Ajuste mecanico para la copula de *Callonychium chilensis* (Hymenoptera: Andrenidae). *Rev. Chil. Entomol.* 12:153–158.

Toro H, Riveros G. 1998. Comportamiento de copula de *Centris mixta* Tamarugalis (Hymenoptera: Anthophoridae). *Rev. Chil. Entomol.* 25:69–75.

Toro H, Rodriguez S. 1997. Correspondencia estructural para la cópula en *Anthidium* (Hymenoptera: Megachilidae). *Rev. Chil. Entomol.* 24:61–80.

Torres A, Brandt J, Lear K, Liu J. 2017. A looming tragedy of the sand commons. *Science* 357(6355):970–971.

Torretta JP, Roig-Alsina A. 2016. First report of *Monoeca* in Argentina, with description of two new species (Hymenoptera: Apidae). *J. Melittology* 59:1–12.

Triponez Y, Arrigo N, Espíndola A, Alvarez N. 2015. Decoupled post-glacial history in mutualistic plant–insect interactions: insights from the yellow loosestrife (*Lysimachia vulgaris*) and its associated oil-collecting bees (*Macropis europaea* and *M. fulvipes*). *J. Biogeogr.* 42(4):630–640.

Trivers RL, Willard DE. 1973. Natural selection of parental ability to vary the sex ratio of offspring. *Science* 179(4068):90–92.

Tropek R, Černá I, Straka J, Cizek O, Konvicka M. 2013. Is coal combustion the last chance for vanishing insects of inland drift sand dunes in Europe? *Biol. Conserv.* 162:60–64.

Trostle G, Torchio PF. 1994. Comparative nesting behavior and immature development of *Megachile rotundata* (Fabricius) and *Megachile apicalis* Spinola (Hymenoptera: Megachilidae). *J. Kans. Entomol. Soc.* 67(1):53–72.

Trunz V, Packer L, Vieu J, Arrigo N, Praz CJ. 2016. Comprehensive phylogeny, biogeography and new classification of the diverse bee tribe Megachilini: can we use DNA barcodes in phylogenies of large genera? *Mol. Phylogenet. Evol.* 103:245–259.

Tuell JK, Ascher JS, Isaacs R. 2009. Wild bees (Hymenoptera: Apoidea: Anthophila) of the Michigan highbush blueberry agroecosystem. *Ann. Entomol. Soc. Am.* 102(2):275–287.

Tuell JK, Isaacs R. 2010. Community and species-specific responses of wild bees to insect pest control programs applied to a pollinator-dependent crop. *J. Econ. Entomol.* 103(3):668–675.

Turner RM. 1990. Long-term vegetation change at a fully protected Sonoran Desert site. *Ecology* 71(2):464–477.

Turner RM, Bowers JE, Burgess TL. 2005. *Sonoran Desert Plants: An Ecological Atlas*. Tucson: University of Arizona Press.

Ulrich W. 1956. Unsere Strepsipteren-Arbeiten. *Zool. Beitr. N. F.* 2:177–255.

Urban D. 1970. As espécies do gênero *Florilegus* Robertson, 1900 (Hymenoptera, Apoidea). *Bol. Universidade Fed. Paraná.* 3:245–280.

Urban D, Graf V. 2000. *Albinapis gracilis* gen. n. e sp. n. e *Hexantheda enneomera* sp. n. do sul do Brasil (Hymenoptera, Colletidae, Paracolletini). *Rev. Bras. Zool.* 17(3):595–601.

van der Pijl L, Dodson CH. 1969. *Orchid Flowers: Their Pollination and Evolution*. Coral Gables: University of Miami Press.

Vanderplanck M, Vereecken NJ, Grumiau L, Esposito F, Lognay G et al. 2017. The importance of pollen chemistry in evolutionary host shifts of bees. *Sci. Rep.* 7:43058.

vanEngelsdorp D, Meixner MD. 2010. A historical review of managed honey bee populations in Europe and the United States and the factors that may affect them. *J. Invertebr. Pathol.* 103:S80–S95.

Vasek FC. 1980. Creosote bush: long-lived clones in the Mojave Desert. *Am. J. Bot.* 67(2): 246–255.

Vásquez A, Olofsson TC. 2009. The lactic acid bacteria involved in the production of bee pollen and bee bread. *J. Apic. Res.* 48(3):189–195.

Vázquez DP, Morris WF, Jordano P. 2005. Interaction frequency as a surrogate for the total effect of animal mutualists on plants. *Ecol. Lett.* 8(10):1088–1094.

Verecken, NJ. 2018. Wallace's Giant Bee for sale: implications for trade regulation and conservation. *J. Insect Conserv.* https://doi.org/10.1007/s10841-018-0108-2.

Vereecken NJ, Dorchin A, Dafni A, Hötling S, Schulz S, Watts S. 2013. A pollinators' eye view of a shelter mimicry system. *Ann. Bot.* 111:1155–1165.

Vereecken NJ, Mahé G. 2007. Larval aggregations of the blister beetle *Stenoria analis* (Schaum) (Coleoptera: Meloidae) sexually deceive patrolling males of their host, the solitary bee *Colletes hederae* Schmidt and Westrich (Hymenoptera: Colletidae). *Ann. Soc. Entomol. Fr. (N. S.)* 43(4):493–496.

Vereecken NJ, McNeil JN. 2010. Cheaters and liars: chemical mimicry at its finest. *Can. J. Zool.* 88(7):725–752.

Vereecken NJ, Schiestl FP. 2008. The evolution of imperfect floral mimicry. *Proc. Natl. Acad. Sci. USA* 105(21):7484–7488.

Vereecken NJ, Wilson CA, Hötling S, Schulz S, Banketov SA, Mardulyn P. 2012. Pre-adaptations and the evolution of pollination by sexual deception: Cope's rule of specialization revisited. *Proc. R. Soc. Lond. B, Biol. Sci.* 1748:4786–4794.

Viana BF, Kleinert AMP. 2006. Structure of bee-flower system in the coastal sand dune of Abaeté, northeastern Brazil. *Rev. Bras. Entomol.* 50(1):53–63.

Vicens N, Bosch J. 2000a. Pollinating efficacy of *Osmia cornuta* and *Apis mellifera* (Hymenoptera: Megachilidae, Apidae) on 'Red Delicious' apple. *Environ. Entomol.* 29(2):235–240.

Vicens N, Bosch J. 2000b. Weather-dependent pollinator activity in an apple orchard, with special reference to *Osmia cornuta* and *Apis mellifera* (Hymenoptera: Megachilidae and Apidae). *Environ. Entomol.* 29(3):413–420.

Vicidomini S. 1996. Biology of *Xylocopa violacea* (Hymenoptera): in-nest ethology. *Ital. J. Zool.* 63(3):237–242.

Vinson SB, Frankie GW, Blum MS, Wheeler JW. 1978. Isolation, identification, and function of the Dufour gland secretion of *Xylocopa virginica texana* (Hymenoptera: Anthophoridae). *J. Chem. Ecol.* 4(3):315–323.

Vinson SB, Frankie GW, Williams HJ. 1996. Chemical ecology of bees of the genus *Centris* (Hymenoptera: Apidae). *Fla. Entomol.* 79(2):109–129.

Vinson SB, Frankie GW, Williams HJ. 2006. Nest liquid resources of several cavity nesting bees in the genus *Centris* and the identification of a preservative, levulinic acid. *J. Chem. Ecol.* 32(9):2013–2021.

Visscher PK, Danforth BN. 1993. Biology of *Calliopsis pugionis* (Hymenoptera: Andrenidae): nesting, foraging, and investment sex ratio. *Ann. Entomol. Soc. Am.* 86(6):822–832.

Visscher PK, Seeley TD. 1982. Foraging strategy of honeybee colonies in a temperate deciduous forest. *Ecology* 63(6):1790–1801.

Visscher PK, Vetter RS, Orth R. 1994. Benthic bees? Emergence phenology of *Calliopsis pugionis* (Hymenoptera: Andrenidae) at a seasonally flooded site. *Ann. Entomol. Soc. Am.* 87(6): 941–945.

Vit P, Pedro SRM, Roubik DW. 2013. *Pot-Honey: A Legacy of Stingless Bees.* New York: Springer.

Vogel ME, Kukuk PF. 1994. Individual foraging effort in the facultatively social halictid bee, *Nomia (Austronomia) australica* (Smith). *J. Kans. Entomol. Soc.* 67(3):225–235.

Vogel S. 1966. Parfümsammelnde Bienen als Bestäuber von und *Gloxinia. Österr. Bot. Z.* 113(3–4):302–361.

Vogel S. 1974. Olblumen und olsammelnde Bienen. *Trop. Subtrop. Pflwelt.* 7:1–267.

Vogel S. 1978. Evolutionary shifts from reward to deception in pollen flowers. In *The Pollination of Flowers by Insects*, Vol. 6, pp. 89–96. London: Academic Press.

Vogel S. 1981. Abdominal oil-mopping—a new type of foraging in bees. *Naturwissenschaften* 68(12):627–628.

Vogel S. 1990. Olblumen und olsammelnde Bienen dritte Folge. *Momordica, Thladiantha* und die Ctenoplectridae. *Trop. Subtrop. Pflwelt* 73:1–186.

Vulliamy B, Potts SG, Willmer PG. 2006. The effects of cattle grazing on plant-pollinator communities in a fragmented Mediterranean landscape. *Oikos* 114(3):529–543.

Wäckers FL, Romeis J, van Rijn P. 2007. Nectar and pollen feeding by insect herbivores and implications for multitrophic interactions. *Annu. Rev. Entomol.* 52:301–323.

Walter DE, Beard JJ, Walker KL, Sparks K. 2002. Of mites and bees: a review of mite-bee associations in Australia and a revision of *Raymentia* Womersley (Acari: Mesostigmata: Laelapidae), with the description of two new species of mites from *Lasioglossum* (*Parasphecodes*) spp. (Hymenoptera: Halictidae). *Aust. J. Entomol.* 41(2):128–148.

Ward JD. 1928. An unrecorded habit of the male of the bee *Anthidium manicatum* L. *Entomologist* 61(787):267–272.

Ward R, Whyte A, James RR. 2010. A tale of two bees: looking at pollination fees for almonds and sweet cherries. *Am. Entomol.* 56(3):170–177.

Warrant EJ. 2008. Seeing in the dark: vision and visual behaviour in nocturnal bees and wasps. *J. Exp. Biol.* 211(11):1737–1746.

Warrant EJ, Kelber A, Gislén A, Greiner B, Ribi W, Wcislo WT. 2004. Nocturnal vision and landmark orientation in a tropical halictid bee. *Curr. Biol.* 14(15):1309–1318.

Watmough RH. 1974. Biology and behaviour of carpenter bees in southern Africa. *J. Entomol. Soc. South Afr.* 37(2):261–281.

Watmough RH. 1983. Mortality, sex ratio and fecundity in natural populations of large carpenter bees (*Xylocopa* spp.). *J. Anim. Ecol.* 52(1):111–125.

Watson JC, Wolf AT, Ascher JS. 2011. Forested landscapes promote richness and abundance of native bees (Hymenoptera: Apoidea: Anthophila) in Wisconsin apple orchards. *Environ. Entomol.* 40(3):621–632.

Wcislo D, Vargas G, Ihle K, Wcislo W. 2012. Nest construction behavior by the orchid bee *Euglossa hyacinthina*. *J. Hymenopt. Res.* 29(1):15–20.

Wcislo WT. 1990. Parasitic and courtship behavior of *Phalacrotophora halictorum* (Diptera: Phoridae) at a nesting site of *Lasioglossum figueresi* (Hymenoptera: Halictidae). *Rev. Biol. Trop.* 38(2):205–209.

Wcislo WT. 1993. Communal nesting in a North American pearly-banded bee, *Nomia tetrazonata*, with notes on nesting behavior of *Dieunomia heteropoda* (Hymenoptera: Halictidae: Nomiinae). *Ann. Entomol. Soc. Am.* 86(6):813–821.

Wcislo WT. 1996. Parasitism rates in relation to nest site in bees and wasps (Hymenoptera: Apoidea). *J. Insect Behav.* 9(4):643–656.

Wcislo WT. 1997. Invasion of nests of *Lasioglossum imitatum* by a social parasite, *Paralictus asteris* (Hymenoptera: Halictidae). *Ethology* 103(1):1–11.

Wcislo WT, Arneson L, Roesch K, Gonzalez V, Smith A, Fernández H. 2004. The evolution of nocturnal behaviour in sweat bees, *Megalopta genalis* and *M. ecuadoria* (Hymenoptera: Halictidae): an escape from competitors and enemies? *Biol. J. Linn. Soc.* 83(3):377–387.

Wcislo WT, Buchmann SL. 1995. Mating behaviour in the bees, *Dieunomia heteropoda* and *Nomia tetrazonata*, with a review of courtship in Nomiinae (Hymenoptera: Halictidae). *J. Nat. Hist.* 29(4):1015–1027.

Wcislo WT, Cane JH. 1996. Floral resource utilization by solitary bees (Hymenoptera: Apoidea) and exploitation of their stored foods by natural enemies. *Annu. Rev. Entomol.* 41:257–286.

Wcislo WT, Danforth BN. 1997. Secondarily solitary: the evolutionary loss of social behavior. *Trends Ecol. Evol.* 12(12):468–474.

Wcislo WT, Engel MS. 1996. Social behavior and nest architecture of nomiine bees (Hymenoptera: Halictidae; Nomiinae). *J. Kans. Entomol. Soc.* 69(4):158–167.

Wcislo WT, Minckley RL, Leschen RAB, Reyes SG. 1994. Rates of parasitism by natural enemies of a solitary bee, *Dieunomia triangulifera* (Hymenoptera, Coleoptera and Diptera) in relation to phenologies. *Sociobiology* 23(3):265–273.

Wcislo WT, Minckley RL, Spangler HC. 1992. Pre-copulatory courtship behavior in a solitary bee, *Nomia triangulifera* Vachal (Hymenoptera: Halictidae). *Apidologie* 23(5):431–442.

Wcislo WT, Schatz B. 2003. Predator recognition and evasive behavior by sweat bees, *Lasioglossum umbripenne* (Hymenoptera: Halictidae), in response to predation by ants, *Ectatomma ruidum* (Hymenoptera: Formicidae). *Behav. Ecol. Sociobiol.* 53(3):182–189.

Wcislo WT, Tierney SM. 2009. Behavioural environments and niche construction: the evolution of dim-light foraging in bees. *Biol. Rev.* 84(1):19–37.

Werner T, Jaenike J. 2017. *Drosophilids of the Midwest and Northeast*. Rochester: River Campus Libraries, University of Rochester.

Westerkamp CH. 1996. Pollen in bee-flower relations some considerations on melittophily. *Bot. Acta* 109(4):325–332.

Westerling AL, Hidalgo HG, Cayan DR, Swetnam TW. 2006. Warming and earlier spring increase western U.S. forest wildfire activity. *Science* 313(5789):940–943.

Westrich P. 1989. *Die Wildbienen Baden-Wurttembergs; Spezieller Teil*. Stuttgart: Eugenne Ulmer.

Westrich P, Knapp A, Berney I. 2015. *Megachile sculpturalis* Smith 1853 (Hymenoptera, Apidae), a new species for the bee fauna of Germany, now north of the Alps. *Eucera* 9:3–10.

Whitten W, Young A, Williams N. 1989. Function of glandular secretions in fragrance collection by male euglossine bees (Apidae, Euglossini). *J. Chem. Ecol.* 15(4):1285–1296.

Whitten WM, Young AM, Stern DL. 1993. Nonfloral sources of chemicals that attract male euglossine bees (Apidae: Euglossini). *J. Chem. Ecol.* 19(12):3017–3027.

Wikelski M, Moxley J, Eaton-Mordas A, López-Uribe MM, Holland R et al. 2010. Large-range movements of neotropical orchid bees observed via radio telemetry. *PLoS One* (5):e10738.

Williams HJ, Strand MR, Elzen GW, Vinson SB, Merritt SJ. 1986. Nesting behavior, nest architecture, and use of Dufour's gland lipids in nest provisioning by *Megachile integra* and *M. mendica mendica* (Hymenoptera: Megachilidae). *J. Kans. Entomol. Soc.* 59(4):588–597.

Williams N, Whitten W. 1983. Orchid floral fragrances and male euglossine bees: methods and advances in the last sequidecade. *Biol. Bull.* 164:355–395.

Williams NM. 2003. Use of novel pollen species by specialist and generalist solitary bees (Hymenoptera: Megachilidae). *Oecologia* 134(2):228–237.

Williams NM, Crone EE, Roulston TH, Minckley RL, Packer L, Potts SG. 2010. Ecological and life-history traits predict bee species responses to environmental disturbances. *Biol. Conserv.* 143(10):2280–2291.

Williams NM, Kremen C. 2007. Resource distributions among habitats determine solitary bee offspring production in a mosaic landscape. *Ecol. Appl.* 17(3):910–921.

Willmer PG. 1988. The role of insect water balance in pollination ecology: *Xylocopa* and *Calotropis*. *Oecologia* 76(3):430–438.

Willmer PG, Stone GN. 2004. Behavioral, ecological, and physiological determinants of the activity patterns of bees. *Adv. Study Behav.* 34:347–466.

Wilson EO. 1971. *The Insect Societies*. Cambridge: Belknap Press of Harvard University Press.

Wilson EO, Hölldobler B. 2005. Eusociality: origin and consequences. *Proc. Natl. Acad. Sci. USA* 102(38):13367–13371.

Wilson JS, Pitts JP, von Dohlen C. 2009. Lack of variation in nuclear genes among isolated populations of the sand dune restricted bee *Colletes stepheni* (Hymenoptera: Colletidae). *J.Kans. Entomol. Soc.* 82(4):316–320.

Winfree R, Aguilar R, Vázquez DP, LeBuhn G, Aizen MA. 2009. A meta-analysis of bees' responses to anthropogenic disturbance. *Ecology* 90(8):2068–2076.

Winfree R, Griswold T, Kremen C. 2007a. Effect of human disturbance on bee communities in a forested ecosystem. *Conserv. Biol.* 21(1):213–223.

Winfree R, Fox JW, Williams NM, Reilly JR, Cariveau DP. 2015. Abundance of common species, not species richness, drives delivery of a real-world ecosystem service. *Ecol. Lett.* 18(7): 626–635.

Winfree R, Williams NM, Dushoff J, Kremen C. 2007b. Native bees provide insurance against ongoing honey bee losses. *Ecol. Lett.* 10(11):1105–1113.

Winfree R, Williams NM, Gaines H, Ascher JS, Kremen C. 2008. Wild bee pollinators provide the majority of crop visitation across land-use gradients in New Jersey and Pennsylvania, USA: crop visitation by wild pollinators. *J. Appl. Ecol.* 45(3):793–802.

Winston ML. 1979. The proboscis of the long-tongued bees: a comparative study. *Univ. Kans. Sci. Bull.* 51(22):631–667.

Winston ML. 1987. *The Biology of the Honey Bee*. Cambridge: Harvard University Press.

Wirtz P, Szabados M, Pethig H, Plant J. 1988. An extreme case of interspecific territoriality: male *Anthidium manicatum* (Hymenoptera: Megachilidae) wound and kill intruders. *Ethology* 78(2):159–167.

Wittmann D, Blochtein B. 1995. Why males of leafcutter bees hold their females antennae with their front legs during mating. *Apidologie* 26(3):181–195.

Wood TJ, Holland JM, Goulson D. 2016. Diet characterisation of solitary bees on farmland: dietary specialisation predicts rarity. *Biodivers. Conserv.* 25(13):2655–2671.

Wood TJ, Roberts SP. 2017. An assessment of historical and contemporary diet breadth in polylectic *Andrena* bee species. *Biol. Conserv.* 215:72–80.

Wright GA, Baker DD, Palmer MJ, Stabler D, Mustard JA et al. 2013. Caffeine in floral nectar enhances a pollinator's memory of reward. *Science* 339(6124):1202–1204.

Wuellner CT. 1999. Nest site preference and success in a gregarious, ground-nesting bee *Dieunomia triangulifera*. *Ecol. Entomol.* 24(4):471–479.

Wynns AA, Jensen AB, Eilenberg J. 2013. *Ascosphaera callicarpa*, a new species of bee-loving fungus, with a key to the genus for Europe. *PLoS One* 8(9):e73419.

Yanega D. 1988. Social plasticity and early-diapausing females in a primitively social bee. *Proc. Natl. Acad. Sci. USA* 85(12):4374–4377.

Yang J-W, Xu H-L, Hu H-Y. 2010. Nesting biology of *Dasypoda hirtipes* (Fabricius) (Hymenoptera: Melittidae). *Acta Entomol. Sin.* 53(4):442–448.

Yeates DK, Greathead D. 1997. The evolutionary pattern of host use in the Bombyliidae (Diptera): a diverse family of parasitoid flies. *Biol. J. Linn. Soc.* 60(2):149–85.

Yoshihara Y, Chimeddorj B, Buuveibaatar B, Lhagvasuren B, Takatsuki S. 2008. Effects of livestock grazing on pollination on a steppe in eastern Mongolia. *Biol. Conserv.* 141(9):2376–2386.

Young AM. 1985. Notes on the nest structure and emergence of *Euglossa turbinifex* Dressler (Hymenoptera: Apidae: Bombinae: Euglossini) in Costa Rica. *J. Kans. Entomol. Soc.* 58(3):538–543.

Zanella FCV, Ferreira AG. 2005. Record of *Austrostelis* Michener and Griswold host (Hymenoptera: Megachilidae) and of its occurrence in the semi-arid caatinga region. *Neotrop. Entomol.* 34(5):857–858.

Zayed A, Constantin ŞA, Packer L. 2007. Successful biological invasion despite a severe genetic load. *PLoS One* 2(9):e868.

Zayed A, Packer L. 2005. Complementary sex determination substantially increases extinction proneness of haplodiploid populations. *Proc. Natl. Acad. Sci. USA* 102(30):10742–10746.

Zayed A, Packer L. 2007. The population genetics of a solitary oligolectic sweat bee, *Lasioglossum* (*Sphecodogastra*) *oenotherae* (Hymenoptera: Halictidae). *Heredity* 99(4):397–405.

Zayed A, Packer L, Grixti JC, Ruz L, Owen RE, Toro H. 2005. Increased genetic differentiation in a specialist versus a generalist bee: implications for conservation. *Conserv. Genet.* 6(6):1017–1026.

Zayed A, Roubik DW, Packer L. 2004. Use of diploid male frequency data as an indicator of pollinator decline. *Proc. R. Soc. Lond. B, Biol. Sci.* 271(Suppl 3):S9–S12.

Zeil J. 1993. Orientation flights of solitary wasps (*Cerceris*; Sphecidae; Hymenoptera). II. Similarities between orientation and return flights and the use of motion parallax. *J. Comp. Physiol. A.* 172(2):207–222.

Zeil J. 2012. Visual homing: an insect perspective. *Curr. Opin. Neurobiol.* 22(2):285–293.

Zeil J, Kelber A. 1991. Orientation flights in ground nesting wasps and bees share a common organization. *Verh. Dtsch. Zool. Ges.* 84:371–372.

Zeil J, Kelber A, Voss R. 1996. Structure and function of learning flights in ground-nesting bees and wasps. *J. Exp. Biol.* 199(1):245–252.

Zhang J, Weirauch C, Zhang G, Forero D. 2016. Molecular phylogeny of Harpactorinae and Bactrodinae uncovers complex evolution of sticky trap predation in assassin bugs (Heteroptera: Reduviidae). *Cladistics* 32(5):538–554.

Zhang W, Kramer EM, Davis CC. 2010. Floral symmetry genes and the origin and maintenance of zygomorphy in a plant-pollinator mutualism. *Proc. Natl. Acad. Sci. USA* 107(14):6388–6393.

Zillikens A, Steiner J. 2004. Nest architecture, life cycle and cleptoparasite of the Neotropical leaf-cutting bee *Megachile* (*Chrysosarus*) *pseudanthidioides* Moure (Hymenoptera: Megachilidae). *J. Kans. Entomol. Soc.* 77(3):193–202.

Ziska LH, Pettis JS, Edwards J, Hancock JE, Tomecek MB et al. 2016. Rising atmospheric CO_2 is reducing the protein concentration of a floral pollen source essential for North American bees. *Proc. R. Soc. Lond. B, Biol. Sci.* 283(1828):20160414.

Zurbuchen A, Cheesman S, Klaiber J, Müller A, Hein S, Dorn S. 2010a. Long foraging distances impose high costs on offspring production in solitary bees. *J. Anim. Ecol.* 79(3):674–681.

Zurbuchen A, Landert L, Klaiber J, Müller A, Hein S, Dorn S. 2010b. Maximum foraging ranges in solitary bees: only few individuals have the capability to cover long foraging distances. *Biol. Conserv.* 143(3):669–676.

Subject Index

NOTE: Page numbers in **bold** indicate boxes. Italic *f* indicates figures, and *t* indicates tables. Footnotes are indicated with n.

Taxonomic Index